$\mathbf{q}$	bulk-phase heat flux vector density (see Table 3.6-1)
$\mathbf{q}^s$	surface-excess heat flux vector density (see Table 3.7-1)
r	spherical polar coordinate in radial direction (see Figure B.1-6)
r_0	radius of a spherical surface (see Figure B.1-6)
r_1, r_2	principal surface curvature radii, Eq. (3.2-15)
R	circular cylindrical coordinate in radial direction (see Figure B.1-5)
R_0	radius of a circular cylindrical surface (see Figure B.1-5)
R_i	bulk-phase reaction rate density of species i, Eq. (5.1-5)
$R^s{}_i$	surface-excess reaction rate density of species i, Eq. (5.2-2)
Re	Reynolds number, Eq. (4.4-92)
S	bulk-phase entropy density (see Table 3.6-1)
S^s	surface-excess entropy density (see Table 3.7-1)
$\mathbf{u}$	surface velocity, Eq. (3.4-5)
U	bulk-phase internal energy density (see Table 3.6-1)
U^s	surface-excess internal energy density (see Table 3.7-1)
$\mathbf{v}$	mass-average velocity, Eq. (3.4-2)
$\mathbf{v}^o$	mass-average (material) surface velocity, Eq. (3.4-6)
$\mathbf{x}$	spatial position vector, Eq. (3.4-1)
$\mathbf{x}_s$	surface position vector, Eq. (3.4-3)
x	Cartesian coordinate (see Figure B.1-4)
x^s	surface-excess mass fraction, Eq. (2.1-5)
y	Cartesian coordinate (see Figure B.1-4)
z	Cartesian coordinate (see Figure B.1-4); circular cylindrical coordinate (see Figure B.1-5)

Greek Letters

δ_{ij}	spatial Kronecker delta, Eq. (A.1-5)
$\delta_{\alpha\beta}$	surface Kronecker delta, Eq. (3.2-8)
γ	shear strain, Eq. (14.2-4)
$\dot{\gamma}$	shear strain rate, Eq. (14.3-8)
ε	unit spatial alternator triadic, Eq. (A.1-21)
ε_s	unit surface alternator dyadic, Eq. (3-2-9)
ε_{ijk}	unit spatial alternator tensor, Eq. (A.1-20)
$\varepsilon_{\alpha\beta}$	unit surface alternator tensor, Eq. (B.1-8)
θ	azimuthal angle in circular cylindrical coordinates measured from the x-direction (see Figure B.1-5); polar angle in spherical polar coordinates measured from the z-direction (see Figure B.1-6); contact angle, Eq. (11.2-10)
κ_1, κ_2	principal surface curvature scalars, Eq. (3.2-12)
κ	bulk dilatational viscosity, Eq. (4.1-16b)
κ^s	interfacial dilatational viscosity, Eq. (4.2-18a)
μ	bulk shear viscosity, Eq. (4.1-16a)
μ^s	interfacial shear viscosity, Eq. (4.2-18a)
π	surface pressure, Eq. (2.1-5)
Π	osmotic pressure, Eq. (2.1-4); ...ning pressure, Eq. (11.2-1)

Interfacial Transport Processes and Rheology

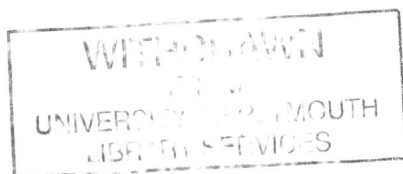

BUTTERWORTH-HEINEMANN SERIES IN CHEMICAL ENGINEERING

SERIES EDITOR

HOWARD BRENNER
Massachusetts Institute of Technology

ADVISORY EDITORS

ANDREAS ACRIVOS
The City College of CUNY

JAMES BAILEY
California Institute of Technology

MANFRED MORARI
California Institute of Technology

E. BRUCE NAUMAN
Rensselear Polytechnic Institute

J.R.A. PEARSON
Schlumberger Cambridge Research

ROBERT K. PRUD'HOMME
Princeton University

SERIES TITLES

Bubble Wake Dynamics in Liquids and Liquid-Solid Suspensions *Liang-Shih Fan and Katsumi Tsuchiya*
Chemical Process Equipment: Selection and Design *Stanley M. Walas*
Chemical Process Structures and Information Flows *Richard S.H. Mah*
Computational Methods for Process Simulations *W. Fred Ramirez*
Constitutive Equations for Polymer Melts and Solutions *Ronald G. Larson*
Fluidization Engineering, 2nd ed. *Daizo Kunii and Octave Levenspiel*
Fundamental Process Control *David M. Prett and Carlos E. Garcia*
Gas-Liquid-Solid Fluidization Engineering *Liang-Shih Fan*
Gas-Separation by Adsorption Processes *Ralph T. Yang*
Granular Filtration of Aerosols and Hydrosols *Chi Tien*
Heterogeneous Reactor Design *Hong H. Lee*
Interfacial Transport Processes and Rheology *David A. Edwards, Howard Brenner and Darsh T. Wasan*
Introductory Systems Analysis for Process Engineers *E. Bruce Nauman*
Microhydrodynamics: Principles and Selected Applications *Sangtae Kim and Seppo J. Karrila*
Modeling With Differential Equations in Chemical Engineering *Stanley M. Walas*
Molecular Thermodynamics of Nonideal Fluids *Lloyd L. Lee*
Phase Equilibria in Chemical Engineering *Stanley M. Walas*
Physicochemical Hydrodynamics: An Introduction *Ronald F. Probstein*
Slurry Flow: Principles and Practice *Clifton A. Shook and Michael C. Roco*
Transport Processes in Chemically Reacting Flow Systems *Daniel E. Rosner*
Viscous Flows: The Practical Use of Theory *Stuart W. Churchill*

REPRINT TITLES

Advanced Process Control *W. Harmon Ray*
Applied Statistical Mechanics *Thomas M. Reed and Keith E. Gubbins*
Elementary Chemical Reactor Analysis *Rutherford Aris*
Kinetics of Chemical Processes *Michel Boudart*
Reaction Kinetics for Chemical Engineers *Stanley M. Walas*

Interfacial Transport Processes and Rheology

David A. Edwards
Department of Chemical Engineering
Massachusetts Institute of Technology

Howard Brenner
Department of Chemical Engineering
Massachusetts Institute of Technology

Darsh T. Wasan
Department of Chemical Engineering
Illinois Institute of Technology

Butterworth-Heinemann
Boston London Oxford Singapore Sydney Toronto Wellington

Recognizing the importance of preserving what has been written, it is the policy of Butterworth-Heinemann to have the books it publishes printed on acid-free paper, and we exert our best efforts to that end.

Library of Congress Cataloging-in-Publication Data

Edwards, David A., 1961-
Interfacial transport processes and rheology / David A. Edwards, Howard Brenner, Darsh T. Wasan.
p. cm. — (Butterworth-Heinemann series in chemical engineering)
Includes bibliographical references and index.
ISBN 0-7506-9185-9
1. Surface chemistry. 2. Rheology 3. Transport theory.
I. Brenner, Howard. II. Wasan, D.T. III. Title. IV. Series
QD506.E38 1991
541.3'3—dc20 91-24665
CIP

British Library Cataloguing in Publication Data

Edwards, David A.
Interfacial transport processes and rheology.
I. Title II. Brenner, Howard
III. Wasan, Darsh T.
531.11

ISBN 0-7506-9185-9

Butterworth-Heinemann
80 Montvale Avenue
Stonheam, MA 02180

10 9 8 7 6 5 4 3 2 1

Printed in the United States of America

Contents

Preface

This textbook is designed to provide the theory, methods of measurement, and principal applications of the expanding field of interfacial hydrodynamics. It is intended to serve the research needs of both academic and industrial scientists, including chemical or mechanical engineers, material and surface scientists, physical chemists, chemical- and bio-physicists, rheologists, physico-chemical hydrodynamicists, and applied mathematicians (especially those with interests in viscous fluid mechanics and continuum mechanics). As a textbook it provides material for a one- or two-semester graduate-level course in interfacial transport processes. It may also be noted that, while separate practical and theoretical subdivisions of material have been introduced, a kind of cross emphasis is often stressed: (i) to the academic scientist, of the importance of understanding major *applications* of interfacial transport; and (ii) to the industrial scientist, of the importance of understanding the underlying *theory*.

Organization of the Textbook

This textbook is divided into two major parts.

Part I: Part I strictly considers the "macroscale" view of fluid interfaces, whereby the latter appear as two-dimensional, singular surfaces, as is the classical idealization. The interface is furthermore regarded in this portion of the text as being *material* in nature, referring to the class of interfacial problems for which no *net* interphase transfer of mass across the interface occurs at any point (or else is sufficiently small to be negligible in its consequences). This mass-transfer restriction does not prevent applications of the resulting theory to the purely *diffusive* transport of solute species across the (material) interface (as such diffusive transport is necessarily measured relative to the mass-average velocity), and is probably applicable to most, though certainly not all, practical cases of interphase transport. As discussed below in the subsection on *Classroom Instruction*, the first segment of the book can usefully serve as an introduction to the general subject of interfacial transport processes, adequate for a one-semester course at the graduate level. Part I is divided into chapters respectively emphasizing the theoretical (Chapters 2 to 5), experimental (Chapters 6 to 9), and applied (Chapters 10 to 14) aspects of interfacial rheology. With the exception of the theoretical portion of this material, which develops an overall physicomathematical structure of interfacial transport processes that begins in Chapter 3 and concludes with Chapter 5, the chapters comprising this portion of the book largely stand alone. As such, they may be read out of sequence, as befits the interests of the reader or instructor.

One of the primary advantages of Part I pertains to the needs and interests of the industrial researcher, who is presumably concerned primarily with the experimental and applications chapters. He or she possesses the opportunity in Part I to acquire from Chapters 1 and 2 a simple theoretical background sufficient for reading those later chapters pertaining to experimental measurements and applications.

The theoretician will be primarily interested in Chapters 3 to 5 of Part I. These introductory chapters may be considered a prerequisite to the more rigorous theoretical chapters of Part II.

Part II: Part II begins with Chapter 15, and concerns the more detailed "microscale" view of a fluid interface, whereby the interface is recognized to be not a singular surface across which discontinuities may occur, but rather is seen to be a thin, diffuse (i.e. continuous), three-dimensional transition zone (characterized by steep spatial physicochemical inhomogeneities) between otherwise relatively homogeneous "bulk"-fluid phases. A Gibbsian *surface-excess* formalism is employed in the context of a rigorous (matched asymptotic expansion) micro-continuum theory with the purpose of 'deriving' the macro-interfacial theory of Part I (Chapters 3 to 5) from the more fundamental microscale perspective. In contrast with Part I, the interface is now no longer necessarily regarded as being material. Rather, the more general case of *nonmaterial* interfaces is treated.

The development of topics provided in Chapters 15 to 17 will be seen to parallel that provided in Chapters 3 to 5. A derivation of interfacial transport equations is supplied in the first three chapters of Part II for nonmaterial interfaces. Each of the nonmaterial relations derived in Part II are shown to reduce to its material counterpart in Part I in circumstances wherein the interface is restricted to being material, albeit with greatly enhanced insight into the physical nature of the purely phenomenological equations of Part I.

Finally, Chapter 18 provides an extension of the generic matched-asymptotic, surface-excess formalism of Chapter 15 to the case of three-phase contact lines. However, only the equilibrium case is considered, thus excluding "line-excess" transport phenomena from consideration in this book.

Part II is intended for students and/or researchers interested in a rigorous (though potentially rewarding) theoretical understanding of interfacial transport processes. It seeks to introduce some of the many fruitful avenues of theoretical research currently open to further investigation in the field, as well as to identify new research areas, e.g. rheological equations of state for nonmaterial interfaces, sources of 'non-Newtonian' interfacial rheological behavior for material interfaces, and "line rheology", to name a few such areas.

Classroom Instruction

A one-semester graduate-level course entitled *Interfacial Transport Processes* was taught during the fall of 1990 in the Chemical Engineering Department at the Massachusetts Institute of Technology (and once again during the spring of 1991 in the Mechanical Engineering Department at the Technion—Israel Institute of Technology) by one of us (D.A.E.) based upon a preliminary version of this text. The first few weeks of the semester began with a brief overview of Chapters 1 and 2; this was followed by a review of basic tensor analysis and the differential geometry of surfaces, employing Appendices A and B and §3.2 of Chapter 3. The remaining portion of the first half of the semester was devoted to a study of the basic theory underlying interfacial transport processes and interfacial rheology, as embodied in Chapters 3 to 5. The second half of the course covered the experimental and applications chapters (6 to 14). The advanced material of Part II was considered only briefly, owing primarily to lack of time in a one-semester introductory course; special attention was given to the kinematics of nonmaterial interfaces, utilizing various examples in Parts I and II pertaining to the kinematics of a nonmaterial spherical gas bubble or droplet interface. Selected homework problems taken from the *Questions* appearing at the end of each chapter were assigned each week, and a 'take-home' midterm and final examination was required of the students.

A possible alternative layout of an interfacial transport processes course would involve two semesters, with the first semester focusing solely upon the purely theoretical material of Chapters 3 to 5 and 15 to 18. As envisioned by the authors, roughly the first half of the semester would be devoted to a consideration of the material provided in Appendices A and B together with Chapters 3 to 5, with the remaining classroom time in the first semester being used for the study of the advanced material of Chapters 15 to 18 (of Part II). The second semester of the course would concentrate upon the measurement (6 to 9) and applications (10 to 14) chapters of the text. (This second semester course could be offered independently of the first by substituting the formal theoretical material covered in the above-described first-semester course with a relatively brief, initial consideration of the less demanding theoretical material of Chapters 1 and 2.) This second semester course would most profitably be taught with the aid of pertinent classroom demonstrations and, ideally, student access to laboratory experiments.

Solutions to the *Questions* appearing at the end of each chapter have been prepared in the form of a *Solutions Manual*. Intended as an aid to those either presenting this material in the classroom or using it for self study, this manual is available from the publisher, *Butterworth-Heinemann*, 80 Montvale Avenue, Stoneham, Massachusetts 02180. Requests should be on official letterheads and over the signature of either a member of a university faculty or the industrial or governmental equivalent.

Contributors to this Textbook

As the quotations preceding each chapter and numerous references throughout this book attest, many researchers have contributed to the early and more recent developments defining the field of interfacial transport processes, some long before it was possible to reconcile their efforts as belonging to an identifiable field of science. Our contribution has largely been one of identifying, collecting, selecting, correlating and finally reconciling pertinent subject material drawn from the prior research efforts of others, to whom much of the credit for the existence of this book must be acknowledged. Many of their names appear in the bibliography at the end of this book. However, since this book has been cast as a textbook rather than as a research monograph, no attempt has been made to provide an exhaustively comprehensive bibliography. Thus, many researchers will have been overlooked, and to these individuals we express our regrets that the restricted scope of our book did not allow explicit identification of their contributions.

In addition to these indirect contributors to the book, a number of individuals have contributed more directly, and we wish to acknowledge them. The evolution of the material in Chapters 15 to 17 of Part II deserves special comment in this regard. The original research culminating in Part II of this book, begun in the late 1970's (Brenner 1979, and Brenner & Leal 1982), was later continued and extended by Li Ting—then a graduate student at the Illinois Institute of Technology with D.T. Wasan in collaboration with H. Brenner. Dr. Ting's early efforts are summarized in his PhD thesis (Ting 1984); his later postdoctoral efforts at MIT contributed immeasurably to the developments outlined in Part II. The doctoral thesis of Dr. Gretchen M. Mavrovouniotis (1989) provides the most recent form of the matched asymptotic, surface-excess transport theory to have appeared in the published literature prior to that outlined in Chapters 15 to 17. Developments subsequent to her thesis owe to the collaborative efforts of D.A. Edwards and H. Brenner.

The authors would like to acknowledge the hospitality of the Department of Chemical Engineering at MIT. Their cooperation in allowing us to develop and teach a new graduate course in *Interfacial Transport Processes* during the Fall 1990 semester afforded an opportunity to fine tune earlier drafts of this book. Moreover, the many questions and helpful suggestions of the graduate and post-graduate students who attended this course contributed greatly to the quality and internal consistency of the text. The authors particularly wish to thank Fuquan Gao, Dave Otis, Chunhai Wang and Alejandro Mendoza-Blanco, each of whose enthusiastic input provided important encouragement during the latter stages of the writing of the book. The *Interfacial and Colloid Phenomena* course taught at IIT in the Fall of 1990 by D.T.W. was also extremely beneficial.

D.A.E. wishes to thank the members of the Faculty of Mechanical Engineering at the Technion, and in particular Dr. Michael Shapiro, for their genuine hospitality and intellectual support during his two periods of residence (1987–89 and Spring, 1991) in their department. Their assistance in permitting the offering of an *Interfacial Transport Processes* course during his second stay in the department proved of value both in the development of a solutions manual to the text and in accomplishing last-minute textual corrections (of which latter efforts Tal Hocherman and Michael Shlyafstein were of particular assistance). Many friends have been made in Haifa,

Jerusalem and Boston, and their support has been of inestimable value; the memories are especially warm and dear of the love, idealism and courage of Miky.

H.B. was aided in the writing of this book by a grant from the *Bernard H. Gordon Engineering Curriculum Development Fund* at MIT, and he is grateful for the encouragement implicit in such an award. He would also like to acknowledge the research support that he and his students have received from the Office of Basic Energy Sciences of the Department of Energy, the Army Research Office, and the National Science Foundation. Such support contributed both directly and indirectly to the writing of this book. Lastly, H.B. would like to acknowledge the award of sabbatical leave grants to him during the academic year 1988–89 in the form of a Fellowship from the John Simon Guggenheim Memorial Foundation and a Chevron Visiting Professorship from the Department of Chemical Engineering of the California Institute of Technology. The hospitality displayed to him at Caltech was valuable in furthering the goals embodied in this book.

D.T.W., who was assisted by grants from the National Science Foundation and the Department of Energy, wishes to express his indebtedness to Professor Robert Schechter for his original interest and collaboration in the early stages of this project. The outline that was cowritten by Robert Schechter and Darsh Wasan in March of 1984 provided an early conceptual nucleus for the book, and proved of particular value in the development of the material in Chapters 7 and 9, where there appears frequent reference to the important research publications of Professor Schechter and his collaborators.

May 26, 1991

D.A. Edwards
H. Brenner
D. T. Wasan

PART I

INTERFACIAL RHEOLOGY: BASIC THEORY, MEASUREMENTS & APPLICATIONS

1

"We live in a world of three dimensions. We measure their length, breadth and thickness. The position of a point can be described by three coordinates, x, y and z. We can not escape from the inside of a spherical surface except by passing through it, but if we are standing in a circle on a surface we escape by stepping over it.

"It is amusing to try to imagine a fourth dimension. We can reason that if we could travel into it, we could escape from the inside of a sphere without going through its surface.

"In the special theory of relativity, Einstein has given us reason for looking upon time as a kind of imaginary fourth dimension which differs from any of the ordinary dimensions of space much as the number one differs from the imaginary number $\sqrt{-1}$.

"In the general theory of relativity, there are suggestions that the effect of gravitation is to warp four-dimensional space-time in a fifth dimension, very much as we have to warp a map of Europe to make it fit onto a globe representing the earth.

"Poincaré in an interesting book, 'Science and Hypothesis,' attempted in 1903 to trace the probable development of science on the earth if it had happened that the earth's atmosphere, like that of Venus, had been perpetually cloudy. Without the ability to observe the stars and sun, mankind would have persisted for long in a belief that the earth is flat. If a pioneer among scientists had made the statement that the surface of the earth has no edge or boundary but yet has a limited area, he would have been disbelieved; for these two statements seemed contradictory to those who believed in a flat earth. . . .

"Many of you perhaps have seen the little book entitled 'Flatland,' written in 1885 by an author who gives the name A. Square, but who is said to be Edwin A. Abbott. . . . I propose to tell you of a real two-dimensional world in which phenomena occur that are analogous to those described in 'Flatland.'"

Irving Langmuir (1936)

CHAPTER 1

Interfacial Rheology and Its Applications

Interfacial rheology, or 'interfacial hydrodynamics', is the field of science that studies the response of mobile interfaces to deformation. First encounters with this field most frequently occur through the study of conventional bulk-phase hydrodynamics, where the need arises at a so-called 'free surface' to specify an appropriate boundary condition upon the normal component of the bulk-phase stress tensor. This boundary condition generally introduces *two-dimensional* tensile forces, as well as viscous and elastic forces, which, aside from their localized action upon a two-dimensional, non-Euclidean, moving and deforming surface, are quite analogous to comparable three-dimensional hydrodynamic forces in bulk-phase fluids. Through the science of interfacial rheology one seeks *inter alia* to determine: (i) the shape of a dynamic free fluid interface; (ii) the nature of interfacial response to deformation; and (iii) the quantitative influence that interfacial stress imparts upon hydrodynamic motion in contiguous fluid phases.

Textbooks concerned with classical hydrodynamics (Lamb 1945, Landau & Lifshitz 1960, Bird *et al.* 1960, Batchelor 1967, Slattery 1981) often omit an explicit consideration of interfacial rheology; rather, they view the normal stress boundary condition at a free fluid interface as a continuity condition imposed upon the normal component of the fluid stress tensor, *modulo* a possible discontinuity arising at a curved interface due to (a homogeneous) interfacial tension. The justification for this first-level hydrodynamic approach to fluid interfaces is that a precise knowledge of the normal stress boundary condition is often unnecessary for a basic understanding of the bulk-fluid motion. Hence, if the fluid surface-to-volume ratio, or *specific surface*, is small for the particular system considered (as is most often the case for introductory-level fluid flow problems), a rigorous formulation of the normal stress condition at fluid boundaries is generally unnecessary.

In circumstances for which the fluid system under study possesses a *large* specific surface, as occurs with many fluid-fluid *colloidal dispersions* (e.g. single bubbles, bubbly liquids, emulsions, and foams), interfacial rheological knowledge may prove indispensable for understanding bulk hydrodynamic behavior; this is illustrated, for example, in processes of dynamic phase mixing, either induced or enhanced through the mechanism of interfacial turbulence. A common feature of many colloidal systems is the presence of molecular or macromolecular *surfactants* at the fluid interface; these adsorbed species tend not only to stabilize the interface in a dispersed state, but also to introduce additional interfacial stresses beyond that stress already contributed by a homogeneous interfacial tension.

This textbook is concerned with the theory, measurement, and practical significance of various interfacial transport processes, with the primary emphasis of Part I focused upon the rheological aspects of fluid interfaces possessing a relatively large specific surface. The chapters comprising Part I of this book are devoted respectively to theory (Chapters 2 to 5), measurement (Chapters 6 to 9), and applications (Chapters 10 to 14). The more theoretical of these chapters (specifically Chapters 3 to 5) employ vector-dyadic notation, together with frequent use of differential geometry, knowledge of which is unnecessary for those readers interested primarily in the measurement and applications chapters of Part I. Accordingly, Chapters 1 and 2 are designed to provide a rheological background sufficient to the task.

This introductory chapter begins with a brief discussion of the historical development of interfacial rheology, followed by descriptive overviews of the nature and practical significance of (i) surfactants, (ii) colloidal dispersions for which the specific surface may be large, and (iii) current engineering processes wherein interfacial rheology may play a significant role.

1.1 Historical Review

Only in recent years has "interfacial rheology", as a unique field of study, achieved a cohesive identity comparable to that existing in the more conventional fields of bulk-phase transport processes. Prior to several key experimental and theoretical observations in the 1950's and early 1960's, the term "interfacial (or 'surface') rheology" had been employed mostly in reference to experimental studies devoted to qualitatively elucidating the rigidity and viscoelasticity of adsorbed monolayers at fluid interfaces (Mouquin & Rideal 1927, Tschoegl 1958, Davies & Rideal 1963, Biswas & Haydon 1963, Joly 1964). Simultaneously, however, there developed other areas of interfacial research that were often of greater scientific interest, relating to observations of bubble and fluid droplet motion, fluid instabilities, thermocapillary migration, interfacial turbulence, the behavior of thin liquid films, coalescence phenomena, etc., variously engaging the attention of leading scientists [Thompson (Lord Kelvin) 1855, Plateau 1869, Rayleigh 1878, Gibbs 1957], often at early stages of their careers. [Thus, for example, the first publications of the youthful Einstein (1901) and Bohr (1909) dealt with interfacial phenomena.]

Until recently, such dynamic interfacial phenomena were often viewed in an independent, *ad hoc* context, often appearing under the heading of 'special topics' within the more established disciplines of hydrodynamics or hydrodynamic stability analysis. For this reason, the phenomena themselves, despite their surprisingly widespread applications, have remained obscure, or perhaps worse — misunderstood, by many scientists and engineers. As discussed in the brief historical review below, these dynamic interfacial phenomena may now be viewed within the context of modern interfacial rheology.

Interfacial Rheology Prior to 1960: The Early Period

It seems plausible that a spherical fluid droplet should settle more rapidly under the influence of gravity than would a comparable solid sphere, since fluid tends to 'slip' at a droplet interface (by establishing an internal circulation), whereas the 'no-slip' condition prevails at a solid surface. Such were the theoretical predictions of Rybczynski (1911) and Hadamard (1911), who generalized Stokes' (1851a) (low-Reynolds number) solution for the settling velocity of a solid sphere to the case of a spherical fluid droplet. However, experiments (Lebedev 1916, Silvey 1916) performed shortly thereafter revealed that fluid droplets of sufficiently small radius settled as if they were solid spheres, obeying Stokes' original formula.

Boussinesq (1913a), in an attempt to resolve this discrepancy between theory and experiment, postulated the existence of a 'surface viscosity', conceived as the two-dimensional equivalent of the conventional three-dimensional viscosity possessed by bulk-fluid phases. Owing to viscous interfacial frictional resistance, this surface viscosity would be expected to 'solidify' the fluid interface by diminishing its mobility, and to scale in such a manner as to become of increasing importance with decreasing droplet radius, all other things being equal. Surface viscosity was not altogether a new concept even then, for Plateau (1869) had earlier suggested such a possibility upon observing the difference between the damping of a needle within a surfactant-adsorbed gas-liquid surface and its comparable damping when submerged within the bulk liquid. Later, Plateau (1871) related the existence of surface viscosity to the stability of foam systems.

Boussinesq (1913b) used his theory of surface viscosity to derive an analytical expression for the settling velocity of a spherical fluid droplet. A 'hardening' of the fluid interface with decreasing radius was indeed predicted, such that for a sufficiently small radius the droplet would behave as a solid sphere, consistent with experiment.

For many years the Boussinesq solution, based upon his surface viscosity postulate, was accepted as the explanation for the anomalous droplet settling-velocity results, encouraging the development of a plethora of instruments for measuring 'surface viscosity', as well as other rheological properties of a fluid interface [see the review by Joly (1964)].

Meanwhile, Rayleigh (1916) was working to establish a quantitative theory of 'Bénard instability', offering circumstantial evidence that the cellular circulation patterns observed by Bénard (1901) in shallow pools of liquid heated from below arise via instabilities created by buoyancy-driven convection. While this explanation of Bénard instability was accepted at the time, subsequent experiments (Low & Brunt 1925) pointed to untenable discrepancies between Rayleigh's theory and the observations of Bénard. Block (1956), who made the crucial observation that instabilities occur also for thin fluid layers heated from *above*, postulated that spatial nonuniformities in surface tension were responsible for the instabilities. This was later confirmed by the analysis of Pearson (1958).

The basic physical process by which a surface-tension gradient produces convection in the bulk phase became known (Marangoni 1871) as the 'Marangoni effect' [though Thompson (Lord Kelvin) (1855), was, in fact, first to observe this phenomenon], referring generally to surface tension variations caused by local inhomogeneities in either surface temperature or surfactant concentration.

Examples of the Marangoni effect in Nature were by then recognized in phenomena such as the formation of droplets (or 'tears') of strong wine on the sides of a wine glass (Thompson 1855, Loewenthal 1931), camphor dance (Bikerman 1958), and crystal climbing (Bikerman 1958), while an additional application to dynamic interfacial processes was soon established by the experiments of Young *et al.* (1959), who showed that gas bubbles could be prevented from gravitationally rising through a fluid by the imposition of a uniform vertical temperature gradient. This constitutes an example of the phenomenon known as 'thermocapillarity'.

Reopening the controversy surrounding the Rybczynski-Hadamard formula for the settling velocity of a fluid droplet, Levich (1962) voiced the opinion that the explanation lay not, as Boussinesq suggested, in the existence of a surface viscosity, but rather in the Marangoni effect. Frumkin & Levich (1947) postulated that the anomalous experimental results revealing that fluid droplets settle as solid spheres were, in fact, due to the presence of surface-active agents, which were swept to the rear of the droplet as it settled. The surface concentration gradients and concomitant interfacial-tension gradients thereby created were thus now assumed to be the agencies responsible for retarding the settling velocity. Experimental findings were soon reported (Gorodetskaya 1949) as qualitatively justifying Levich's theory.

As both effects were understood to arise from the presence of surface-active agents, the controversy over "surface viscosity effect" vs "surface-tension gradient effect" was regarded as being nontrivial. This controversy was further rekindled as interest developed in the classical problem of wave damping by surface-active agents.

Ever since early observations indicated that oil spread over the surface of the sea causes a damping of ripples [Pliny the Elder, in Book 2 of his *Natural History*, noted that divers found damping of waves by this mechanism useful for improved underwater vision, whose observations were later expanded upon by Benjamin Franklin (1773)], researchers have sought to explain this practically useful phenomenon. Whereas Thompson (Lord Kelvin) (1871) was the first theoretician to consider the capillary wave problem in an inviscid medium, it was not until the studies of Levich (1941) and Lamb (1945) that the damping effect of a thin liquid film or surfactant monolayer at the fluid surface was explained through the existence of inhomogeneities in surface tension, i.e. as resulting from the Marangoni effect.

For many years thereafter, studies of capillary (Brown 1940, Levich 1941, Hansen & Mann 1964, Lucassen & Hansen 1966) and longitudinal (Platikanov *et al.* 1966, Lucassen 1968, Lucassen & van den Tempel 1972, Garrett & Joos 1976, Maru & Wasan 1979, Ting *et al.* 1985) waves at fluid interfaces revealed that a surface viscous effect, if indeed present, would impart a damping influence identical to that caused by the Marangoni effect. As this synchronous behavior had also been observed earlier with falling droplets, the existence and/or importance of surface viscosity as a unique physical property of a fluid interface [in the sense that Plateau (1869) and Boussinesq (1913a) had seemingly intended] became a questionable issue, particularly were one to dismiss the (admittedly qualitative) experiments that had for many years been invoked to document the 'viscous' nature of adsorbed monolayers.

A novel phenomenon, commonly subsumed under the appellation 'interfacial turbulence', was initially observed experimentally in the thesis of Wei (1955). The basic phenomenon he sought to investigate concerned the spontaneous mixing that arises when two unequilibrated and immiscible fluids, at least one of which contains a solute that is soluble in both phases, are brought into mutual contact. The interfacial activity generated as the solute underwent interphase transport was observed to range between relatively minor rippling and twitching motions at the interface, to the creation of streams of liquid reaching into the second phase, such streamers breaking finally into dispersed emulsion droplets. The most intense activity was observed when contact of the phases occasioned a rapid, exothermic chemical reaction between two solutes in the neighborhood of the interface (Wei 1955, Sherwood & Wei 1957).

Once again, the Marangoni effect was soon shown to be the primary cause of interfacial turbulence (Sternling & Scriven 1959), though the theoretical analysis, which employed the original Boussinesq theory, demonstrated that interfacial viscosity effects were in this case clearly distinguishable from Marangoni effects. Thus, whereas the Marangoni effect was observed to *create* the instability, owing to solute inhomogeneities across the interface, the interfacial viscosity effect was observed to *damp* the interfacial instability.

The work by Sternling & Scriven (1959), together with Scriven's (1960) paper, wherein Boussinesq's theory was generalized to 'material' interfaces of arbitrary curvature, marked a turning point toward today's "modern" understanding of interfacial rheology.

Interfacial Rheology After 1960: The Modern Period

Since the work of Scriven (1960), consistent theoretical analyses of existing experimental devices have been developed to allow reproducible, quantitative measurements of interfacial rheological properties, such as interfacial viscosity and elasticity (this latter property being related to interfacial-tension gradients, i.e. Marangoni effects). These will be discussed in Chapters 6 to 9. Analyses of Bénard and Rayleigh instabilities, and interfacial turbulence, have both undergone further refinement, often with an explicit accounting of the generally unique contributions of interfacial-tension gradients and interfacial viscosities, as will be discussed in Chapter 10. Rational explanations for the similar influences of interfacial viscosity and interfacial-tension gradients encountered in certain interfacial flows, such as surface wave propagation and the sedimentation of spherical droplets, have also been developed in the modern, post-1960 period, as will be discussed in Chapters 4 and 5.

The importance of interfacial rheology to the drainage and stability of thin liquid films has been established in recent years; moreover, through this single phenomenon, the importance of interfacial rheology to important industrial processes involving the dynamics of foams and emulsions has been identified. These topics will be addressed in Chapters 10 to 14.

Many novel theoretical developments have transpired since 1960. These are reviewed in Chapter 15, which initiates Part II of this text, the latter being concerned with the current state of these new theoretical developments.

Presently, interfacial rheology is understood to play a significant role in many natural and industrial processes involving dynamic fluid interfaces, particularly when the specific surface is large, and surfactants are adsorbed to the interface. In the remainder of Chapter 1, this basic physical context of interfacial rheology is further elaborated, as too are several current engineering applications.

1.2 Surfactants

Surface-active materials possess various names, such as detergent, wetting agent, emulsifier, demulsifier or dispersing agent, depending upon their function. Generally, they are large molecules (the classical surfactant molecule has a molecular weight of 200-2000) possessing a bipolar structure composed respectively of *hydrophobic* (water-'hating') and *hydrophilic* (water-'loving') segments, such molecules being 'attracted' to an interface separating aqueous and nonaqueous phases. This dual physicochemical nature causes the molecule to seek the bi-phase environment of the phase transition zone.

Fatty acids and alcohols are typical surfactant materials, serving as the active ingredient for many commercial products, including soaps, detergents, corrosion and rust inhibitors, dispersing agents, demulsifiers, germicides, and fungicides. The hydrocarbon "tail" of the fatty acid or alcohol is the *lypophilic* (oil-'loving') or hydrophobic part of the surfactant, and the polar

—COOH or —OH "head" the hydrophilic part (see Figure 1.2-1). Short-chain fatty acids and alcohols are generally soluble in the aqueous phase; however, by increasing the hydrocarbon chain length, or alternatively by reducing the polarity of the polar head groups (e.g. by the addition of electrolyte to the aqueous phase), these surfactants are ultimately rendered *insoluble*, and may form an *insoluble monolayer* at the fluid interface.

Aside from fatty acids and alcohols, the role of surface-active agents may also be played by polymeric molecules, proteins, and small solid particles (Schwartz *et al.* 1977).

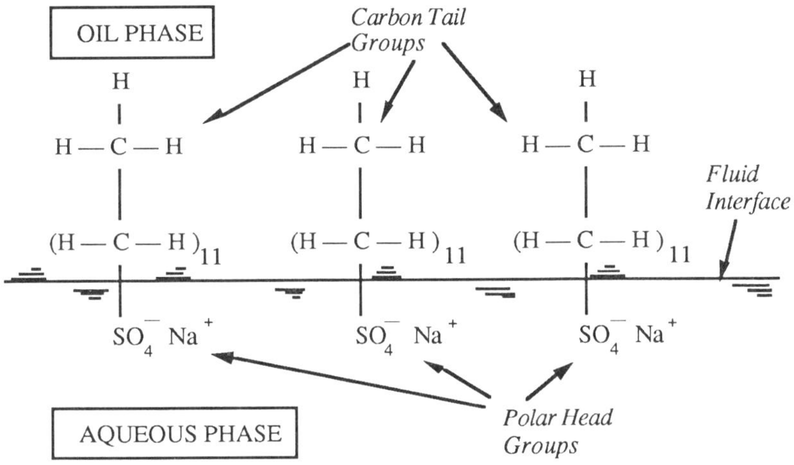

Figure 1.2-1 Sodium dodecyl sulfate molecules are depicted adsorbed to a fluid interface, with their polar heads in the aqueous phase and their hydrophobic carbon tails in the nonaqueous phase.

Synthetic and natural polymers often exhibit a significant surfactant tendency, particularly when the hydrophobic and hydrophilic parts of the molecule are sufficiently separated, as with block and graph copolymers. Carboxylic polymers, sulfonated polymers, phenolic polymers and polyvinyl polymers constitute further examples of polymeric surfactant molecules.

Protein, a natural polymeric substance, is a linear polymer formed of amino acid groups connected by peptide linkages to form a polypeptide chain, and may be surface active by virtue of the existence of hydrophobic/hydrophilic segments. In a homogeneous bulk phase, proteins often assume a helical configuration; however, when adsorbed to an interface, this helical structure is destroyed as the molecule arranges itself according to its hydrophilic/hydrophobic constitution. In the process the protein is denatured (i.e. bonds are broken and the polypeptide chain altered) and hence may become insoluble in the aqueous phase. As a result, an insoluble monolayer is formed, which, when subject to deformation may exhibit peculiar gel-like interfacial rheological behavior.

Submicron solid particles may display surface activity owing to the intrinsic physicochemical hydrophobicity of their surfaces. This hydrophobicity is manifested by a finite *three-phase equilibrium contact angle* at the solid/water/air (or oil) contact line (see Figure 1.2-2). The particle in the vicinity of the interface is then forced by the concomitant interfacial forces to adopt an equilibrium position *at* the interface, for which this contact angle requirement is met. This action of surfactancy is useful in the engineering process of ore flotation, as is discussed in §1.3.

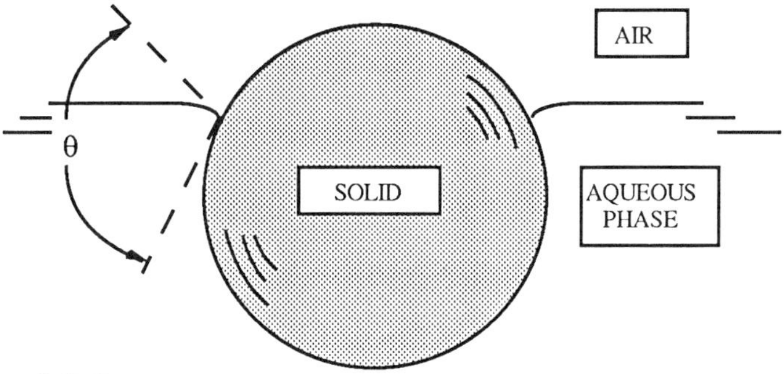

Figure 1.2-2 A nonzero contact angle θ is shown between an aqueous phase/air surface and a solid particle. It arises from the hydrophobicity of the particle surface. A completely hydrophilic particle exhibits a zero contact angle.

A surfactant is generally classified according to the polarity of its hydrophilic head; thus, the primary classifications are *anionic* (e.g. carboxylic acids, sulfuric esters and sulfonates), *nonionic* (e.g. polyethenoxy and polyhydroxy surfactants), or *cationic* (e.g. fatty nitriles and amines). Anionic surfactants are those most commonly encountered in practice.

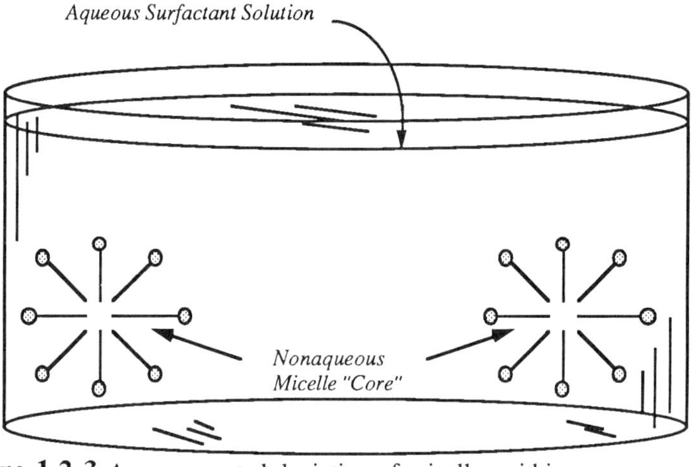

Figure 1.2-3 An exaggerated depiction of micelles within an aqueous solution. Each micelle is composed of individual ionic surfactants, which aggregate in such a way that the hydrophobic surfactant tails are sheltered within nonaqueous cores.

At very low concentrations, molecular surfactants behave as normal (perhaps electrolytic) solutes; however, as a certain bulk-phase concentration known as the *critical micelle concentration* (CMC) is approached, these solutes often form aggregates in the bulk phase. Such agglomerates are termed *micelles*, or *association colloids*, as depicted in Figure 1.2-3. Micelle formation in the bulk phase may have a significant effect upon interfacial and bulk-phase physicochemical and rheological properties.

1.3 Colloidal Dispersions

When surfactant adsorbs onto a fluid interface, interfacial properties, such as interfacial tension, are significantly altered; moreover, new *dynamic* interfacial properties may arise. A further consequence of surfactant adsorption is that fluid-fluid *colloidal dispersions,* by which is meant suspensions of liquid or gas within a continuous dispersing liquid, are stabilized. Such dispersions may exist for extended periods of time. These colloidal dispersions, called *bubbly liquids* (a dilute suspension of gas bubbles within a liquid), *foams* (large gas-phase volume fraction within a dispersing liquid), or *emulsions* (liquid droplets dispersed within a second immiscible liquid), generally possess a large specific surface, often exhibiting gross hydrodynamic behavior that is significantly influenced by these interfacial rheological attributes. The largest specific surfaces arise in foams or liquid-liquid systems containing very small emulsion droplets. The latter, termed *microemulsions*, possess droplet sizes in the range of 10-100 nm.

1.4 Interfacial Hydrodynamics in Engineering Processes

Engineering processes that capitalize upon the unique properties of liquid dispersions find wide application in industry, particularly in the area of chemical engineering separation processes. In this final section we provide a short, qualitative description of several current liquid dispersion technologies wherein interfacial hydrodynamics possesses a recognized role. References to each of these applications may be found in the 'additional reading' citations provided at the conclusion of this chapter.

Processing/Flow of Emulsions

Flows of emulsified systems are widely encountered in engineering applications; examples are provided by the following technologies: hydraulic fluid, food, health care, waste processing, cosmetic, herbicide, paint, paper making, oil recovery, liquid membranes. As the bulk-phase macroscale rheological properties of an emulsion (which determine whether the emulsion will be Newtonian or non-Newtonian, highly viscous or relatively inviscid, etc.) are often significantly related to the interfacial rheological properties of the surfactant-adsorbed fluid interface, microscale interfacial hydrodynamics may largely influence the composite emulsion flow behavior. As will be demonstrated in Chapter 4 (cf. example 6 of §4.4), a dilute emulsion of (spherical) droplets dispersed within an otherwise continuous Newtonian fluid is rendered *elastic* by virtue of interfacial tension, even leaving aside the additional rheological possibility of interfacial-tension gradients; moreover, the bulk-phase viscosity of the emulsion is enhanced by the existence of interfacial viscosity at the droplet interfaces. Interfacial rheology may also indirectly influence the nature of emulsions formed in dispersion processes by impacting on the phenomenon of droplet-droplet coalescence (see Chapter 13).

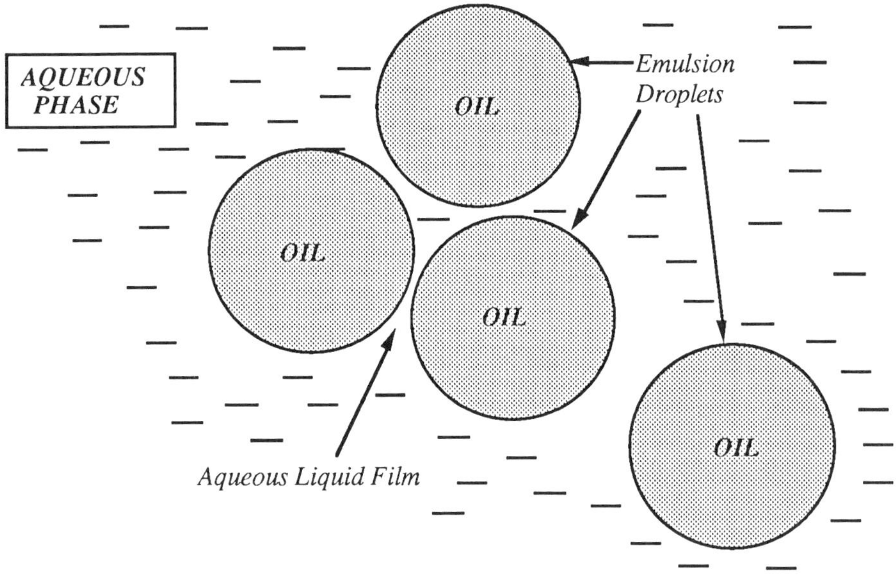

Figure 1.4-1 Liquid films between dispersed droplets of an emulsion.

A knowledge of interfacial hydrodynamic theory may be useful in suggesting the proper choice of surfactant (e.g. to influence bulk-phase emulsion rheological properties), power requirements or process design. Alternatively, interfacial rheological properties may contribute adversely to processing characteristics, such as unwanted flocculation or agglomeration of droplets.

Processing/Flow of Foams

Owing to the large interfacial area-to-volume ratios encountered with foams, the rheological properties of foams are generally more strongly influenced by interfacial rheology than are the comparable properties of emulsions. Thus, as we will learn in Chapter 14, the primary aspects of foam response to deformation (e.g. the elastic or highly viscous nature of foams and the existence of yield stress phenomena) may be directly correlated with its static and dynamic interfacial properties.

Foam flows may be encountered in froth flotation technologies, fire fighting, well drilling, enhanced oil recovery and 'clean-out' operations in the oil industry, foam fracturing and particle transport by foams, in addition to many of the technologies already cited for emulsion flows.

Stability of Emulsions and Foams

A common problem encountered in the treatment of waste streams is the need to 'break' an emulsion that contains an unwanted phase (typically the organic phase). The process of *demulsification* depends primarily upon the stability of the liquid films which form between droplets, as depicted in Figure 1.4-1. The stability of this liquid film is in part related to the rate of film drainage from a *thick* to a *thin* film (discussed at length in Chapter 11), and in part to the stability of the final equilibrium thin film to mechanical or thermal disturbances (discussed in Chapter 12): both processes may depend strongly upon interfacial properties. Likewise, referring to Figure 1.4-2, the process of *defoaming* involves the breaking of thin liquid films, formed between adjacent gas bubbles. The importance of interfacial rheological properties to foam and emulsion stability will be discussed in Chapter 13.

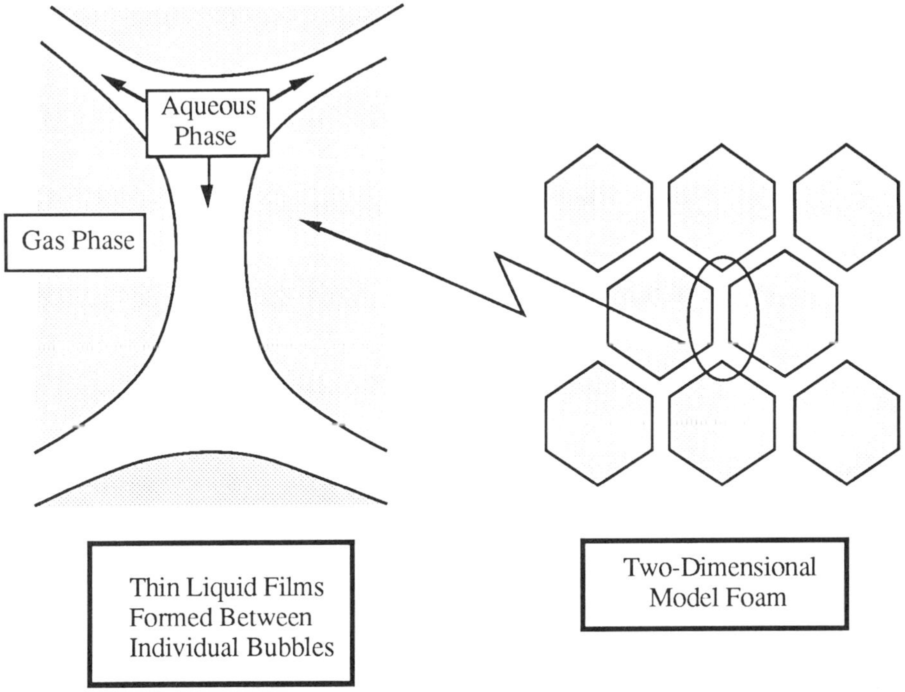

Figure 1.4-2 Thin liquid films are shown, formed between individual bubble "cells" of a model two-dimensional hexagonal foam. The drainage and stability of these films determines the stability of the foam.

Spraying and Atomization

Ink-jet printing involves the creation of a stream of ink droplets ejected from a narrow-bore nozzle which is made to vibrate at a frequency near the resonance frequency of the continuous liquid jet exiting the tube. This causes the jet to break up into essentially uniform-size droplets (see Figure 1.4-3), according to the phenomenon of Rayleigh instability (discussed in Chapter 10). The carrier fluid (liquid developer) is generally water, and the water-insoluble ink is dispersed either in the form of an emulsion or as small solid particles, though in either event various surfactants will be present. These surfactants, as well as other additives (which may themselves possess a limited surface activity), have the effect of creating interfacial rheological stresses at the ink-droplet surfaces created from the ejecting liquid jet; hence, the size, shape, and behavior of the ink droplets will depend at least to some degree upon dynamic interfacial effects.

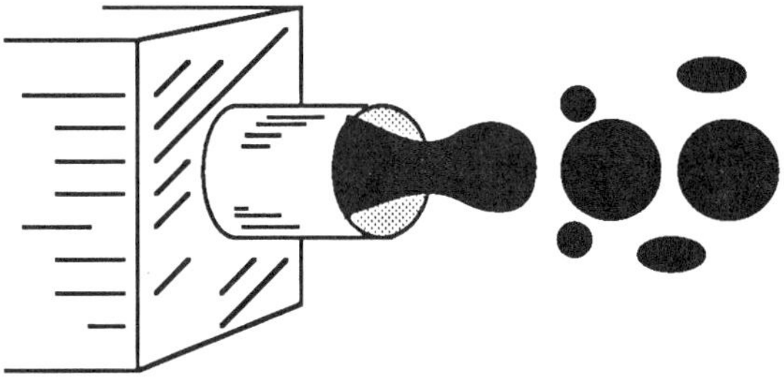

Figure 1.4-3 Ink droplets ejected from a narrow nozzle. A uniform distribution of droplet sizes is created by vibrating the nozzle near the resonance frequency of the initially continuous liquid jet issuing from the tube.

Ink-jet printing is only one example of *atomization*, which is the process whereby droplets smaller than about 100 micron are created. Other examples of atomization arise during fuel injection in the internal combustion process, jet engines, and rockets. Similarly, operations which produce larger droplets, generally in the range of 100–1000 micron, are known as *spraying* operations, and include processes such as evaporation, humidification, paint-lacquer or soluble plastic spraying, firefighting, and agricultural spraying. In most of these operations, as with ink-jet printing, droplets of a particular size range are desired: this range may be influenced by interfacial hydrodynamical effects.

Distillation

Distillation is a common chemical engineering separation process, whereby various volatile chemical components are separated from a liquid solvent solution on the basis of partial vapor pressure differences. In conventional distillation processes, a solution is heated within a sieve-plate or packed-bed distillation column, as depicted in Figure 1.4-4. Either in a discontinuous fashion (as is the case for the sieve-plate column) or continuously (as with the packed-bed column), vapor is intimately contacted with the liquid phase, enabling the more volatile components present in the parent solvent to be successively *stripped.*

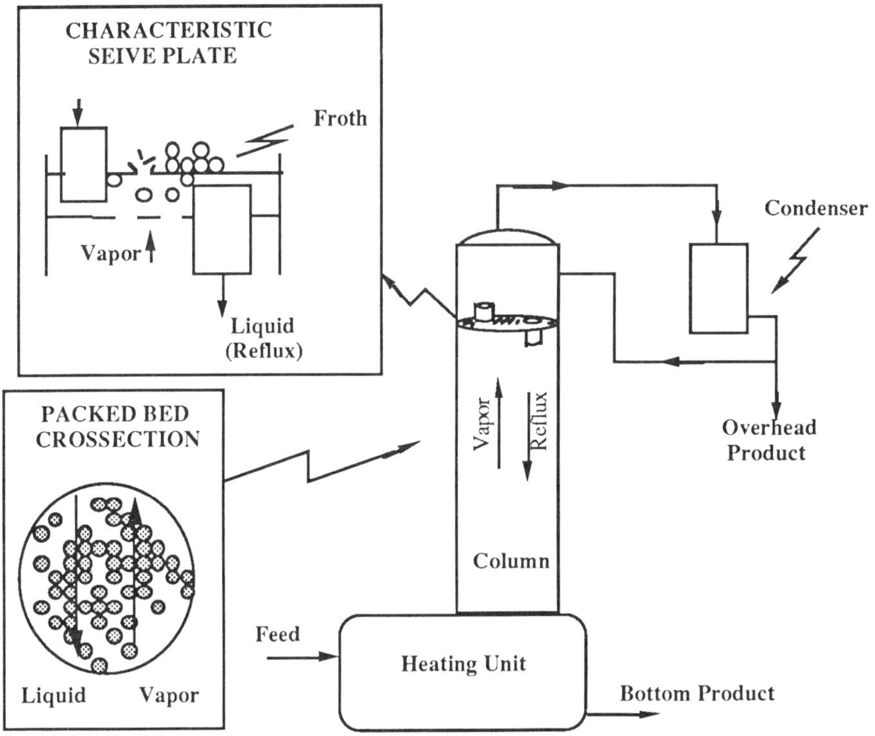

Figure 1.4-4 A typical distillation column, with cross-sectional representations given for sieve-plate and packed-bed columns.

For the sieve-plate column, a bubbly liquid is created on each sieve plate, which in the presence of surfactant additionally creates a froth situated immediately above the bubbling liquid. This froth, if not so severe as to cause column flooding, produces an even greater contact area between vapor and liquid than exists in the bubbly liquid phase, thereby producing a potentially significant increase in the efficiency of the distillation process. The choice of surfactant that might be used to effect this enhanced efficiency will depend, in part, upon the interfacial rheological properties existing at the bubble surfaces — particularly in relation to the stability of bubbles created.

For the packed-bed column, interfacial rheological properties are important in determining the stability of the thin, continuous liquid films which form over the surfaces of the bed packing, and which, if such films prove unstable, break up into rivulets and small droplets to expose patches of the packing material to the vapor phase. Naturally, the resulting dry area does not contribute to the interphase transfer process.

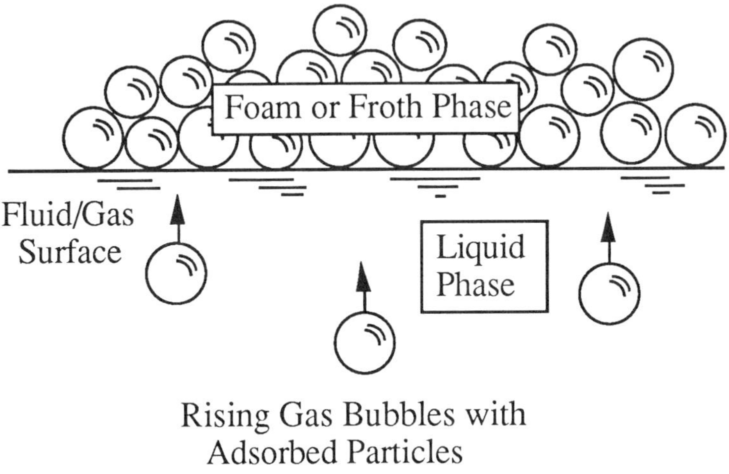

Figure 1.4-5 The basic phenomenon of flotation is depicted. Rising gas bubbles entrain surfactant particles distributed throughout the continuous liquid phase, and remove the particles from the latter, ultimately to a foam or froth phase, which gathers at the liquid/gas surface.

Flotation

A common method of surfactant-aided separation occurs during the process of flotation. This process is used to separate both undissolved (particulate flotation) and dissolved (colligend flotation) matter from a liquid phase by bubbling gas through the liquid. The matter is separated by its inclination to adsorb onto bubble surfaces, which then carry the matter to a separate froth phase (see Figure 1.4-5). The most widely used flotation process is that of ore flotation, whereby suspended mineral particles are removed from mineral slurries. Other flotation processes include ore flotation, precipitate flotation, protein separation, surfactant flotation, and hydrolyzable metal flotation.

Interfacial hydrodynamics may play a role in flotation processes by affecting the rate of bubble rise through the liquid (as is discussed via examples in Chapters 4 and 5) as well as by affecting the stability of the froth formed.

Microemulsion and Micellar Engineering Processes

Microemulsion and micellar colloidal dispersions are of particular utility for enhancing interphase mass transfer exchange rates owing to the extremely large specific surfaces available in these systems. They may also be useful as high-interfacial area "microreactors". Examples of the former are micellar-enhanced ultrafiltration for treatment of waste water, reversed-micelle solvent extraction for bioseparation of proteins and admicellar chromatography; examples of the latter are provided by chemical reactions occuring within microemulsion droplets; emulsion polymerization, photochemical reactions, enzymatic reactions, and precipitation phenomena.

The role of interfacial rheology *per se* in microcolloidal dispersed systems may ultimately prove to be rather small; in fact, intrinsic interfacial viscous and elastic stresses are likely so large that the dispersed liquid droplets may effectively be regarded as solid-like, at least with respect to their hydrodynamic behavior.

Coating Processes

Coating processes, such as those encountered in the photographic industry, may be significantly influenced by interfacial rheological properties. Coating fluids generally contain molecular species which readily adsorb to the surface of a spread fluid layer. Owing to the rheological properties imparted to the interface by surfactant adsorption, both the shape and thickness of the single- or multi-layered coating may be anticipated to depend upon intrinsic dynamic interfacial phenomena. Thus, experimental knowledge of the magnitude of interfacial rheological properties of the particular surfactant-adsorbed monolayers encountered, together with a theoretical understanding of the hydrodynamics of the coating process, will facilitate a quantitative understanding of coatability.

1.5 Summary

Interfacial rheology addresses the hydrodynamic boundary conditions that obtain at an interfacial boundary separating two immiscible fluids, on which boundary a surfactant may be adsorbed. This field of rheological science is of particular importance to those multiphase fluid systems possessing a large specific surface, such as occurs with many liquid-fluid colloidal dispersions. Dispersions of this kind, namely bubbly liquids, emulsions and foams, are often encountered in engineering processes; hence, an understanding of these technological processes often requires a quantitative knowledge of interfacial rheology.

Colloidal dispersions are typically stabilized by the presence of surface-active materials. Surfactants stabilize such systems largely by imparting an intrinsic rheological behavior to the fluid interface, with respect to both its viscous and elastic interfacial natures. In Chapter 2 we will consider those basic interfacial properties that arise by virtue of the adsorption of surfactants onto a fluid interface.

Questions for Chapter 1

1.1 What are colloidal dispersions? Why is surfactant generally present in a fluid-fluid colloidal dispersion?

1.2 Under which two physicochemical conditions should interfacial rheology be considered in the theoretical description of an interfacial flow?

1.3 What was the principle controversy in the early (pre-1960) period of interfacial rheology?
1.4 Cite two physicochemical causes for the 'Marangoni effect'.

Additional Reading for Chapter 1

§1.2 Surfactants

Schwartz, A.M., Perry, J.W. & Berch, J. 1977 *Surface Active Agents and Detergents*. New York: Krieger.

Gaines, G.L. 1966 *Insoluble Monolayers at Liquid-Gas Interfaces*. New York: Interscience Publishers.

§1.3 Colloidal Dispersions

Davies, J.T. & Rideal, E.K. 1963 *Interfacial Phenomena*. New York: Academic Press.

Vold, R.D. & Vold, M.J. 1983 *Colloid & Interface Chemistry*. London: Addison-Wesley.

§1.4 Interfacial Hydrodynamics in Engineering Processes

Processing/Flow of Emulsions

Becher, P. 1983 *Encyclopedia of Emulsion Technology*. New York: Marcel Dekker.

Lissant, K.J. 1974 *Emulsions and Emulsion Technology: Part I*. Surfactant Science Series Vol. 6. New York: Marcel Dekker.

Processing/Flow of Foams

Kraynik, A. 1988 Foam Flows. *Ann. Rev. Fluid Mech.* **20**, 325–357.

Stability of Emulsions and Foams

Becher, P. 1983 *Encyclopedia of Emulsion Technology*. New York: Marcel Dekker.

Spraying and Atomization

Schwartz, A.M., Perry, J.W. & Berch, J. 1977 *Surface Active Agents and Detergents*. New York: Krieger.

Torrey, S. 1984 *Emulsions and Emulsifier Applications*. Park Ridge, New Jersey: Noyes Data Corporation.

Distillation

Berg, J.C. 1988 The effect of surface-active agents on distillation processes. In *Surfactants in Chemical/Process Engineering*, vol. 28 (eds. D.T. Wasan, M.E. Ginn & D.O. Shah), pp. 29–76. New York: Marcel Dekker.

Flotation

Lemlich, R. 1972 *Adsorptive Bubble Separation Techniques*. New York: Academic Press.

Somasundaran, P. & Ramachandran, R. 1988 Surfactants in flotation. In *Surfactants in Chemical/Process Engineering*, vol. 28 (eds. D.T. Wasan, M.E. Ginn & D.O. Shah), pp. 195–235. New York: Marcel Dekker.

Microemulsion and Micellar Engineering Processes

Leung, R., Hou, M.J. & Shah, D.O. 1988 Microemulsions: formation, structure, properties, and novel applications. In *Surfactants in Chemical/Process Engineering*, Vol. 28 (eds. D.T. Wasan, M.E. Ginn & D.O. Shah), pp. 315–367. New York: Marcel Dekker.

Robb, I.D. 1982 *Microemulsions*. New York: Plenum Press.

Coating Processes

Valentini, J.E., Thomas, W.R., Sevenhysen, P., Jiang, T.S., Liu, Y. & Yen, S.C. 1991 Materials and interfaces. Role of dynamic surface tension in slide coating. *Ind. Eng. Chem.* **30**, 453–461.

2

"I begin by the study of an element, the influence of which must be regarded as self-sufficient, namely, the tension of liquid surfaces, a curious property whose existence has long remained a mere hypothesis . . .. I draw the following conclusions:

"1st, tension really exists in every liquid surface and consequently in every liquid film; 2nd, this tension is independent of the curvature of the surface of the film; it is the same throughout the whole extent of the same surface, or of the same film, and at each point is the same in all tangential directions; 3rd, it is independent of the thickness of the film, at least so long as this thickness is not less than twice the radius of molecular attraction; 4th, it varies with the nature of the liquid; 5th, in the same liquid it varies in the opposite direction to the temperature, but at ordinary temperatures it undergoes only small alterations; 6th, we possess a great number of processes for measuring this tension."

J. Plateau (1869)

CHAPTER 2

Basic Properties of Interfacial Rheology

In conventional circumstances, a fluid interface in equilibrium exhibits at the "macroscale" (i.e. a characteristic experimental length scale much larger than the 'thickness' of the interfacial transition zone itself) an intrinsic state of tension, constituting the two-dimensional analog of the intrinsic state of compression (quantified by the thermodynamic pressure) in "bulk"-phase, three-dimensional fluids at equilibrium. The magnitude of this interfacial tension varies with temperature, bulk-phase pressure, and areal surfactant concentration within the fluid interface; thus, should a nonuniformity (in surfactant concentration, say) develop within the fluid interface, an interfacial-tension gradient will result, subsequently inducing both areal and volumetric fluid motions. This gradient-driven flow is the basic physical mechanism underlying the so-called 'Marangoni effect'.

In addition to the possible existence of interfacial-tension gradients at surfactant-adsorbed fluid interfaces, other interfacial rheological stresses of a viscous nature may arise, such as those relating to interfacial shear and dilatational viscosities. Such interfacial viscosities pose a damping influence, similar to the action of bulk-phase viscosity, upon any three-dimensional fluid motion proximate to the fluid interface; moreover, in certain circumstances the existence (and magnitude) of these areal viscosity coefficients may appreciably influence the interfacial motion itself.

While other interfacial stresses, such as interfacial shear elasticity and yield stress, may also be observed to occur at fluid interfaces, the three types of interfacial stresses cited above are those most commonly encountered. In Chapter 2 we focus attention upon these interfacial stresses, pursuing obvious physical analogies existing between interfacial and bulk-phase fluid properties; this entré into the field is pursued within a simple mathematical context, without recourse to vector-dyadic notation or differential geometry (which, however, constitute the more mature context of later chapters). A sufficient, intuitive understanding of interfacial rheology is thereby provided in this chapter for those readers interested specifically in the practical, experimental measurement and applications chapters of Part I, which begin with Chapter 6. For the more general reader, a brief perusal of Chapter 2 may prove sufficient, as all of this material will be covered more thoroughly in Chapters 3 to 5.

We begin in §2.1 with a brief consideration of the physics of equilibrium interfacial tension, followed by a discussion of dynamic interfacial stresses in §2.2. Simple illustrations of the Marangoni effect, interfacial shear viscosity, and interfacial dilatational viscosity are then respectively provided in subsequent sections. A summary in §2.6 concludes the chapter.

2.1 Interfacial Tension

The equilibrium property of interfacial tension is inherent in many accepted norms of interfacial behavior; included therein are the tendency for bubbles and droplets to assume a spherical shape, the beading-up of droplets on a solid surface, and the rise of liquids through narrow capillaries. Additionally, interfacial turbulence, Rayleigh instabilities, the elastic nature of foams, and the coalescence behavior of droplets are each strongly influenced by the property of interfacial tension.

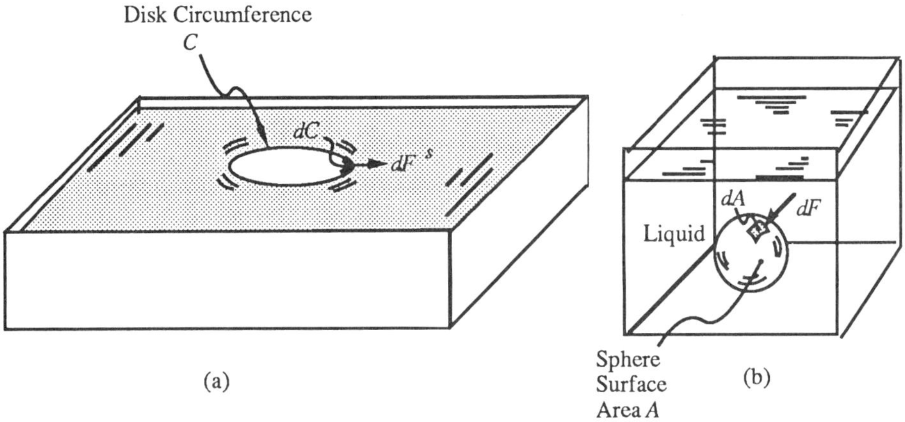

Figure 2.1-1 (a) A tensile force dF^s acts upon the differential line element dC within a fluid interface. This two-dimensional force arises by the action of interfacial tension σ. (b) The bulk-fluid analog of the interfacial tension is the hydrostatic pressure p, which acts as a three-dimensional compressive stress (force per unit area), producing a compressive force dF acting upon a differential element of area dA immersed within the bulk liquid.

Thus, imagine (as depicted in Figure 2.1-1) a thin, circular disk of circumference C and negligible thickness, immersed within a planar fluid interface. Acting outwardly along a differential element dC of the perimeter of such an immersed disk is a tensile force dF^s exerted by the (shaded) fluid upon the disk, and lying in the tangent plane of the interface, as shown in the figure. This contact force may be expressed quantitatively as

$$dF^s = \sigma dC\,, \qquad (2.1\text{-}1)$$

where σ is the interfacial tension. The physical dimensions of Eq. (2.1-1) are such that interfacial tension may be regarded as a force per unit length acting within a (macroscale) fluid interface. [A more general vector-force definition of dF^s is provided in §3.3.]

Interfacial tension is the two-dimensional counterpart of the three-dimensional hydrostatic pressure p. As illustrated in the figure, a *compressive* force dF, directed inwardly from the volumetric fluid, will act upon a differential area element dA of a sphere immersed within a bulk fluid. This contact force arises from the action of the hydrostatic pressure, and may be expressed as

$$dF = pdA. \qquad (2.1\text{-}2)$$

The intimate relationship between p and σ, as suggested by Eqs. (2.1-1) and (2.1-2), will recur in various contexts throughout this text.

Interfacial tension forces become particularly manifest when the interface is placed in a state of torsion or bending, such as generally occurs when the interface is curved. In such circumstances the interfacial tension then produces a net vector force that is normal to the interface, thereby influencing the magnitude of the bulk-phase pressure forces. Early in the 19th century, Laplace (1806) and Young (1805), considering the equilibrium state of contiguous fluid phases separated by a curved fluid interface, independently discovered the basic equation of capillarity, namely, the existence of a pressure difference

$$\Delta p = 2H\sigma\,, \qquad (2.1\text{-}3)$$

across any curved fluid interface, with H the mean curvature of the surface. It is as a consequence of this relationship (known as the "Young-Laplace" or "Laplace" equation) that gas bubbles and droplets adopt a spherical shape at equilibrium and that curved fluid interfaces rise through narrow-bore vertical capillaries.

The thermodynamics of interfacial tension, its significance in relation to electrocapillarity, wetting phenomena at solid surfaces, and nucleation — as well as a great many additional physical contexts and applications, several of which are not explicitly addressed within this text — are subjects discussed in detail in a number of classical treatises concerned with surface phenomena (Adam, 1948, Davies & Rideal 1963, Adamson 1982).

Interestingly, the magnitude of surface tension forces generally diminishes significantly when a surface-active material accumulates at the interface. Thus, imagine that a *monomolecular film* of surfactant is spread uniformly over an entire fluid interface, subsequent to which a knife-edge barrier is used to sweep away a portion of this monolayer from the remainder. This action creates a highly concentrated monolayer on one side of the barrier, together with a less concentrated monolayer on the other, thereby giving rise to a force on the barrier in the direction of the less concentrated monolayer. Per unit length of barrier, this force is referred to in the literature as *surface pressure*, and its measurement in this manner displays the principle of the *Langmuir balance* (Langmuir 1917).

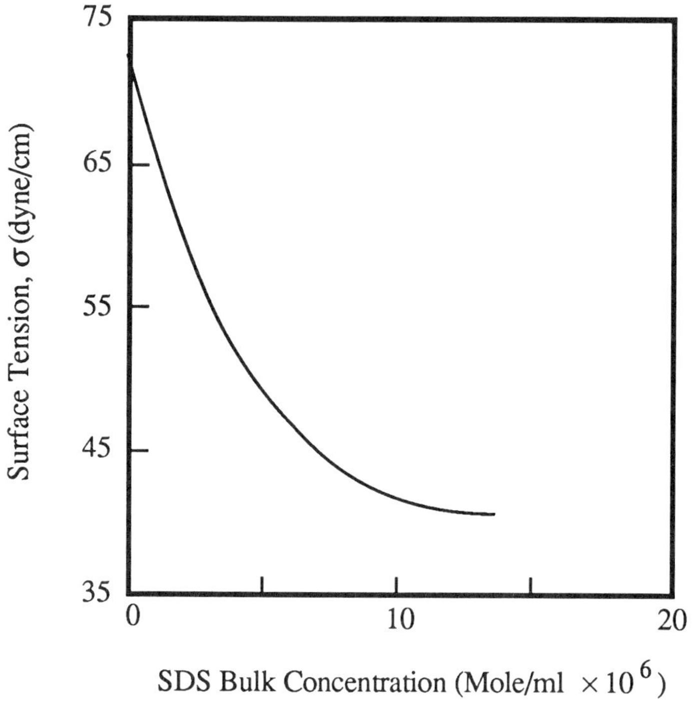

Figure 2.1-2 The equilibrium surface tension of sodium dodecyl sulfate (SDS) vs bulk-phase surfactant concentration. Values taken from Elsworthy & Mysels (1966).

The lowering of interfacial tension by the presence of an adsorbed surfactant may be regarded as the interfacial analog of the phenomenon of osmotic pressure, created by the presence of a dissolved solute in a bulk-fluid phase. According to Van't Hoff's relation, a dilute (i.e. ideal) solution of solute in equilibrium with the pure solvent across a semi-permeable membrane requires the maintenance of an *excess* pressure on the solution side, which exceeds that of the pure solvent by an amount

$$\Pi = p - p_o = kTx \, . \qquad (2.1\text{-}4)$$

Here Π, is the osmotic pressure, p_o the hydrostatic pressure in the absence of solute, k the Boltzmann constant, T the absolute temperature, and x the volumetric number density of solute molecules. (This osmotic pressure is necessary to prevent the solvent molecules from crossing the semi-permeable membrane and entering the solution phase, thereby diluting the latter.) Similarly, according to the Gibbs adsorption equation (cf. §5.6), a dilute layer of adsorbed surfactant molecules at an interface causes a diminution in the interfacial tension by the amount

$$\pi = \sigma_o - \sigma = kTx^s \tag{2.1-5}$$

with π the so-called *surface pressure*, σ_o the interfacial tension in the absence of surfactant, and x^s the surface-excess areal number density of surfactant molecules. [Thus, a lineal barrier in the plane of the interface separating (and enclosing) a surfactant monolayer from the pure solvent interface, requires the maintenance of a surface pressure on the side of the pure solvent as indicated by Eq. (2.1-5).] Equation (2.1-5) is only strictly appropriate in the dilute, ideal behavior limit, $x^s \to 0$; other more general adsorption equations, valid for nonideal systems, will be discussed in Chapter 5 (cf. §5.5).

Experimentally, one may determine the functional dependence of interfacial tension upon surfactant concentration for particular surfactant systems. These data are generally displayed graphically in the form of an *adsorption isotherm*, consisting of a plot of equilibrium interfacial tension vs bulk-phase surfactant concentration (at a given temperature). A typical adsorption isotherm is depicted in Figure 2.1-2.

Owing to the dependence of interfacial tension upon surfactant concentration, in a dynamic, disequilibrium state, instantaneous inhomogeneities in surfactant concentration created at the interface may produce interfacial-tension gradients. In turn, these result in an interfacial stress in the tangential plane of the interface, animating fluid motion in the neighborhood of the phase interface. Such dynamic interfacial behavior, induced or influenced by dynamical interfacial properties, such as interfacial-tension gradients, interfacial viscosity, and interfacial elasticity, constitutes the subject of the remaining sections of this chapter.

2.2 Dynamic Interfacial Properties

To simplify our initial examination of interfacial rheological stresses, the remaining sections of this chapter focus on a wholly *intrinsic* view of a macroscopically planar, dynamic fluid surface. According to this (asymptotically limiting) view, the interface is regarded as a two-dimensional entity, or material body in its own right, independent of the physical existence of the surrounding three-dimensional fluids in which it is, in fact, embedded. Practically, such an interface might correspond to a highly-viscous insoluble monolayer. This special limiting class of surfactant-adsorbed interfaces (which may be observed with certain mixed surfactant monolayers, or denatured proteins at a fluid interface), when deformed within the plane of the interface, responds dynamically in a manner that is essentially independent of any ensuing motion existing within the contiguous, three-dimensional, bulk phases. In other words, the interfacial rheological stresses acting within such surfactant monolayers are sufficiently large in comparison with the bulk-fluid stresses acting across the interface, to enable examination of the intrinsic nature of interfacial response to deformation, without simultaneously directing attention to any volumetric motion existing in the proximate bulk phases.

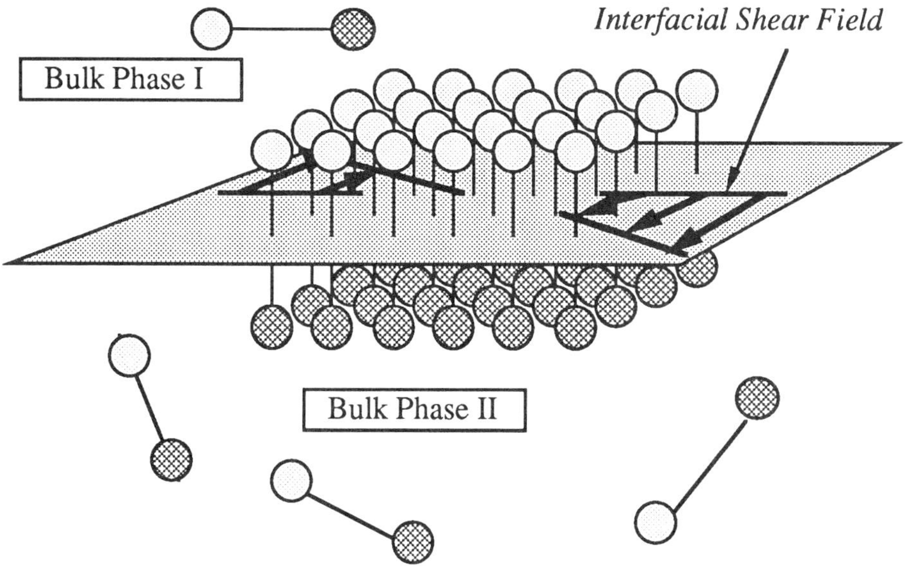

Figure 2.2-1 Idealized representation of a monolayer of surfactant "dumbells" (possessing hydrophobic/ hydrophilic extremities) at a fluid-fluid interface separating bulk phases I and II, and subject to an interfacial shear field. An excess viscous stress occurs in the region of the phase interface owing to the (volumetric) excess of surfactant particles in this zone, relative to the comparable concentration of particles in bulk phases I and II.

As was discussed briefly in Chapter 1, surfactant adsorption at a fluid interface is often a necessary condition for the existence of intrinsic interfacial rheological stresses, such as interfacial-tension gradients. Other interfacial stresses observed in the dynamic state are of a *viscous* nature. The basic physical interpretation of such viscous stresses may be understood, for example, via reference to Figure 2.2-1, which depicts a shear field imposed upon a surfactant monolayer at a fluid interface.

Owing to the presence of the monolayer, interfacial response to deformation is altered. Since the shear field imposed across the interface results in a large viscous interaction between adjacent surfactant particles, their close proximity in the interfacial zone results in a viscous stress that is much larger than that existing in the surrounding fluid, where the concentration of surfactant particles is considerably less. As the sizes of the surfactant particles are beyond resolution by a macroscale experimental probe, the local microstructure of the interfacial layer is not explicitly recognized by a macroscale observer. Rather, such an observer views the interface as being a two-dimensional surface at the intersection of otherwise homogeneous bulk phases. To account for those viscous stresses generated by the hydrodynamically interacting surfactant particles within the interfacial region, the macroscale observer is compelled to assign a *surface-excess stress* to the (two-dimensional) interface, which he or she then considers to be an intrinsic rheological property of the interface.

2.3 Interfacial Shear Viscosity

Imagine a simple shear field imposed upon a planar fluid interface, as depicted in Figure 2.3-1. Acting upon the lineal boundaries of the two-dimensional fluid element shown in the figure are two-dimensional surface stresses, represented here by the notation P^s_{ij}. (In subsequent chapters we will frequently use the equivalent dyadic notation $\mathbf{P}^s$). The two subscripts i and j may each adopt independent values (i.e. 1 and 2, x and y, etc.), corresponding to any pair of mutually orthogonal directions within the (two-dimensional) interface. In general circumstances there then exist four independent interfacial stress components: in the figure, a net force is delivered upon the boundaries of the surface element by the single surface stress component P^s_{xy}.*

In the absence of any external forces, the net force acting upon the surface element of Figure 2.3-1 may be expressed as

$$\left(P^s_{xy}\Big|_{x+\Delta x} - P^s_{xy}\Big|_{x} \right)\Delta y = 0. \qquad (2.3\text{-}1)$$

Note in the above that the dimensions of P^s_{ij} are force per unit length. Divide Eq. (2.3-1) by the areal element $\Delta x \Delta y$, and allow this element to become vanishingly small, as is the conventional three-dimensional continuum approach. This process yields the differential relation

$$\frac{\partial P^s_{xy}}{\partial x} = 0, \qquad (2.3\text{-}2)$$

representing the balance of interfacial linear momentum for a highly-viscous insoluble monolayer subjected to a simple shearing motion.

It remains to supply a constitutive relation between the interfacial pressure tensor P^s_{ij} and the rate of surface shear S^o_{ij}, the latter being given in the present case by the expression

---

* To distinguish among the several classes of interfacial variables that will appear in the subsequent theory (e.g. whether the variables are *surface-excess* properties, such as the surface-excess stress tensor, or merely properties *assigned* to the interface), we shall throughout the text consistently follow the postscripting convention indicated in Table 2.3-1.

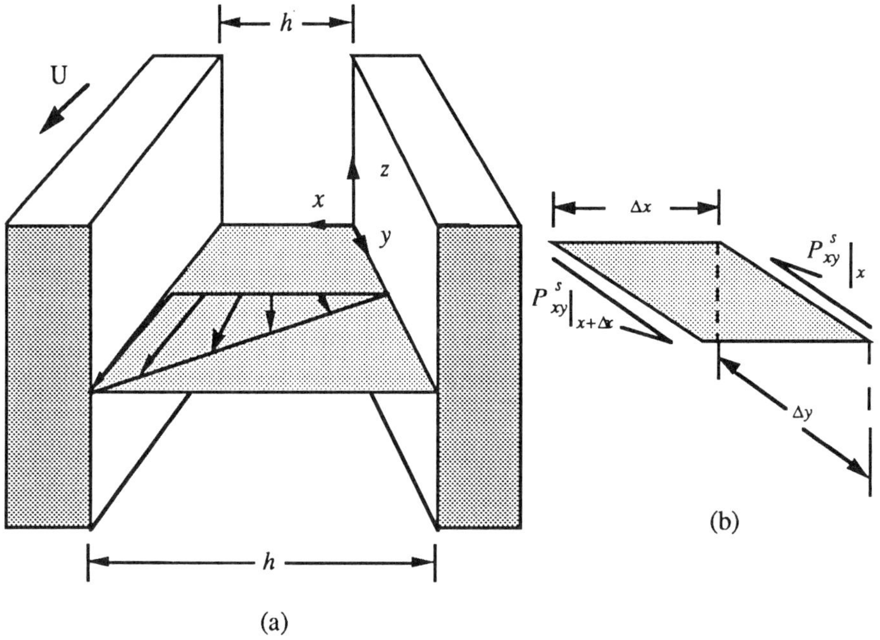

Figure 2.3-1 (a) Interfacial shear field imposed upon a planar fluid interface by the motion of the left-hand wall; (b) stresses acting upon a differential surface element $\Delta x \Delta y$.

$$S^o_{xy} = \frac{\partial v^o_y}{\partial x}. \tag{2.3-3}$$

Throughout much of the text this constitutive assumption will consist of a simple proportionality between interfacial shear stress and rate of shear,* whence

$$\boxed{P^s_{xy} = \mu^s \frac{\partial v^o_y}{\partial x}}, \tag{2.3-4}$$

where μ^s is the *surface shear viscosity* or, by analogy with bulk-fluid hydrodynamics, the *Newtonian* surface viscosity.

The surface shear viscosity μ^s (normally reported in units of surface poise, sp, equivalent to 1 g/s) is the most extensively investigated of all rheological properties of surfactant-adsorbed interfaces. As will be discussed in later chapters, the existence of a surface shear viscosity has been found to

---

* The actual constitutive relation between P^s_{xy} and S^o_{xy} must, of course, be established through experiments performed upon the particular surfactant-adsorbed interface being considered. Several alternative *generic* constitutive relations are discussed in §4.3 of Chapter 4. Chapter 9 surveys experimental methods for establishing interfacial viscous stress constitutive relations.

be an important stabilizing property of emulsion-droplet and foam-bubble interfaces; its presence influences not only the flow behavior of the interface itself, but also that of the bulk fluid in its proximity.

Table 2.3-1
POSTSCRIPTING OF INTERFACIAL VARIABLES

H^s	Surface-excess property*
H_s	Property mathematically assigned to the interface, or a geometrical property of the mathematical dividing surface
H^o	Bulk-phase property evaluated at the interface
H_o	Equilibrium value of H

Surface shear viscosity is negligible for a surfactant-free interface; its magnitude increases monotonically with adsorbed surfactant concentration, normally until the critical micelle concentration (CMC) is reached, at which point a more complex behavior ensues. Figure 2.3-2 illustrates the typical functional dependence of μ^s upon the area occupied by a single surfactant molecule in the low surfactant concentration range (below the CMC). The abrupt increase in surface shear viscosity observed with diminishing area-per-molecule may be attributed to the crowding of surfactant molecules, and concomitant interaction of their long paraffin chains.

The measurement and chemical composition dependence of surface shear viscosity will be considered in detail in Chapter 7.

In the absence of bulk-phase viscous stresses, the interfacial linear momentum equation (2.3-2) in combination with Eq. (2.3-4) adopts the simple form

$$\frac{d^2 v_y^o}{dx^2} = 0. \tag{2.3-5}$$

* The terminology 'surface-excess' will throughout the text be applied either to *extensive* variables of the type defined in Eqs. (3.6-7a) and (3.6-10a) or *intensive field* variables of the type defined in Eqs. (3.6-7b) and (3.6-10b). Exceptions to this rule will arise for those intensive 'surface-excess' quantities (e.g. μ^s) which [though not satisfying either of the relations (3.6-7b) or (3.6-10b)] are loosely termed 'surface excess' owing to the possibility of expressing these two-dimensional fields (later, in the matched asymptotic surface-excess theory of Part II) as quadratures of corresponding three-dimensional fields of the type [cf. Eq. (16.3-21)] which applies to 'true' surface-excess quantities [cf. Eqs. (15.6-5) and (15.6-9)].

Subject to the no-slip boundary conditions (see Figure 2.3-1)

$$v_y^o = 0 \quad @ \; x = 0 \tag{2.3-6}$$

and

$$v_y^o = U \quad @ \; x = h, \tag{2.3-7}$$

Eq. (2.3-5) possesses the linear solution

$$v_y^o = Gx\,, \tag{2.3-8}$$

as illustrated in the figure, where $G \equiv U/h$ is the shear rate.

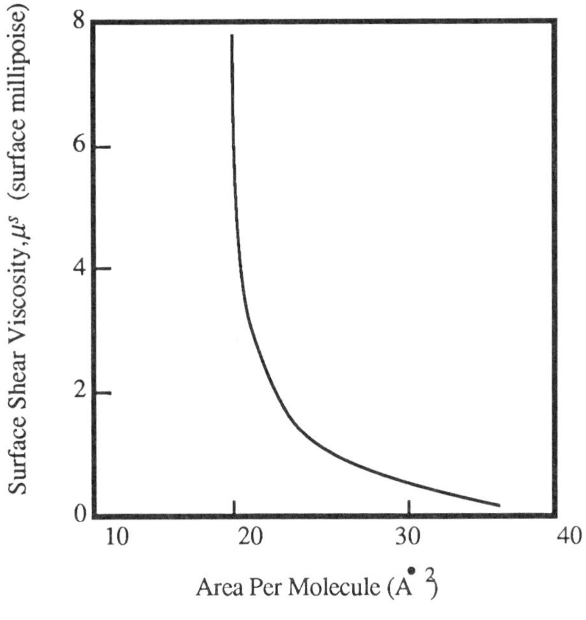

Figure 2.3-2 Surface shear viscosity of a stearic acid monolayer at an air-water interface vs the area-per-molecule adsorbed. Values taken from Goodrich *et al.* (1975).

2.4 Interfacial-Tension Gradient

The Marangoni effect is the source of many interesting, often aesthetically pleasing phenomena in interfacial hydrodynamics, involving interfacially driven bulk-phase flows, possibly together with bizarre interfacial instabilities (as will be discussed in Chapter 10). In many circumstances the interfacial-tension gradients underlying these flows represent the most practically important interfacial rheological stresses.

Marangoni effects arise from the existence of interfacial-tension gradients within a fluid interface creating bulk fluid motions through the mechanism of conservation of interfacial linear momentum, as in Eq. (2.3-2). While the bulk-phase flows generated by this interfacial phenomenon are frequently the principal *raison dé être* for interest in the Marangoni effect, the basic physical mechanism of gradient-driven flows is again more easily perceived within the intrinsic context of the highly-viscous insoluble monolayer model, independently of any eventual volumetric-flow consequences.

In this context, by analogy with classical pressure gradient-driven Poiseuille flow in a rectilinear channel, imagine a time-independent, fully-developed "surface pressure" gradient acting within the plane of an interface between two parallel plates, as depicted in Figure 2.4-1.

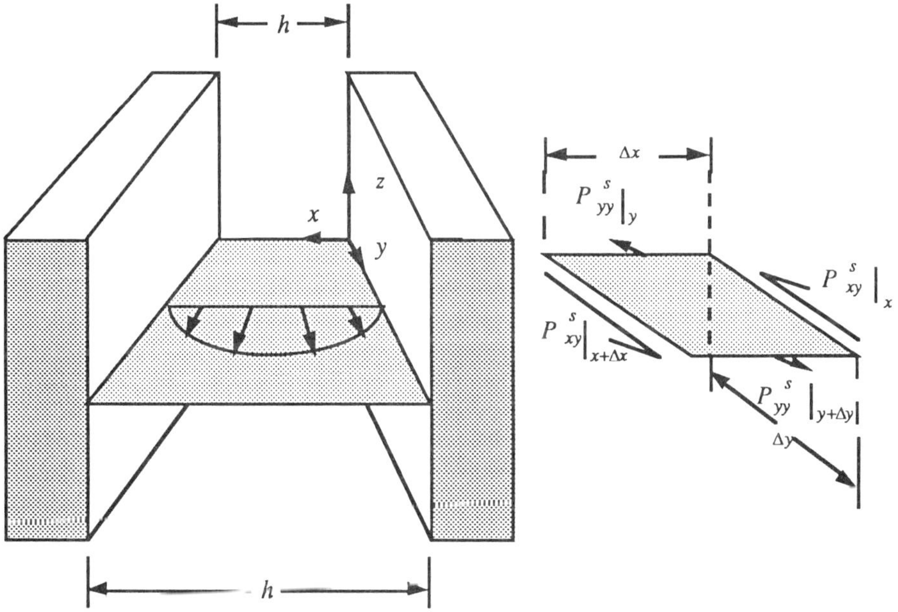

Figure 2.4-1 A "surface pressure" (or surface tension) gradient is imposed across a fluid-fluid interface between two parallel plates, generating a two-dimensional 'Poiseuille flow' in the plane of the interface.

The balance of interfacial forces acting upon the boundaries of the surface element shown in the figure may be expressed as

$$\left(P^s_{xy}\Big|_{x+\Delta x} - P^s_{xy}\Big|_x\right)\Delta y + \left(P^s_{yy}\Big|_{y+\Delta y} - P^s_{yy}\Big|_y\right)\Delta x = 0. \quad (2.4\text{-}1)$$

Division by the areal element $\Delta x \Delta y$ while allowing this element to become vanishingly small (as in §2.3) thereby yields

$$\frac{\partial P^s_{xy}}{\partial x} + \frac{\partial P^s_{yy}}{\partial y} = 0, \qquad (2.4\text{-}2)$$

expressing the balance of interfacial linear momentum for the surface pressure gradient-driven flow. The constitutive expression for the surface stress component P^s_{yy} is simply

$$\boxed{P^s_{yy} = \sigma,} \qquad (2.4\text{-}3)$$

where σ is the interfacial tension.

As interfacial tension is highly sensitive to surfactant adsorption, any disturbance within the interface that disturbs the homogeneity of the surfactant surface concentration thereby raises or lowers the interfacial tension, thus creating a dynamic interfacial tension gradient at the location of the inhomogeneity. Equation (2.4-2) in combination with Eq. (2.4-3) indicates that the existence of an interfacial-tension gradient $\partial\sigma/\partial y$ creates an interfacial stress P^s_{xy}. In turn, this results during interfacial motion in the course of the system's attempt to re-establish an equilibrium state of uniform surfactant concentration.

Together, Eqs. (2.4-2), (2.4-3), and (2.3-4) combine to yield

$$\frac{d\sigma}{dy} + \mu^s \frac{d^2 v^o_y}{dx^2} = 0, \qquad (2.4\text{-}4)$$

where the interfacial tension gradient is a constant of the interfacial flow owing to the fully developed nature of the interfacial motion. Surfactant concentration inhomogeneities are not the only source of such gradients; rather, temperature gradients constitute another source owing to the sensitive dependence of interfacial tension upon temperature (see Questions 5.4–5.6 of Chapter 5).

Boundary conditions imposed upon Eq. (2.4-4) again correspond to the no-slip conditions

$$v^o_y = 0 \quad @\ x = 0 \qquad (2.4\text{-}5)$$

and

$$v^o_y = 0 \quad @\ x = h \qquad (2.4\text{-}6)$$

at the plate surfaces in Figure 2.4-1. Thereby, one obtains the quadratic field

$$v_y^o = \frac{h^2}{2\mu^s}\left(\frac{d\sigma}{dy}\right)\left[\frac{x}{h} - \left(\frac{x}{h}\right)^2\right], \tag{2.4-7}$$

corresponding to the two-dimensional 'Poiseuille flow' shown in the figure.

2.5 Interfacial Dilatational Viscosity

Material compression and expansion is a more commonplace occurrence with interfaces than with bulk fluids, owing in part to the ability of surface matter to flow to and from the bulk phase. Thus, whereas the dilatational or bulk viscosity coefficient κ of the classical Newtonian model for three-dimensional viscous fluids (Liebermann 1949) is seldom encountered in practice, its two-dimensional counterpart — the interfacial dilatational viscosity κ^s — is an important rheological property of Newtonian interfaces.

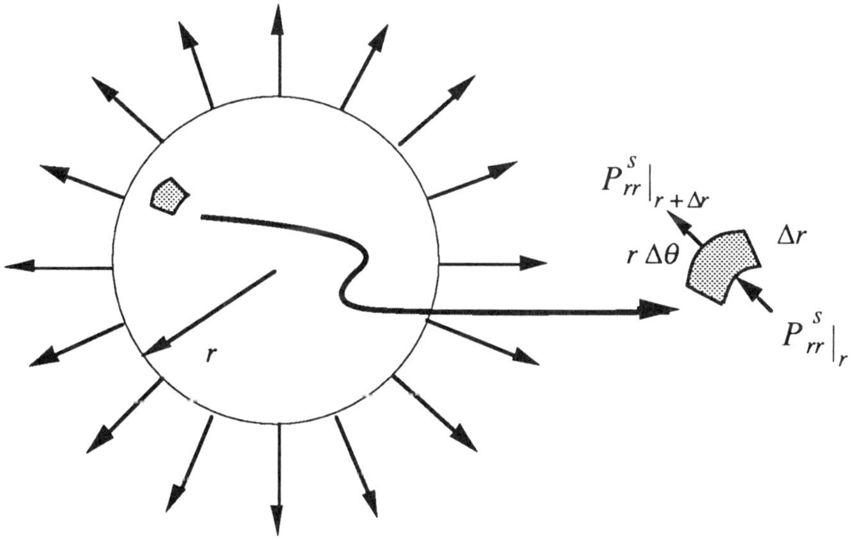

Figure 2.5-1 Uniform dilatation of a planar interface. An expanded view of the stresses acting upon the surface element of area $r\Delta\theta\Delta r$ is shown.

As an introduction to the rheological manifestation of the interfacial dilatational viscosity, consider the uniform expansion of a circular, flat interface at constant rate* $\Delta = A^{-1}dA/dt$ (with $A = \pi r^2$ the area of the circle),

---

* Δ here should not be confused with the differential delta, such as in $\Delta\theta$.

as shown in Figure 2.5-1. A balance of forces acting upon the surface element shown yields

$$\left(P^s_{rr}\Big|_{r+\Delta r} - P^s_{rr}\Big|_{r} \right) r\,\Delta\theta = 0. \tag{2.5-1}$$

Upon dividing by $r\Delta\theta\Delta r$ and subsequently allowing this differential area element to become vanishingly small, we obtain

$$\frac{dP^s_{rr}}{dr} = 0. \tag{2.5-2}$$

The constitutive expression for P^s_{rr} in the present case is given by [cf. Eqs. (4.2-15) and (4.2-17) of Chapter 4: see also Eq. (4.2-19)]

$$\boxed{P^s_{rr} = \sigma + \kappa^s \Delta,} \tag{2.5-3}$$

where a linear constitutive relation has again been assumed between stress and deformation rate, as in Eq. (2.3-4).

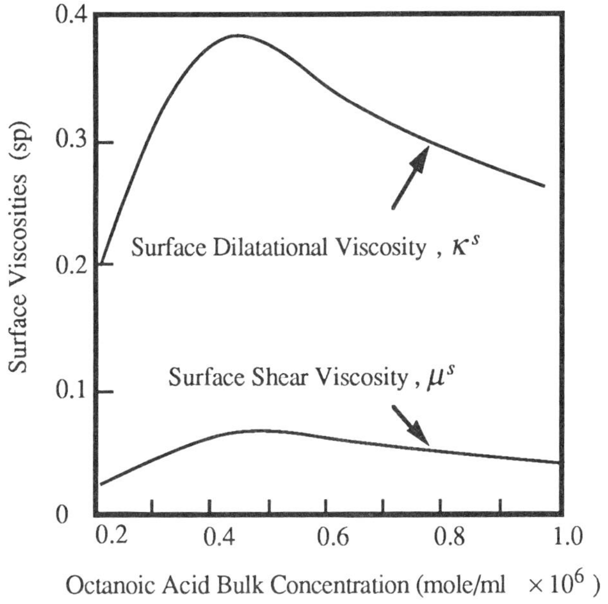

Figure 2.5-2 Surface dilatational and shear viscosities of an aqueous octanoic acid solution/air interface versus octanoic acid concentration. Values taken from Ting *et al.* (1984).

The second term of Eq. (2.5-3), which was not present in the previous example [cf. Eq. (2.4-3)] owing to the absence of interfacial dilatation, may constitute a significant interfacial rheological contribution, inasmuch as the magnitude of κ^s for most adsorbed surfactant interfaces appears to be at least as large as the interfacial shear viscosity μ^s. This is illustrated in Figure 2.5-2, which compares the respective surface dilatational and shear viscosities of an aqueous octanoic acid solution/air interface.

The force required to expand the highly-viscous insoluble monolayer may be obtained simply as $2\pi r\, P^s_{rr}$, with r the instantaneous radius of the expanding interface, and P^s_{rr} given by Eq. (2.5-3). Upon removal of the external surface force giving rise to this expansion, interfacial tension will act to contract the interface (the largest tension existing at the center of the expanding circle of Figure 2.5-1). Opposing this contraction is the dissipative resistance offered by the interfacial dilatational viscosity. The balance of the two determines the rate at which the interface returns to its equilibrium state.

2.6 Summary

Fluid interfaces containing adsorbed surfactants may exhibit dynamic behavior that differs considerably from that of *clean*, surfactant-free fluid interfaces.

In the current chapter we have provided an introduction to the primary interfacial rheological properties of interfaces. This has been accomplished by pursuing analogies existing between three-dimensional bulk fluids and two-dimensional interfaces, albeit within the limiting context of a planar, highly-viscous, surfactant-adsorbed monolayer. The latter constitutes an idealized model of interfacial behavior, one whose simplicity derives from its *intrinsic* response, independent of that of the three-dimensional substrate(s) to which it is irrevocably affixed.

Since the highly-viscous insoluble monolayer responds to interfacial deformation in a manner that is (to a leading approximation) independent of the hydrodynamic stresses exerted upon the interface by the contiguous bulk-fluid phases, the various limiting forms of the surface linear momentum balance developed in this chapter have appeared to be independent of these bulk-phase stresses. This independence, while greatly simplifying the analysis of dynamic interfacial stresses, nevertheless fails to reveal the fundamental role such stresses play in an interfacial linear momentum equation that serves as the normal stress boundary condition imposed upon the bulk-phase hydrodynamical equations. As such, the model largely obscures the importance of interfacial rheological properties to bulk-phase flows.

The *extrinsic* rheology of fluid interfaces will be discussed in Chapters 3 to 5, beginning with the more general conceptual framework of interfacial transport processes. The influence of interfacial rheology upon bulk-phase hydrodynamics will be developed theoretically, followed by several illustrative examples. Those readers interested primarily in the *practical* aspects of interfacial rheology, may wish at this point to turn directly to Chapters 6 to 14.

Questions for Chapter 2

2.1 A three-phase contact line forms at the lineal boundary between solid, fluid and gas phases. The contact angle at which the fluid-gas surface approaches the solid phase (see Figure 1.2-2) is determined by the *wettability* of the solid. (The contact angle θ at a perfectly wetting solid surface is 0°.) Let the spherical solid particle depicted in Figure 1.2-2 exhibit a 90° contact angle with the air-water surface. The particle is assumed to possess a mass m and radius r_0. Denote by σ the surface tension of the water-air surface. Determine a general expression for the equilibrium distance d of the solid particle center below the plane of the surface owing to the force of gravity (with g the gravitational constant).

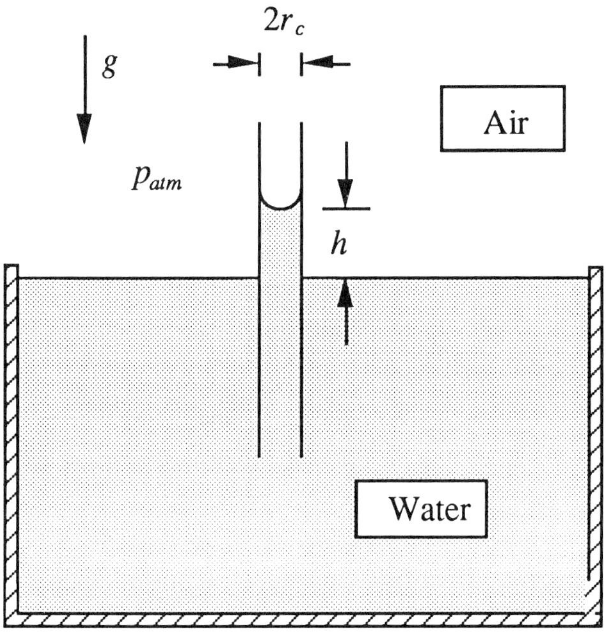

2.2 Determine the equilibrium height h of water (σ=73 dyne/cm, ρ=1 g/cm^3) that will rise through a narrow circular-cylindrical capillary of radius r_c=100 microns under the influence of gravity (see the figure below). Assume the water perfectly wets the walls of the capillary, hence, the shape of the meniscus is hemispherical ($H=-1/R$, with R the radius of the meniscus). Here, g=980 cm/s^2.

2.3 In the example problem of §2.3 (see Figure 2.3-1): *i*. What is the stress (force/length) exerted upon the boundary at $x = h$? *ii*. What would be the velocity profile if the boundary at $x = 0$ were moving in the y direction with a velocity $-U$? *iii*. What would be the stress on the boundary $x = h$ in this case? *iv*. Does interfacial stress at the channel walls increase or decrease with a diminishing separation distance h? In what terms would you explain this?

2.4 In the example problem of §2.4 (see Figure 2.4-1): *i*. Express the surface velocity, Eq. (2.4-7), in terms of the average surface velocity. *ii*. What is the rate of surface flow between the channel walls? *iii*. What is the stress acting on the boundary $x = h$?

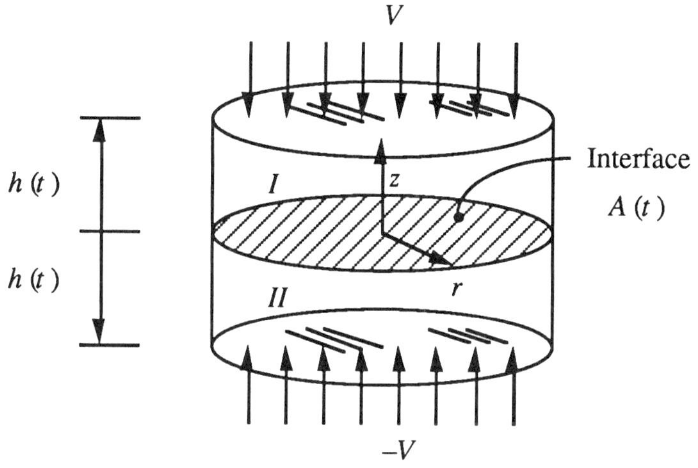

2.5 As an explicit example of the expanding surface problem of §2.5, consider the figure shown below, in which a tube of incompressible liquid, bisected by a planar fluid interface of area A, is compressed along its axial coordinate. Given the following velocity field in each of the two phases I and II;

$$v_r = \frac{V}{2h} r \qquad v_z = -\frac{V}{h} z,$$

show that

$$\Delta = \frac{1}{r}\frac{\partial}{\partial r}(r v_r)$$

is equivalent to the rate of surface expansion

$$\Delta = \frac{1}{A}\frac{dA}{dt}.$$

Additional Reading for Chapter 2

§2.1 Interfacial Tension

Defay, R. & Prigogine, I. 1966 *Surface Tension and Adsorption*, New York: Wiley.

§2.2 Dynamic Interfacial Properties

Levich, V.G. 1962 *Physicochemical Hydrodynamics*, Englewood Cliffs, New Jersey: Prentice-Hall.

§2.3 Interfacial Shear Viscosity

Joly, M. 1964, Surface viscosity. In *Recent Progress in Surface Science*, vol. I (eds. J.F. Danielli, K.G.A. Pankhurst, and A.C. Riddiford), pp. 1–50. New York: Academic Press.

Mannheimer, R.J. & Schechter, R.S. 1970 An improved apparatus and analysis for surface rheological measurements. *J. Colloid Interface Sci.* **32**, 195–211.

Jiang, T.S., Chen, J.D. & Slattery, J.C. 1983 Nonlinear interfacial stress-deformation behavior measured with several interfacial viscometers. *J. Colloid Interface Sci.* **96**, 7–19.

§2.4 Interfacial Tension Gradient

Levich, V.G. & Krylov, V.S. 1969 Surface-tension-driven phenomena. In, *Annual Review of Fluid Mechanics*, vol. 1, pp. 293–316. California: Annual Reviews, Inc.

Sternling, C.V. & Scriven, L.E. 1959 Interfacial turbulence: hydrodynamic instability and the Marangoni effect. *A.I.Ch.E. J.* **5**, 514–523.

§2.5 Interfacial Dilatational Viscosity

Hansen, R.S. & Ahmad, J. 1971 Waves at interfaces. In, *Progress in Surface and Membrane Science*, vol. 4 (eds. J.F. Danielli, M.D. Rosenberg and D.A. Cadenhead), pp. 1–55. New York: Academic Press.

Ting, L., Wasan, D.T., Miyano, K. & Xu, S.Q. 1984 Longitudinal surface waves for the study of dynamic properties of surfactant systems II. gas-liquid surface. *J. Colloid Interface Sci.* **102**, 248–258.

3

"Again, everybody is aware that all springs are colder in summer than in winter, as well as the following miracles of nature; that . . . all sea water is made smooth by oil, and so divers sprinkle oil from their mouth because it calms the rough element and carries light down with them . . .."

Gaius Plinius Secundus (Pliny the Elder) (AD 77)

CHAPTER 3

Interfacial Transport Processes

Interfacial transport processes often bear an analogy to volumetric transport processes occurring within bulk-fluid phases. Such areal processes may influence the nature of the volumetric transport phenomena, particularly in the vicinity of the fluid interface, the latter being conventionally viewed in analyses of bulk transport phenomena as a two-dimensional surface partially bounding two or more three-dimensional bulk phases.

A fluid interface is not, of course, truly a two-dimensional material entity. Nevertheless, as an interface is generally observed only over an experimental length scale (the *macroscale*) that is quite large relative to the 'thickness' of the interfacial transition zone, an interface is conveniently idealized as a two-dimensional, singular 'surface' possessing a macroscopically defined location, configuration and orientation between a pair of contiguous, three-dimensional, immiscible bulk-fluid phases.

Clearly, there are two levels or scales at which interfacial transport processes may be studied. The macroscale description of interfaces, which is the view that will be adopted here in Part I of this book, is founded upon the assumption that a fluid interface between immiscible fluid phases may reasonably be regarded as a two-dimensional, singular surface. This is the conventional view of a fluid interface, and in virtually all circumstances is the most practically useful of such interfacial models. At this macroscale level, one seeks to develop transport laws governing the response of the fluid interface; these phenomenological relations function as boundary conditions imposed upon comparable volumetric transport fields at the interface. There is certainly more than one methodology for deriving these laws, but conventionally one proceeds simply by analogy with existing conservation and constitutive transport laws for three-dimensional fluid continua, making due allowance for the generally non-Euclidean 'metric' nature of curved, two-dimensional domains. In Part II of this book, a second and more fundamental path is taken, proceeding from a more basic microscale, three-dimensional interfacial description, accounting for the 'diffuse' transition region existing between the two fully developed bulk fluid phases on either side. While the conventional (macroscale) conclusions resulting from these two alternative paths (regarding interfacial behavior at the macroscale) are mutually consistent, the microscale view affords the more rational means for deriving interfacial constitutive laws, additionally offering the advantage of insight into the origin of "two-dimensional" interfacial properties.

Chapter 3 provides a general foundation for the macroscale theory of interfacial transport processes. This generic theory is specifically adapted in Chapters 4 and 5 to the basic dynamical transport processes underlying interfacial rheology. We begin in §3.1 by considering the macroscale perspective of interfacial transport, whereby the interfacial transition zone is collapsed onto a singular two-dimensional surface separating contiguous bulk phases. Interfacial geometrical preliminaries are considered in §3.2, followed by interfacial statics in §3.3. Section 3.4 is devoted to an examination of the kinematics of *material* fluid interfaces. The remaining sections of Chapter 3 develop a single, generic, macroscale transport equation for such a material fluid interface; eventually, this generic development is employed in Chapters 4 and 5 to derive the basic equations respectively governing interfacial momentum, mass, and species transport.

3.1 The Macroscale View of a Fluid Interface*

The geometrical configuration of a fluid interface, its movement, and the static or dynamic state of the contiguous bulk phases bounding it, are each strongly dependent upon steep, fine-scale inhomogeneities in volumetric physical properties extant across the transition zone between bulk phases. These volumetric inhomogeneities include, for example, nonuniformities in the overall mass density of the fluid, species mass density, and shear viscosity fields, each rapidly varying in a direction normal to the macroscopically perceived phase interface. The consequences of such steep, physical property gradients are dramatically illustrated by the example of a rapidly varying pressure field across a curved interface. This microscale pressure inhomogeneity occasions short-range, attractive, microscale forces within the interfacial transition zone, which are finally responsible at the coarser, macroscale, for the existence of a state of tension at the interface.

Though the ultimate physical origin of, say, the tensile force existing at a fluid interface cannot be fully understood without a thorough micromechanical analysis of the interfacial transition region (as provided in Chapter 15), for strictly operational purposes, empirical knowledge of macromechanical phenomenological properties such as interfacial tension, interfacial viscosity, and interfacial elasticity, may often prove sufficient to describe the pertinent phenomena. Indeed, in the resolution of most scientific and technological problems, one conventionally views the interface from this strictly macroscale point of view, where the interface, *per se*, is regarded as a

* The reader may refer to §15.1 for a discussion of the corresponding "microscale" view of a fluid interface.

singular two-dimensional boundary separating contiguous bulk-fluid phases (Figure 3.1-1). Macroscale surface conservation and constitutive laws embodying appropriate macroscale phenomenological coefficients (such as interfacial viscosities) are ascribed to these two-dimensional boundaries, serving ultimately as (macroscale) boundary conditions imposed upon the field variables in the bulk-phase transport equations. In the following chapters we will develop these macroscopic interfacial laws, or boundary conditions for fluid interfaces, subsequently applying them to practical dynamic interfacial problems. In Part II of the book, a microscale rationale for each of these interfacial equations will be derived from the perspective of a detailed, rigorous, micromechanical theory.

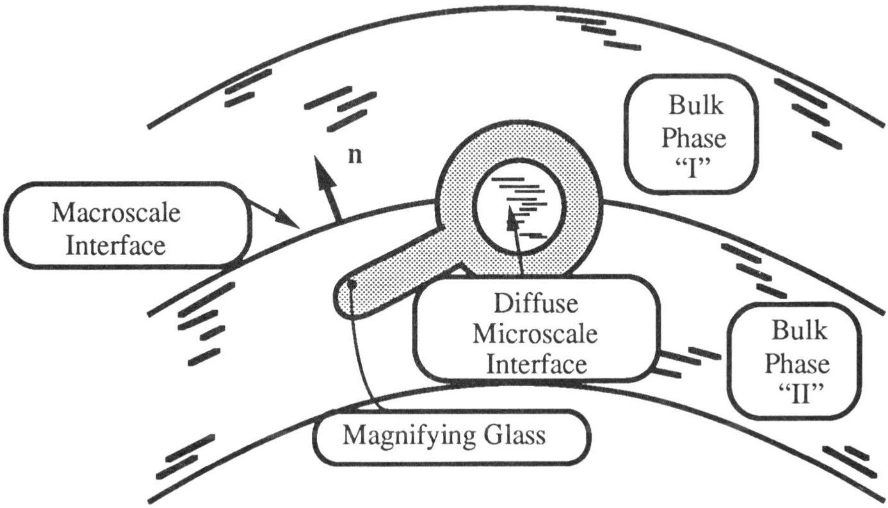

Figure 3.1-1 A depiction of the contrasts between the microscale (l, say) and the macroscale (L, say) views of the interface. From the macroscale perspective, the interface appears to be a two-dimensional surface, whereas from the microscale perspective (achieved by viewing the interfacial region through a magnifying glass), the interface appears as a highly inhomogeneous three-dimensional transition region.

Throughout Part I, however, our perspective of a fluid interface will be that of a strictly macroscale observer, wherein the interface is viewed simply as a two-dimensional phase boundary, possessing its own intrinsic material surface properties, and which moves and deforms between immiscible bulk-fluid phases. Even at this 'elementary' macroscale level the dynamical behavior of fluid interfaces may appear quite complex, involving as it does, deformable (and hence time-dependent) non-Euclidean spaces. For a proper understanding of interfacial transport processes, one accordingly needs to be familiar with the basic geometrical description of a two-dimensional surface, to which description we turn in the following section. For a clarification of unfamiliar terminology or notation, the reader may wish to consult any of the standard texts of tensor analysis and differential geometry cited at the conclusion of this chapter, in addition to Appendices A and B of this book.

3.2 Interfacial Geometry

Intrinsic and Extrinsic Views

Surface geometry may be considered either from an *intrinsic* or *extrinsic* point of view. The former, intrinsic perspective constitutes a two-dimensional description of the surface geometry that can be achieved entirely by an observer constrained to lie within the surface itself (independently of the three-dimensional space in which the surface is embedded). The intrinsic perspective, to which brief consideration is given in Question 3.27, corresponds to the interfacial perspective adopted in Chapter 2. The extrinsic perspective constitutes a description of the surface geometry by an 'external' observer situated outside of the surface itself, and lying within the three-dimensional space in which the surface is embedded. The latter, extrinsic view is the most practically useful of the two for the purposes of describing interfacial transport phenomena (as it emphasizes the connection between the two-dimensional surface 'phase' and the three-dimensional bulk phases surrounding it). As such, we shall adopt the extrinsic perspective in the remainder of Part I.

Extrinsic Surface Geometry

The differential geometrical properties to be discussed in this subsection will be encountered repeatedly throughout the theoretical chapters of this text. Appendix B may be consulted for a further elucidation of these surface geometrical concepts in semi-orthogonal* curvilinear coordinates. Supplementing these brief reviews are standard texts, cited in the Additional Reading list at the conclusion of the chapter.

Let the vector $\mathbf{x}_s$ denote a two-point vector field, specifying the position of a point P_s in a two-dimensional Riemannian surface relative to a space-fixed origin O.† Furthermore, let the duo of (generally nonorthogonal) curvilinear coordinates $q^\alpha \equiv (q^1, q^2)$ denote specified functions of position $\mathbf{x}_s \equiv \mathbf{x}_s (q^1, q^2)$ at each point P_s on the surface [which surface may be regarded as a particular q^3-coordinate surface in the general parameterization of three-dimensional space by the semi-orthogonal curvilinear coordinate system (q^1, q^2, q^3), as further discussed in §B.1]. Then,

$$\begin{aligned} d\mathbf{x}_s &= \frac{\partial \mathbf{x}_s}{\partial q^\alpha} dq^\alpha \\ &\equiv \mathbf{a}_\alpha dq^\alpha \end{aligned} \tag{3.2-1}$$

* The property of semi-orthogonality is explained in §B.1.

† The analogous three-dimensional parameterization of Euclidean space is provided in Appendix A [particularly note the discussion immediately preceding and following Eq. (A.1-1)].

provides a relation between the two-point (bilocal) differential displacement $d\mathbf{x}_s$ between neighboring interfacial points $\mathbf{x}_s$ and $\mathbf{x}_s+d\mathbf{x}_s$ and the comparable differential displacement dq^i of the coordinates q^i. The 'one-point' vectors

$$\mathbf{a}_\alpha \overset{\text{def}}{=} \frac{\partial \mathbf{x}_s}{\partial q^\alpha} \tag{3.2-2}$$

comprise a system of *surface base vectors*, defined locally at the point P_s $[\equiv \mathbf{x}_s \equiv (q^1, q^2)]$, 'induced' by the specific choice of the functional dependence of the q^i upon position $\mathbf{x}_s$. In terms of these, the *surface metric tensor* is defined as

$$a_{\alpha\beta} = \mathbf{a}_\alpha \cdot \mathbf{a}_\beta \qquad (\alpha,\ \beta = 1,\ 2), \tag{3.2-3}$$

whose determinant is

$$a = \det\ a_{\alpha\beta} \equiv |\mathbf{a}_1 \times \mathbf{a}_2|^2. \tag{3.2-4}$$

(By way of example, explicit expressions for $\mathbf{a}_\alpha$, $a_{\alpha\beta}$ and a are calculated for a spherical surface in Question 3.18 and for a prolate spheroidal surface in example 1 of §3.7.)

The local *unit normal vector* at a point (q^1, q^2) of the interface is defined by the expression

$$\mathbf{n} = \mathbf{a}_1 \times \mathbf{a}_2 / \sqrt{a}. \tag{3.2-5}$$

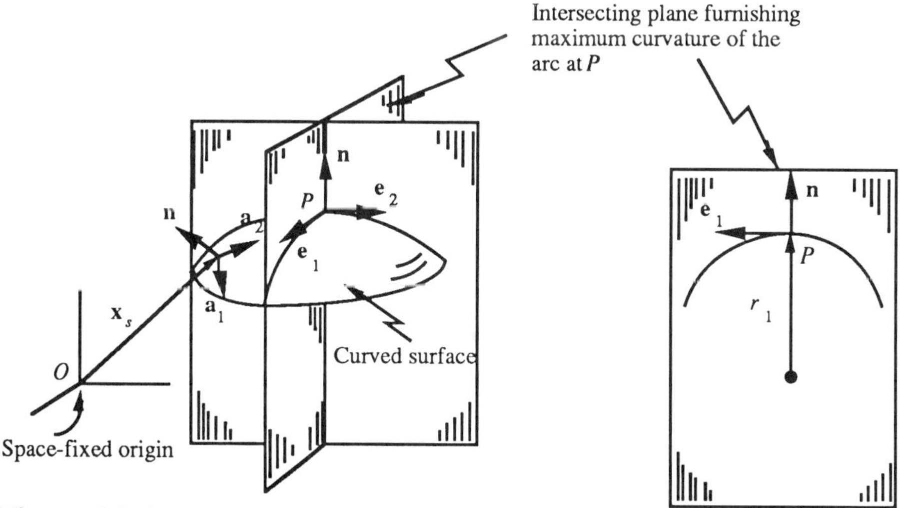

Figure 3.2-1 A planar surface containing the unit normal **n** to a curved surface at P is rotated until the curvature of the arc (formed by the intersection of the plane with the surface) at the locally defined unit normal vector is maximized. The radius of curvature of this arc at P is r_1. A second plane, also containing **n**, is then drawn perpendicular to the first. The radius of curvature of this second arc at P is r_2. The algebraically signed scalars r_1 and r_2 are the two principal radii of curvature of the surface at P. (See Questions 3.8 and 3.9.)

This yields the following representation of the *dyadic surface idemfactor*:

$$\mathbf{I}_s \stackrel{\text{def}}{=} \mathbf{I} - \mathbf{nn}$$
$$\equiv \mathbf{a}^1\mathbf{a}_1 + \mathbf{a}^2\mathbf{a}_2, \qquad (3.2\text{-}6)$$

with the dyadic **I** the (three-dimensional) spatial idemfactor. Here,

$$\mathbf{a}_\alpha \cdot \mathbf{a}^\beta = \delta_\alpha^\beta \qquad (\alpha,\ \beta = 1, 2) \qquad (3.2\text{-}7)$$

provides a defining relation [cf. Eq. (B.1-4b)] for the *reciprocal* surface base vectors $(\mathbf{a}^1, \mathbf{a}^2)$ used in the representation (3.2-6) of the surface idemfactor. In Eq. (3.2-7),

$$\delta_\alpha^\beta = \begin{cases} 1 & \text{if} \quad \alpha = \beta \\ 0 & \text{if} \quad \alpha \neq \beta \end{cases} \qquad (3.2\text{-}8)$$

is the surface (two-dimensional) Kronecker delta.

Likewise, we obtain the following representation of the *dyadic surface unit alternator*:

$$\boldsymbol{\varepsilon}_s \stackrel{\text{def}}{=} \boldsymbol{\varepsilon} \cdot \mathbf{n}$$
$$\equiv (\mathbf{a}^1\mathbf{a}^2 - \mathbf{a}^2\mathbf{a}^1)\sqrt{a}. \qquad (3.2\text{-}9)$$

Here, $\boldsymbol{\varepsilon}$ ($\equiv -\mathbf{I}\times\mathbf{I}$) is the triadic spatial unit alternator [cf. Eq. (A.1-21)].

Local curvature of a surface may be expressed in terms of the variation of the unit surface normal **n** with position (q^1, q^2) in the surface. Thus, the *surface curvature dyadic* **b** possesses the representation

$$\mathbf{b} \equiv -\nabla_s \mathbf{n}, \qquad (3.2\text{-}10)$$

where the *surface gradient operator* $\nabla_s \equiv \partial / \partial \mathbf{x}_s$ is defined as

$$\nabla_s \stackrel{\text{def}}{=} \mathbf{I}_s \cdot \nabla \equiv \mathbf{a}^\alpha \frac{\partial}{\partial q^\alpha}, \qquad (3.2\text{-}11)$$

in which $\nabla \equiv \partial / \partial \mathbf{x}$ is the spatial gradient operator. The surface curvature dyadic is positive-definite* when the surface is concave in the direction of the unit surface normal **n**.

The surface curvature dyadic is symmetric. It may be expressed in 'canonical' form as†

---

* A symmetric dyadic **D** is positive-definite if for all nonzero vectors **u**,

$$\mathbf{u} \cdot \mathbf{D} \cdot \mathbf{u} > 0.$$

It is non-negative definite if for all vectors **u**

$$\mathbf{u} \cdot \mathbf{D} \cdot \mathbf{u} \geq 0.$$

† The arbitrary choices of the index labels "1" and "2" appearing in Eq. (3.2-12a) derive from the invariant nature of the curvature dyadic and have nothing whatsoever to do with the explicit choice of the curvilinear coordinates (q^1, q^2) nor their indices. Nevertheless, in illustrative problems throughout the text, where q^1 and q^2 are *orthogonal* curvilinear

$$\mathbf{b} = \kappa_1 \mathbf{e}_1 \mathbf{e}_1 + \kappa_2 \mathbf{e}_2 \mathbf{e}_2 \qquad (3.2\text{-}12\text{a})$$

in terms of the pair of mutually perpendicular unit vectors ($\mathbf{e}_1$, $\mathbf{e}_2$) in the respective directions of the principal axes of curvature, together with the principal curvatures κ_1 and κ_2 of the surface. These 'principal' quantities are respectively defined as the pair of eigenvectors $\mathbf{e}_\alpha$ (normalized such that $|\mathbf{e}_\alpha| = 1$) and eigenvalues κ_α of $\mathbf{b}$ (see Questions 3.8 and 3.9):

$$\mathbf{b} \cdot \mathbf{e}_\alpha = \kappa_\alpha \mathbf{e}_\alpha \quad (\text{no sum on } \alpha) \quad (\alpha = 1, 2). \qquad (3.2\text{-}12\text{b})$$

The two (algebraically signed) curvature scalars may themselves be expressed in terms of two more commonly used scalar measures of curvature, namely the mean surface curvature

$$\begin{aligned} H &\overset{\text{def}}{=} -\tfrac{1}{2}\nabla_s \cdot \mathbf{n} \\ &\equiv \tfrac{1}{2}\mathbf{I}_s : \mathbf{b} \\ &= \tfrac{1}{2}(\kappa_1 + \kappa_2), \end{aligned} \qquad (3.2\text{-}13)$$

and the Gaussian or total curvature

$$\begin{aligned} K &\overset{\text{def}}{=} -\frac{1}{2}\,\boldsymbol{\varepsilon}_s : (\mathbf{b} \cdot \boldsymbol{\varepsilon}_s \cdot \mathbf{b}) \\ &\equiv \kappa_1 \kappa_2 . \end{aligned} \qquad (3.2\text{-}14)$$

In the preceding pair of equations the double-dot product follows the nesting convention of Chapman & Cowling (1961) [e.g. $\mathbf{mn}:\mathbf{pq} = (\mathbf{n}\cdot\mathbf{p})(\mathbf{m}\cdot\mathbf{q})$ with $\mathbf{m}$, $\mathbf{n}$, $\mathbf{p}$, $\mathbf{q}$ any vectors], rather than that of Wilson & Gibbs (1929).

Yet another mode of expressing the principal curvatures κ_1 and κ_2 is in terms of the principal radii of curvature r_1 and r_2:

$$\kappa_1 = -\frac{1}{r_1} \qquad \kappa_2 = -\frac{1}{r_2}. \qquad (3.2\text{-}15)$$

Figure 3.2-1 depicts the locally defined principal directions ($\mathbf{e}_1$, $\mathbf{e}_2$) and principal radii of curvature (r_1, r_2) at a point P (q^1, q^2) on the surface.

Orthogonal Curvilinear Surface Coordinates (q^1, q^2)

In circumstances wherein (q^1, q^2) represent *orthogonal* curvilinear coordinates (as in every example considered in the subsequent text) the pair of surface base vectors ($\mathbf{a}_1$, $\mathbf{a}_2$) are orthogonal, whence

$$\mathbf{a}_1 \cdot \mathbf{a}_2 = 0. \qquad (3.2\text{-}16)$$

---

coordinates [cf. the subsection following Eq. (3.2-15)], we shall always make this identification. In such circumstances κ_1 and κ_2 are then the same as the curvatures of the respective q^1 and q^2 coordinate curves [cf. Eqs. (B.1-20) to (B.1-22)].

Comparing this relation with Eq. (3.2-3) reveals that in the present case of orthogonal surface coordinates (q^1, q^2), the surface metric tensor becomes simply (in matrix form)

$$a_{\alpha\beta} = \begin{bmatrix} a_{11} & 0 \\ 0 & a_{22} \end{bmatrix}. \tag{3.2-17}$$

For such orthogonal systems the normalized surface base vectors are *self reciprocal*, in the sense that

$$\frac{1}{|\mathbf{a}_1|}\mathbf{a}_1 = \frac{1}{|\mathbf{a}^1|}\mathbf{a}^1, \qquad \frac{1}{|\mathbf{a}_2|}\mathbf{a}_2 = \frac{1}{|\mathbf{a}^2|}\mathbf{a}^2 \tag{3.2-18}$$

as is readily shown (see Question 3.19). When this is the case, it is often convenient to introduce in their stead comparable *unit* vectors

$$\mathbf{i}_\alpha = \mathbf{a}_\alpha / |\mathbf{a}_\alpha| \qquad (\alpha = 1, 2) \quad \text{(no sum on } \alpha\text{)}; \tag{3.2-19}$$

explicitly,

$$\mathbf{i}_1 = \frac{\mathbf{a}_1}{\sqrt{a_{11}}}, \qquad \mathbf{i}_2 = \frac{\mathbf{a}_2}{\sqrt{a_{22}}}. \tag{3.2-20}$$

Frequent use will be made throughout the text of the orthonormal system $(\mathbf{i}_1, \mathbf{i}_2)$ of unit vectors.

3.3 Interfacial Statics

This section addresses the basic nature of macroscale interfacial force balances at an arbitrarily curved fluid interface in a state of hydrostatic equilibrium. This static state serves as a standard from which nonequilibrium interfacial transport processes depart.

Analogous to three-dimensional fluid continua, macroscale fluid interfaces may be acted upon by two fundamental types of forces: (i) (areal) surface *body* force densities $\mathbf{f}^s$ resulting from the action of forces originating outside of the two-dimensional interface itself; and (ii) (lineal) surface *contact* stresses $\mathbf{p}^s_\nu$ resulting from the action of contiguous two-dimensional interfacial elements upon one another.

Surface Body Forces

Body force surface vector densities, $\mathbf{f}^s$ (possessing the units of force per unit area), are supposed continuously distributed over each areal fluid element on the interface. These 'action-at-a-distance' forces are transmitted without necessity of physical contact (within the two-dimensional interfacial domain) between the surface fluid element and the source of the force.

Such surface body force densities may be expressed as a sum:

$$\mathbf{f}^s = \mathbf{F}^s + \mathbf{F}^b, \tag{3.3-1}$$

where $\mathbf{F}^s$ and $\mathbf{F}^b$ are, respectively, the *surface-excess body force* and the *bulk-phase body force* vector densities. The surface-excess contribution, $\mathbf{F}^s$ [cf. Eqs. (16.1-14) and the first subsection of §17.2], is the two-dimensional analog of continuum body forces in three-dimensional fluids. Such surface forces may arise, for example, through gravitational or electromagnetic forces exerted upon the fluid interface by means of distant external sources lying outside of the interface [e.g., see its relation to the 'disjoining pressure' acting within a thin, plane-parallel, liquid film, as indicated in Eq. (11.2-7)]. On the other hand, the bulk-phase contribution, $\mathbf{F}^b$, has no counterpart for three-dimensional fluids. This force (which in the state of hydrostatic equilibrium derives from the bulk-phase pressure) is exerted upon the two-dimensional areal element by the contiguous bulk-fluid phases. Thus, $\mathbf{F}^b$ is in actuality a (net) contact force, applied intimately at the interface by the surrounding three-dimensional bulk phases; nevertheless, from the standpoint of an intrinsic, surface-fixed observer, necessarily unaware of the presence and intimacy of these three-dimensional bulk phases, it indeed appears to act as a conventional 'distant' areal body force, since its origin does not reside in the two-dimensional interfacial domain itself.

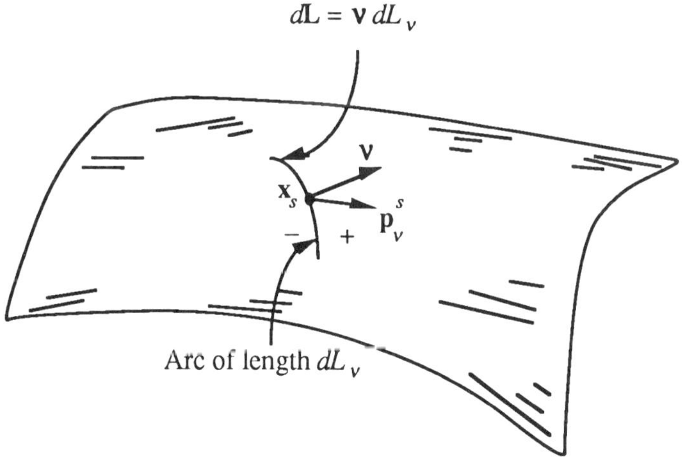

Figure 3.3-1 A directed line element $d\mathbf{L}=\boldsymbol{\nu}dL_\nu$ is drawn within the interface. By convention, the unit normal $\boldsymbol{\nu}$ is drawn from the negative to the positive side of the line element. A differential surface contact force $d\mathbf{F}^s = \mathbf{p}_\nu^s\, dL_\nu$ is exerted upon the portion of interfacial fluid lying on the negative side of the line element by the interfacial fluid lying on the positive side of this element.

Surface Contact Forces

A surface stress vector, $\mathbf{p}_\nu^s$ (possessing the units of force per unit length), acts linealy within the interface by virtue of the intimate contact between contiguous two-dimensional interfacial fluid elements. Per unit length, the vector corresponds to a frictional, pressure, traction, or contact force, and is defined such that, if $d\mathbf{L}=\nu dL_\nu$ is a directed element of arc length with unit normal ν lying in the tangent plane of the interface (see Figure 3.3-1), drawn such that it points from the negative (–) to the positive (+) side of the arc, then $\mathbf{p}_\nu^s\, dL_\nu$ gives the infinitesimal vector force $d\mathbf{F}^s$ exerted by the interfacial material lying on the plus side of the line element upon the material lying on the negative side. This two-dimensional definition and sign convention is the analog of that for the comparable three-dimensional stress vector in terms of a directed element of surface area $d\mathbf{S}$.

The lineal surface contact stress vector $\mathbf{p}_\nu^s$ is not itself an interfacial field variable, as it depends upon the orientational sense in which one draws the lineal element within the interface. Explicitly, the functional dependence,

$$\mathbf{p}_\nu^s = \mathbf{p}_\nu^s(\mathbf{x}_s, \nu), \tag{3.3-2}$$

of $\mathbf{p}_\nu^s$ at the interfacial point $\mathbf{x}_s$ is such that it depends not only upon $\mathbf{x}_s$ but also upon ν. As demonstrated in Appendix 3.A at the conclusion of this chapter, the surface contact stress vector may be expressed in terms of a dyadic field variable $\mathbf{P}^s \equiv \mathbf{P}^s(\mathbf{x}_s)$ as

$$\mathbf{p}_\nu^s = \nu \cdot \mathbf{P}^s, \tag{3.3-3}$$

with $\mathbf{P}^s$ the surface-excess pressure tensor. The latter dyadic, which is an interfacial field quantity, constitutes the fundamental mode via which the state of contact stress within a fluid interface is expressed. Its basic nature is such that $d\mathbf{L}\cdot\mathbf{P}^s$ gives the force $d\mathbf{F}^s$ exerted across the line element $d\mathbf{L}$ in the sense described at the beginning of this subsection.

For an *isotropic* interface existing in a state of hydrostatic equilibrium the surface-excess pressure tensor is given by

$$\mathbf{P}^s = \sigma \mathbf{I}_s, \tag{3.3-4}$$

where the scalar σ is the interfacial tension. It is a macroscale property of the physicochemical system, fundamentally dependent only upon (macroscale) pressure, temperature and interfacial composition at the point $\mathbf{x}_s$.

As discussed in Chapter 2, interfacial tension is the primary mechanical property of equilibrium fluid interfaces, and may be considered as the two-dimensional analog of the thermodynamic pressure p for three-dimensional continua. Unlike these pressure forces, which are compressive in nature, interfacial tension is tensile. As such, it provides fluid interfaces with their familiar tensile nature, whereby they tend to contract such as to minimize interfacial area.

In nonequilibrium circumstances the surface-excess pressure tensor $\mathbf{P}^s$ will generally be a nonisotropic (and nonsymmetric) tensor, possessing in the most general circumstances six independent components (as will be demonstrated in Part II): thus, assuming an orthogonal parameterization of the surface, with $(\mathbf{i}_1, \mathbf{i}_2)$ the orthonormal base vectors defined in Eq. (3.2-20),

$$\mathbf{P}^s = (\mathbf{i}_1\ \mathbf{i}_2)\begin{pmatrix} P^s_{11} & P^s_{12} & P^s_{13} \\ P^s_{21} & P^s_{22} & P^s_{23} \end{pmatrix}\begin{pmatrix} \mathbf{i}_1 \\ \mathbf{i}_2 \\ \mathbf{n} \end{pmatrix}. \tag{3.3-5a}$$

Alternatively, written out explicitly in component form,

$$\begin{aligned} \mathbf{P}^s &= \mathbf{i}_1\mathbf{i}_1 P^s_{11} + \mathbf{i}_2\mathbf{i}_2 P^s_{22} \\ &+ \mathbf{i}_1\mathbf{i}_2 P^s_{12} + \mathbf{i}_2\mathbf{i}_1 P^s_{21} \\ &+ \mathbf{i}_1\mathbf{n}\, P^s_{13} + \mathbf{i}_2\mathbf{n}\, P^s_{23}. \end{aligned} \tag{3.3-5b}$$

The pair of scalar components (P^s_{11}, P^s_{22}) bearing repeated indices are surface pressures, or *normal* surface contact forces per unit area (analogous to 'normal' stresses in three-dimensional fluids); (P^s_{12}, P^s_{21}) are likewise surface traction forces, identifiable with frictional or viscous forces (analogous to shearing stresses in three-dimensional fluids); (P^s_{13}, P^s_{23}) are surface *bending* forces. The latter have no analog in the ordinary three-dimensional stress case, as they reflect effects arising from the nature of the (two-dimensional) interfacial pressure tensor as deriving from a quadrature of corresponding three-dimensional pressure tensor fields [cf. Eq. (16.1-13)].

Hydrostatic Force Balance at a Fluid Interface

At equilibrium, a force balance over an element of area A lying on a fluid interface, such as depicted in Figure 3.3-2, may be expressed as

$$\int_A dA\,\mathbf{f}^s + \oint_C d\mathbf{L}\cdot\mathbf{P}^s = \mathbf{0}, \tag{3.3-6}$$

where $d\mathbf{L}$ is an outwardly directed, differential line element to the closed contour C of the areal domain A. According to the surface divergence theorem (Question 3.10)

$$\oint_C d\mathbf{L}\,\mathfrak{R} = \int_A dA\,\nabla_s\cdot(\mathbf{I}_s\,\mathfrak{R}), \tag{3.3-7}$$

where $\mathfrak{R} \equiv \mathfrak{R}(\mathbf{x}_s)$ is a generic field of arbitrary tensorial order. Equation (3.3-7), together with the general identity [cf. Eqs. (3.2-6), (3.2-10), (3.2-11), (3.2-13)],

$$\nabla_s\cdot\mathbf{I}_s \equiv 2H\,\mathbf{n}, \tag{3.3-8}$$

provides the relation

$$\oint_C d\mathbf{L}\,\mathfrak{R} = \int_A dA(\nabla_s\mathfrak{R} + 2H\,\mathbf{n}\,\mathfrak{R}), \tag{3.3-9}$$

since $\mathbf{I}_s^{\dagger} = \mathbf{I}_s$ and $\mathbf{I}_s \cdot \nabla_s \equiv \nabla_s$. [This generic identity may be specialized by inserting a dot- or cross-product operational sign into identical tensorial positions on both sides of the equality sign, e.g. with use of a dot, $d\mathbf{L}\cdot \mathfrak{R}$, $\nabla_s \cdot \mathfrak{R}$ and $\mathbf{n}\cdot\mathfrak{R}$ replace the respective 'noncontracted' terms in Eq. (3.3-9).] With use of the preceding relation [also noting from Eq. (3.3-5) that $\mathbf{n}\cdot\mathbf{P}^s = \mathbf{0}$], Eq. (3.3-6) becomes

$$\int_A dA(\mathbf{f}^s + \nabla_s \cdot \mathbf{P}^s) = \mathbf{0}. \qquad (3.3\text{-}10)$$

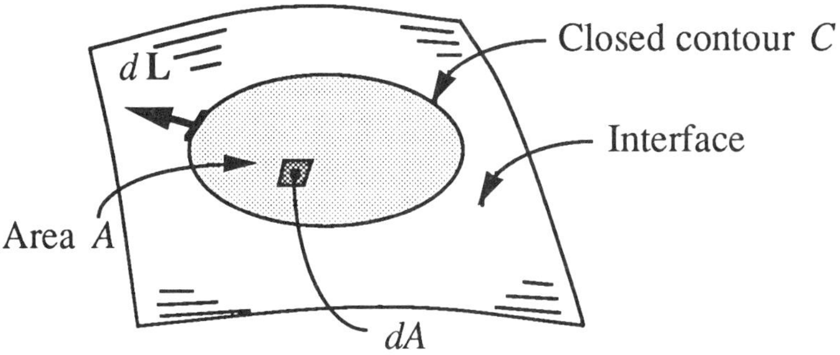

Figure 3.3-2 An interfacial area element A bounded by the closed contour C. The vector $d\mathbf{L}$ denotes an outwardly directed differential line element along the curve C, whereas dA represents a differential areal element of A.

As the choice of interfacial domain A is arbitrary, and since the field variables $\mathbf{f}^s(\mathbf{x}_s)$ and $\mathbf{P}^s(\mathbf{x}_s)$ are independent of this choice, this requires at each point $\mathbf{x}_s$ that

$$\mathbf{f}^s + \nabla_s \cdot \mathbf{P}^s = \mathbf{0}, \qquad (3.3\text{-}11)$$

which constitutes the local surface force balance at a static fluid interface.*

Substitute Eqs. (3.3-4) and (3.3-1) into Eq. (3.3-11), together with Eq. (3.3-8) to obtain the expression

$$\mathbf{F}^b + \mathbf{F}^s + \nabla_s\sigma + 2H\sigma\mathbf{n} = \mathbf{0}. \qquad (3.3\text{-}12)$$

This reduces to the Laplace condition (2.1-3) in the absence of interfacial tension gradients and the surface-excess body force density $\mathbf{F}^s$ (with $\mathbf{F}^b\cdot\mathbf{n} \equiv -\Delta p$).

---

* This is the analog of the corresponding hydrostatic equation

$$\mathbf{F} + \nabla \cdot \mathbf{P} = \mathbf{0}$$

for a three-dimensional fluid continuum, where typically $\mathbf{F}(\mathbf{x})$ is the gravity force ρg and $\mathbf{P} = -\mathbf{I}p$, with $p \equiv p(\mathbf{x})$ the thermodynamic pressure.

3.4 Interfacial Kinematics

A net transfer of mass between the interface and surrounding bulk phases often accompanies solute motion normal to a fluid interface. In this context, nonequilibrium fluid interfaces are respectively distinguished as being either *nonmaterial* or *material*, according as mass does or does not pass to or from the interface at any point $\mathbf{x}_s$. Here, in Part I, where attention is exclusively directed to the *macroscale* interfacial perspective, the fluid interface will for conceptual simplicity always be regarded as being (kinematically) material, even if this is only approximately rather than exactly true. Eventually, in Part II, where attention will be broadened to consider the relationships existing between the micro- and macro-interfacial perspectives, and where more rigorous (albeit conceptually more demanding) foundations will be laid for interfacial transport processes, the purely macroscale transport results of Part I will be generalized and extended to include *inter alia* those additional transport mechanisms arising from the possibly nonmaterial nature of fluid interfaces.

Material Fluid Interfaces

A material interface does not exchange mass* (at any point $\mathbf{x}_s$) with surrounding bulk phases, whence by definition such an interface moves and deforms with the mass-averaged motion of the neighboring fluid phases. In particular, the normal component of the *mass-averaged* vector velocity evaluated at some point $\mathbf{x}_s$ of a material fluid interface, is identical to the actual normal surface velocity at that point. Even when surfactant or solute mass transfer occurs between interfacial and bulk phases, the kinematical treatment of the interface as being a material interface often represents a sufficiently accurate approximation to the true state of affairs, as is illustrated by way of example in Chapter 17 of Part II.

Material Surface Velocity

Imagine, as in Figure 3.4-1a, the trajectory of a particular fluid particle P (i.e. a material point) moving within a three-dimensional fluid continuum. Initially, at time $t = 0$, this material particle instantaneously occupies the position $\mathbf{x}(0)$ relative to a space-fixed origin O. This same material particle

---

* It is nevertheless possible for net solute and solvent molecular exchanges to occur by molecular diffusion without violating the condition of no *mass* exchange. Thus, our restriction to a material interface, which is tantamount to no *convective* transport across the interface, does not rule out the possibility of a diffusive transport (since molecular diffusion is always measured relative to the mass-average velocity.)

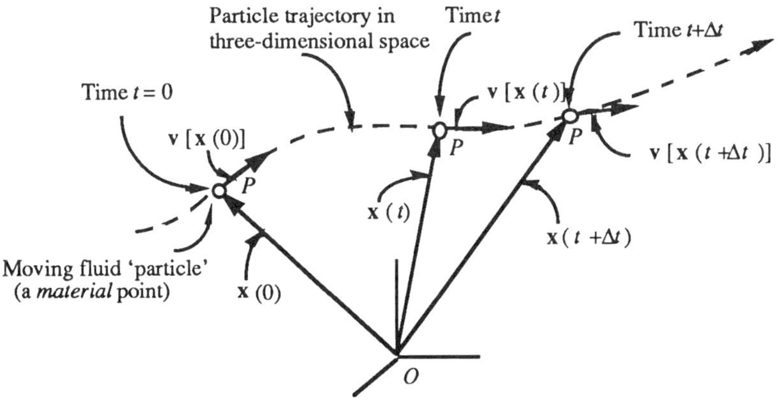

(a)

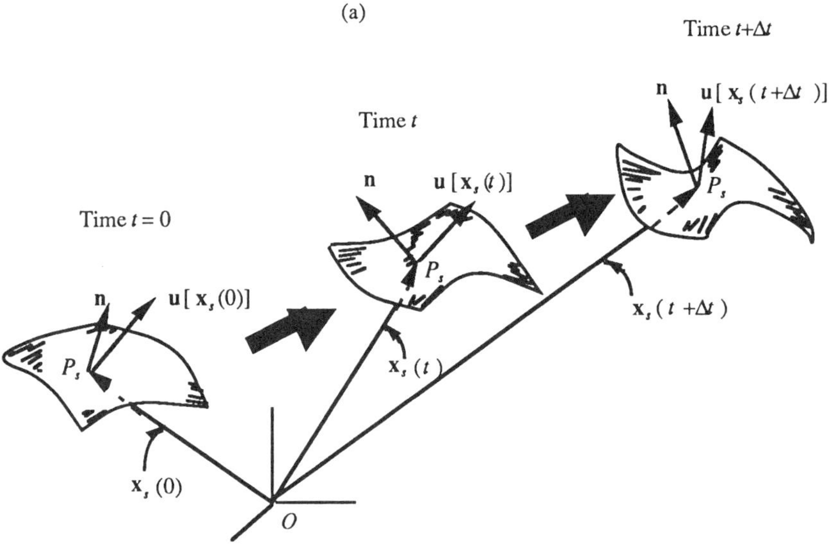

(b)

Figure 3.4-1 (a) The trajectory of a fluid particle (material point) P moving in space is depicted; the instantaneous position $\mathbf{x}$ of the material particle in space is shown at three successive instants of time; (b) A surface is depicted as moving in space; the surface is shown at three successive instants of time, as is the same point P_s lying permanently on the moving surface.

$\mathbf{x}(0)$ pursues a spatial trajectory in time that is defined by the parameterization

$$\mathbf{x} \equiv \mathbf{x}[\mathbf{x}(0),\ t]\,; \tag{3.4-1}$$

this functional relation gives the instantaneous position $\mathbf{x}$ at time t of the particular mass point P that originally occupied the position $\mathbf{x}(0)$ at time t=0. The material (or *mass-averaged*) velocity $\mathbf{v}$ of the convected particle $\mathbf{x}(0)$ may thus be defined in the standard manner (Aris 1962) as

$$\mathbf{v}(\mathbf{x},t) \overset{\text{def}}{=} \lim_{\Delta t \to 0} \left. \frac{\mathbf{x}(t+\Delta t) - \mathbf{x}(t)}{(t+\Delta t) - t} \right|_{\mathbf{x}(0)}$$

$$\equiv \left(\frac{\partial \mathbf{x}}{\partial t}\right)_{\mathbf{x}(0)}. \tag{3.4-2}$$

Imagine now a two-dimensional surface moving through three-dimensional space, as depicted in Figure 3.4-1b. At time t=0, the position relative to a space-fixed origin O of a particular interfacial point P_s lying on this surface may be specified by its instantaneous spatial position vector $\mathbf{x}_s(0)$. At later moments in time this same point P_s locked onto the interface [which point may be identified literally (rather than symbolically by P_s) for all times by the generic label $\mathbf{x}_s(0)$, referring to its original location at t=0] will occupy different positions relative to O, according to the parameterization

$$\mathbf{x}_s \equiv \mathbf{x}_s[\mathbf{x}_s(0),\ t], \tag{3.4-3}$$

where $\mathbf{x}_s$ denotes the current spatial position vector of the point P_s at time t. Assuming a nonvanishing Jacobian (see §B.1), this functional relation may in principle be inverted to provide the convected coordinates $\mathbf{x}_s(0)$ of the surface-fixed point P_s:

$$\mathbf{x}_s(0) \equiv \mathbf{x}_s(0)[\mathbf{x}_s,\ t]. \tag{3.4-4}$$

Explicitly, this functional relation identifies the particular 'interfacial point' $\mathbf{x}_s(0)$, i.e. P_s, which happens to be situated at t at the spatial point $\mathbf{x}_s$. The instantaneous interfacial velocity $\mathbf{u}$ of the convected surface point $P_s \equiv \mathbf{x}_s(0)$ at time t may thus be defined as

$$\mathbf{u}(\mathbf{x}_s,t) \overset{\text{def}}{=} \lim_{\Delta t \to 0} \left. \frac{\mathbf{x}_s(t+\Delta t) - \mathbf{x}_s(t)}{(t+\Delta t) - t} \right|_{\mathbf{x}_s(0)}$$

$$\equiv \left(\frac{\partial \mathbf{x}_s}{\partial t}\right)_{\mathbf{x}_s(0)}, \tag{3.4-5}$$

analogous to the definition of the spatial material velocity provided in Eq. (3.4-2). Equation (3.4-5) may be viewed as a *Lagrangian*-like definition of the instantaneous velocity $\mathbf{u}$ of a point P_s lying permanently on the interface, defined following the motion of the interface in time.

Let the symbol

$$\mathbf{v}^o(\mathbf{x}_s, t) \overset{\text{def}}{=} \mathbf{v}(\mathbf{x}, t)\big|_{\mathbf{x}=\mathbf{x}_s} \tag{3.4-6}$$

denote the velocity through space of the particular material particle that happens to be occupying the spatial point $\mathbf{x}_s$ at time t. For the circumstances of interest to us here in Part I, where the interface is a *material* surface, we have that

$$\mathbf{u}(\mathbf{x}_s, t) = \mathbf{v}^o(\mathbf{x}_s, t). \tag{3.4-7}$$

That is, for a material interface, the velocity of a point lying permanently on the interface is trivially the same as the mass-average velocity $\mathbf{v}^o$ of the two three-dimensional fluid continua lying on either side of the interface. Later on, in Part II (Chapter 15), we will learn that the velocity $\mathbf{u}(\mathbf{x}_s,t)$ of a point lying permanently on the interface may differ from the material velocity $\mathbf{v}^o(\mathbf{x}_s,t)$ at that point; indeed, in such circumstances the material velocities* $\mathbf{v}^+$ and $\mathbf{v}^-$ [or more precisely their normal components $\mathbf{n}\cdot\mathbf{v}^+$ and $\mathbf{n}\cdot\mathbf{v}^-$] lying respectively on the two 'sides' of the interface will generally differ, so that $\mathbf{u}(\mathbf{x}_s,t) \neq \mathbf{v}^+(\mathbf{x}_s,t) \neq \mathbf{v}^-(\mathbf{x}_s,t)$.

As the position vector $\mathbf{x}_s(0)$ denotes a point P_s lying permanently on the interface, the derivative

$$\frac{D_s}{Dt} \overset{\text{def}}{=} \left(\frac{\partial}{\partial t}\right)_{\mathbf{x}_s(0)}, \tag{3.4-8}$$

is known as the *convected surface derivative*. This derivative may be related to the conventional (surface-fixed) partial time derivative $(\partial/\partial t)_{\mathbf{x}_s}$ by applying the chain rule:

$$\left(\frac{\partial}{\partial t}\right)_{\mathbf{x}_s(0)} = \left(\frac{\partial}{\partial t}\right)_{\mathbf{x}_s} + \left(\frac{\partial \mathbf{x}_s}{\partial t}\right)_{\mathbf{x}_s(0)} \cdot \left(\frac{\partial}{\partial \mathbf{x}_s}\right)_t; \tag{3.4-9}$$

equivalently,

$$\frac{D_s}{Dt} = \frac{\partial}{\partial t} + \mathbf{u}\cdot\nabla_s. \tag{3.4-10}$$

Henceforth, following convention, the partial time derivative $\partial/\partial t$ appearing without an identifying subscript $\mathbf{x}_s$ (or $\mathbf{x}$) indicating what is being kept constant in the differentiation will always be understood to refer to a

---

* Here, $\mathbf{v}^+$ and $\mathbf{v}^-$ are explicitly defined such that

$$\mathbf{v}^{\pm}(\mathbf{x}_s, t) \overset{\text{def}}{=} \mathbf{v}(\mathbf{x}^{\pm}, t),$$

where $\mathbf{x}^+$ and $\mathbf{x}^-$ denote points lying just 'above' and just 'below' the interfacial point $\mathbf{x}_s$; physically, these represent the respective material velocities at the interface of the two immiscible fluids whose common boundary forms the interface. (See Question 3.3 at the end of the Chapter.)

point *fixed* either in three-dimensional space, i.e. $(\partial/\partial t)_{\mathbf{x}}$, or in the surface, i.e. $(\partial/\partial t)_{\mathbf{x}_s}$. In particular, the terminology 'surface-fixed point' refers to a point on the surface characterized by constant, intrinsic surface coordinates (q^1, q^2). Thus [cf. Eqs. (15.3-1)–(15.3-5)], *

$$\left(\frac{\partial}{\partial t}\right)_{\mathbf{x}_s} \equiv \left(\frac{\partial}{\partial t}\right)_{q^1,\, q^2} . \tag{3.4-10a}$$

For a material interface, Eq. (3.4-7) permits us to replace (3.4-10) by the identity

$$\frac{D_s}{Dt} = \frac{\partial}{\partial t} + \mathbf{v}^o \cdot \nabla_s . \tag{3.4-11}$$

As Part I will always be concerned with material (or approximately material) interfaces, this latter equation will represent the preferred representation of the convected surface derivative in this portion of the text. [The advantage residing in the use of (3.4-11) over (3.4-10) is that only material velocities, namely $\mathbf{v}^o$ and $\mathbf{v}$, will then appear, and these are related by Eq. (3.4-6). Indeed, on occasion we may even unambiguously write $\mathbf{v}(\mathbf{x}_s,t)$ in place of $\mathbf{v}^o$, without ambiguity. In such circumstances only the (material) velocity $\mathbf{v}$ will appear in our equations.]

The (material) surface velocity $\mathbf{v}^o$ may be decomposed into normal and tangential components by means of the surface idemfactor $\mathbf{I}_s$ (which acts as a projection operator):

---

* An alternative form of Eq. (3.4-4) may be expressed as

$$\mathbf{x}_s(0) = \mathbf{x}_s(0)[q^1, q^2, t],$$

wherein the argument q_0^3 has been omitted with the understanding that the surface coordinates (q^1, q^2) pertain to this particular q^3-coordinate surface. Thus, Eq. (3.4-9) may be expressed as

$$\begin{aligned}
\left.\frac{\partial}{\partial t}\right)_{\mathbf{x}_s(0)} &= \left.\frac{\partial}{\partial t}\right)_{q^1,\, q^2} + \left.\frac{\partial q^\alpha}{\partial t}\right)_{\mathbf{x}_s(0)} \left.\frac{\partial}{\partial q^\alpha}\right)_t \\
&= \left.\frac{\partial}{\partial t}\right)_{q^1,\, q^2} + \left.\frac{\partial q^\alpha}{\partial t}\right)_{\mathbf{x}_s(0)} \mathbf{a}_\alpha \cdot \mathbf{a}^\beta \left.\frac{\partial}{\partial q^\beta}\right)_t \\
&= \left.\frac{\partial}{\partial t}\right)_{q^1,\, q^2} + \mathbf{u} \cdot \mathbf{I}_s \cdot \nabla_s \\
&\equiv \left.\frac{\partial}{\partial t}\right)_{q^1,\, q^2} + \mathbf{u} \cdot \nabla_s .
\end{aligned}$$

Comparision of Eqs. (3.4-9) and the above relation verifies the relation (3.4-10a).

$$\begin{aligned}\mathbf{v}^o(\mathbf{x}_s, t) &= \mathbf{I}\cdot\mathbf{v}^o(\mathbf{x}_s, t)\\ &\equiv \mathbf{I}_s\cdot\mathbf{v}^o + \mathbf{n}\mathbf{n}\cdot\mathbf{v}^o\\ &= \mathbf{v}_s(\mathbf{x}_s, t) + \mathbf{v}_n(\mathbf{x}_s, t);\end{aligned} \tag{3.4-12}$$

here, the vectors $\mathbf{v}_s \overset{\text{def}}{=} \mathbf{I}_s\cdot\mathbf{v}^o$ and $\mathbf{v}_n \overset{\text{def}}{=} \mathbf{n}\mathbf{n}\cdot\mathbf{v}^o$ are, respectively, the tangential and normal surface velocities. As implied by the choice of names, the former vector has no normal component, i.e. $\mathbf{n}\cdot\mathbf{v}_s=0$, whereas the latter has no tangential component, only a normal component.

Surface Reynolds Transport Theorem

A kinematical theorem that proves useful in the derivation of surface conservation laws is the two-dimensional analog of the Reynolds transport theorem. The *surface* Reynolds transport theorem may be derived in a manner entirely analogous to its three-dimensional counterpart (for a derivation of the latter, see Slattery 1981) by considering a surface moving through three-dimensional space, such as is depicted in Figure 3.4-2. At time t=0, a differential areal element on this surface may be denoted by dA_0, whereas at time t it is given by dA: a geometrical relation exists between these two differential areal elements, namely

$$dA = j\,dA_0, \tag{3.4-13}$$

where $j \equiv j[\mathbf{x}_s(0), t] \equiv j[\mathbf{x}_s, t]$ is the *surface Jacobian*, representing the amount by which the original area dA_0 is magnified at time t. The surface Jacobian j is the two-dimensional analog of the standard volumetric Jacobian J of three-dimensional space [which latter quantity may be viewed analogously to Eq. (3.4-13) as the scaling factor $J \equiv dV/dV_0$ between differential volumetric elements dV and dV_0].

As demonstrated in Question 3.4, the convected time derivative of the surface Jacobian is given as

$$\frac{1}{j}\frac{D_s j}{Dt} = \nabla_s\cdot\mathbf{u}\,. \tag{3.4-14}$$

The surface Reynolds transport theorem may now be derived by considering the convected time derivative of a surface (scalar, vector or polyadic) field $\Re \equiv \Re(\mathbf{x}_s, t)$ integrated over the instantaneous configuration of a moving surface $A \equiv A(t)$:

$$\frac{d}{dt}\int_A \Re\, dA \tag{3.4-15}$$

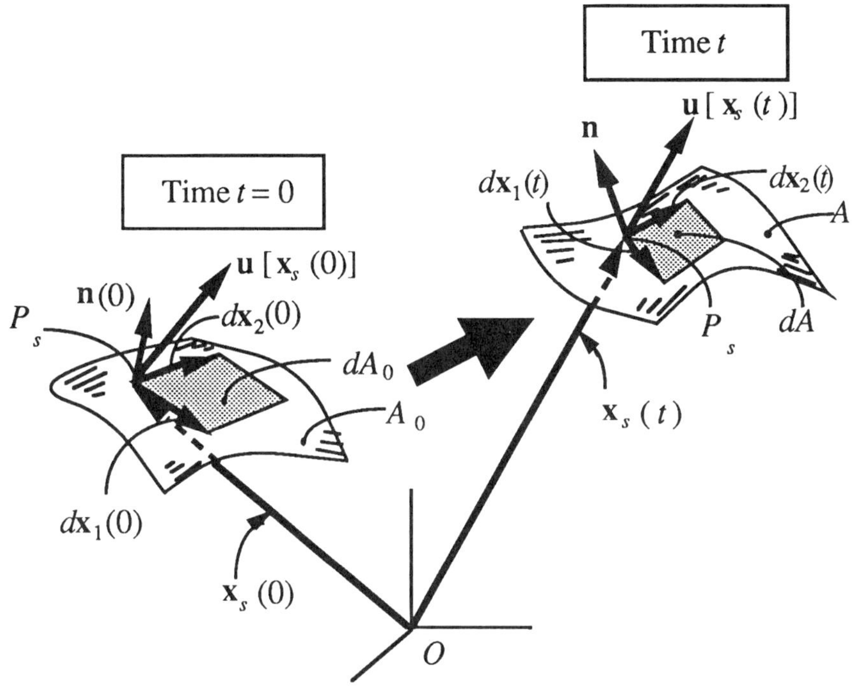

Figure 3.4-2 A surface is depicted as moving through three-dimensional space. At time t=0, a differential areal element $dA_0 \equiv \mathbf{n}(0) \cdot [d\mathbf{x}_1(0) \times d\mathbf{x}_2(0]$ may be defined in terms of the differential length vectors $[d\mathbf{x}_1(0), d\mathbf{x}_2(0]$ originating at the surface-fixed point P_s, which point occupies the spatial position $\mathbf{x}_s(0)$. At time t the convected surface point P_s is located at the spatial position $\mathbf{x}_s(t)$, and the areal element is given by $dA \equiv \mathbf{n}(t) \cdot [d\mathbf{x}_1(t) \times d\mathbf{x}_2(t)]$.

According to Eq. (3.4-13) this time derivative may be expressed as*

---

* It is easy to follow the logic implied in the intermediate steps of Eq. (3.4-16) by inserting the explicit arguments of the various functions composing the integrand [while resorting to the functional dependencies (3.4-3) and its inverse (3.4-4)]. Thus, temporarily writing $A(0)$ in place of A_0 for clarity, we have for the first step of (3.4-16)

$$\frac{d}{dt}\int_{A(t)} \Re[\mathbf{x}_s, t]dA = \frac{d}{dt}\int_{A(0)} \Re[\mathbf{x}_s(0), t]\, j\, [\mathbf{x}_s(0), t]dA(0).$$

The integration domain $A(0)$, being time invariant, is independent of time t, whence the integration and convected time differentiation steps in the preceding operation may be interchanged to obtain

$$\int_{A(0)} \left.\frac{\partial}{\partial t}\right)_{\mathbf{x}_s(0)} \{\Re[\mathbf{x}_s(0), t]\, j\, [\mathbf{x}_s(0), t]\}dA(0)$$

$$\frac{d}{dt}\int_A \Re dA = \frac{d}{dt}\int_{A(0)} \Re j dA_0 = \int_A \frac{1}{j}\frac{D_s}{Dt}(\Re j)\,dA\,. \quad (3.4\text{-}16)$$

Upon employing Eq. (3.4-14), this yields

$$\frac{d}{dt}\int_A \Re dA = \iint_A \left[\frac{D_s\Re}{Dt} + \frac{\Re}{j}\frac{D_s j}{Dt}\right] dA$$

$$= \iint_A \left[\frac{D_s\Re}{Dt} + \Re\nabla_s\cdot\mathbf{u}\right] dA\,, \quad (3.4\text{-}17)$$

which is the surface Reynolds transport theorem. Alternatively, with use of Eq. (3.4-10), this theorem can be written as

$$\boxed{\frac{d}{dt}\int_A \Re dA = \iint_A \left[\frac{\partial\Re}{\partial t} + \nabla_s\cdot(\mathbf{u}\Re)\right] dA\,.} \quad (3.4\text{-}18)$$

For a material surface, Eq. (3.4-18) adopts the equivalent form

$$\frac{d}{dt}\int_A \Re dA = \iint_A \left[\frac{\partial\Re}{\partial t} + \nabla_s\cdot(\mathbf{v}^o\Re)\right] dA\,. \quad (3.4\text{-}19)$$

Equations (3.4-13) and (3.4-14) jointly reveal the nature of the property $\nabla_s\cdot\mathbf{u}$ as the *fractional rate of areal expansion* (of the moving surface), namely,

$$\frac{1}{\delta A}\frac{D_s}{Dt}\delta A = \nabla_s\cdot\mathbf{u} \quad (3.4\text{-}20)$$

(or its counterpart with $\nabla_s\cdot\mathbf{v}^o$ for the material surfaces considered here in Part I), with δA an infinitesimal area. This is the analog of the comparable volumetric relation (Aris 1962)

$$\frac{1}{\delta V}\frac{D}{Dt}\delta V = \nabla\cdot\mathbf{v},$$

which interprets the divergence of the vector velocity field $\mathbf{v}$ in three dimensions as being the fractional rate of volumetric expansion of a

---

for the right-hand term. Upon reverting to the standard convected derivative notation (3.4-8) and converting back from the original surface domain to the current surface domain variables via Eq. (3.4-4) we thereby obtain the last term on the right-hand side of Eq. (3.4-16), which may be written out explicitly as

$$\int_{A(t)} \frac{D_s}{Dt}\{\Re[\mathbf{x}_s, t]\, j[\mathbf{x}_s, t]\}\frac{dA}{j}\,.$$

convected infinitesimal volume element δV. Using Eq. (3.4-9), the rate of material surface expansion may be decomposed into contributions arising from tangential and normal surface velocities, as

$$\nabla_s \cdot \mathbf{v}^o = \nabla_s \cdot \mathbf{v}_s + \nabla_s \cdot \mathbf{v}_n . \qquad (3.4\text{-}21)$$

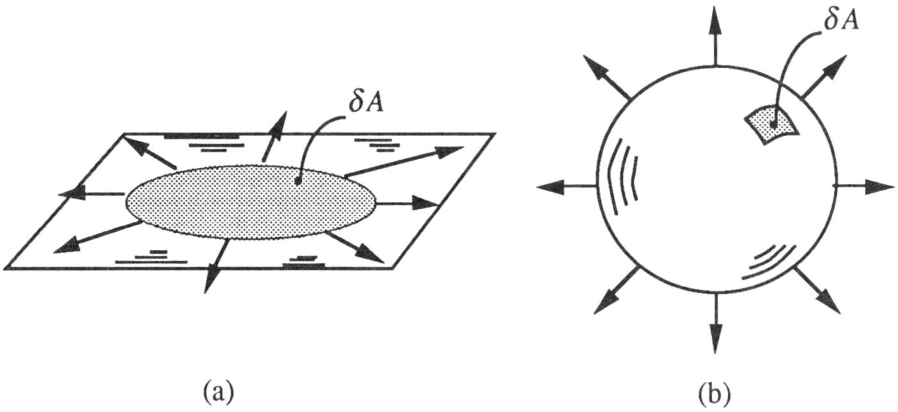

Figure 3.4-3 The fractional rate of surface expansion $\frac{1}{\delta A}\frac{D_s}{Dt}\delta A = \nabla_s \cdot \mathbf{v}^o$ may be viewed as the sum of: (a) a surface stretching motion, quantified by $\nabla_s \cdot \mathbf{v}_s$; and (b) a surface inflation, quantified by $\nabla_s \cdot \mathbf{v}_n$.

The first term in Eq. (3.4-20) represents the rate of surface expansion that accompanies a "stretching" motion in the interface, whereas the second term, the surface expansion that accompanies an "inflation" of the surface. These two motions are illustrated in Figure 3.4-3.

3.5 The Generic Volumetric Transport Equation in Continuous Three-Dimensional Media

Areal transport processes within and normal to fluid interfaces are generally accompanied by simultaneous volumetric transport processes within the two contiguous bulk-phase fluids, each of which is separately a three-dimensional fluid continuum. These areal and volumetric forms of transport are necessarily interdependent, at least in proximity to the interface. The volumetric bulk-phase physical properties (e.g. viscosity, density, solute diffusivity, etc.) and fields (velocity, pressure, solute concentration, etc.), though separately continuous within each phase, are generally discontinuous across the phase boundary separating them. Accordingly, one view of the role of interfacial fields and interfacial physical properties (and the view that will be adopted throughout this text) is that of furnishing physically appropriate boundary conditions imposed upon these bulk fields at their common boundary.

Rather than consider the individual transport processes separately governing the balance of mass, momentum, species, etc., in the remaining sections of this chapter, we focus instead upon a single, abstract, *generic* conservation law governing the transport of *all* extensive physical properties, firstly in continuous three-dimensional media, then in discontinuous media, and finally within a fluid interface. Ultimately, the generic balance equations will be applied in later chapters to specific physical circumstances. (This same generic procedure is followed in Part II; cf. §§15.6 and 15.7.)

Let $\Psi \equiv \Psi(t)$ be the amount of some generic, extensive physical property $\mathcal{P}$ (e.g. mass of some species i, linear momentum, energy, etc.) instantaneously contained within a domain of fluid (i.e., a *material volume*, or volume that convects with the fluid motion) $V \equiv V(t)$ (see Figure 3.5-1) forming some portion of a three-dimensional fluid continuum. An elementary generic overall balance expresses the time-rate of accumulation $d\Psi/dt$ of the property $\mathcal{P}$ within V as

$$\frac{d\Psi}{dt} = -\Phi + \Pi + Z, \qquad (3.5\text{-}1)$$

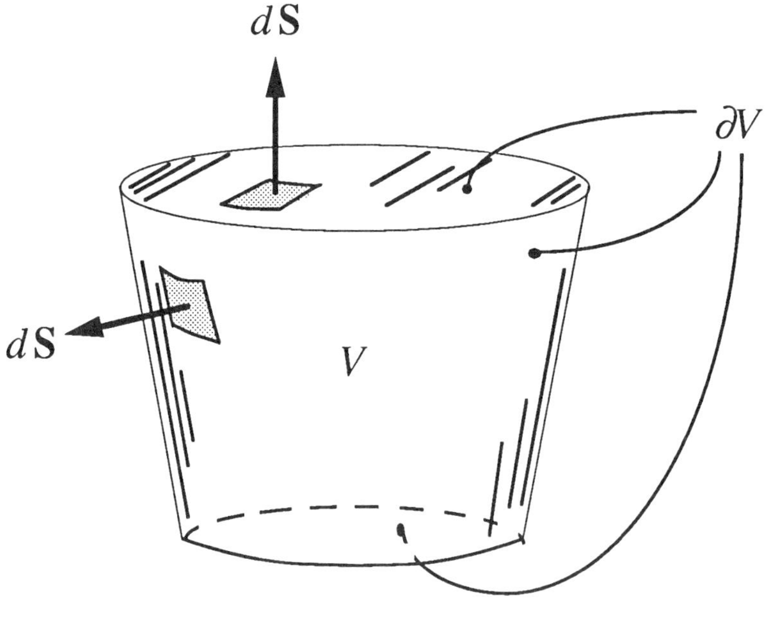

Figure 3.5-1 A material domain $V(t)$ bounded by the closed surface $\partial V(t)$. The latter consists of the sides, top, and bottom of the pillbox; $d\mathbf{S}$ denotes an outwardly drawn directed element of surface area on ∂V.

where Φ is the 'molecular' or 'diffusive' flux of $\mathcal{P}$ *out* of V through its surface ∂V, Π is the time-rate of production or creation of $\mathcal{P}$ within V (e.g. by chemical reaction), and Z the rate of supply of the property $\mathcal{P}$ from sources external to V acting upon each individual element of the volume V (e.g. momentum may be supplied to the contents of V by the action of a 'long-range' external force). Depending upon the tensorial nature of the property $\mathcal{P}$ being considered (e.g. scalar in the case of mass, vector in the case of momentum, etc.), Ψ may be a scalar, vector, etc. It is not necessary that we explicitly specify the tensorial rank of this generic quantity, except, of course, to recognize that all extensive quantities appearing in Eq. (3.5-1) must be of the same rank.

We assume that the total amount Ψ of $\mathcal{P}$ in V may be expressed in terms of a continuous volumetric density field $\psi(\mathbf{x}, t)$ (amount of the property $\mathcal{P}$ per unit volume at a point $\mathbf{x}$ of V at time t) such that

$$\Psi = \int_V dV \;\psi(\mathbf{x}, t)\,, \qquad (3.5\text{-}2)$$

where the tensorial rank of the field ψ is the same as that of Ψ. According to the Reynolds transport theorem (for a derivation of this theorem, refer to Question 3.5), it follows upon differentiating Eq. (3.5-2) that

$$\frac{d\Psi}{dt} = \int_{V(t)} \left\{ \frac{\partial}{\partial t}\psi(\mathbf{x}, t) + \nabla\cdot[\mathbf{v}\,\psi(\mathbf{x}, t)]\right\} dV\,. \qquad (3.5\text{-}3)$$

Likewise, the rate Π of production of $\mathcal{P}$ within V may be expressed in terms of the volumetric production rate density field $\pi(\mathbf{x}, t)$ as

$$\Pi = \int_V dV \;\pi(\mathbf{x}, t)\,. \qquad (3.5\text{-}4)$$

Similarly, with the field $\zeta(\mathbf{x}, t)$ the volumetric rate of supply density,

$$Z = \int_V dV \;\zeta(\mathbf{x}, t)\,. \qquad (3.5\text{-}5)$$

With $\phi(\mathbf{x}, t)$ the areal flux density (amount per unit time per unit area) of the property $\mathcal{P}$ at a point $\mathbf{x}$ lying on ∂V, the time-rate Φ at which the property $\mathcal{P}$ is transferred *out* of the volume V through its surface is

$$\Phi = \oint_{\partial V} d\mathbf{S}\cdot\phi\,, \qquad (3.5\text{-}6)$$

with $d\mathbf{S}$ the outward-drawn normal on ∂V. (As $d\mathbf{S}$ is a vector, the tensorial rank of the density ϕ is one order higher than that of Φ; thus, if the property

$\mathcal{P}$ is a vector, e.g. momentum, ϕ is a dyadic.) Equation (3.5-6) may be alternatively expressed as a volume integral by using the divergence theorem (Reddy & Rasmussen 1982) to obtain

$$\Phi = \int_V dV\, \nabla \cdot \phi \,. \tag{3.5-7}$$

Substitution of the above relations into Eq. (3.5-1) yields

$$\int_V dV \left[\frac{\partial \psi}{\partial t} + \nabla \cdot (\mathbf{v}\, \psi) + \nabla \cdot \phi - \pi - \zeta \right] = \mathbf{0}\,. \tag{3.5-8}$$

As the choice of the domain V is arbitrary the integrand must vanish at each point $\mathbf{x}$, whence

$$\boxed{\frac{\partial \psi}{\partial t} + \nabla \cdot (\mathbf{v}\, \psi + \phi) - \pi - \zeta = \mathbf{0}\,.} \tag{3.5-9}$$

This fundamental relation constitutes the generic, volumetric balance equation for continuous three-dimensional media at each point $\mathbf{x}$ of the continuum. The sum $\mathbf{v}\psi+\phi$ appearing in Eq. (3.5-9), consisting of the convective flux $\mathbf{v}\psi$ and diffusive flux ϕ of the property $\mathcal{P}$, represents the total flux relative to a stationary observer

Some examples illustrating the respective choices of generic variables in Eq. (3.5-9) for common physical properties $\mathcal{P}$ are provided in Table 3.5-1. Thus, in the case of mass, Eq. (3.5-9) adopts the form

$$\frac{\partial \rho}{\partial t} + \nabla \cdot (\mathbf{v}\, \rho) = 0, \tag{3.5-10}$$

corresponding to the well-known continuity equation, expressing the law of conservation of mass.

Table 3.5-1
IDENTIFICATION OF GENERIC VOLUMETRIC FIELDS IN EQ. (3.5-9)

Extensive Property $\mathcal{P}$	Volumetric Property Density ψ	Areal Molecular Flux ϕ	Internal Production Rate π	External Supply Rate ζ
Mass	ρ	0	0	0
Lin. momentum	$\rho\mathbf{v}$	$-\mathbf{P}$	0	$\mathbf{F}$
Ang. momentum	$\mathbf{x}\times(\rho\mathbf{v})+\rho\mathbf{a}$	$\mathbf{P}\times\mathbf{x}-\mathbf{C}$	0	$\mathbf{x}\times\mathbf{F}+\mathbf{G}$
Mass of species i	ρ_i	$\mathbf{j}_i$	R_i	0
Energy	$U+(1/2)\rho\mathbf{v}\cdot\mathbf{v}$	$\mathbf{q}$	$-\nabla\cdot(\mathbf{v}\cdot\mathbf{P})$	$\mathbf{F}\cdot\mathbf{v}$
Entropy	S	$\mathbf{q}/T$	Φ	0

$\mathbf{a}$ = internal angular momentum pseudovector

$\mathbf{C}$ = couple-stress pseudodyadic

$\mathbf{F}$ = external body force density vector

$\mathbf{G}$ = body couple density pseudovector

$\mathbf{j}_i$ = species diffusive mass flux vector measured relative to $\mathbf{v}$

$\mathbf{P}$ = pressure dyadic

$\mathbf{q}$ = heat flux vector

R_i = species volumetric reaction rate

T = absolute temperature

U = internal energy density

$\mathbf{v}$ = mass-average velocity vector

$\mathbf{x}$ = space-fixed position vector

ρ = mass density

ρ_i = species mass density

Φ = entropy production density rate

3.6 The Generic Volumetric Transport Equation in Discontinuous Three-Dimensional Media and the Generic Surface-Excess Transport Law

When the specific field variables indicated in Table 3.5-1 are substituted into Eq. (3.5-11), familiar (Bird *et al.* 1960) balance equations are obtained governing the transport of mass, momentum, etc. for continuous media. In the presence of a physical (macroscale) discontinuity, however, such as generally occurs across a fluid-fluid phase boundary for each of the fields indicated in Table 3.5-1, these equations must be supplemented by an appropriate equation quantifying the "jump" in the (normal component of the) field across the discontinuity. (This jump expresses the *magnitude* of the discontinuity.) For fluid interfaces, this quantification adopts the form of a surface-excess balance equation [see Eq. (3.6-16)] operative at the macroscopically singular phase interface, which then functions as a boundary condition imposed upon the (discontinuous) bulk-field densities.

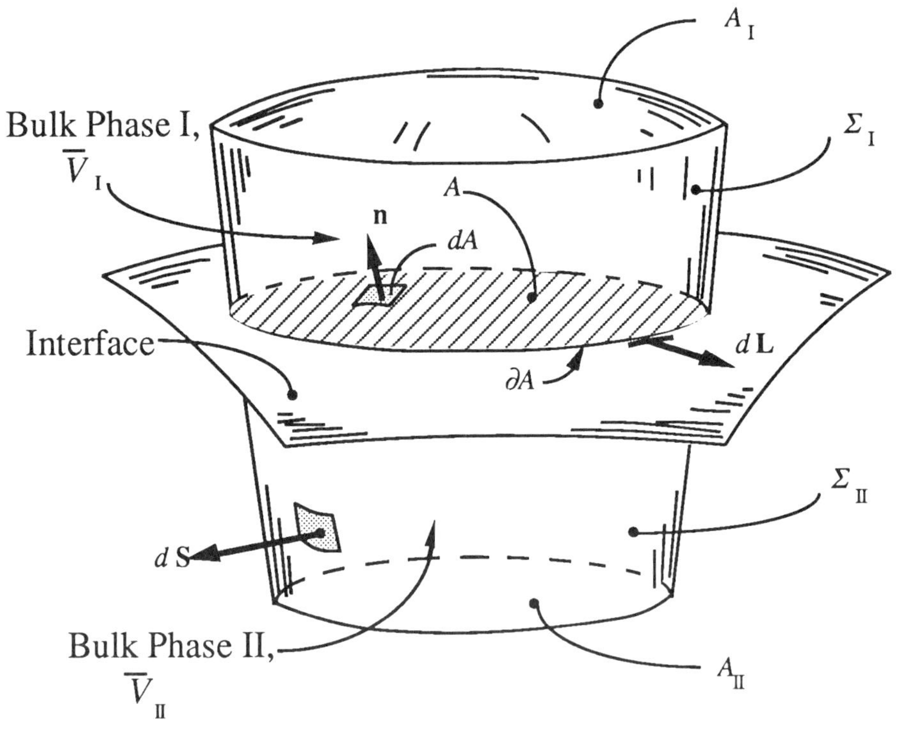

Figure 3.6-1 A macroscopic material-fixed pillbox straddles a singular (material) fluid interface which separates bulk fluid phases I and II.

Consider the material pillbox $V \equiv V(t)$ depicted in Figure 3.6-1. This domain straddles a moving and deforming fluid interfacial domain $A \equiv A(t)$, internal to the pillbox, and instantaneously represented by the shaded area in Figure 3.6-1. The domain V may be decomposed as*

$$V = \bar{V} \oplus A, \tag{3.6-1}$$

with

$$\bar{V} = \bar{V}_{\mathrm{I}} \oplus \bar{V}_{\mathrm{II}}, \tag{3.6-2}$$

where the overbar consistently denotes in this section a bulk-phase quantity. The closed surface ∂V bounding the volume V may similarly be decomposed as

$$\partial V = \partial \bar{V} \oplus \partial A, \tag{3.6-3a}$$

in which

$$\partial \bar{V} = A_{\mathrm{I}} \oplus A_{\mathrm{II}} \oplus \Sigma_{\mathrm{I}} \oplus \Sigma_{\mathrm{II}}. \tag{3.6-3b}$$

Here A_{I} and A_{II} respectively represent the top and bottom end caps of the pillbox.

As depicted in Figure 3.6-2, points within the pillbox V may be identified by the space-fixed position vector $\mathbf{x}$, or, alternatively, by the surface position vector $\mathbf{x}_s$ together with an algebraically signed normal coordinate n originating at the interface and reckoned positive in the direction of the unit surface normal $\mathbf{n}$:

$$n \stackrel{\text{def}}{=} (\mathbf{x} - \mathbf{x}_s) \cdot \mathbf{n}\,, \tag{3.6-4a}$$

such that

$$\mathbf{x} = \mathbf{x}_s + n\mathbf{n}\,. \tag{3.6-4b}$$

where $\mathbf{x}$ and $\mathbf{x}_s$ denote positions along identical normal coordinate trajectories (as shown in the figure). Throughout this text, $\mathbf{n}$ will always be drawn from phase II into phase I [occasionally identified as being drawn from the – to the + phase as, for example, in Eq. (3.6-12)].

A generic volumetric variable λ may thus be parameterized within V as being of the equivalent functional form

$$\lambda(\mathbf{x};t) \equiv \lambda(\mathbf{x}_s, n;\ t) \tag{3.6-5}$$

---

* Since we are particularly interested here in examining the case of volumetric domains V whose interior contains a singular surface A, we use the symbol $\bar{V}$ to exclude from V the set of points lying on the surface A. The distinction is unimportant except in a purely technical sense since the points excluded from V constitute a "set of measure zero". Similar remarks apply to $\partial \bar{V}$ in Eq. (3.6-3), which is distinguished from ∂V by the set of points lying on the closed contour ∂A in Figure 3.6-1.

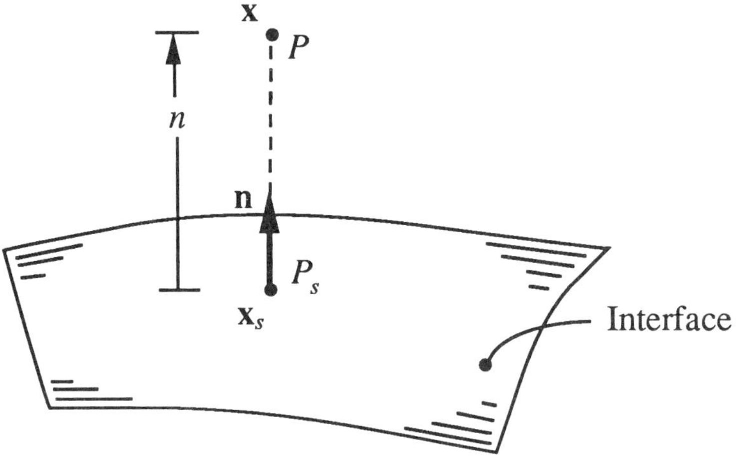

Figure 3.6-2 The relation between space- and surface-fixed position vectors, **x** and $\mathbf{x}_s$ respectively, is illustrated. The points P and P_s are chosen so as to lie along the same normal coordinate n trajectory, in the direction of the unit surface normal **n**.

As discussed in §3.1, the fundamental physical perspective of interfaces adopted in this text (formally elaborated upon in Chapter 15 of Part II), admits of two disparate length scales — the macro and microscales. From the standpoint of a macroscopic observer, unable to resolve distances of the order of the 'thickness' of the interfacial transition region, the total amount $\bar{\Psi}$ of $\mathcal{P}$ contained within $\bar{V}$ may be expressed in terms of the generally discontinuous volumetric density field $\bar{\psi}(\mathbf{x}, t)$ [amount of the property $\mathcal{P}$ per unit volume at a point **x** of $\bar{V}$ (technically, $\mathbf{x} \in \bar{V}$, where the symbol $\in$ designates "included in")] such that

$$\bar{\Psi} = \int_{\bar{V}} dV \ \bar{\psi}(\mathbf{x},t) \,. \tag{3.6-6}$$

On the other hand, from the finer *microscale* viewpoint (see Figure 3.6-3), a *true* field density $\psi(\mathbf{x}, t)$ becomes apparent, this latter microscale field density being fully continuous throughout the continuous domain V. From the standpoint of a microscale observer, the *true* total amount Ψ of $\mathcal{P}$ in V may be expressed in terms of the continuous volumetric density field $\psi(\mathbf{x}, t)$ as in Eq. (3.5-2), viz.

$$\Psi = \int_{V} dV \ \psi(\mathbf{x}, t) \,.$$

It is only in the 'vicinity' of the fluid interface that differences between the two volumetric densities $\psi(\mathbf{x}, t)$ and $\bar{\psi}(\mathbf{x}, t)$ emerge, such differences being attributable to the large normal gradients $\partial\psi/\partial n$ in ψ existing within the interfacial transition zone. Thus, as a means of reconciling the usual (discontinuous) macroscale view of fluid interfaces with the true (continuous) microscale view, the residual difference

$$\Psi^s \overset{\text{def}}{=} \Psi - \overline{\Psi} \tag{3.6-7a}$$

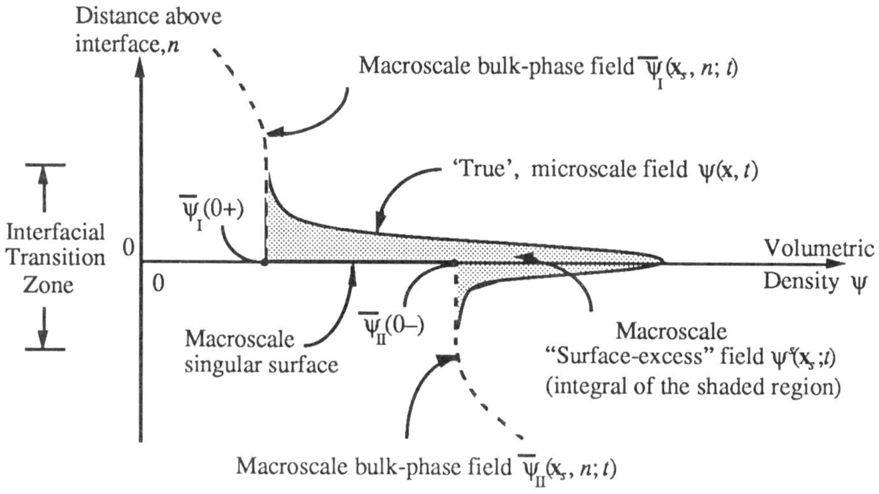

Figure 3.6-3 The macroscale bulk and surface-excess fields at the point $\mathbf{x}_s$ from the ('true', continuous, microscale) length scale l perspective of the interfacial transition zone. The surface-excess field represents the difference between the macro and microscale fields, assigned to the macroscale interface.

is *assigned* to the macroscale surface A as representing the total 'surface-excess' amount of the property $\mathcal{P}$ in A. In turn, assuming all subsequent intensive areal fields to be continuous within the interface, this definition of the (areally) extensive variable permits us to define the surface-excess areal density field $\psi^s \equiv \psi^s(\mathbf{x}_s, t)$ at each point $\mathbf{x}_s$ of A by the expression

$$\Psi^s = \int_A dA\, \psi^s(\mathbf{x}_s, t). \tag{3.6-7b}$$

The generic areal field density $\psi^s(\mathbf{x}_s, t)$ corresponds to the (excess) amount of the property $\mathcal{P}$ per unit area at the point P_s of the two-dimensional *interfacial continuum*, just as the volumetric field density $\psi(\mathbf{x}, t)$ corresponds to the amount of the property $\mathcal{P}$ per unit volume at a point P of the three-dimensional *volumetric continuum*. The reader should note (as follows from the above) that the areal field density ψ^s is *not* to be regarded as the actual amount of the property $\mathcal{P}$ per unit area at the point P_s (in which case ψ^s might better be regarded as a "surface" rather than "surface-excess" field), but rather represents the amount *assigned* to the interface as the intensive field representation of the extensive residual difference, Eq. (3.6-7a).

With use of the preceding equations, the corollary to Eq. (3.5-2) for the macroscopically discontinuous domain depicted in Figure 3.6-1 may be expressed as

$$\Psi = \int_{\bar{V}} dV \ \bar{\psi} + \int_{A} dA \ \psi^s \ . \qquad (3.6\text{-}8)$$

Explanatory comments similar to the above may be applied to the total molecular efflux Φ of the property $\mathcal{P}$ through the boundaries of the macroscopically discontinuous volume element V depicted in Figure 3.6-1. Thus, from a purely macroscale perspective, the efflux $\bar{\Phi}$ through the boundary $\partial\bar{V}$ may be expressed in terms of the generally discontinuous areal flux density $\bar{\phi}(\mathbf{x}, t)$ (L-scale amount of the property $\mathcal{P}$ per unit area per unit time flowing across a directed element of surface centered at a point $\mathbf{x}$ lying on the bounding surface $\partial\bar{V}$) as

$$\bar{\Phi} = \oint_{\partial\bar{V}} d\mathbf{S} \cdot \bar{\phi} . \qquad (3.6\text{-}9\text{a})$$

A continuous flux density $\phi(\mathbf{x}, t)$ is perceived at the microscale level, this latter *true* flux density differing substantially from $\bar{\phi}(\mathbf{x}, t)$ particularly in the vicinity of the phase transition zone, in a manner closely analogous to that depicted in Figure 3.6-3 for the volumetric density field ψ. The *true* efflux Φ through the material boundary ∂V as perceived by a microscale observer may be expressed [similar to Eq. (3.5-6)] as

$$\Phi = \oint_{\partial V} d\mathbf{S} \cdot \phi . \qquad (3.6\text{-}9\text{b})$$

As in Eq. (3.6-7a), the disparity between these micro- and macro-scale flux perspectives is reconciled by assigning the residual difference

$$\Phi^s \overset{\text{def}}{=} \Phi - \bar{\Phi} \qquad (3.6\text{-}10\text{a})$$

to the macroscale contour ∂A as representing the total 'surface-excess' (molecular) flux of the property out of A through the closed contour ∂A bounding A; (the contour ∂A represents the intersection of the surface ∂V with A). The analog of Eq. (3.6-7b) follows from our assumption that all areal fields are continuous within the interface, whence we are led to define the *surface-excess lineal density field* $\phi^s(\mathbf{x}_s, t)$ at each point $\mathbf{x}_s$ of the interface as

$$\Phi^s = \oint_{\partial A} d\mathbf{L} \cdot \phi^s \ , \qquad (3.6\text{-}10\text{b})$$

where $d\mathbf{L}$ is an outwardly directed differential lineal element along the contour ∂A. The generic field $\phi^s(\mathbf{x}_s, t)$ corresponds to the (excess) amount of the property $\mathcal{P}$ per unit time per unit arc length flowing (in a 'positive' sense) across the line element $d\mathbf{L}$ in Figure 3.6-1 at the point P_s of the two-dimensional interfacial continuum. [It represents the two-dimensional analog of the areal flux field $\phi(\mathbf{x}, t)$ corresponding to the flux of the property $\mathcal{P}$ at a point P of the three-dimensional volumetric continuum; compare Eq. (3.6-9b) with (3.6-10b).] The field quantity ϕ^s possesses a tensorial rank that is one order higher than that of the tensor Φ (as $d\mathbf{L}$ is a vector).

Formal proofs of (the physically primitive and strictly intuitive) Eqs. (3.6-8) and (3.6-10) will be given in Part II [cf. Eqs. (15.6-5), (15.6-18), (15.6-22) and (15.6-36)], where the microscopically continuous and macroscopically discontinuous views of a fluid interface will be mathematically reconciled within the framework of a matched asymptotic expansion scheme.

Similar to Eq. (3.6-8), other volumetric relations corresponding to Eqs. (3.5-4) and (3.5-5) adopt the forms

$$\Pi = \int_{\bar{V}} dV\ \bar{\pi} + \int_A dA\ \pi^s\ , \tag{3.6-11a}$$

$$Z = \int_{\bar{V}} dV\ \bar{\zeta} + \int_A dA\ \zeta^s\ , \tag{3.6-11b}$$

with π^s and ζ^s respectively, the surface-excess areal production and supply densities. With use of Eqs. (3.6-9) and (3.6-10), the flux analog of Eq. (3.6-8) becomes

$$\Phi = \int_{\partial \bar{V}} d\mathbf{S} \cdot \bar{\phi} + \oint_{\partial A} d\mathbf{L} \cdot \phi^s\ . \tag{3.6-11c}$$

The form of the divergence theorem appropriate to a convected material domain V possessing in its interior a convected material surface of discontinuity A (see Question 3.2 for the derivation of this theorem, obtained by applying the divergence theorem separately to each of the continuous domains $\bar{V}_{\mathrm{I}}$ and $\bar{V}_{\mathrm{II}}$) may be applied to Eq. (3.6-11c) together with Eq. (3.3-7) to obtain

$$\Phi = \int_{\bar{V}} dV\ \nabla \cdot \bar{\phi} + \int_A dA\ \mathbf{n} \cdot \|\bar{\phi}\| + \int_A dA\ \nabla_s \cdot (\mathbf{I}_s \cdot \phi^s)\ . \tag{3.6-12a}$$

Here, generically

$$\|\Re\| \overset{\text{def}}{=} \Re_{\mathrm{I}}(0+) - \Re_{\mathrm{II}}(0-) \tag{3.6-12b}$$

denotes the 'jump' in a discontinuous generic tensor field $\Re$ across a singular surface, with $\Re_{\mathrm{I}}(0+)$ and $\Re_{\mathrm{II}}(0-)$ denoting the respective values of $\Re$ on the two 'sides' of the interface, n=0+ and n=0–.

The volumetric Reynolds transport theorem (3.5-3) permits us to write

$$\frac{d}{dt}\int_{\bar{V}_{\mathrm{I}}(t)\oplus\bar{V}_{\mathrm{II}}(t)} dV\ \bar{\psi} = \int_{\bar{V}(t)} dV\left[\frac{\partial}{\partial t}\bar{\psi} + \nabla\cdot(\mathbf{v}\ \bar{\psi})\right], \qquad (3.6\text{-}13\text{a})$$

where the absence of an overbar on the macroscale mass-average velocity vector $\mathbf{v}$ corresponds to the fact that $\mathbf{v}$ is continuous across the macroscale interface [cf. Eq. (3.4-6)]. Together with the surface Reynolds transport theorem (3.4-19) for a convected material surface, namely

$$\frac{d}{dt}\int_A dA\ \psi^s = \int_A dA\left[\frac{\partial}{\partial t}\psi^s + \nabla_s\cdot(\mathbf{v}^o\ \psi^s)\right], \qquad (3.6\text{-}13\text{b})$$

Eq. (3.6-13a) may be employed to obtain the expression

$$\frac{d\Psi}{dt} = \int_V dV\left[\frac{\partial}{\partial t}\bar{\psi} + \nabla\cdot(\mathbf{v}\ \bar{\psi})\right] + \int_A dA\left[\frac{\partial}{\partial t}\psi^s + \nabla_s\cdot(\mathbf{v}^o\ \psi^s)\right] \qquad (3.6\text{-}14)$$

for the time rate of change of the amount Ψ of the property $\mathcal{P}$ within the material volume V (containing the material interface A).

Substitute Eqs. (3.6-11), (3.6-12) and (3.6-14b) into Eq. (3.5-1) to obtain

$$\int_V dV\left\{\frac{\partial}{\partial t}\bar{\psi} + \nabla\cdot(\mathbf{v}\ \bar{\psi} + \bar{\phi}) - \bar{\pi} - \bar{\zeta}\right\}$$
$$+\int_A dA\left\{\frac{\partial}{\partial t}\psi^s + \nabla_s\cdot(\mathbf{v}^o\ \psi^s + \mathbf{I}_s\cdot\phi^s) - \pi^s - \zeta^s + \mathbf{n}\cdot\|\bar{\phi}\|\right\} = 0.$$

Upon recognizing that V and A have both been arbitrarily chosen, there results*

---

* The form of the pointwise three-dimensional conservation equation (3.6-15) appropriate to the entire domain $V = \bar{V} \oplus A$ which now includes those points $\mathbf{x}_s$ lying on the interface, is given by the expression

$$\frac{\partial\ \bar{\psi}}{\partial\ t} + \nabla\cdot(\bar{\mathbf{v}}\ \bar{\psi} + \bar{\phi}) - \bar{\pi} - \bar{\zeta} = -\mathbf{n}\cdot\|\bar{\phi}\|\delta(\mathbf{x} - \mathbf{x}_s),$$

$$\frac{\partial \bar{\psi}}{\partial t} + \nabla \cdot (\mathbf{v}\,\bar{\psi} + \bar{\boldsymbol{\phi}}) - \bar{\pi} - \bar{\zeta} = \mathbf{0} \tag{3.6-15}$$

and

$$\boxed{\frac{\partial \psi^s}{\partial t} + \nabla_s \cdot (\mathbf{v}^o\,\psi^s + \mathbf{I}_s \cdot \boldsymbol{\phi}^s) - \pi^s - \zeta^s = -\,\mathbf{n} \cdot \|\boldsymbol{\phi}\|.} \tag{3.6-16}$$

Equation (3.6-15) represents the pointwise generic bulk-phase conservation equation, valid at each point $\mathbf{x}$ of the volumetric regions $\bar{V}_{\mathrm{I}}$ and $\bar{V}_{\mathrm{II}}$ [compare Eq. (3.5-9)], excluding those points $\mathbf{x}_s$ lying on the interface A. Likewise, Eq. (3.6-16) represents the pointwise generic interfacial equation, valid at each point $\mathbf{x}_s$ of A. Recall from our convention [cf. Eq. (3.4-10a) and the discussion thereof] that $\partial/\partial t$ in Eq. (3.6-15) represents $\partial/\partial t)_{\mathbf{x}}$ whereas $\partial/\partial t$ in Eq. (3.6-16) represents $\partial/\partial t)_{\mathbf{x}_s}$.

Owing to the fact that the interface is taken here in Part I to be a material interface, across which no mass thus flows, it follows that $\mathbf{v}_{\mathrm{I}}(0+) = \mathbf{v}_{\mathrm{II}}(0-) \equiv \mathbf{v}(0)$, say; that is, the bulk-phase velocity is continuous across the interface. The symbol $\mathbf{v}^o$ in Eq. (3.6-16), representing the velocity at a point $\mathbf{x}_s$ of the material interface, is identical to this common velocity. Explicitly, as in Eq. (3.4-6),

$$\boxed{\mathbf{v}^o(\mathbf{x}_s, t) \equiv \mathbf{v}(\mathbf{x}, t)\big|_{\mathbf{x}=\mathbf{x}_s} .} \tag{3.6-17}$$

As will be shown in Part II (Chapters 16 and 17), the surface-excess properties appearing in the latter equation, which correspond to the *physical* properties defined in Table 3.5-1, are those given in Table 3.6-1.

Equation (3.6-16) represents a generic surface-excess balance equation, appropriate to a material fluid interface. Comparison with Eq. (3.6-15) reveals a nearly complete analogy between the respective volumetric and areal terms appearing in each of the equations. The jump in the normal component of the bulk-phase molecular flux, which appears on the right side of Eq. (3.6-16), is one term which quite clearly has no analog in the volumetric conservation equation (3.6-15). It represents a "source" (or "sink") term in the surface conservation equation, corresponding to diffusive

---

where $\delta(\mathbf{x} - \mathbf{x}_s)$ is the Dirac delta function (Morse & Feshbach 1953). For continuous three-dimensional fluids, $\mathbf{n} \cdot \|\boldsymbol{\phi}\| = \mathbf{0}$, thereby reducing this equation to the form of Eq. (3.5-9).

Table 3.6-1
IDENTIFICATION OF GENERIC SURFACE-EXCESS FIELDS IN EQ. (3.6-16)

Extensive Surface Property $\mathcal{P}$	Areal Property Density ψ^s	Lineal Molecular Flux ϕ^s	Areal Production Rate π^s	Areal Supply Rate ζ^s
Mass	ρ^s	0	0	0
Lin. momentum	$\rho^s \mathbf{v}^o$	$-\mathbf{P}^s$	0	$\mathbf{F}^s$
Ang. momentum	$\mathbf{x}_s \times \rho^s \mathbf{v}^o + \rho^s \mathbf{a}^s$	$\mathbf{P}^s \times \mathbf{x}_s - \mathbf{C}^s$	0	$\mathbf{x}_s \times \mathbf{F}^s + \mathbf{G}^s$
Mass of species *i*	ρ_i^s	$\mathbf{j}_i^s$	R_i^s	0
Energy	$U^s + 1/2 \rho^s \mathbf{v}^o \cdot \mathbf{v}^o$	$\mathbf{q}^s$	$-\nabla_s \cdot (\mathbf{v}^o \cdot \mathbf{P}^s)$	$\mathbf{F}^s \cdot \mathbf{v}^o$
Entropy	S^s	$\mathbf{q}^s / T$	Φ^s	0

$\mathbf{C}^s$ = surface-excess couple stress pseudodyadic

$\mathbf{F}^s$ = surface-excess external body force density vector

$\mathbf{G}^s$ = surface-excess body couple density pseudovector

$\mathbf{j}_i^s$ = surface-excess species diffusive mass flux vector measured relative to $\mathbf{v}^o$

$\mathbf{P}^s$ = surface-excess pressure dyadic

$\mathbf{q}^s$ = surface-excess heat flux vector

R_i^s = surface-excess species areal reaction rate

S^s = surface-excess entropy density

T = temperature

U^s = surface-excess internal energy density

$\mathbf{v}^o$ = mass-average velocity vector at the material interface

$\mathbf{x}_s$ = space-fixed position vector in the interface

ρ^s = surface-excess mass density

ρ_i^s = surface-excess species mass density

$\mathbf{a}^s$ = surface-excess internal angularmomentum pseudovector

Φ^s = surface-excess rate of entropy production density

transport of the property in question between the bulk and interfacial phases; that is, viewed from the perspective of an observer situated in the two-dimensional interface, the source (or sink) term represents a 'mysterious' appearance (or disappearance) of the property from (or into) the third dimension. A second, though lesser distinction occurs between the molecular flux or respective "ϕ" terms in Eqs. (3.6-15) and (3.6-16). As three-dimensional space is 'flat', the idemfactor $\mathbf{I}$ that would have otherwise appeared in Eq. (3.6-16) (in the flux term as $\mathbf{I}\cdot\phi$) is constant throughout space, and hence independent of $\mathbf{x}$. In contrast, the curvature of the interface is such that $\mathbf{I}_s$ is generally not a constant (except for a *flat* interface), but is rather a function of $\mathbf{x}_s$ [through the dependence of $\mathbf{n}$ upon $\mathbf{x}_s$; cf. Eq. (3.2-6)].

3.7 Examples

Examples are provided in this section to illustrate geometrical (§3.2) and kinematical (§3.4) formulas presented previously in the chapter. Illustrative examples of interfacial transport processes (§3.6) will be provided later in §§4.4 and 5.6, in which latter sections attention is addressed to explicit representations of the generic equations (3.6-15) and (3.6-16) for the cases interfacial linear momentum and species transport respectively.

1. Surface Geometry of a Prolate Spheroid

Consider the orthogonal curvilinear coordinate parameterization of the prolate spheroidal surface $\xi \equiv \xi_0$ depicted in Figure 3.7-1. [The reader may refer to Appendix A of Happel & Brenner (1986) for further details of this curvilinear coordinate system.]

Utilizing Eq. (B.1-12), with the definitions of the scale factors provided in Figure 3.7-1, the explicit choice of surface coordinates $(q^1, q^2) \equiv (\eta, \phi)$ is seen to induce the following surface base vectors $(\mathbf{a}_\eta, \mathbf{a}_\xi)$:

$$\mathbf{a}_\eta = \mathbf{i}_\eta \frac{1}{h_\eta},$$

$$\equiv \mathbf{i}_\eta \, c\left(\sinh^2 \xi_0 + \sin^2 \eta\right)^{1/2}, \tag{3.7-1}$$

and

$$\mathbf{a}_\phi = \mathbf{i}_\phi \frac{1}{h_\phi},$$

$$\equiv \mathbf{i}_\eta \, c \sinh \xi_0 \sin \eta. \tag{3.7-2}$$

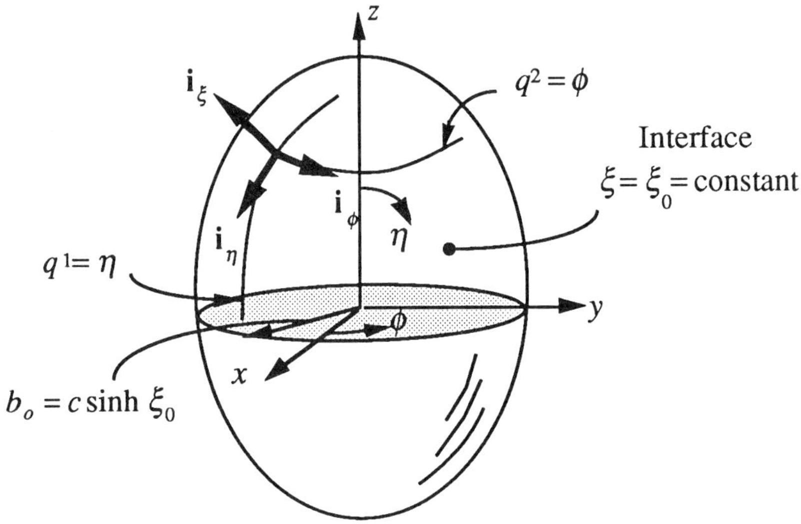

Prolate spheroid scale factors
$h_\eta = h_\xi = \dfrac{1}{c(\sinh^2\xi + \sin^2\eta)^{1/2}}$
$h_\phi = \dfrac{1}{c \sinh \xi \sin \eta}$

Figure 3.7-1 A prolate spheroidal surface.

Similarly, Eq. (B.1-13) may be used to provide the following expressions for the reciprocal surface base vectors:

$$\mathbf{a}^\eta = \mathbf{i}_\eta \, h_\eta \,,$$

$$\equiv \mathbf{i}_\eta \frac{1}{c\left(\sinh^2 \xi_0 + \sin^2 \eta\right)^{1/2}}, \tag{3.7-3}$$

and

$$\mathbf{a}^\phi = \mathbf{i}_\phi \, h_\phi \,,$$

$$\equiv \mathbf{i}_\phi \frac{1}{c \sinh \xi_0 \sin \eta}. \tag{3.7-4}$$

The two nonzero independent components of the surface metric $a_{\alpha\beta}$ [cf. Eq. (3.2-17)] may be determined directly from the above by Eq. (3.2-3). This gives

$$a_{\eta\eta} = c^2\left(\sinh^2 \xi_0 + \sin^2 \eta\right) \tag{3.7-5}$$

and

$$a_{\phi\phi} = c^2 \sinh^2 \xi_0 \sin^2 \eta, \qquad (3.7\text{-}6)$$

whence it follows that the determinant of the surface metric (3.2-4) is given by

$$\begin{aligned} a &= a_{\eta\eta} a_{\phi\phi} \\ &\equiv c^4 \left(\sinh^2 \xi_0 + \sin^2 \eta \right) \sinh^2 \xi_0 \sin^2 \eta. \end{aligned} \qquad (3.7\text{-}7)$$

From Eq. (3.2-5) the unit normal $\mathbf{n}(\eta, \phi)$ to the surface $\xi \equiv \xi_0$ is given by

$$\begin{aligned} \mathbf{n} &= \frac{1}{\sqrt{a}} \mathbf{a}_\eta \times \mathbf{a}_\phi \\ &= \mathbf{i}_\eta \times \mathbf{i}_\phi \\ &= \mathbf{i}_\xi, \end{aligned} \qquad (3.7\text{-}8)$$

whence, from Eq. (3.2-6), the unit surface idemfactor $\mathbf{I}_s$ possesses the representation

$$\mathbf{I}_s = \mathbf{i}_\eta \mathbf{i}_\eta + \mathbf{i}_\phi \mathbf{i}_\phi .$$

According to Eq. (B.1-20), the surface curvature dyadic $\mathbf{b}$ [cf. Eq. (3.2-10)] for the prolate spheroid surface is

$$\mathbf{b} = -\mathbf{i}_\eta \mathbf{i}_\eta \frac{\sinh \xi_0 \cosh \xi_0}{c \left(\sinh^2 \xi_0 + \sin^2 \eta \right)^{3/2}} - \mathbf{i}_\phi \mathbf{i}_\phi \frac{\cosh \xi_0 / \sinh \xi_0}{c \left(\sinh^2 \xi_0 + \sin^2 \eta \right)^{1/2}}, \qquad (3.7\text{-}9)$$

yielding [cf. Eqs. (B.1-21) and (B.1-22)]

$$\kappa_\eta = -\frac{\sinh \xi_0 \cosh \xi_0}{c \left(\sinh^2 \xi_0 + \sin^2 \eta \right)^{3/2}}, \qquad \kappa_\phi = -\frac{\cosh \xi_0 / \sinh \xi_0}{c \left(\sinh^2 \xi_0 + \sin^2 \eta \right)^{1/2}} \qquad (3.7\text{-}10)$$

for the two principal surface curvatures. It may be noted that the ϕ-coordinate curve along the surface at $\eta = \pi/2$ is circular with radius b_o (the minor semiaxis of the prolate spheroid surface). This is revealed by the surface curvature κ_ϕ which, for $\eta = \pi/2$, yields

$$\kappa_\phi\Big|_{\eta=\pi/2} = -\frac{\cosh\,\xi_0/\sinh\,\xi_0}{c\left(1+\sinh^2\,\xi_0\right)^{1/2}},$$

$$= -\frac{1}{c\,\sinh\,\xi_0}, \tag{3.7-11a}$$

or with $b_o = c\,\sinh\,\xi_0$ (Figure 3.7-1),

$$\kappa_\phi\Big|_{\eta=\pi/2} = -\frac{1}{b_o}. \tag{3.7-11b}$$

In addition, the curvature dyadic **b** of the prolate spheroidal surface is seen to be isotropic at the poles of the spheroid ($\eta = 0, \pi$):

$$\kappa_\eta\Big|_{\eta=0,\ \pi} = \kappa_\phi\Big|_{\eta=0,\ \pi} = -\frac{\cosh\,\xi_0}{c\,\sinh^2\,\xi_0}, \tag{3.7-11c}$$

as anticipated on account of the symmetry of the surface about all planes possessing the z-axis.

Finally, the mean and Gaussian curvatures may be determined respectively from Eqs. (3.2-13) and (3.2-14), utilizing Eqs. (3.7-10), as

$$H = -\frac{1}{2}\left(\frac{\sinh\,\xi_0\cosh\,\xi_0}{c\left(\sinh^2\,\xi_0+\sin^2\,\eta\right)^{3/2}} + \frac{\cosh\,\xi_0/\sinh\,\xi_0}{c\left(\sinh^2\,\xi_0+\sin^2\,\eta\right)^{1/2}}\right) \tag{3.7-12}$$

and

$$K = \frac{\cosh^2\,\xi_0}{c^2\left(\sinh^2\,\xi_0+\sin^2\,\eta\right)^2}. \tag{3.7-13}$$

2. The Kinematics of a Translating, Rotating, Expanding Circular Cylindrical Surface

Consider the example of a circular cylindrical interface of infinite axial length, simultaneously undergoing rigid-body translation and rotation, in addition to a radial expansion (Figure 3.7-2).

Relative to an observer fixed at the point O along the z axis, a surface point P_s lying upon the cylindrical surface may be located at time t by the position vector

$$\mathbf{x}_s = Ut\,\mathbf{i}_z + R_0\mathbf{i}_R, \tag{3.7-14}$$

where U is a constant, $R_0 \equiv R_0(t)$, and the (time-varying) radial unit vector $\mathbf{i}_R$ satisfies the kinematic relation

$$\left(\frac{\partial \mathbf{i}_R}{\partial t}\right)_{\mathbf{x}_s(0)} = \boldsymbol{\Omega} \times \mathbf{i}_R, \tag{3.7-15}$$

where

$$\boldsymbol{\Omega} = \mathbf{i}_z \Omega \tag{3.7-16}$$

represents the angular velocity of the cylindrical surface about the z axis. Here,

$$\mathbf{x}_s(0) = R_0(0)\mathbf{i}_R(0) \tag{3.7-17}$$

denotes the initial location of the interfacial point P_s.

The interfacial velocity $\mathbf{v}^o$ [cf. Eqs. (3.4-5) and (3.4-7)] may be determined from the above relations as

$$\mathbf{v}^o = \left(\frac{\partial \mathbf{x}_s}{\partial t}\right)_{\mathbf{x}_s(0)} = \mathbf{i}_z U + R_0 \Omega \mathbf{i}_\phi + \mathbf{i}_R \dot{R}_0, \tag{3.7-18}$$

representing a rigid translation, rigid rotation, and radial expansion of the interface. Here, $\dot{R}_0$ represents the time derivative of R_0.

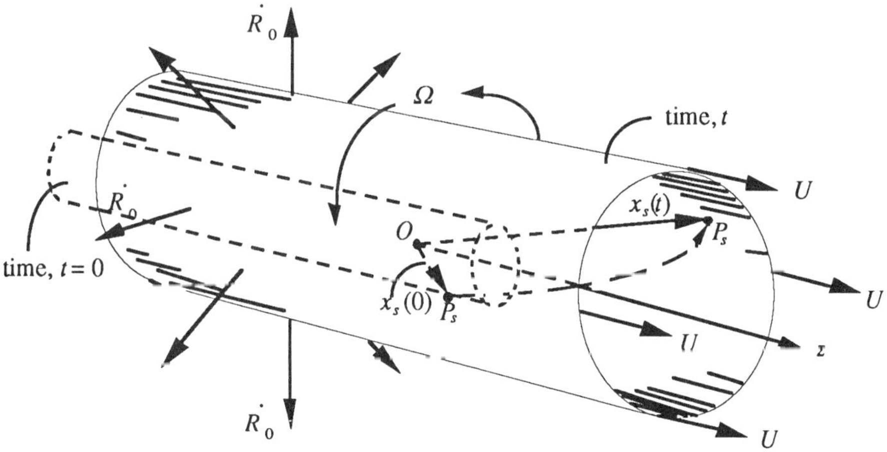

Figure 3.7-2 A translating, rotating, expanding circular cylindrical interface ($R = R_0$).

Employing the chain rule upon the surface convected time derivative in Eq. (3.7-18) [as in (Eq. (3.4-9)], provides

$$\mathbf{v}^o = \left(\frac{\partial \mathbf{x}_s}{\partial t}\right)_{\mathbf{x}_s(0)} = \mathbf{i}_z U + R_0 \Omega \mathbf{i}_\phi + \mathbf{i}_R \dot{R}_0,$$

$$= \left(\frac{\partial \mathbf{x}_s}{\partial t}\right)_{q^1,\, q^2} + \left(\frac{\partial \mathbf{x}_s}{\partial t}\right)_{\mathbf{x}_s(0)} \cdot \left(\frac{\partial \mathbf{x}_s}{\partial \mathbf{x}_s}\right)_t .$$

Noting that

$$\mathbf{I}_s = \mathbf{i}_z \mathbf{i}_z + \mathbf{i}_\phi \mathbf{i}_\phi ,$$

together with the representation of the surface idemfactor in Question 3.14, yields

$$\left(\frac{\partial \mathbf{x}_s}{\partial t}\right)_{q^1,\, q^2} = \mathbf{i}_R \dot{R}_0 . \tag{3.7-19}$$

As discussed in §3.4, the time derivative $\partial/\partial t)_{q^1,\, q^2} \equiv \partial/\partial t)_{\mathbf{x}_s}$ represents time differentiation relative to an observer 'fixed' at a point (q^1, q^2) within the two-dimensional surface space. Equation (3.7-19) therefore illustrates that $\partial/\partial t)_{q^1,\, q^2}$ is nonzero only in the special case wherein the surface is moving normal to itself in three-dimensional space. Interfacial motion *within* the two-dimensional interface (corresponding in the present example to rigid-body translations and rotations) is entirely intrinsic; hence, as it does not impart an extrinsic motion of the surface in three-dimensional space, transverse surface motion does not appear in Eq. (3.7-19).

Of the three interfacial motions represented in (3.7-18), it is naturally only the surface expansion that causes a deformation of the interface. This deformation is easily determined by noting that the instantaneous area of the cylinder surface (per unit axial length) is given by

$$A = 2\pi R_o, \tag{3.7-20}$$

whence,

$$\frac{\dot{A}}{A} = \frac{2\dot{R}_0}{R_o} = 2H\dot{R}_0 = 2Hv^o_R = 2H\, \mathbf{n} \cdot \mathbf{v}^o = \nabla_s \cdot \mathbf{v}^o, \tag{3.7-21}$$

confirming $\nabla_s \cdot \mathbf{v}^o$ as the rate of surface expansion [cf. Eq. (3.4-20)].

3. A Nonmaterial, Spherical Droplet Interface

A spherical droplet possesses an instantaneous total mass $M(t)$ of liquid phase II, which liquid possesses a constant mass density $\rho_{\text{II}} \equiv \rho$. Liquid II is absorbed at an areal rate f by the surrounding liquid phase I (which latter liquid possesses the same mass density as liquid II; i.e. $\rho_{\text{I}} \equiv \rho$). Owing to the identical bulk-phase densities, the material velocity $\mathbf{v}$ is continuous across the interface (cf. Question 15.5), thereby satisfying Eq. (3.4-6). In these circumstances,*

$$f = \rho \, \mathbf{n} \cdot (\mathbf{v}^o - \mathbf{u}) \qquad (3.7\text{-}22)$$

expresses the net flux of liquid II across the spherical interface (Figure 3.7-3).

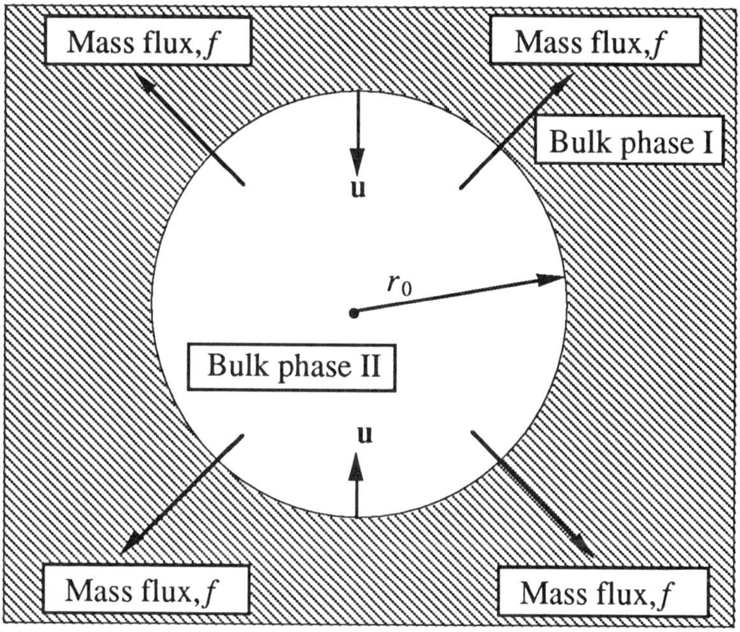

Figure 3.7-3 A nonmaterial, spherical interface through which there is a net mass flux of gas $\rho \, \mathbf{n} \cdot (\mathbf{v}^o - \mathbf{u})$ from phase II to phase I. The bulk phases possess an identical mass density ρ.

The total volume V of liquid II at any instant of time t is given by

$$V = \frac{M}{\rho},$$

whence,

$$\begin{aligned} \dot{V} &= \frac{\dot{M}}{\rho}, \\ &= -\frac{fA}{\rho} \\ &= -\mathbf{n} \cdot (\mathbf{v}^o - \mathbf{u})A \qquad (3.7\text{-}23) \end{aligned}$$

---

* Refer to Question 17.3 for the generalization of this example to the case where the mass densities of the respective bulk phases differ.

represents the time rate of volumetric change owing to the net mass flux out of the droplet. Here, $A = 4\pi r_0^2$ and $\mathbf{n} = \mathbf{i}_r$. Thus, the radius of the droplet diminishes at a rate governed by:

$$\dot{r}_0 = \frac{\dot{V}}{A},$$
$$= -\mathbf{n} \cdot (\mathbf{v}^o - \mathbf{u}),$$
$$\equiv \mathbf{n} \cdot \mathbf{u}. \tag{3.7-24}$$

(The last equivalence in the above expression corresponds to the definition of **u** as the surface velocity.) It follows directly from Eq. (3.7-24) that the mass-averaged velocity at the interface is zero; explicitly,

$$\mathbf{n} \cdot \mathbf{v}^o = 0. \tag{3.7-25}$$

Thus, the absorption of liquid II by the surrounding liquid I results in the inward collapse of the droplet surface, at a rate governed by

$$\mathbf{n} \cdot \mathbf{u} = -\frac{f}{\rho}, \tag{3.7-26}$$

as follows from a comparison of Eqs. (3.7-23) and (3.7-25).

The preceding example demonstrates that when a net mass transfer occurs across an interface, the (normal components of the) mass-average and surface velocities will always differ. The significance of this difference to interfacial transport will depend upon factors such as the magnitude of the mass flux (3.7-22) and the surface-to-volume ratio [cf. Eq. (3.7-24)].

3.8 Summary

Chapter 3 has provided the basic physical and mathematical foundations of interfacial transport processes, particularly for the case of material interfaces (as opposed to phase interfaces), where no mass crosses the interface at any point. Attention was focused entirely upon the macroscale view of a fluid interface, from which vantage point the interface appears to be a continuous, two-dimensional, moving and deforming boundary between two contiguous three-dimensional 'bulk' fluid phases (each of which is separately continuous). The basic geometrical properties and kinematics of such singular surfaces have been discussed, as too has been the mechanical description of static interfaces. The surface-excess pressure dyadic, characterizing the state of stress in an interface, was introduced via Eqs. (3.3-3) and (3.3-5). For an isotropic interface in equilibrium, this tensor is represented by a single (positive) scalar, namely the interfacial tension, as in Eq. (3.3-4). Section 3.4 briefly discussed material vs nonmaterial interfaces; it was indicated that here in Part I, attention will be confined exclusively to material fluid interfaces. Finally, generic conservation laws governing the transport of extensive physical properties were developed for both the bulk-phase fluids [cf. Eq. (3.6-15)], and the interface [cf. Eq. (3.6-16)]. In the latter case an important distinction was drawn between surface quantities and surface-excess quantities, and the generic surface-excess transport equation observed to play the role of a boundary condition between otherwise discontinuous bulk-phase fields.

In the following chapters the latter equations will be developed for the particular cases of momentum and mass transfer, and subsequently applied to several practical, illustrative problems pertaining to interfacial rheology.

Appendix 3.A

The Surface-Excess Pressure Tensor: Derivation of Eq. (3.3-3)

In §3.3 the surface-excess contact stress vector $\mathbf{p}_\nu^s$ was expressed in terms of a dyadic field property $\mathbf{P}^s = \mathbf{P}^s(\mathbf{x}_s)$., as

$$\mathbf{p}_\nu^s = \nu \cdot \mathbf{P}^s . \tag{3. A- 1}$$

A proof of the existence of such a dyadic field is provided in this appendix.

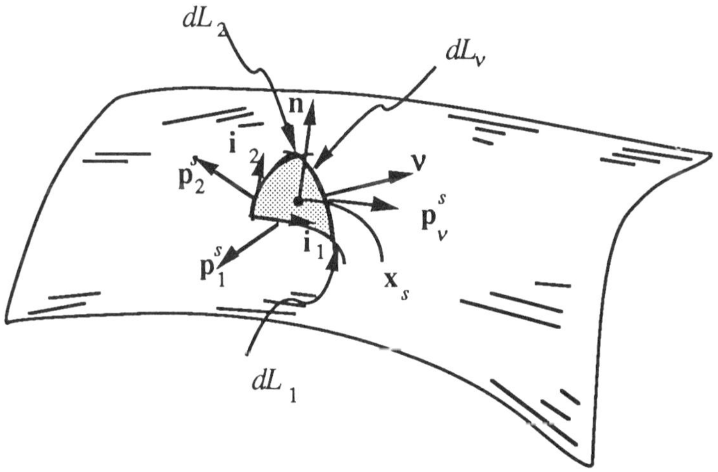

Figure 3.A-1 An infinitesimal curvilinear triangle of sides dL_1, dL_2 and dL_ν, with $dL_\alpha = \mathbf{i}_\alpha \cdot \nu dL_\nu$ (α=1,2) representing differential lengths measured along orthogonal surface coordinates (q^1, q^2), is drawn within the interface. Surface lineal stress vectors $\mathbf{p}_1^s$, $\mathbf{p}_2^s$, and $\mathbf{p}_\nu^s$, act on the three sides of this triangle. When the normal components (i.e. perpendicular to the line element) of the vector forces $\mathbf{p}_1^s dL_1$, $\mathbf{p}_2^s dL_2$, and $\mathbf{p}_\nu^s dL_\nu$ point outward from the surface element they represent tensile forces. When directed inward they are compressive.

Consider the curvilinear triangle drawn in Figure 3.A-1. A balance of forces upon this surface element may be expressed as

$$\int_{\delta A} dA\,\mathbf{f}^s + \oint_{\delta C} dL\,\mathbf{p}^s = \mathbf{0}, \tag{3.A-2}$$

where δA is the area of the surface element and $\delta C = dL_\nu + dL_1 + dL_2$ is its closed, lineal boundary. Taking the limit as the area δA becomes vanishingly small, the first integral of Eq. (3.A-2) diminishes most rapidly, at a rate $O(l^2)$, whereas the second integral diminishes at a rate $O(l)$, whence

$$\lim_{\delta A \to 0} \oint_{\delta C} dL\,\mathbf{p}^s = \mathbf{0}. \tag{3.A-3}$$

In this limit, Eq. (3.A-3) suggests that

$$dL_\nu\,\mathbf{p}^s_\nu + dL_1\mathbf{p}^s_{(-1)} + dL_2\mathbf{p}^s_{(-2)} = \mathbf{0}, \tag{3.A-4}$$

where [note here that the unit normals ($\mathbf{i}_1$, $\mathbf{i}_2$) to the differential lengths (dL_1, dL_2) respectively, are inwardly directed to the area element δA (though tangent to the interface at $\mathbf{x}_s$)]

$$\mathbf{p}^s_{(-\alpha)} = -\,\mathbf{p}^s_{(\alpha)} \tag{3.A-5}$$

As noted in the figure caption,

$$dL_\alpha = \mathbf{i}_\alpha \cdot \nu\; dL_\nu\,, \tag{3.A-6}$$

hence, Eq. (3.A-4) may be expressed with Eq. (3.A-5) as

$$dL_\nu\,\mathbf{p}^s_\nu = dL_\nu\;\nu\cdot\left[\mathbf{i}_1\mathbf{p}^s_1 + \mathbf{i}_2\mathbf{p}^s_2\right]. \tag{3.A-7}$$

The stress vectors ($\mathbf{p}^s_1$, $\mathbf{p}^s_2$) may be decomposed into three spatial components as

$$\mathbf{p}^s_1 = \mathbf{i}_1 P^s_{11} + \mathbf{i}_2 P^s_{12} + \mathbf{n}\,P^s_{13},$$

$$\mathbf{p}^s_2 = \mathbf{i}_1 P^s_{21} + \mathbf{i}_2 P^s_{22} + \mathbf{n}\,P^s_{23}, \tag{3.A-8}$$

which yields, upon substitution into Eq. (3.A-7),

$$\mathbf{p}^s_\nu = \nu\cdot\left[\mathbf{i}_1\mathbf{i}_1 P^s_{11} + \mathbf{i}_1\mathbf{i}_2 P^s_{12} + \mathbf{i}_1\mathbf{n}\,P^s_{13} + \mathbf{i}_2\mathbf{i}_1 P^s_{21} + \mathbf{i}_2\mathbf{i}_2 P^s_{22} + \mathbf{i}_2\mathbf{n}P^s_{23}\right]. \tag{3.A-9}$$

Equation (3.A-9) may finally be expressed as

$$\mathbf{p}^s_\nu = \nu\cdot\,\mathbf{P}^s\,, \tag{3.A-10}$$

where $\mathbf{P}^s$ is the dyadic defined in Eq. (3.3-5).

Questions for Chapter 3

3.1 Using Eqs. (3.4-12), (3.2-6) and (A.1-11) show that $\mathbf{n}\cdot\mathbf{v}_s=0$.

3.2 Derive the divergence theorem for a fluid volume possessing a surface of discontinuity:

$$\int_{\bar{V}} dV\, \nabla \cdot \mathfrak{R} = \int_{\partial\bar{V}} d\mathbf{S} \cdot \mathfrak{R} - \int_{A} dA\,\mathbf{n} \cdot \|\mathfrak{R}\|\,,$$

where $\bar{V}$ is the fluid volume, $\mathfrak{R}$ is a generic polyadic or tensorial field quantity, $\partial\bar{V}$ is the bounding surface of the volume with outward directed surface element $d\mathbf{S}$, and A is the surface of discontinuity. The jump operator is defined in Eq. (3.6-12b). [Hint: Use the divergence theorem [cf. Eqs. (3.5-6) and (3.5-7)] for each of the two closed volumes (which do not include those points of the respective volumes lying on A) bounding the surface of discontinuity. Follow the notation of Figure 3.6-1.]

3.3 A spherical liquid droplet of radius r_o is suspended within a continuous fluid at rest. A (purely geometrical) surface point P_s rotates within the surface about the z axis (see Figure B.1-6 of Appendix B) with angular speed Ω. What is the material surface velocity $\mathbf{v}^o$? What is the surface velocity $\mathbf{u}$?

3.4 Derivation of Eq. (3.4-14):

a. Beginning with the definition of the surface Jacobian provided in Eq. (3.4-13), use Eq. (3.2-5), the definitions of the differential surface area elements provided in the caption to Figure 3.4-2 [while choosing the differential length elements such that

$$d\mathbf{x}_{\alpha} = \mathbf{a}_{\alpha} dq^{\alpha} \qquad \text{(no sum on } \alpha\text{)} \qquad (\alpha = 1, 2)],$$

to prove

$$j = \sqrt{\frac{a}{a_o}}\,.$$

Here, a is the determinant of the surface metric on the (time-dependent) surface $A(t)$, and a_o is the surface metric determinant on the surface A_o;

b. use Eqs. (3.2-1), (3.2-3) and (3.4-5) to show that

$$\frac{D_s}{Dt}(a_{\alpha\beta}) = (\mathbf{a}_{\alpha}\mathbf{a}_{\beta} + \mathbf{a}_{\beta}\mathbf{a}_{\alpha}) : \nabla_s \mathbf{u}\,.$$

c. utilize the preceding relation, together with the identity (which follows the Einstein summation convention—see §B.1)

$$a = \frac{1}{2} e^{\alpha\gamma} e^{\beta\delta} a_{\alpha\beta} a_{\gamma\delta} ,$$

to show that

$$\frac{1}{a} \frac{D_s a}{Dt} = 2\nabla_s \cdot \mathbf{u} ;$$

d. hence, derive Eq. (3.4-14).

3.5 Use the volumetric relation

$$dV = J dV_o ,$$

analogous to the areal relation (3.4-13), together with

$$\frac{1}{J} \frac{DJ}{Dt} = \nabla \cdot \mathbf{v} ,$$

analogous to Eq. (3.4-14), to:

a. derive the volumetric Reynolds transport theorem for a moving volume $V(t)$

$$\frac{d}{dt} \int_V \Re dV = \int_V \left[\frac{D\Re}{Dt} + \Re \nabla \cdot \mathbf{v} \right] dV ,$$

analogous to Eq. (3.4-17);

b. derive the alternative forms

$$\frac{d}{dt} \int_V \Re dV = \int_V \left[\frac{\partial \Re}{\partial t} + \nabla \cdot (\mathbf{v}\, \Re) \right] dV$$

and

$$\frac{d}{dt} \int_V \Re dV = \int_V \frac{\partial \Re}{\partial t} dV + \oint_S d\mathbf{S} \cdot \mathbf{v} \Re ,$$

where S is a closed surface bounding the volume V.

c. If $\Re = \psi$ is the volumetric density of some property $\mathcal{P}$, how would you physically interpret the last of the above relations.

3.6 Starting with Eq. (3.4-8), derive the relation

$$\frac{d}{dt} \int_A \Re dA = \int_A \frac{\partial \Re}{\partial t} dA + \oint_C d\mathbf{L} \cdot \mathbf{u} \Re ,$$

which is the areal analog of the volumetric relation given in the final equation of Question 3.5. Provide a physical interpretation of the terms appearing on the right-hand side of this equation.

3.7 The field density ψ in Eq. (3.5-9) represents the amount of some property $\mathcal{P}$ per unit volume. Derive the following form of this equation in terms of the field density $\hat{\psi} \equiv \psi/\rho$, which represents the amount of $\mathcal{P}$ per unit *mass*:

$$\rho \frac{D\hat{\psi}}{Dt} + \nabla \cdot \phi - \pi - \zeta = \mathbf{0}\,,$$

where

$$\frac{D}{Dt} \overset{\text{def}}{=} \frac{\partial}{\partial t} + \mathbf{v} \cdot \nabla$$

is the material or convected derivative. (To distinguish between the two densities, ψ and $\hat{\psi}$ may respectively be termed the *volumetric* and *mass* densities of the property $\mathcal{P}$.)

3.8 Determination of the principal directions of curvature $\mathbf{e}_1$ and $\mathbf{e}_2$: Construct a tangent plane S to the surface A at a point O on its surface, as depicted in the figure below. Locate a cartesian coordinate system at O with components (x_1, x_2, x_3) and corresponding unit basis vectors $(\mathbf{i}_1, \mathbf{i}_2, \mathbf{i}_3)$. The unit vector $\mathbf{i}_3$ is chosen to lie normal to the tangent plane, whereas the unit (tangent) vectors $\mathbf{i}_1$ and $\mathbf{i}_2$, respectively, are to be determined as coinciding with the principal directions of curvature ($\mathbf{e}_1$ and $\mathbf{e}_2$) of A at O.

Background for Question 3.8

Let the surface S be defined by the equation

$$F(x_1, x_2, x_3) = 0, \tag{a}$$

or, alternatively,

$$x_3 = f(x_1, x_2) \tag{b}$$

where

$$F(x_1, x_2, x_3) = x_3 \quad f(x_1, x_2). \tag{c}$$

Perform a Taylor series expansion of (b) about the point of tangency O to obtain

$$x_3 = f_o + \mathbf{r} \cdot (\nabla_2 f)_o + \frac{1}{2!}\mathbf{r}\,\mathbf{r}{:}(\nabla_2 \nabla_2 f)_o + O(r^3), \tag{d}$$

where

$$\mathbf{r} = \mathbf{i}_1 x_1 + \mathbf{i}_2 x_2,$$

$$\nabla_2 \equiv \frac{\partial}{\partial \mathbf{r}} = \mathbf{i}_1 \frac{\partial}{\partial x_1} + \mathbf{i}_2 \frac{\partial}{\partial x_2},$$

$$r = |\mathbf{r}| = \left(x_1^2 + x_2^2\right)^{1/2}.$$

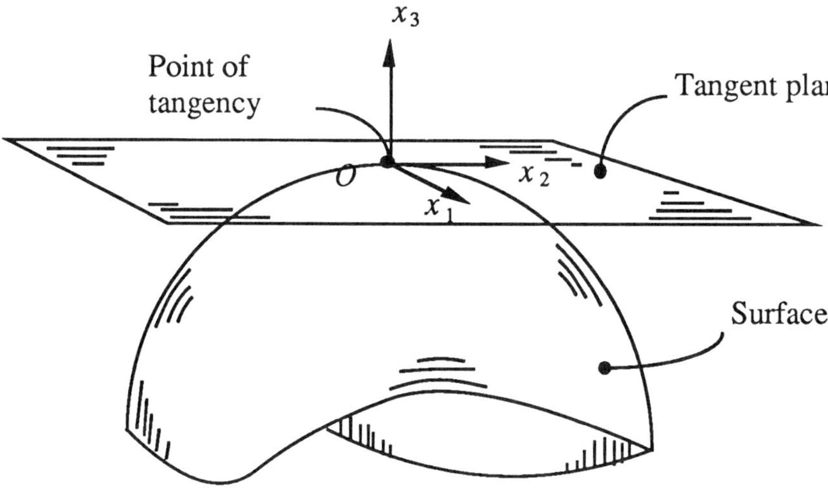

Forming the gradient of (c) yields

$$(\nabla F)_o = \mathbf{i}_3 - (\nabla_2 f)_o \, .$$

The left-hand side of this equation is normal to the surface F=constant, thus making it parallel to the surface A at O. This makes

$$(\nabla_2 f)_o = \mathbf{0}$$

since $(\nabla_2 f)_o$ lies in the plane S wholly tangent to the surface A at O. In addition,

$$f_o = 0$$

since $x_3 = 0$ at the point of tangency O. It then follows from (d) that

$$x_3 = \frac{1}{2}\mathbf{r}\,\mathbf{r}\!:\!(\nabla_2 \nabla_2 f)_o + O(r^3). \qquad \text{(e)}$$

Show that equation (e) is equivalent to the invariant form

$$x_3 = \frac{1}{2}\mathbf{r} \cdot \mathbf{b} \cdot \mathbf{r} + O(r^3). \qquad \text{(f)}$$

By choosing ($\mathbf{i}_1$, $\mathbf{i}_2$) to coincide with ($\mathbf{e}_1$, $\mathbf{e}_2$), show that there follows from (e) the relation

$$x_3 = \frac{1}{2}\left(\kappa_1 x_1^{\,2} + \kappa_2 x_2^{\,2}\right) + O(r^3), \qquad \text{(g)}$$

where (κ_1, κ_2) are the principal curvature scalars.

3.9 Consider an osculating plane (of the type shown in Figure 3.2-1) containing the x_3 axis and making an angle ϕ relative to the x_1 axis, as depicted in the figure below. A circle of radius R is drawn in the osculating plane ϕ=constant, whose center lies at the point (0, 0, $-R$). This circle passes through O and is tangent to the plane S. Its equation is (maintaining the notation of the Question 3.8)

$$(x_3 + R)^2 + r^2 = R^2.$$

i. Show that for $r/R \ll 1$, there obtains

$$x_3 = -\frac{r^2}{2R} + O\left(\frac{r^4}{R^3}\right).$$

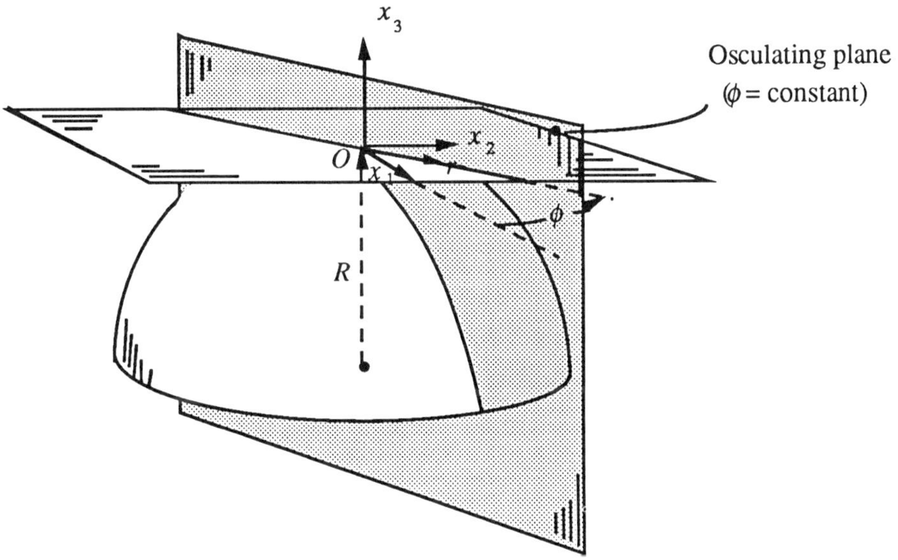

ii. Compare the previous result with Eq. (g) of Question 3.8 to show that

$$-\frac{1}{R} = \kappa_1 \cos^2\phi + \kappa_2 \sin^2\phi .$$

Hence, conclude from the preceding relation that the maximum and minimum curvatures correspond to the angles ϕ=0 and $\pi/2$ (within the range $0 \le \phi \le \pi$), and thereby confirm the physical interpretation of ($\mathbf{e}_1$, $\mathbf{e}_2$).

Refer to section B.1 of Appendix B in connection with each of the following questions.

3.10 Prove the validity of the surface divergence theorem, Eq. (3.3-7), in orthogonal curvilinear coordinates. (Hint: Begin with Stokes' theorem

$$\int_S \mathbf{n} \cdot \nabla \times \mathbf{A}\, dS = \oint_C \mathbf{t} \cdot \mathbf{A}\, dL ,$$

where S is a surface bounded by the closed curve C. Here, the vectors $\mathbf{n}$ and $\mathbf{t}$ respectively represent the unit normal to the surface S and the unit tangent vector to the curve C. Make the substitution $\mathbf{A} \equiv \boldsymbol{\varepsilon}_s \mathcal{R}$.)

3.11 Consider a planar surface undergoing a small perturbation f away from its equilibrium shape ($z = z_o$=constant), such that the equation of the surface may be expressed as

$$\bar{z} \stackrel{\text{def}}{=} z/z_o = 1 + f(x, y) ,$$

where $|f| \ll 1$. Derive an expression for the mean surface curvature $H \equiv H(f)$ to terms of the first order in the small parameter f.

3.12 Consider a circular cylindrical surface undergoing a small perturbation f away from its equilibrium shape ($r = R$=constant), such that the equation of the surface may be expressed as

$$\bar{r} \stackrel{\text{def}}{=} r/R = 1 + f(\theta, z) ,$$

where $|f| \ll 1$. Derive an expression for the mean surface curvature $H \equiv H(R, f)$ to terms of the first order in the small parameter f.

3.13 Prove that the mean and total curvatures may be expressed in terms of the two curvature scalars, as shown in Eqs. (3.2-13) and (3.2-14). [Begin with the vector-invariant definitions provided on the first lines of Eqs. (3.2-13) and (3.2-14).]

3.14 Show from Eqs. (3.2-2) and (B.1-4a) that the surface idemfactor possesses the representation

$$\mathbf{I}_s = \frac{\partial \mathbf{x}_s}{\partial \mathbf{x}_s} \quad (\equiv \nabla_s \mathbf{x}_s),$$

analogous to the corresponding representation

$$\mathbf{I} = \frac{\partial \mathbf{x}}{\partial \mathbf{x}} \quad (\equiv \nabla \mathbf{x})$$

of the three-dimensional idemfactor. Observe that the 'invariant' representation of the dyadic surface idemfactor $\mathbf{I}_s$ appearing in the first equation of this problem is the counterpart of the comparable tensor relation

$$\delta_\alpha^\beta = \frac{\partial q^\beta}{\partial q^\alpha} .$$

3.15 Using Eqs. (B.1-5) through (B.1-7), show that the surface idemfactor possesses the additional 'covariant' representation

$$\mathbf{I}_s = a_{\alpha\beta}\,\mathbf{a}^\alpha \mathbf{a}^\beta ,$$

and 'contravariant' representation

$$\mathbf{I}_s = a^{\alpha\beta}\,\mathbf{a}_\alpha \mathbf{a}_\beta .$$

3.16 Define a surface metric dyadic $\mathbf{a}$ as

$$\mathbf{a} = \mathbf{a}^\alpha \mathbf{a}^\beta a_{\alpha\beta}$$

and its conjugate surface metric dyadic as

$$\mathbf{A} = \mathbf{a}_\alpha \mathbf{a}_\beta a^{\alpha\beta} .$$

Pursuing the analogy between the surface dyadics **a** and **A** and the comparable space dyadics **g** and **G** [defined respectively in Eqs. (A.1-13) and (A.1-15)]:

i. Show that the dyadics **a** and **A** are symmetric

$$\mathbf{a}^\dagger = \mathbf{a} \qquad \mathbf{A}^\dagger = \mathbf{A} .$$

ii. Show that these dyadics are positive definite [see the footnote below Eq. (3.2-7)].

iii. Show that

$$\mathbf{a} \cdot \mathbf{A} = \mathbf{I}_s .$$

iv. Given the previous relation and assuming a nonvanishing surface Jacobian, what is the inverse of the dyadic **a** (i.e. what is the dyadic $\mathbf{a}^{-1}$)?

v. Show that

$$a\,A = 1 ,$$

where a and A are the determinants of **a** and **A** respectively.

3.17 Show that, in addition to Eq. (A.1-11), the following relation holds:

$$\mathbf{v} \cdot \mathbf{I} = \mathbf{v} ,$$

with **v** an arbitrary (generally three-dimensional) vector.

3.18 Using the expressions provided in section B.1 for the scale factors h_α of a spherical-polar coordinate system, determine (in terms of the unit base vectors $\mathbf{i}_\alpha$, the radius of the sphere r_0, and/or the surface coordinates θ and/or ϕ) explicit expressions for: (a) the surface base vectors $\mathbf{a}_\alpha$ (and their reciprocals $\mathbf{a}^\alpha$); (b) the surface metric $a_{\alpha\beta}$ (and the reciprocal metric $a^{\alpha\beta}$); and (c) the determinant a of the surface metric.

3.19 Prove that normalized surface base vectors, satisfying $\mathbf{a}_\alpha / |\mathbf{a}_\alpha| = \mathbf{a}^\alpha / |\mathbf{a}^\alpha|$, arise only in the case of *orthogonal* curvilinear coordinates (q^1, q^2).

3.20 An alternative representation of Eq. (B.1-8a) for the unit surface alternator is given by

$$\boldsymbol{\varepsilon}_s = \mathbf{a}_\alpha \mathbf{a}_\beta \varepsilon^{\alpha\beta} .$$

Find a relation between the respective covariant and contravariant tensors $\varepsilon_{\alpha\beta}$ and $\varepsilon^{\alpha\beta}$.

3.21 Find a relation between the tensorial components v^j of a vector **v** [see Eq. (A.1-6)] and the *physical* components $v(i)$ of the vector **v**, where

$$v(i) \equiv \mathbf{i}_i \cdot \mathbf{v},$$

such that

$$\mathbf{v} \equiv \mathbf{i}_i v(i) = \mathbf{i}_1 v(1) + \mathbf{i}_2 v(2) + \mathbf{i}_3 v(3).$$

[As noted at the beginning of §A.2, the notation v_j, rather than $v(i)$, is used throughout this text for the physical components of a vector **v**, as attention herein is primarily limited to orthogonal curvilinear coordinate systems. Thus, the representation provided in Eq. (A.2-1) pertains.]

3.22 Show that Eq. (A.1-28) together with $|\mathbf{i}_i| = 1$ provides an alternative to Eq. (A.1-30) for determining the metrical coefficients; namely as

$$\frac{1}{h_j} = \sqrt{\left(\frac{\partial x}{\partial q^j}\right)^2 + \left(\frac{\partial y}{\partial q^j}\right)^2 + \left(\frac{\partial z}{\partial q^j}\right)^2} \qquad (j = 1, 2, 3).$$

3.23 Show that the metric dyadic **g** and the conjugate metric dyadic **G** are both positive-definite forms.

3.24 Show from Eq. (A.1-16) that the conjugate metric dyadic is, in fact, the inverse of the metric dyadic, i.e. $\mathbf{G}=\mathbf{g}^{-1}$.

3.25 Using the invariant definition

$$df = d\mathbf{x} \cdot \nabla f$$

of the gradient operator for any function $f \equiv f(q^1, q^2, q^3)$, say, together with Eqs. (A.1-1) and (A.1-26), confirm the validity of the representation (A.3-2) of the gradient operator in orthogonal curvilinear coordinates.

3.26 Using the invariant definition of the Laplacian operator $\nabla^2 \equiv \nabla \cdot \nabla$ together with Eq. (A.3-2) and Eqs. (A.3-1) for the derivatives of the unit vectors, confirm Eq. (A.3-3).

3.27 *i.* Given the surface vector $\mathbf{A}_s = \mathbf{i}_1 A_1 + \mathbf{i}_2 A_2$, show (for simplicity, in orthogonal curvilinear coordinates) using the definition of the surface gradient operator provide by Eq. (4.B-1) [see also Eq. (B.1-13)], that

$$\nabla_s \nabla_s \mathbf{A}_s - (\nabla_s \nabla_s)^{\dagger} \mathbf{A}_s = \mathbf{Riemann}_s \cdot \mathbf{A}_s,$$

where $\mathbf{Riemann}_s$ is a surface tetradic, expressible in tensorial form as

$$\mathbf{Riemann}_s = \mathbf{a}^\alpha \mathbf{a}^\beta \mathbf{a}^\gamma \mathbf{a}^\delta R_{\alpha\beta\gamma\delta} .$$

Here, the Riemann tensor $R_{\alpha\beta\gamma\delta}$ may be represented as

$$R_{\alpha\beta\gamma\delta} \equiv \mathbf{a}_\alpha \cdot \left(\frac{\partial^2 \mathbf{a}_\beta}{\partial q^\gamma \partial q^\delta} - \frac{\partial^2 \mathbf{a}_\beta}{\partial q^\delta \partial q^\gamma} \right).$$

[Note in the above that the surface is *Euclidean* (i.e., the surface gradient operation is commutative) only when $\mathbf{Riemann}_s = \mathbf{0}$. The vanishing of the Riemann tetradic is, therefore, an intrinsic indication that the surface is 'flat'.]

ii. Show that

$$\mathbf{Riemann}_s = \varepsilon_s \, \varepsilon_s \, K ,$$

where the total curvature K is defined in Eq. (3.2-14). (The total curvature K is, therefore, an *intrinsic* curvature scalar.)

iii. Show that

$$K = \frac{1}{4} \varepsilon_s : \mathbf{Riemann}_s : \varepsilon_s .$$

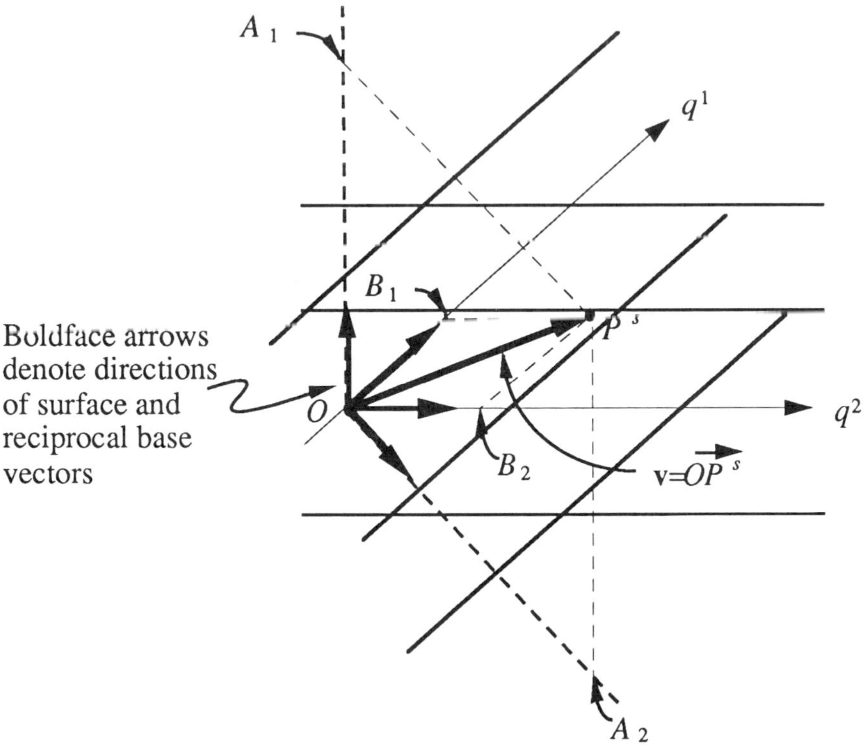

3.28 Consider the two-dimensional nonorthogonal rectilinear coordinate system depicted on the preceding page. Give the relationship between the covariant (v_1, v_2) and contravariant (v^1, v^2) components of the surface vector $\mathbf{v}$ ($\equiv v_1\mathbf{a}^1+v_2\mathbf{a}^2\equiv v^1\mathbf{a}_1+v^2\mathbf{a}_2$) and the resolutions ($A_1$, A_2, B_1, B_2) of the vector $\mathbf{v}$ on the (q^1, q^2) oblique coordinates shown. Use Eqs. (3.2-2) and (3.2-7) for this purpose.

Additional Reading for Chapter 3

§3.2 Interfacial Geometry

Wilson, E.B. & J.W. Gibbs 1929 *Vector Analysis.* New York: Dover.

Brand, L. 1947 *Vector and Tensor Analysis.* New York: Wiley.

Aris, R. 1962 *Vectors, Tensors, and the Basic Equations of Fluid Mechanics.* Englewood Cliffs, New Jersey: Prentice-Hall.

§3.3 Interfacial Statics

Buff, F.P. 1960 The theory of capillarity. In *Handbuch der Physik*, vol. X, pp. 281–304. Berlin: Springer-Verlag.

Brenner, H. 1979 A micromechanical derivation of the differential equation of interfacial statics. *J. Colloid Interface Sci.* **3**, 422–439.

§3.4 Interfacial Kinematics

Moeckel, G.P. 1975 Thermodynamics of an interface. *Arch. Rat. Mech. Anal.* **57**, 255–280.

§3.6 The Generic Transport Equation in Discontinuous Media and the Generic Surface-Excess Transport Law

Moeckel, G.P. 1975 Thermodynamics of an interface. *Arch. Rat. Mech. Anal.* **57**, 255–280.

Deemer, A.R. & Slattery, J.C. 1978 Balance equations and structural models for phase interfaces. *Int. J. Multiphase Flow* **4**, 171–192.

4

"Let us imagine ourselves inhabitants of a surface. Like the Lines and Polygons who populate Abbott's Flatland we are not aware of a third dimension except as an abstract mathematical notion. Unlike the inhabitants of Flatland we find that our surface may at times have curvature—curvature which changes with time. As hydrodynamicists and engineers we seek to describe mathematically the motion of material in our world; at the moment we wish to do this for a certain class of (two-dimensional) fluids which we choose to call Newtonian."

L.E. Scriven (1960)

CHAPTER 4

Interfacial Transport of Momentum

Bulk-phase macroscale fluid fields often exhibit discontinuities across interfacial boundaries, generally owing to comparable discontinuities in bulk fluid physical properties, e.g. density or viscosity. These field discontinuities, which may, for example, arise for mass, momentum, or energy density fields, require the specification of appropriate boundary conditions stating the magnitude of the interfacial discontinuity. Two examples of such boundary conditions are normal stress and solute concentration boundary conditions, corresponding to discontinuities in the normal component of the respective bulk-phase linear momentum density and species concentration fields at an interface.

In Chapter 3, a generic, two-dimensional, macroscale surface-excess balance equation was developed for material interfaces. This equation is adapted in the current chapter to the respective cases of surface-excess linear and angular momentum, thereby furnishing the basic dynamical equations underlying interfacial rheology. We begin in §4.1 with a brief discussion of the equations of mass and momentum transport in continuous three-dimensional media, focusing upon nonpolar, incompressible, Newtonian fluids. Subsequently, in §4.2, we examine the equations governing the transport of surface-excess linear and angular momentum; these arise as boundary conditions imposed upon the contiguous bulk-phase fluid stress fields at a fluid-fluid interface. (In the equilibrium case, the well-known Laplace equation of capillarity [cf. Eq. (2.1-3)] for the bulk-phase pressure difference across a curved interface constitutes an elementary example of this sort of boundary condition.) As in the case of three-dimensional transport, we focus upon nonpolar, (two-dimensional) Newtonian fluid interfaces. In §4.3 we examine non-Newtonian surface rheological constitutive equations. Use of the equations of interfacial rheology are illustrated in §4.4 through solutions of several hydrodynamic problems involving surfactant-adsorbed fluid interfaces. These detailed solutions include: (i) streaming flow past a spherical droplet; (ii) the settling of a droplet at the center of a hollow, fluid-filled spherical container; (iii) an expanding gaseous spherical bubble; (iv) the dilatational viscosity of a dilute bubbly liquid; (v) Newtonian as well as non-Newtonian interfacial flows within a deep-channel surface shear viscometer; (vi) a (spherical) fluid droplet suspended in a homogeneous shear field; (vii) the shear viscosity and viscoelasticity of a dilute emulsion.

As the mathematical development of this chapter is largely performed using dyadic notation (also employed in the previous chapter), this may cause difficulties for the reader uncomfortable with this notation. Such readers are nevertheless encouraged to bear through (since the hydrodynamic equations at a free fluid boundary may otherwise assume decidedly more unwieldy forms when such a concise notation is not followed). However, the important equations gleaned from this chapter are subsequently presented (cf. Tables 4.1 and 4.2) in planar, cylindrical, and spherical coordinate systems for quick and practical reference.

4.1 The Equations of Mass and Momentum Transport in Continuous Three-Dimensional Media

The equations of motion for continuous fluid media may be obtained by substituting into the generic, nonspecific Eq. (3.5-9) the appropriate variables corresponding to mass and momentum transport fields. These important equations are discussed thoroughly in many popular texts (Bird *et al.* 1960, Batchelor 1967), particularly when the bulk-phase fluid may be classified as incompressible, nonpolar, and Newtonian. In such circumstances this pair of equations of motion respectively constitute the *continuity* and *Navier-Stokes equations*. The interfacial stress boundary conditions developed in this chapter will generally be applicable to *any* rheological class of contiguous bulk fluid, although the bulk-phase fluids with which we subsequently deal will generally be Newtonian, and hence obey the Navier-Stokes equations. As the reader is assumed to be generally familiar with classical hydrodynamics, our review in this section is brief.

The mass balance equation for a fluid continuum, in the form of the well-known continuity equation, may be obtained by substituting into Eq. (3.5-9) the appropriate definitions appearing in Table 3.5-1 for the property of mass [$\psi \equiv \rho$]. This yields

$$\frac{\partial \rho}{\partial t} + \nabla \cdot (\rho \mathbf{v}) = 0, \qquad (4.1\text{-}1)$$

applicable at each point $\mathbf{x}$ of the three-dimensional fluid continuum. Equivalently (cf. Question 3.7),

$$\frac{D\rho}{Dt} + \rho \nabla \cdot \mathbf{v} = 0. \qquad (4.1\text{-}2)$$

Similarly, upon substituting into Eq. (3.5-9) the definitions of the generic fields indicated in Table 3.5-1 appropriate to the property of linear momentum [$\psi \equiv \rho \mathbf{v}$], one obtains

$$\frac{\partial}{\partial t}(\rho\,\mathbf{v}) + \nabla\cdot(\rho\,\mathbf{vv}) - \nabla\cdot\mathbf{P} - \mathbf{F} = \mathbf{0} \qquad (4.1\text{-}3)$$

for the balance of linear momentum. Equivalently, from Eq. (4.1-1),

$$\rho\frac{D\mathbf{v}}{Dt} - \nabla\cdot\mathbf{P} - \mathbf{F} = \mathbf{0}. \qquad (4.1\text{-}4)$$

The angular momentum equation [with $\psi\equiv\mathbf{x}\times(\rho\mathbf{v})+\rho\mathbf{a}$ the total moment of momentum at a point $\mathbf{x}$ of the fluid continuum] follows likewise from Eq. (3.5-9) as

$$\frac{\partial}{\partial t}(\rho\,\mathbf{x}\times\mathbf{v}) + \frac{\partial}{\partial t}(\rho\,\mathbf{a}) + \nabla\cdot(\rho\,\mathbf{vx}\times\mathbf{v}) + \nabla\cdot(\rho\,\mathbf{va})$$
$$- \nabla\cdot(\mathbf{P}\times\mathbf{x}) - \nabla\cdot\mathbf{C} - \mathbf{x}\times\mathbf{F} - \mathbf{G} = \mathbf{0}. \qquad (4.1\text{-}5)$$

The pseudo-vector invariant $\mathbf{P}_\times$ of the antisymmetric portion $\frac{1}{2}(\mathbf{P}-\mathbf{P}^\dagger)$ of the pressure tensor is defined as

$$\mathbf{P}_\times = -\,\varepsilon\!:\frac{1}{2}(\mathbf{P}-\mathbf{P}^\dagger)$$
$$\equiv -\,\varepsilon\!:\mathbf{P}, \qquad (4.1\text{-}6)$$

where ε is the unit alternator [cf. Eq. (A.1-21)]. In terms of this pseudovector, Eq. (4.1-5) may be rearranged into the form

$$\mathbf{x}\times\left[\frac{\partial}{\partial t}(\rho\,\mathbf{v}) + \nabla\cdot(\rho\,\mathbf{vv}) - \nabla\cdot\mathbf{P} - \mathbf{F}\right]$$
$$+ \frac{\partial}{\partial t}(\rho\,\mathbf{a}) + \nabla\cdot(\rho\,\mathbf{va}) - \mathbf{P}_\times - \nabla\cdot\mathbf{C} - \mathbf{G} = \mathbf{0}.$$

As the term in brackets vanishes in consequence of Eq. (4.1-3), we thereby obtain the standard form of the angular momentum balance equation as (Dahler & Scriven 1963)

$$\frac{\partial}{\partial t}(\rho\,\mathbf{a}) + \nabla\cdot(\rho\,\mathbf{va}) - \mathbf{P}_\times - \nabla\cdot\mathbf{C} - \mathbf{G} = \mathbf{0}. \qquad (4.1\text{-}7)$$

Equivalently, from Eq. (4.1-1),

$$\rho\frac{D\mathbf{a}}{Dt} - \mathbf{P}_\times - \nabla\cdot\mathbf{C} - \mathbf{G} = \mathbf{0}\,, \qquad (4.1\text{-}8)$$

which constitutes the angular momentum analog of the linear momentum equation (4.1-4).

Nonpolar Fluids

Nonpolar fluids correspond to fluid continua for which

$$\mathbf{a} = \mathbf{0}, \qquad \mathbf{C} = \mathbf{0}, \qquad \mathbf{G} = \mathbf{0}. \qquad (4.1\text{-}9)$$

These properties arise for fluids devoid of internal microstructure [or if such a microstructure does exist, it does not support microcouples, such as might arise if the fluid structure were dipolar and subjected to an external field (Brenner 1970, 1984)]. From Eq. (4.1-7), Eqs. (4.1-9) require that

$$\mathbf{P}_{\times} = \mathbf{0}, \tag{4.1-10}$$

and hence, from the identity (see Question 4.7)

$$\mathbf{P} - \mathbf{P}^{\dagger} = \boldsymbol{\varepsilon} \cdot \mathbf{P}_{\times} \tag{4.1-11}$$

that

$$\mathbf{P} = \mathbf{P}^{\dagger}. \tag{4.1-12}$$

Thus, the pressure tensor is symmetric for nonpolar fluids.

The pressure tensor **P** may generally be decomposed into isotropic and deviatoric parts, as

$$\mathbf{P} = -\,p\,\mathbf{I} + \boldsymbol{\tau}, \tag{4.1-13}$$

where p is the thermodynamic pressure and τ the viscous stress tensor. At a point $\mathbf{x}$ of the fluid continuum the former is defined as that hypothetical pressure which would appear in an equilibrium equation of state governing the temperature T, specific volume (or mass density ρ) and composition (if the fluid is a solution) x_i existing at the point $\mathbf{x}$ (i=1,2, . . . N; $x_i = \rho_i/\rho$ = mass fraction of species i; N = number of chemical species i); explicitly,

$$p \equiv p(\rho, T, x_1, x_2, \ldots), \tag{4.1-13 a}$$

with $\sum_{i=1}^{N} x_i = 1$. This thermodynamic pressure must be distinguished from the *mean pressure*,

$$\bar{p} \overset{\text{def}}{=} -\tfrac{1}{3}\mathbf{I}:\mathbf{P}, \tag{4.1-13 b}$$

defined at the point $\mathbf{x}$. In general, these two pressures differ, although they are the same for incompressible Newtonian fluids [cf. Eq. (4.1-17)] and, presumably, for all incompressible fluids, Newtonian or not. In general, it is assumed to be the mean pressure $\bar{p}$ (rather than p) that is measured by a pitot-static tube immersed into the flowing fluid (Prandtl & Tietjiens 1934).

For nonpolar fluids, it follows from Eqs. (4.1-12) and (4.1-13) that the viscous stress tensor is symmetric:

$$\boldsymbol{\tau} = \boldsymbol{\tau}^{\dagger}. \tag{4.1-14}$$

The Newtonian Stress Tensor

The constitutive assumption for isotropic Newtonian fluids is that the viscous stress tensor τ may be expressed as (Bird *et al.* 1960)

$$\boldsymbol{\tau} = \left(\kappa - \tfrac{2}{3}\mu\right)(\mathbf{I}:\mathbf{D})\,\mathbf{I} + 2\mu\,\mathbf{D}, \qquad (4.1\text{-}15)$$

where μ and κ are the shear and dilatational viscosities respectively, and $\mathbf{D}$ is the rate of deformation tensor, defined as (see Question 4.11)

$$\mathbf{D} = \tfrac{1}{2}\left(\nabla\mathbf{v} + \nabla\mathbf{v}^{\dagger}\right). \qquad (4.1\text{-}16)$$

The scalars μ and κ are *material properties* of the fluid, and hence of the functional forms

$$\mu \equiv \mu(\rho, T, x_1, x_2, \ldots), \quad \kappa \equiv \kappa(\rho, T, x_1, x_2, \ldots) \qquad (4.1\text{-}16a)$$

at each point $\mathbf{x}$ of the fluid.

It follows upon forming the trace of Eq. (4.1-13) and using Eq. (4.1-13b) together with Eq. (4.1-15) that for a compressible Newtonian fluid the thermodynamic and mean pressures are related by the expression

$$\bar{p} - p = -\kappa\nabla\cdot\mathbf{v}, \qquad (4.1\text{-}17)$$

involving only the dilatational viscosity coefficient κ.

Incompressible Fluids

Incompressible fluids correspond to fluid continua for which the mass density ρ is constant in space and time:

$$\rho = \text{const.}, \qquad (4.1\text{-}18)$$

yielding

$$\frac{D\rho}{Dt} = \frac{\partial\rho}{\partial t} + \mathbf{v}\cdot\nabla\rho = 0; \qquad (4.1\text{-}19)$$

equivalently, from Eq. (4.1-2),

$$\nabla\cdot\mathbf{v} = 0. \qquad (4.1\text{-}20)$$

Thus, for an incompressible fluid, $\mathbf{I}:\mathbf{D} \equiv \nabla\cdot\mathbf{v} = 0$, whence Eq. (4.1-15) simplifies to the form

$$\boldsymbol{\tau} = 2\mu\,\mathbf{D}. \qquad (4.1\text{-}21)$$

Moreover, for incompressible Newtonian fluids, Eq. (4.1-17) shows that the thermodynamic and mean pressures become identical:

$$\bar{p} = p. \qquad (4.1\text{-}22)$$

The Navier-Stokes Equations for Incompressible, Newtonian Fluids

In combination, Eqs. (4.1-3), (4.1-13), (4.1-15) and (4.1-20), yield

Table 4.1-1
THE NAVIER-STOKES EQUATIONS IN CARTESIAN COORDINATES (x,y,z)

x − comp:
$$\rho\left(\frac{\partial v_x}{\partial t} + v_x\frac{\partial v_x}{\partial x} + v_y\frac{\partial v_x}{\partial y} + v_z\frac{\partial v_x}{\partial z}\right) = -\frac{\partial p}{\partial x} + \mu\left(\frac{\partial^2 v_x}{\partial x^2} + \frac{\partial^2 v_x}{\partial y^2} + \frac{\partial^2 v_x}{\partial z^2}\right),$$

y − comp:
$$\rho\left(\frac{\partial v_y}{\partial t} + v_x\frac{\partial v_y}{\partial x} + v_y\frac{\partial v_y}{\partial y} + v_z\frac{\partial v_y}{\partial z}\right) = -\frac{\partial p}{\partial y} + \mu\left(\frac{\partial^2 v_y}{\partial x^2} + \frac{\partial^2 v_y}{\partial y^2} + \frac{\partial^2 v_y}{\partial z^2}\right),$$

z − comp:
$$\rho\left(\frac{\partial v_z}{\partial t} + v_x\frac{\partial v_z}{\partial x} + v_y\frac{\partial v_z}{\partial y} + v_z\frac{\partial v_z}{\partial z}\right) = -\frac{\partial p}{\partial z} + \mu\left(\frac{\partial^2 v_z}{\partial x^2} + \frac{\partial^2 v_z}{\partial y^2} + \frac{\partial^2 v_z}{\partial z^2}\right).$$

Continuity equation:
$$\frac{\partial v_x}{\partial x} + \frac{\partial v_y}{\partial y} + \frac{\partial v_z}{\partial z} = 0.$$

THE NEWTONIAN PRESSURE TENSOR IN CARTESIAN COORDINATES

$$P_{xx} = -p + 2\mu\frac{\partial v_x}{\partial x},$$

$$P_{yy} = -p + 2\mu\frac{\partial v_y}{\partial y},$$

$$P_{zz} = -p + 2\mu\frac{\partial v_z}{\partial z},$$

$$P_{xy} = P_{yx} = \mu\left(\frac{\partial v_x}{\partial y} + \frac{\partial v_y}{\partial x}\right),$$

$$P_{yz} = P_{zy} = \mu\left(\frac{\partial v_y}{\partial z} + \frac{\partial v_z}{\partial y}\right),$$

$$P_{zx} = P_{xz} = \mu\left(\frac{\partial v_z}{\partial x} + \frac{\partial v_x}{\partial z}\right).$$

Table 4.1-2
THE NAVIER-STOKES EQUATIONS IN CYLINDRICAL COORDINATES (R, ϕ, z)*

R − comp: $$\rho\left(\frac{\partial v_R}{\partial t} + v_R\frac{\partial v_R}{\partial R} + \frac{v_\phi}{R}\frac{\partial v_R}{\partial \phi} - \frac{v_\phi^{\,2}}{R} + v_z\frac{\partial v_R}{\partial z}\right) = -\frac{\partial p}{\partial R} + \mu\left(\nabla^2 v_R - \frac{v_R}{R^2} - \frac{2}{R^2}\frac{\partial v_\phi}{\partial \phi}\right),$$

φ − comp: $$\rho\left(\frac{\partial v_\phi}{\partial t} + v_R\frac{\partial v_\phi}{\partial R} + \frac{v_\phi}{R}\frac{\partial v_\phi}{\partial \phi} + \frac{v_R v_\phi}{R} + v_z\frac{\partial v_\phi}{\partial z}\right) = -\frac{1}{R}\frac{\partial p}{\partial \phi} + \mu\left(\nabla^2 v_\phi - \frac{v_\phi}{R^2} + \frac{2}{R^2}\frac{\partial v_R}{\partial \phi}\right),$$

z − comp: $$\rho\left(\frac{\partial v_z}{\partial t} + v_R\frac{\partial v_z}{\partial R} + \frac{v_\phi}{R}\frac{\partial v_z}{\partial \phi} + v_z\frac{\partial v_z}{\partial z}\right) = -\frac{\partial p}{\partial z} + \mu\nabla^2 v_z$$

Continuity equation: $$\frac{1}{R}\frac{\partial}{\partial R}(R v_R) + \frac{1}{R}\frac{\partial v_\phi}{\partial \phi} + \frac{\partial v_z}{\partial z} = 0.$$

THE NEWTONIAN PRESSURE TENSOR IN CYLINDRICAL COORDINATES

$$P_{RR} = -p + 2\mu\frac{\partial v_R}{\partial R},$$

$$P_{\phi\phi} = -p + 2\mu\left(\frac{1}{R}\frac{\partial v_\phi}{\partial \phi} + \frac{v_R}{R}\right),$$

$$P_{zz} = -p + 2\mu\frac{\partial v_z}{\partial z},$$

$$P_{R\phi} = P_{\phi R} = \mu\left[R\frac{\partial}{\partial R}\left(\frac{v_\phi}{R}\right) + \frac{1}{R}\frac{\partial v_R}{\partial \phi}\right],$$

$$P_{\phi z} = P_{z\phi} = \mu\left(\frac{\partial v_\phi}{\partial z} + \frac{1}{R}\frac{\partial v_z}{\partial \phi}\right),$$

$$P_{zR} = P_{Rz} = \mu\left(\frac{\partial v_z}{\partial R} + \frac{\partial v_R}{\partial z}\right).$$

---

* The Laplace operator in circular cylindrical coordinates is

$$\nabla^2 = \frac{1}{R}\frac{\partial}{\partial R}\left(R\frac{\partial}{\partial R}\right) + \frac{1}{R^2}\frac{\partial^2}{\partial \phi^2} + \frac{\partial^2}{\partial z^2}.$$

Table 4.1-3
THE NAVIER-STOKES EQUATIONS IN SPHERICAL COORDINATES $(r, \theta, \phi)^*$

$$r-comp\colon\ \rho\left(\frac{\partial v_r}{\partial t}+v_r\frac{\partial v_r}{\partial r}+\frac{v_\theta}{r}\frac{\partial v_r}{\partial \theta}-\frac{v_\phi^2+v_\theta^2}{r}+\frac{v_\phi}{r\sin\theta}\frac{\partial v_r}{\partial \phi}\right)$$

$$=-\frac{\partial p}{\partial r}+\mu\left(\nabla^2 v_r-\frac{2}{r^2}v_r-\frac{2}{r^2}\frac{\partial v_\theta}{\partial \theta}-\frac{2}{r^2}\cot\theta\, v_\theta-\frac{2}{r^2\sin\theta}\frac{\partial v_\phi}{\partial \phi}\right),$$

$$\theta-comp\colon\ \rho\left(\frac{\partial v_\theta}{\partial t}+v_r\frac{\partial v_\theta}{\partial r}+\frac{v_\theta}{r}\frac{\partial v_\theta}{\partial \theta}+\frac{v_\phi}{r\sin\theta}\frac{\partial v_\theta}{\partial \phi}+\frac{v_r v_\theta}{r}-\frac{\cot\theta\, v_\phi^2}{r}\right)$$

$$=-\frac{1}{r}\frac{\partial p}{\partial \theta}+\mu\left(\nabla^2 v_\theta+\frac{2}{r^2}\frac{\partial v_r}{\partial \theta}-\frac{v_\theta}{r^2\sin^2\theta}-\frac{2\cos\theta}{r^2\sin^2\theta}\frac{\partial v_\phi}{\partial \phi}\right),$$

$$\phi-comp\colon\ \rho\left(\frac{\partial v_\phi}{\partial t}+v_r\frac{\partial v_\phi}{\partial r}+\frac{v_\theta}{r}\frac{\partial v_\phi}{\partial \theta}+\frac{v_r v_\phi}{r}+\frac{v_\phi}{r\sin\theta}\frac{\partial v_\phi}{\partial \phi}\right.$$

$$\left.+\frac{\cot\theta\, v_\theta v_\phi}{r}\right)=-\frac{1}{r\sin\theta}\frac{\partial p}{\partial \phi}+\mu\Big(\nabla^2 v_\phi$$

$$+\frac{2}{r^2\sin^2\theta}\frac{\partial v_r}{\partial \phi}-\frac{v_\phi}{r^2\sin^2\theta}+\frac{2\cos\theta}{r^2\sin^2\theta}\frac{\partial v_\theta}{\partial \phi}\Big).$$

$$Continuity\ eqn\colon\ \frac{1}{r^2}\frac{\partial}{\partial r}\left(r^2 v_r\right)+\frac{1}{r\sin\theta}\frac{\partial}{\partial \theta}\left(\sin\theta\, v_\theta\right)+\frac{1}{r\sin\theta}\frac{\partial v_\phi}{\partial \phi}=0.$$

THE NEWTONIAN PRESSURE TENSOR IN SPHERICAL COORDINATES

$$P_{rr}=-p+2\mu\frac{\partial v_r}{\partial r},$$

$$P_{\theta\theta}=-p+2\mu\left(\frac{1}{r}\frac{\partial v_\theta}{\partial \theta}+\frac{v_r}{r}\right),$$

$$P_{\phi\phi}=-p+2\mu\left(\frac{1}{r\sin\theta}\frac{\partial v_\phi}{\partial \phi}+\frac{v_r}{r}+\frac{\cot\theta\, v_\theta}{r}\right),$$

$$P_{\theta r}=P_{\theta r}=\mu\left[r\frac{\partial}{\partial r}\left(\frac{v_\theta}{r}\right)+\frac{1}{r}\frac{\partial v_r}{\partial \theta}\right],$$

$$P_{\theta\phi}=P_{\phi\theta}=\mu\left[\frac{\sin\theta}{r}\frac{\partial}{\partial \theta}\left(\frac{v_\phi}{\sin\theta}\right)+\frac{1}{r\sin\theta}\frac{\partial v_\theta}{\partial \phi}\right],$$

$$P_{\phi r}=P_{r\phi}=\mu\left[\frac{1}{r\sin\theta}\frac{\partial v_r}{\partial \phi}+r\frac{\partial}{\partial r}\left(\frac{v_\phi}{r}\right)\right].$$

---

* The Laplace operator in spherical coordinates is

$$\nabla^2=\frac{1}{r^2}\frac{\partial}{\partial r}\left(r^2\frac{\partial}{\partial r}\right)+\frac{1}{r^2\sin\theta}\frac{\partial}{\partial \theta}\left(\sin\theta\frac{\partial}{\partial \theta}\right)+\frac{1}{r^2\sin^2\theta}\frac{\partial^2}{\partial \phi^2}.$$

$$\boxed{\rho\left(\frac{\partial \mathbf{v}}{\partial t} + \mathbf{v} \cdot \nabla \mathbf{v}\right) = -\nabla p + \mu \nabla^2 \mathbf{v}} \tag{4.1-23}$$

and

$$\boxed{\nabla \cdot \mathbf{v} = 0.} \tag{4.1-24}$$

These respectively represent the *Navier-Stokes* and *continuity* equations for a three-dimensional, incompressible, nonpolar, Newtonian fluid continuum for cases wherein the material properties ρ and μ are constant throughout the fluid. Component forms of these equations are provided for Cartesian, circular cylindrical, and spherical coordinate systems in Tables 4.1-1 through 4.1-3.

4.2 The Equations of Interfacial Mass and Momentum Transport

At a fluid boundary, Eqs. (4.1-22) and (4.1-23) require the specification of kinematical and dynamical interfacial conditions. *Kinematical* conditions appropriate to (macroscale) material fluid interfaces have been previously described in §3.4. Requisite *dynamical* conditions at a macroscale material fluid interface are provided by the two-dimensional analogs of the linear and angular momentum equations developed in the previous section [cf. Eqs. (4.1-3) and (4.1-7)]. These dynamical conditions are obtained from the generic surface-excess balance equation (3.6-16) by substituting from Table 3.6-1 the appropriate definitions of the particular momental properties concerned.

The surface-excess mass balance equation may be obtained by substituting into Eq. (3.6-16) the appropriate definitions appearing in Table 3.6-1 for the property of surface-excess mass [$\psi^s \equiv \rho^s$]. This yields

$$\frac{\partial \rho^s}{\partial t} + \nabla_s \cdot (\rho^s \mathbf{v}^o) = 0, \tag{4.2-1}$$

applicable at each point $\mathbf{x}_s$ of the two dimensional interfacial continuum. Here, $\mathbf{v}^o(\mathbf{x}_s)$ is the mass-averaged velocity of the interface, defined in Eq. (3.4-6), and ρ^s is the surface-excess mass density. Equivalently, from Eq. (3.4-10),

$$\frac{D_s \rho^s}{Dt} + \rho^s \nabla_s \cdot \mathbf{v}^o = 0. \tag{4.2-2}$$

As the interface has been assumed here to be a *material* interface, no transfer of mass between the interface and the contiguous bulk fluids occurs. This explains the absence of a jump boundary condition on the right-hand side of Eq. (4.2-1). [See Eq. (16.1-8) for the case where the interface is *nonmaterial*.]

Similarly to the above, upon substituting into Eq. (3.6-16) the definitions of the generic fields indicated in Table 3.6-1 appropriate to the property of surface-excess linear momentum [$\psi^s \equiv \rho^s \mathbf{v}^o$], one obtains

$$\frac{\partial}{\partial t}(\rho^s \mathbf{v}^o) + \nabla_s \cdot (\rho^s \mathbf{v}^o \mathbf{v}^o) - \nabla_s \cdot \mathbf{P}^s - \mathbf{F}^s = \mathbf{n} \cdot \|\bar{\mathbf{P}}\|, \quad (4.2\text{-}3)$$

where $\mathbf{P}^s$ the surface-excess pressure tensor* and $\mathbf{F}^s$ the surface-excess force density vector. Equivalently, from Eq. (4.2-2),

$$\boxed{\rho^s \frac{D_s \mathbf{v}^o}{Dt} - \nabla_s \cdot \mathbf{P}^s - \mathbf{F}^s = \mathbf{n} \cdot \|\bar{\mathbf{P}}\|.} \quad (4.2\text{-}4)$$

The surface-excess angular momentum equation follows similarly [with $\psi^s \equiv \mathbf{x}_s \times (\rho^s \mathbf{v}^o) + \rho^s \mathbf{a}^s$] from Eq. (3.6-16) as:

$$\begin{aligned}&\frac{\partial}{\partial t}(\rho^s \mathbf{x}_s \times \mathbf{v}^o) + \frac{\partial}{\partial t}(\rho^s \mathbf{a}^s) + \nabla_s \cdot (\rho^s \mathbf{v}^o \mathbf{x}_s \times \mathbf{v}^o) \\ &+ \nabla_s \cdot (\rho^s \mathbf{v}^o \mathbf{a}^s) - \nabla_s \cdot (\mathbf{P}^s \times \mathbf{x}_s) - \nabla_s \cdot \mathbf{C}^s \\ &\quad - \mathbf{x}_s \times \mathbf{F}^s - \mathbf{G}^s - \mathbf{x}_s \times \mathbf{n} \cdot \|\bar{\mathbf{P}}\| - \mathbf{n} \cdot \|\bar{\mathbf{C}}\| = 0, \end{aligned} \quad (4.2\text{-}5)$$

where $\mathbf{a}^s$ is the surface-excess internal angular momentum density pseudovector, $\mathbf{C}^s$ the surface-excess couple stress tensor, and $\mathbf{G}^s$ the surface-excess body couple density vector. The pseudovector invariant $\mathbf{P}^s_\times$ of the antisymmetric portion $\frac{1}{2}(\mathbf{P}^s - \mathbf{P}^{s\dagger})$ of the surface-excess stress tensor is defined as [cf. Eq. (4.1-6)]

$$\mathbf{P}^s_\times = -\,\varepsilon : \mathbf{P}^s. \quad (4.2\text{-}6)$$

In terms of this pseudovector, Eq. (4.2-5) may be rearranged into the form

---

* The surface-excess pressure tensor possesses the property $\mathbf{I}_s \cdot \mathbf{P}^s = \mathbf{P}^s$, as is demonstrated in Appendix 3.A [cf. Eqs. (3.A-9) and (3.A-10): see also Eq. (3.6-11c)]. This property, which is utilized in obtaining Eq. (4.2-3), may be attributed (adhering to the physical context of Appendix 3.A) to the fact that the surface-excess contact force $\mathbf{p}^s = \mathbf{v} \cdot \mathbf{P}^s$ given by Eq. (3.A-10) acts upon a lineal boundary dL situated in the interface in a direction ν that is normal to the line element and lies in the *tangent plane* to the interface. Such is the commonly observed property of interfacial tension. Nevertheless, $\mathbf{P}^s \cdot \mathbf{n}$ is generally *nonzero*. This latter fact, which apparently has no analogy with the pressure tensor $\mathbf{P}$ in three dimensions, owes to the fact that (two-dimensional) interfacial forces may, in general, act directly upon contiguous (three-dimensional) bulk phases, normal to the interface [see Edwards (1987)].

$$\mathbf{x}_s \times \left[\frac{\partial}{\partial t}(\rho^s \mathbf{v}^o) + \nabla_s \cdot (\rho^s \mathbf{v}^o \mathbf{v}^o) - \nabla_s \cdot \mathbf{P}^s - \mathbf{F}^s - \mathbf{n} \cdot \|\bar{\mathbf{P}}\|\right]$$
$$+ \frac{\partial}{\partial t}(\rho^s \mathbf{a}^s) + \nabla_s \cdot (\rho^s \mathbf{v}^o \mathbf{a}^s) - \nabla_s \cdot \mathbf{C}^s - \mathbf{P}^s_{\times} - \mathbf{G}^s - \mathbf{n} \cdot \|\bar{\mathbf{C}}\| = \mathbf{0}.$$

As the terms in brackets vanish in consequence of Eq. (4.2-3), we thereby obtain the surface-excess angular momentum equation

$$\frac{\partial}{\partial t}(\rho^s \mathbf{a}^s) + \nabla_s \cdot (\rho^s \mathbf{v}^o \mathbf{a}^s) - \nabla_s \cdot \mathbf{C}^s - \mathbf{P}^s_{\times} - \mathbf{G}^s = \mathbf{n} \cdot \|\bar{\mathbf{C}}\|. \tag{4.2-7}$$

Equivalently, from Eq. (4.2-2),

$$\boxed{\rho^s \frac{D_s \mathbf{a}^s}{Dt} - \nabla_s \cdot \mathbf{C}^s - \mathbf{P}^s_{\times} - \mathbf{G}^s = \mathbf{n} \cdot \|\bar{\mathbf{C}}\|,} \tag{4.2-8}$$

which constitutes the surface-excess angular momentum analog of the surface-excess linear momentum equation (4.2-4).

Equations (4.2-1) through (4.2-8) may be observed to provide the respective two-dimensional analogs of Eqs. (4.1-1) through (4.1-8). The only qualitatively new terms appearing in the analogous surface equations [cf. Eqs. (4.2-4) and (4.2-8)] correspond to 'source' terms, via which linear and angular momentum may be supplied to the two-dimensional interface from the three-dimensional contiguous bulk phases, and conversely. [See Eqs. (16.1-16) and (16.2-8) for the respective generalizations of the latter two interfacial conditions for *nonmaterial* interfaces.]

In all practical circumstances, $\rho^s << \bar{\rho}$ (owing to the extreme thinness of the interfacial transition zone; cf. Eqs. (16.1-2) and (16.1-10) and the discussion thereof).* In such circumstances,

$$\rho^s = 0, \tag{4.2-8a}$$

whence Eqs. (4.2-4) and (4.2-8) reduce to the respective forms

$$-\mathbf{n} \cdot \|\bar{\mathbf{P}}\| = \mathbf{F}^s + \nabla_s \cdot \mathbf{P}^s \tag{4.2-9}$$

and

$$-\mathbf{n} \cdot \|\bar{\mathbf{C}}\| = \nabla_s \cdot \mathbf{C}^s + \mathbf{P}^s_{\times} + \mathbf{G}^s. \tag{4.2-10}$$

* Note, however, this does not exclude the important possibility that $\rho_i^s \neq 0$, such as occurs when species i is a surfactant. The reader may refer to §5.2 for a fuller discussion.

Nonpolar Fluids

As will be more fully discussed in Chapter 19, circumstances may exist wherein antisymmetric surface-excess stresses (i.e. $\mathbf{P}^s_\times \neq \mathbf{0}$) may arise at a surfactant-adsorbed interface. Such surface stresses would arise, for example, should the adsorbed surfactant particles be subjected to body couples originating from the action of an external electromagnetic field upon permanent internal dipoles locked into the surfactant molecules. Such couples may be balanced by orientational, adsorptive couples (Edwards 1987) acting upon surfactant particles in the vicinity of the fluid interface. It may therefore be possible for a nonpolar bulk fluid satisfying, *inter alia*, the bulk-phase equality $\overline{\mathbf{P}}_\times = \mathbf{0}$] to be contiguous to a *polar* interface, for which $\mathbf{P}^s_\times \neq \mathbf{0}$!

For the present, however, we will henceforth assume that

$$\mathbf{C}^s = \mathbf{0}, \quad \mathbf{G}^s = \mathbf{0}, \tag{4.2-11}$$

in addition to Eqs. (4.1-9). These conditions correspond to the absence of internal microstructure in both the interfacial and bulk phases. From Eq. (4.2-10) it therefore follows that

$$\mathbf{P}^s_\times = \mathbf{0}, \tag{4.2-12}$$

and hence from the identity (see Question 4.8)

$$\mathbf{P}^s - \mathbf{P}^{s\dagger} = \boldsymbol{\varepsilon} \cdot \mathbf{P}^s_\times \tag{4.2-13}$$

that

$$\mathbf{P}^s = \mathbf{P}^{s\dagger}. \tag{4.2-14}$$

Thus, both the bulk and interfacial pressure tensors are symmetric for nonpolar phases.

A general decomposition of the surface-excess pressure tensor may be made in terms of isotropic and deviatoric parts, analogous to the decomposition of the bulk-phase pressure tensor in Eq. (4.1-13); thus, we write

$$\boxed{\mathbf{P}^s = \mathbf{I}_s \sigma + \boldsymbol{\tau}^s,} \tag{4.2-15}$$

where σ is the *thermodynamic* interfacial tension and $\boldsymbol{\tau}^s$ the surface-excess stress tensor. At a point $\mathbf{x}_s$ of the interfacial continuum the former is defined as that hypothetical surface tension which would appear in an equilibrium equation of state [cf. Eqs. (5.5-1)–(5.5-3)] governing the temperature T, surface-excess mass density ρ^s, and composition (if surfactant is adsorbed to the interface) x_i^s adsorbed at the point $\mathbf{x}_s$ (i=1,2, . . . N; $x_i^s = \rho_i^s/\rho^s$= surface-excess mass fraction of species i; N = number of chemical species i); explicitly,

$$\sigma \equiv \sigma(\rho^s, T, x_1^s, x_2^s, \ldots), \tag{4.2-15a}$$

where $\sum_{i=1}^{N} x_i^s = 1$. This thermodynamic interfacial tension must be distinguished from the *mean interfacial tension*, [note the analogy with Eq. (4.1-13b)],

$$\bar{\sigma} \overset{\text{def}}{=} \tfrac{1}{2}\mathbf{I}_s : \mathbf{P}^s, \tag{4.2-15b}$$

defined at the interfacial point $\mathbf{x}_s$. This latter tension is more commonly known as the *dynamic interfacial tension.* Insofar as interfacial phases are generally *not* 'incompressible', in the sense that $\nabla_s \cdot \mathbf{v}_s \neq 0$ (even if $\nabla \cdot \mathbf{v} = 0$), a difference will generally exist between the 'dynamic interfacial tension' $\bar{\sigma}$ defined above, and the thermodynamic interfacial tension σ [cf. Eq. (4.4-32)]. The consequences of this difference are further elaborated upon in Chapters 6 and 8. From Eqs. (4.2-14) and (4.2-15), it follows that the surface-excess stress tensor is symmetric for a nonpolar interface

$$\boldsymbol{\tau}^s = \boldsymbol{\tau}^{s\dagger}. \tag{4.2-16}$$

The Newtonian Surface-Excess Stress Tensor

To obtain an explicit expression for the surface-excess pressure tensor under dynamic conditions, a specific constitutive law for the surface-excess stress tensor $\boldsymbol{\tau}^s$ [in Eq. (4.2-15)] must be supplied. In this context, we now consider the *Boussinesq*, or *Boussinesq-Scriven* constitutive law for a 'Newtonian' interface, which is the constitutive relation most widely assumed in practice. Later, in §4.3, we will consider other constitutive relations that have been proposed for non-Newtonian interfaces.

The Boussinesq-Scriven constitutive expression, appropriate to so-called *Newtonian interfaces*, is of the form (Scriven 1960)

$$\boldsymbol{\tau}^s = (\kappa^s - \mu^s)(\mathbf{I}_s : \mathbf{D}_s)\mathbf{I}_s + 2\mu^s\mathbf{D}_s, \tag{4.2-17}$$

where μ^s and κ^s are respectively the interfacial shear and dilatational viscosities at a point $\mathbf{x}_s$ of the interface, and

$$\mathbf{D}_s = \tfrac{1}{2}[(\nabla_s \mathbf{v}^o)\cdot \mathbf{I}_s + \mathbf{I}_s \cdot (\nabla_s \mathbf{v}^o)^\dagger] \tag{4.2-18}$$

is the surface rate of deformation tensor (see Appendix 4.C at the end of this chapter). This expression for $\boldsymbol{\tau}^s$ may be compared with the comparable bulk-phase stress tensor (4.1-15). As with the bulk-phase material properties of

Eq. (4.1-16a), the scalars μ^s and κ^s are *material properties* of the fluid interface, and hence of the functional forms

$$\mu^s \equiv \mu^s(\rho^s, T, x_1^s, x_2^s, \ldots), \quad \kappa^s \equiv \kappa^s(\rho^s, T, x_1^s, x_2^s, \ldots) \quad (4.2\text{-}18\text{a})$$

at each point $\mathbf{x}_s$ of the interface.

The mean interfacial tension (4.2-15b) here adopts the following form for Newtonian interfaces

$$\bar{\sigma} - \sigma = \kappa^s \nabla_s \cdot \mathbf{v}^o, \qquad (4.2\text{-}19)$$

which should be compared with the comparable bulk-phase expression (4.1-17). As noted following Eq. (4.2-15b), whereas incompressible *bulk fluids* are common in practice, 'incompressible' *interfaces* (i.e. $\nabla_s \cdot \mathbf{v}^o = 0$) are rare, if not nonexistent. Thus, the interfacial dilatational viscosity contribution (4.2-19) to the mean interfacial tension $\bar{\sigma}$ will often be sizeable in practice, especially as κ^s is often of the order of μ^s in magnitude [cf. Figure 2.5-2 and Eq. (16.3-26)]. Illustrations of this phenomenon are numerous throughout this text. Included are examples 1-4, as well as example 6 of §4.4, example 2 of §5.6, each of the three classes of interfacial dilatational phenomena cited in Chapter 8, the interfacial turbulence example discussed in §10.4, the thin film drainage and stability examples of Chapters 11 and 12, and each of the foam rheology problems of Chapter 14.

The Newtonian Surface Stress Boundary Condition

Upon substitution of Eqs. (4.2-15) and (4.2-13) into Eq. (4.2-9), and regarding μ^s and κ^s as constants, independent of interfacial position $\mathbf{x}_s$, we obtain

$$\begin{aligned} -\mathbf{n} \cdot \|\bar{\mathbf{P}}\| &= \mathbf{F}^s + 2H\sigma\mathbf{n} + \nabla_s\sigma + (\kappa^s + \mu^s)\nabla_s\nabla_s \cdot \mathbf{v}^o \\ &+ 2\mu^s\mathbf{n}(\mathbf{b} - 2H\,\mathbf{I}_s) : \nabla_s\mathbf{v}^o + 2H\,\mathbf{n}(\kappa^s + \mu^s)\nabla_s \cdot \mathbf{v}^o \\ &+ \mu^s\{\mathbf{n} \times \nabla_s[(\nabla_s \times \mathbf{v}^o) \cdot \mathbf{n}] - 2(\mathbf{b} - 2H\,\mathbf{I}_s) \cdot (\nabla_s\mathbf{v}^o) \cdot \mathbf{n}\}, \end{aligned} \qquad (4.2\text{-}20)$$

as demonstrated in Appendix 4.A at the end of this chapter. This vector equation may be resolved into vectors respectively normal and tangential to the interface via the generic scheme (4.B-2) applicable to any vector **f**. This yields

$$\mathbf{n} \cdot \|\bar{\mathbf{P}}\| = \|\bar{\mathbf{P}}_{nn}\| + \|\bar{\mathbf{P}}_{ns}\|,$$

where

$$\left\|\overline{\mathbf{P}}_{nn}\right\| \overset{\text{def}}{=} \mathbf{n}\cdot\left\|\overline{\mathbf{P}}\right\|\cdot\mathbf{n}\,\mathbf{n} \equiv -\mathbf{F}^{s}\cdot\mathbf{n}\,\mathbf{n} - 2H\sigma\,\mathbf{n}$$
$$- \mu^{s}\mathbf{n}(\mathbf{b} - 2H\,\mathbf{I}_{s}):\nabla_{s}\mathbf{v}^{o} - 2H\,\mathbf{n}(\kappa^{s} + \mu^{s})\nabla_{s}\cdot\mathbf{v}^{o}$$

(4.2- 20 a)

and

$$\left\|\overline{\mathbf{P}}_{ns}\right\| \overset{\text{def}}{=} \mathbf{n}\cdot\left\|\overline{\mathbf{P}}\right\|\cdot\mathbf{I}_{s} \equiv -\mathbf{F}^{s}\cdot\mathbf{I}_{s} - \nabla_{s}\sigma - (\kappa^{s} + \mu^{s})\nabla_{s}\nabla_{s}\cdot\mathbf{v}^{o}$$
$$- \mu^{s}\{\mathbf{n}\times\nabla_{s}[(\nabla_{s}\times\mathbf{v}^{o})\cdot\mathbf{n}] + 2(\mathbf{b} - 2H\,\mathbf{I}_{s})\cdot(\nabla_{s}\mathbf{v}^{o})\cdot\mathbf{n}\}.$$

(4.2- 20 b)

This pair of equations is often referred to respectively as the *normal stress* and *tangential stress* boundary conditions. Together, they provide the explicit interfacial stress boundary condition for a Newtonian interface. In Tables 4.2-1, 4.2-2 and 4.2-3 component forms of the vector equation (4.2-20) are provided for planar, circular cylindrical, and spherical interfaces (see Appendix 4.B).

We offer here a comment on the application of the boundary condition (4.2-20), and specifically (4.2-20a) to fluid-mechanical problems. Bulk flow problems involving deformable interfacial boundaries often arise wherein the geometrical configuration of the interface is assumed to remain constant in time, or at any rate is known *a priori* as a function of time (e.g. an expanding spherical gas bubble; cf. example 2 of §4.4). In such cases, specifying the interface shape prior to solution of the relevant hydrodynamic problem is tantamount to specification of the normal velocity $\mathbf{n}\cdot\mathbf{v}^{o}$ at each point $\mathbf{x}_{s}$ of the interface. In such circumstances, imposing the normal stress $\left\|\overline{\mathbf{P}}_{nn}\right\|$ boundary condition overspecifies the problem formulation. *Either* the geometry of the interface is known *a priori* and a boundary condition upon normal velocity at the fluid interface is imposed, or else the interfacial condition $\left\|\overline{\mathbf{P}}_{nn}\right\|$ is imposed. Thus, the normal stress boundary condition (4.2-20a) is, in fact, responsible for defining the geometrical configuration of the fluid interface.

This point is aptly illustrated by the example of a closed, bounded, hydrostatic interface in the absence of externally applied forces. The normal stress $\left\|\overline{\mathbf{P}}_{nn}\right\|$ boundary condition reduces in this case to the Laplace condition, Eq. (2.1-3). In this limit of mechanical equilibrium, pressures and interfacial tension are each constant, independent of respective spatial and interfacial positions, $\mathbf{x}$ and $\mathbf{x}_{s}$. The $\left\|\overline{\mathbf{P}}_{nn}\right\|$ condition thereby requires the existence of a *constant* mean surface curvature H at each point $\mathbf{x}_{s}$ of the interface. This geometrical requirement can only be met by a *spherical* interface. It therefore follows that bubbles and droplets dispersed in a homogeneous pressure field necessarily seek a spherical shape.

Table 4.2-1
THE NORMAL STRESS BOUNDARY CONDITION FOR PLANAR NEWTONIAN INTERFACES (z = const.)

$$-\mathbf{n} \cdot \|\bar{\mathbf{P}}\| = \mathbf{F}^s + \nabla_s \cdot \mathbf{P}^s$$

Tangential Components

$$x - comp: \quad -\|\bar{P}_{zx}\| = F_x^s + \frac{\partial \sigma}{\partial x} + (\kappa^s + \mu^s)\frac{\partial}{\partial x}\left(\frac{\partial v_x^o}{\partial x} + \frac{\partial v_y^o}{\partial y}\right) + \mu^s \frac{\partial}{\partial y}\left(\frac{\partial v_x^o}{\partial y} - \frac{\partial v_y^o}{\partial x}\right),$$

$$y - comp: \quad -\|\bar{P}_{zy}\| = F_y^s + \frac{\partial \sigma}{\partial y} + (\kappa^s + \mu^s)\frac{\partial}{\partial y}\left(\frac{\partial v_x^o}{\partial x} + \frac{\partial v_y^o}{\partial y}\right) - \mu^s \frac{\partial}{\partial x}\left(\frac{\partial v_x^o}{\partial y} - \frac{\partial v_y^o}{\partial x}\right).$$

Normal Component

$$z - comp: \quad -\|\bar{P}_{zz}\| = F_z^s .$$

Table 4.2-2
THE NORMAL STRESS BOUNDARY CONDITION FOR CYLINDRICAL NEWTONIAN INTERFACES [$R = R_0(t)$, say]

$$-\mathbf{n}\cdot\|\bar{\mathbf{P}}\| = \mathbf{F}^s + \nabla_s \cdot \mathbf{P}^s$$

Tangential Components

$$\phi - comp:\quad -\left\|\bar{P}_{R\phi}\right\| = F^s_\phi + \frac{1}{R_0}\frac{\partial\sigma}{\partial\phi} + \mu^s\frac{\partial}{\partial z}\left(\frac{\partial v^o_\phi}{\partial z} - \frac{1}{R_0}\frac{\partial v^o_z}{\partial\phi}\right) + \frac{\kappa^s + \mu^s}{R_0}\frac{\partial}{\partial\phi}\left(\frac{1}{R_0}\frac{\partial v^o_\phi}{\partial\phi} + \frac{\partial v^o_z}{\partial z}\right),$$

$$z - comp:\quad -\left\|\bar{P}_{Rz}\right\| = F^s_z + \frac{\partial\sigma}{\partial z} - \frac{\mu^s}{R_0}\frac{\partial}{\partial\phi}\left(\frac{\partial v^o_\phi}{\partial z} - \frac{1}{R_0}\frac{\partial v^o_z}{\partial\phi}\right) + (\kappa^s + \mu^s)\frac{\partial}{\partial z}\left(\frac{1}{R_0}\frac{\partial v^o_\phi}{\partial\phi} + \frac{\partial v^o_z}{\partial z}\right).$$

Normal Component

$$R - comp:\quad -\left\|\bar{P}_{RR}\right\| = F^s_R - \frac{\sigma}{R_0} + \frac{2\mu^s}{R_0}\frac{\partial v^o_z}{\partial z} - \frac{\kappa^s v^o_R}{R_0^2} - \frac{\kappa^s + \mu^s}{R_0}\left(\frac{1}{R_0}\frac{\partial v^o_\phi}{\partial\phi} + \frac{\partial v^o_z}{\partial z}\right).$$

Table 4.2-3
THE NORMAL STRESS BOUNDARY CONDITION FOR SPHERICAL NEWTONIAN INTERFACES [$r = r_0(t)$, say]

$$-\mathbf{n}\cdot\left\|\bar{\mathbf{P}}\right\| = \mathbf{F}^s + \nabla_s \cdot \mathbf{P}^s$$

Tangential Components

$$\theta - comp:\quad -\left\|\bar{P}_{r\theta}\right\| = F_\theta^s + \frac{1}{r_0}\frac{\partial\sigma}{\partial\theta}$$
$$+\,\mu^s\left\{\frac{2v_\theta^o}{r_0^2} + \frac{1}{r_0\sin\theta}\frac{\partial}{\partial\phi}\left[\frac{1}{r_0\sin\theta}\left(\frac{\partial v_\theta^o}{\partial\phi} - \frac{\partial}{\partial\theta}(\sin\theta\, v_\phi^o)\right)\right]\right\}$$
$$+\,\frac{\kappa^s+\mu^s}{r_0}\frac{\partial}{\partial\theta}\left[\frac{1}{r_0\sin\theta}\left(\frac{\partial}{\partial\theta}(\sin\theta\, v_\theta^o) + \frac{\partial v_\phi^o}{\partial\phi}\right)\right],$$

$$\phi - comp:\quad -\left\|\bar{P}_{r\phi}\right\| = F_\phi^s + \frac{1}{r_0\sin\theta}\frac{\partial\sigma}{\partial\phi}$$
$$+\,\mu^s\left\{\frac{2v_\phi^o}{r_0^2} - \frac{1}{r_0}\frac{\partial}{\partial\theta}\left[\frac{1}{r_0\sin\theta}\left(\frac{\partial v_\theta^o}{\partial\phi} - \frac{\partial}{\partial\theta}(\sin\theta\, v_\phi^o)\right)\right]\right\}$$
$$+\,\frac{\kappa^s+\mu^s}{r_0\sin\theta}\frac{\partial}{\partial\phi}\left[\frac{1}{r_0\sin\theta}\left(\frac{\partial}{\partial\theta}(\sin\theta\, v_\theta^o) + \frac{\partial v_\phi^o}{\partial\phi}\right)\right].$$

Normal Component

$$r - comp:\quad -\left\|\bar{P}_{rr}\right\| = F_r^s - \frac{2\sigma}{r_0} - \frac{4\kappa^s v_r^o}{r_0^2}$$
$$-\,\frac{2\kappa^s}{r_0}\left[\frac{1}{r_0\sin\theta}\left(\frac{\partial}{\partial\theta}(\sin\theta\, v_\theta^o) + \frac{\partial v_\phi^o}{\partial\phi}\right)\right].$$

4.3 Non-Newtonian Interfacial Rheological Behavior

Alternative forms of the interfacial stress boundary condition (4.2-20) have been proposed to describe experimentally observed non-Newtonian interfacial rheological behavior. Several of these non-Newtonian constitutive models are outlined in this section.

Nonlinear Viscous Models

The Boussinesq-Scriven constitutive model (4.2-17) poses a linear relation between interfacial stress $\boldsymbol{\tau}^s$ and interfacial deformation rate $\mathbf{D}_s$, wherein the interfacial viscosities μ^s and κ^s are assumed to be independent of the deformation rate. While this independence is to be expected at small rates of interfacial deformation, at sufficiently large rates most surfactant-adsorbed fluid interfaces exhibit a shear-thinning or dilatation-thinning behavior. Such shear thinning is evidenced, for example, by the deformation-rate dependence of the interfacial viscosity μ^s at high shear rates, shown in Figure 4.3-1.

In general, the interfacial viscosities μ^s and κ^s are functions of the principal invariants I_s and II_s of the interfacial rate-of-deformation tensor $\mathbf{D}_s$ defined in Eq. (4.2-18),* these invariants being defined as the coefficients appearing in the quadratic equation

$$\det(\mathbf{D}_s - D_s\mathbf{I}_s) = 0 = D_s^{\,2} - \mathrm{I}_s\, D_s + \mathrm{II}_s \tag{4.3-1}$$

governing the pair of eigenvalues D_{s1} and D_{s2} of the symmetric dyadic $\mathbf{D}_s$ (see Question 4.14). The first invariant, namely

$$\mathrm{I}_s \overset{\text{def}}{=} \operatorname{tr}\mathbf{D}_s \equiv \mathbf{I}_s\!:\mathbf{D}_s, \tag{4.3-2}$$

corresponds to the rate of surface dilatation, whereas the second invariant,

$$\mathrm{II}_s \overset{\text{def}}{=} \det\mathbf{D}_s \equiv \tfrac{1}{2}\left[(\mathbf{I}_s\!:\mathbf{D}_s)^2 - \mathbf{I}_s\!:(\mathbf{D}_s\cdot\mathbf{D}_s)\right], \tag{4.3-3}$$

is related to the rate of surface shear. In particular, when $\mathrm{I}_s = 0$ the magnitude $\dot{\gamma}$ of the shear rate is given by $\dot{\gamma} = \sqrt{\mathrm{II}_s}$ (see Question 4.13).

---

* This is analogous to the comparable dependence of the bulk-phase shear viscosity μ upon the three principal invariants (I, II, III) of the three-dimensional deformation-rate dyadic $\mathbf{D}$, defined in Eq. (4.1-16) [see, e.g., Bird *et al.* 1960)].

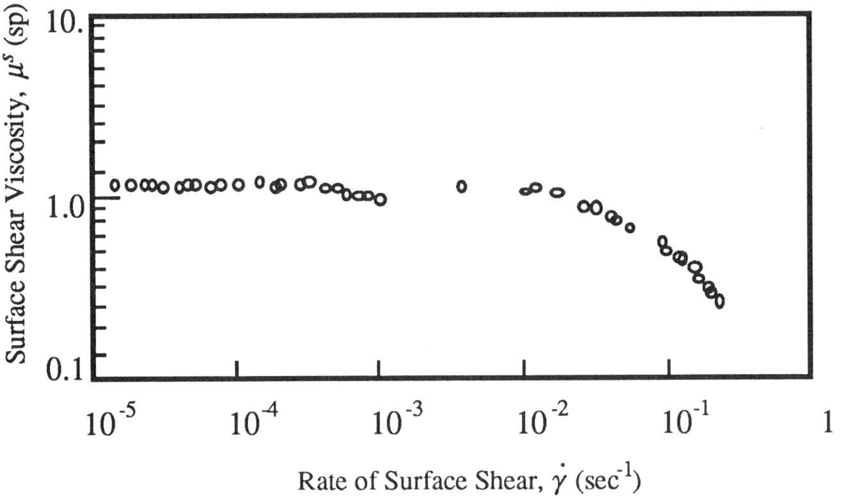

Figure 4.3-1 Apparent interfacial shear viscosity as a function of the rate o interfacial shear at the interface between air and a 0.1% aqueous solution of sodium dodecy sulfate. Data are taken from Jiang *et al.* (1983).

One simple nonlinear model (Hedge & Slattery 1971b) of interfacia rheological behavior seeks a generalization of the Boussinesq-Scriven mode (4.2-17) by allowing a functional dependence of interfacial shear anc dilatational viscosities upon the two principal invariants of the surface deformation tensor, namely,

$$\mu^s = \mu^s(\mathrm{I}_s, \mathrm{II}_s) \tag{4.3-4}$$

and

$$\kappa^s = \kappa^s(\mathrm{I}_s, \mathrm{II}_s). \tag{4.3-5}$$

An explicit functional form for such nonlinear deformation-rate dependence has been proposed by Pintar *et al.* (1971). Their model is the two-dimensional analog of the Powell-Erying model for bulk fluids [Bird *e al.* (1960)]. In particular, the following dependence was postulated for the material parameter μ^s:

$$\mu^s = \mu_0^s + \frac{\alpha \sinh^{-1}\left(\beta\sqrt{\mathrm{II}_s}\right)}{\sqrt{\mathrm{II}_s}}, \tag{4.3-6}$$

where α and β are material constants. This relation has been shown to appl at low interfacial shear rates to certain polymeric-surfactant solutions, a illustrated in Figure 4.3-2.

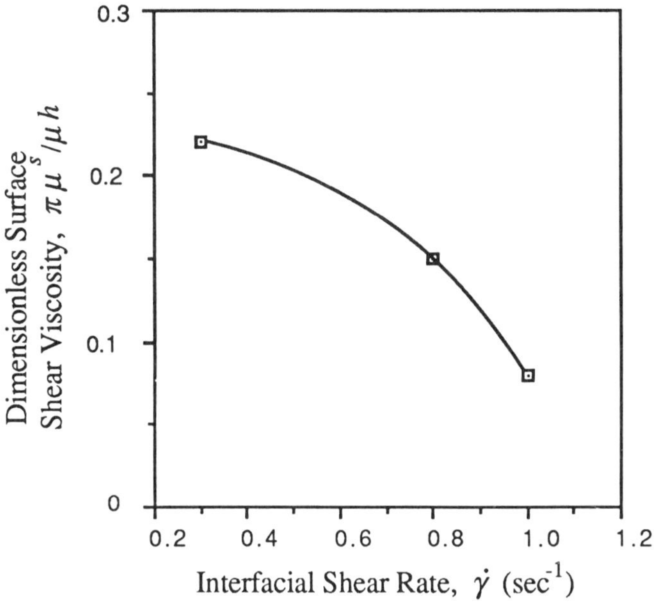

Figure 4.3-2 Surface shear viscosity of a 1 % solution of Duponol RA in 1 % carboxymethyl cellulose (CMC) solution as a function of shear rate. The solid line represents the predictions of the Powell-Erying model. Data are taken from Pintar *et al.* (1971).

The 'network-rupture' model, discussed at the end of this section, is a second and more comprehensive model; it predicts an explicit dependence of both the interfacial rheological properties μ^s and κ^s upon the rate of surface deformation.

Bingham Plastic Model

The development of a plastic-like film at an aged fluid interface has been observed for certain bulk solutions. Such an interfacial 'film' exhibits both viscous and yield properties.

The Bingham-fluid model of interfacial rheological behavior (Mannheimer & Schechter 1967) poses the existence of a (scalar) yield stress τ_0^s, say. A single Cartesian component of the interfacial stress tensor $\boldsymbol{\tau}^s$ would, according to this model, be represented as

$$\tau_{xy}^s = \mu^s \frac{\partial v_x^o}{\partial y} \pm \tau_0^s \qquad \left(\left|\tau_{xy}^s\right| > \tau_0^s\right),$$

$$\frac{\partial v_x^o}{\partial y} = 0 \qquad \left(\left|\tau_{xy}^s\right| < \tau_0^s\right). \tag{4.3-7}$$

This model has been employed successfully by Mannheimer & Schechter (1970) to account for the rheological behavior of highly viscous surfactant solutions.

Linear Viscoelastic Models

The interfacial adsorption of proteins (or other large, flexible surfactant molecules) may result in a dependence of the interfacial stress tensor τ^s upon the deformational *history* of the interface [in addition to its (viscous) dependence upon the *instantaneous* interfacial deformation rate $\mathbf{D}_s$]. Such interfacial rheological behavior is accounted for by the existence of shear and dilatational *elasticities*. [As further discussed in Chapters 5 and 6, interfacial dilatational elasticity can arise from interfacial-tension gradients. Though such elastic behavior does not necessarily violate the Newtonian model (4.2-17) of interfacial rheological behavior, it is nevertheless included briefly in the following discussion since it constitutes an important example of *interfacial viscoelasticity*.]

Shear and dilatational elasticity effects are often observed in practice in circumstances for which the interfacial deformation rate $\mathbf{D}_s$ at a surfactant-adsorbed interface is time-dependent. The simplest example of this phenomenon arises when a sinusoidally oscillating shear or dilatational field is imposed upon a linear viscoelastic interface, whose dynamic response may be expressed in terms of two independent coefficients (Ferry 1958).

The *dilatational modulus* E^* may be represented for a sinusoidally oscillating surface dilatation [$\mathrm{I}_s \equiv \Delta_0 \exp(i\omega t)$] as [cf. Eq. (5.6-60)]

$$E^* \overset{\text{def}}{=} \frac{d\sigma}{d\ln\ \delta A} \equiv E' + i\ E''. \qquad (4.3\text{-}8)$$

Here, E' and E'' respectively represent 'in-phase' and 'out-of-phase' components (the former being known as the *dilatational elasticity*); ω is the frequency of oscillation and $i = \sqrt{-1}$ is the imaginary unit.*

In circumstances wherein interfacial expansion occurs very slowly, the dilatational elasticity E' is often called the *Gibbs* elasticity. [See, however, Eq. (5.5-6) and Question 5.7 for the definition of the Gibbs elasticity used in this text.] Conversely, when measured at very high rates of interfacial expansion, the dilatational elasticity is called the *Marangoni* elasticity .

---

* As discussed in §8.1, E'' represents the apparently 'dissipative' contribution to the dilatational modulus [owing to its being the purely imaginary contribution]. As such, a direct identification of E'' with the interfacial dilatational viscosity is often made [see the introductory discussion of Chapter 8] by assuming that $E''= \omega\kappa^S$. This formula is, however, incorrect inasmuch as E'' embodies the relaxational effect of surfactant adsorption/desorption processes upon the thermodynamic interfacial tension σ.

Likewise, for a sinusoidally oscillating shear field [$\dot{\gamma} \equiv \dot{\gamma}_0 \exp(i\omega t)$], the interfacial shear viscosity of a linear viscoelastic interface may be expressed as the complex quantity

$$(\mu^s)^* = \mu_s' - \frac{i\,G_s}{\omega}, \tag{4.3-10}$$

where μ_s' is termed the *dynamic shear viscosity* and G_s the *shear elasticity*.

A model of linear viscoelasticity, useful in predicting the parameters μ_s' and G_s, is provided by the *Maxwell-Voight surface viscoelasticity model.* It postulates the functional form

$$(\mu^s)^* = \frac{\mu_s''}{\left[1 + \dfrac{(\mu_s''/\eta_s)(\omega\tau_o)^2}{1+(\omega\tau_o)^2} + i\left\{\omega t_o + \dfrac{(\mu_s''/\eta_s)(\omega\tau_o)^2}{1+(\omega\tau_o)^2}\right\}\right]}. \tag{4.3-11}$$

A mechanical interpretation of the various parameters appearing in this equation is given in Figure 4.3-3. [Note that Eq. (4.3-11) applies only in the case of a sinusoidally oscillating interfacial shear field.] The two characteristic time constants appearing in Eq. (4.3-11) are defined as follows: (i) the characteristic *retardation time* of the Voight element,

$$\tau_o = \frac{\eta_s}{E_s}; \tag{4.3-12}$$

and (ii) the characteristic *relaxation time* of the Maxwell element,

$$t_o = \frac{\mu_s''}{G_s''}. \tag{4.3-13}$$

In these expressions the parameters G_s'' and E_s are respectively the *instantaneous* and *retarded* shear elasticities; μ_s'' and η_s are respectively the dynamic *Maxwell shear viscosity* and internal *Voight shear viscosity.*

The Maxwell-Voight model represents the two-dimensional analog of a three-dimensional model (Turner & Gurnee 1958) of viscoelasticity, combining both the *Maxwell* and *Voight* models of (bulk-fluid) elasticity. Both models employ springs (corresponding to ideal Hookean elasticity) and dashpots (corresponding to ideal Newtonian viscosity) to represent viscoelastic behavior, with the Maxwell model (spring and dashpot in *series*) representing the ideal behavior of a viscoelastic liquid and the Voight model (spring and dashpot in *parallel*) representing the ideal response of a viscoelastic solid. This two-dimensional model has been used to characterize the non-Newtonian interfacial rheological behavior of various interfacial surfactant films, as is further discussed in §9.2.

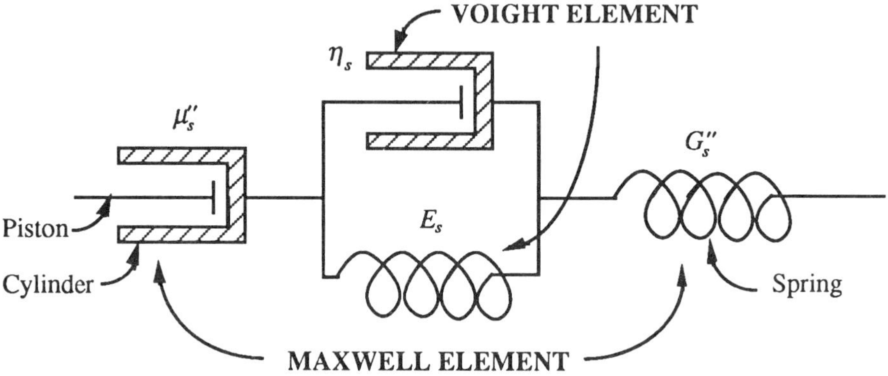

Figure 4.3-3 Spring-dashpot elements of the Maxwell-Voight model.

Network Rupture Model of Non-Newtonian Interfacial Behavior Both nonlinear and viscoelastic behavior (including dilatational elasticity effects) are incorporated into the network rupture model of a fluid interface, proposed by Gardner *et al.* (1978) by analogy with the comparable three-dimensional network rupture model.

At low rates of interfacial deformation this model may be expressed for an incompressible interface as

$$\tau^s = \int_0^\infty m(t)\mathbf{G}(t)dt, \tag{4.3-14}$$

with $m(t)$ a time-dependent material function, t the time of deformation, and the dyadic **G** an interfacial strain tensor [the two–dimensional analog of the three-dimensional Cauchy strain tensor; see Bird *et al.* (1977)]. Gardner *et al.* (1978) suggest the following constitutive form for $m(t)$:

$$m(t) = \frac{\delta(t)}{t}\eta_\infty + \sum_{n=1}^{\infty}\frac{\mu_n}{\lambda_n^2}\exp(-t/\lambda_n) \;, \tag{4.3-15}$$

$$\mu_n = (\eta_0 - \eta_\infty)\frac{\left(\dfrac{2}{n+1}\right)^\alpha}{\sum_{p=1}^{\infty}\left(\dfrac{2}{p+1}\right)^\alpha} \tag{4.3-16}$$

and

$$\lambda_n = \lambda\left(\frac{2}{n+1}\right)^\beta . \tag{4.3-17}$$

The parameters α, β, λ, η_0 and η_∞ are five (of the six) intrinsic parameters of the model, each being dependent upon only the physicochemical nature of the interface.

For the case of simple surface shear, this "memory–dependent" interfacial rheological model yields

$$\mu^s(\dot{\gamma}) = \eta_\infty + \sum_{n=1}^{\infty} \mu_n \left\{ 1 - \left(1 + \frac{B}{\lambda_n \dot{\gamma}}\right) \exp\left(-\frac{\beta}{\lambda_n \dot{\gamma}}\right) \right\} \qquad (4.3\text{-}18)$$

for the shear-rate dependent surface shear viscosity μ^s, where B is the sixth coefficient of the model. Similarly, for the case of oscillating shear flows, we obtain

$$\mu_s'(\omega) = \eta_\infty + \sum_{n=1}^{\infty} \frac{\mu_n}{1 + \lambda_n^2 \omega^2} \qquad (4.3\text{-}19)$$

and

$$G_s(\omega) = \omega \sum_{n=1}^{\infty} \frac{\mu_n \lambda_n \omega}{1 + \lambda_n^2 \omega^2} \qquad (4.3\text{-}20)$$

for the dynamic viscoelastic properties defined in Eq. (4.3-10).

The network rupture model also predicts dilatational elasticity effects; however, an explicit model, such as is provided by Eqs. (4.3-15) to (4.3-17) for the viscous contribution to the stress, does not yet exist. At large rates of interfacial deformation the model predicts 'network rupture', in the sense that the interface is no longer capable of permanently resisting flow.

This model has been used by Gardner *et al.* (1978) in attempting to account for the observed viscoelastic behavior of polymeric monolayers adsorbed to an air/water surface.

4.4 Examples

Use of the (Newtonian) interfacial stress boundary condition (4.2-20) is illustrated in this section by solving several interfacial hydrodynamic examples. Case II of example 4 illustrates a non-Newtonian interface problem. Interfacial tension gradients are deliberately avoided in the following examples, as they require simultaneous consideration of interfacial and bulk-phase species transfer. Such phenomena will be addressed and illustrated in Chapter 5.*

* The overbars introduced in §3.6 for bulk-phase field variables that are generally discontinuous across the macroscale fluid interface [e.g. appearing in Eqs. (4.4-20) and

1. Creeping Flow Past an Emulsion Droplet

Imagine as in Figure 4.4-1 an emulsion droplet whose surface exhibits Newtonian rheological behavior, and which sediments with a constant velocity **U** parallel to the gravity field **g** through a quiescent, Newtonian fluid. We suppose the Reynolds number to be very small, i.e.

$$\mathrm{Re} = \frac{2a\rho U}{\mu} << 1, \tag{4.4-1}$$

such that the low-Reynolds number, Stokes, or creeping-flow form of the Navier-Stokes equations (4.1-23) obtains in both the continuous fluid ($a < r < \infty$) and droplet ($0 < r < a$) phases:

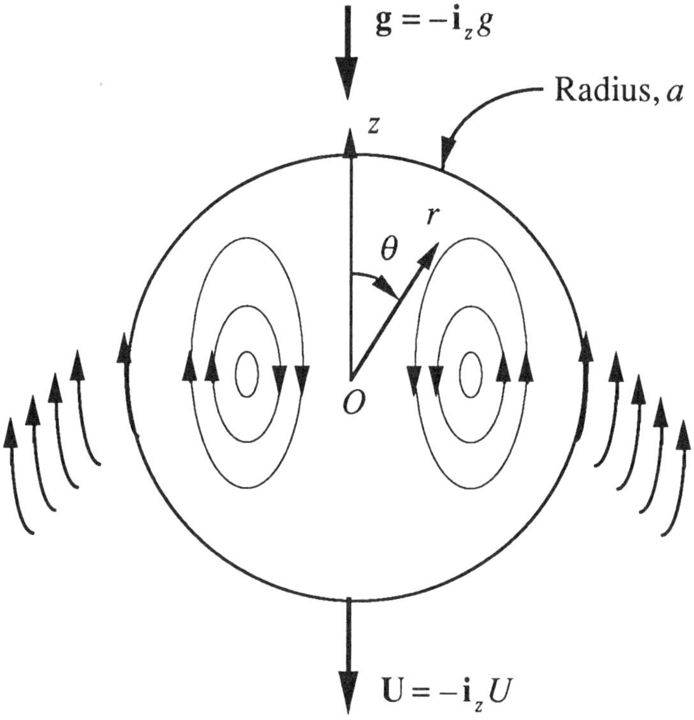

Figure 4.4-1 A cross-sectional meridional-plane view (ϕ = constant) of a spherical droplet settling within a quiescent fluid under the influence of gravity.

$$\mu \nabla^2 \mathbf{v} = \nabla p^* \tag{4.4-2}$$

and

Tables 4.2-1–4.2-3] are omitted for notational simplicity in the examples of this section; moreover, they are also omitted in remaining examples in the text.

$$\tilde{\mu}\nabla^2\tilde{\mathbf{v}} = \nabla\tilde{p}^*, \tag{4.4-3}$$

together with the corresponding equations of continuity (4.1-24) in each phase, namely

$$\nabla \cdot \mathbf{v} = 0 \tag{4.4-4}$$

and

$$\nabla \cdot \tilde{\mathbf{v}} = 0. \tag{4.4-5}$$

In the above, (~) is used to denote droplet-phase variables. The modified pressures

$$p^* = p + \rho gr \cos\theta \tag{4.4-6}$$

and

$$\tilde{p}^* = \tilde{p} + \tilde{\rho} gr \cos\theta \tag{4.4-7}$$

have been introduced so as to combine the hydrostatic pressures ($p_h = \rho gz$, $\tilde{p}_h = \tilde{\rho} gz$) with the thermodynamic pressures (p, $\tilde{p}$). In the representations (4.4-6) and (4.4-7), polar coordinates r and θ are employed for reasons discussed below.

In our solution of this problem we will assume a spherical surface shape *a priori*. This assumption will be confirmed *a posteriori* [cf. Eqs. (4.4-21)–(4.4-23)] by showing that the resulting velocity and pressure fields identically (and fortuitously) satisfy the normal stress boundary condition (4.2-20a) for this particular shape.

Referring to Table 4.1-3, Eqs. (4.4-2)–(4.4-5) may be expressed in spherical polar coordinates as:

$$\mu\left(\nabla^2 v_r - \frac{2}{r^2}v_r - \frac{2}{r^2}\frac{\partial v_\theta}{\partial\theta} - \frac{2}{r^2}\cot\theta\, v_\theta\right) = \frac{\partial p^*}{\partial r}, \tag{4.4-8 a}$$

$$\mu\left(\nabla^2 v_\theta + \frac{2}{r^2}\frac{\partial v_r}{\partial\theta} - \frac{v_\theta}{r^2\sin^2\theta}\right) = \frac{1}{r}\frac{\partial p^*}{\partial\theta}, \tag{4.4-8 b}$$

$$\frac{1}{r^2}\frac{\partial}{\partial r}(r^2 v_r) + \frac{1}{r\sin\theta}\frac{\partial}{\partial\theta}(\sin\theta\, v_\theta) = 0 \tag{4.4-8 c}$$

and

$$\tilde{\mu}\left(\nabla^2 \tilde{v}_r - \frac{2}{r^2}\tilde{v}_r - \frac{2}{r^2}\frac{\partial \tilde{v}_\theta}{\partial\theta} - \frac{2}{r^2}\cot\theta\, \tilde{v}_\theta\right) = \frac{\partial \tilde{p}^*}{\partial r}, \tag{4.4-9 a}$$

$$\tilde{\mu}\left(\nabla^2 \tilde{v}_\theta + \frac{2}{r^2}\frac{\partial \tilde{v}_r}{\partial\theta} - \frac{\tilde{v}_\theta}{r^2\sin^2\theta}\right) = \frac{1}{r}\frac{\partial \tilde{p}^*}{\partial\theta}, \tag{4.4-9 b}$$

$$\frac{1}{r^2}\frac{\partial}{\partial r}(r^2 \tilde{v}_r) + \frac{1}{r\sin\theta}\frac{\partial}{\partial\theta}(\sin\theta\, \tilde{v}_\theta) = 0, \tag{4.4-9 c}$$

upon supposing by symmetry that $v_\phi = 0$ and that the fields v_r, v_θ and p^* are each independent of the ϕ coordinate (along with similar assumptions pertaining to the internal droplet fields). The relevant form of the ϕ-independent Laplacian operator in spherical polar coordinates is

$$\nabla^2 = \frac{1}{r^2}\frac{\partial}{\partial r}\left(r^2\frac{\partial}{\partial r}\right) + \frac{1}{r^2 \sin\theta}\frac{\partial}{\partial\theta}\left(\sin\theta\frac{\partial}{\partial\theta}\right). \tag{4.4-9d}$$

These equations are to be solved subject to the condition that far from the droplet the fluid velocity (relative to the translating droplet center O) is

$$\mathbf{v} \to \mathbf{i}_z U \qquad \text{as} \quad r \to \infty. \tag{4.4-10}$$

In addition, upon utilizing Eq. (3.6-17), the condition of continuity of velocity across the interface requires that

$$v_r = \tilde{v}_r = 0 = v_r^o, \text{ say} \qquad \text{at } r = a, \tag{4.4-11}$$

$$v_\theta = \tilde{v}_\theta \equiv v_\theta^o, \text{ say} \qquad \text{at } r = a. \tag{4.4-12}$$

The tangential stress boundary condition (4.2-20b) follows from Table 4.2-3 as*

$$\tilde{\mu} a\frac{\partial}{\partial r}\left(\frac{\tilde{v}_\theta}{r}\right) - \mu a\frac{\partial}{\partial r}\left(\frac{v_\theta}{r}\right) = \left(\frac{\kappa^s + \mu^s}{a}\right)\frac{\partial}{\partial\theta}\left[\frac{1}{a\sin\theta}\frac{\partial}{\partial\theta}\left(\sin\theta\, v_\theta^o\right)\right]$$

$$+ 2\mu^s\frac{v_\theta^o}{a^2} \qquad \text{at } r = a. \tag{4.4-13}$$

The normal stress boundary condition (4.2-20a) also follows from Table 4.2-3 as

---

* In §5.6, a similar settling droplet problem is considered that includes the hydrodynamic effects of interfacial-tension gradients. Such interfacial tension inhomogeneities develop along the interface through the convective 'sweeping' of surfactant toward the trailing end of the droplet interface, this phenomenon acting in combination with the functional dependence of interfacial tension upon surfactant concentration. The present problem may be regarded as corresponding to the case of small surface Peclèt number ($\text{Pe}_s \equiv Ua/D_s \ll 1$), for which case surface diffusion of surfactant (as quantified by the surface diffusivity D_s) is able to counteract the convective transport of surfactant, thereby maintaining an essentially *uniform* surfactant concentration over the surface of the droplet.

$$-\tilde{p}+2\tilde{\mu}\frac{\partial \tilde{v}_r}{\partial r}+p-2\mu\frac{\partial v_r}{\partial r}=-\frac{2\sigma}{a}$$
$$-\frac{2\kappa^s}{a^2 \sin\theta}\left[\frac{\partial}{\partial\theta}(\sin\theta\, v_\theta^o)\right] \quad \text{at } r=a \qquad (4.4\text{-}14)$$

Owing to the axisymmetric nature of the fluid motion, Eqs. (4.4-8) and (4.4-9) may be solved using a Stokes stream function formulation (Happel & Brenner 1986) for both the internal and external fluid motions. Alternatively, subject to *a posteriori* verification, we suppose the required fields are of the respective forms

$$\tilde{v}_r=-2U\cos\theta\,(A+Br^2), \qquad (4.4\text{-}15)$$

$$\tilde{v}_\theta=2U\sin\theta\,(A+2Br^2), \qquad (4.4\text{-}16)$$

$$v_r=U\cos\theta\left(1-\frac{2E}{r^3}-\frac{2C}{r}\right), \qquad (4.4\text{-}17)$$

$$v_\theta=U\sin\theta\left(-1-\frac{E}{r^3}+\frac{C}{r}\right), \qquad (4.4\text{-}18)$$

together with

$$p^*=K-2\mu U\frac{C}{r^2}\cos\theta, \qquad (4.4\text{-}19)$$

$$\tilde{p}^*=\tilde{K}-20\tilde{\mu}UBr\cos\theta. \qquad (4.4\text{-}20)$$

These velocity fields satisfy the respective creeping-flow and continuity equations as well as the boundary condition (4.4-10). The coefficients A, B, C and E are to be determined by imposing the conditions specified by the remaining boundary conditions (4.4-11)–(4.4-13) on the droplet surface. (See Question 4.4 for the calculation of these coefficients.)

The spherical droplet surface assumption may be verified by showing that the velocity and pressure field solutions provided above satisfy the normal stress boundary condition (4.4-14). Substituting Eqs. (4.4-15)–(4.4-20), (4.4-6) and (4.4-7) into Eq. (4.4-14) furnishes the following algebraic relation:

$$\tilde{K}-K-\frac{2\sigma}{a}+U\cos\theta\left[Ba\left(12\tilde{\mu}+16\frac{\kappa^s}{a}\right)\right.$$
$$\left.-\frac{6\mu C}{a^2}-\frac{12\mu E}{a^4}+\frac{8A\kappa^s}{a^2}+\frac{(\tilde{\rho}-\rho)ga}{U}\right]=0.$$
$$(4.4\text{-}21)$$

As this equation must hold at all surface positions θ, the normal stress boundary condition may be identically satisfied for the spherical droplet shape by separately requiring that

$$\tilde{K} - K - \frac{2\sigma}{a} = 0 \tag{4.4-22}$$

and

$$\frac{(\tilde{\rho} - \rho)ga}{U} - \mu\left[\frac{6C}{a^2} + \frac{12E}{a^4} - \frac{8A\kappa^s}{\mu a^2} - Ba\left(12\frac{\tilde{\mu}}{\mu} + 16\frac{\kappa^s}{\mu a}\right)\right] = 0. \tag{4.4-23}$$

The former equation represents the equilibrium Laplace condition (2.1-3). The latter relation, (4.4-23), may be shown (see Question 4.4) to furnish a relation between the settling speed U of the droplet and the viscous/buoyancy terms (as obtained, alternatively, by balancing the net hydrodynamic drag force acting upon the droplet surface with buoyancy and gravitational forces). From Eq. (4.4-23), together with the explicit algebraic expressions obtained (Question 4.4) for the coefficients A, B, C and E, it follows that

$$U = \frac{2}{9}a^2\frac{(\tilde{\rho} - \rho)}{\mu}g\left[1 + \frac{1}{2}\left(1 + \frac{\kappa^s}{a\mu} + \frac{3}{2}\frac{\tilde{\mu}}{\mu}\right)^{-1}\right]. \tag{4.4-24}$$

The absence of an interfacial shear viscosity term in Eq. (4.4-24) reveals that the interfacial motion in this case is purely dilatational. The absence of a μ^s term may be traced to the fact that the interfacial velocity field v_θ lacks any ϕ dependence.

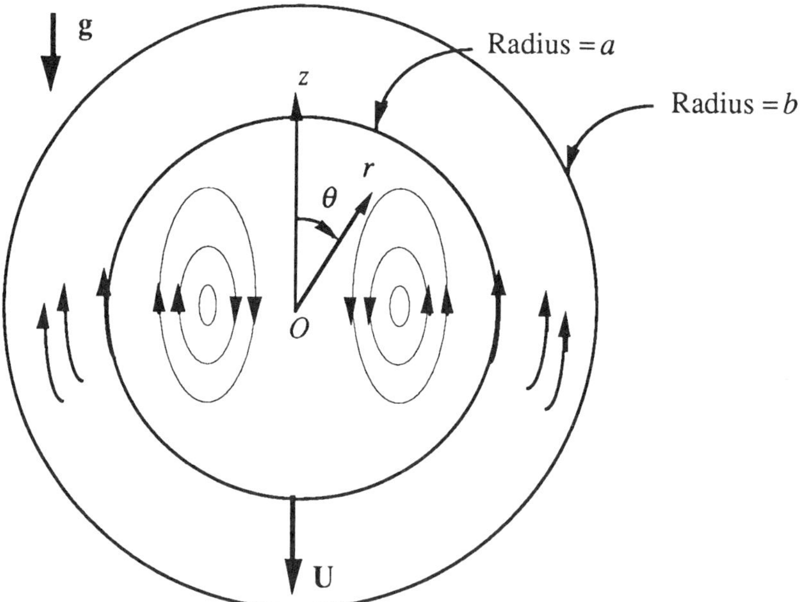

Figure 4.4-2 Cross-sectional meridional-plane view (ϕ = constant) of a spherical droplet of radius a contained within a rigid spherical enclosure of radius b.

Closely related to the above problem is that of a spherical droplet settling quasistatically within a Newtonian fluid contained within a hollow spherical shell, at the instant that the droplet passes the center of the shell, as in Figure 4.4-2. Following arguments comparable to those used to obtain Eq. (4.4-24), it may be shown that (Question 4.5)

$$U = \frac{1}{K_r} U_\infty, \tag{4.4-25}$$

where U_∞ now refers to the velocity of the settling droplet in an unbounded fluid as given by Eq. (4.4-24), and K_r is a dimensionless wall-effect coefficient, the latter being defined by

$$K = H^{-1}\{1 - [1 - 5(3\Lambda - 1)^{-1}]\lambda^5\}, \tag{4.4-26}$$

in which $\lambda = a/b$ and

$$H = 1 - \frac{9}{4}\left(1 - \frac{5}{3}\Lambda^{-1}\right)\lambda + \frac{5}{2}(1 - \Lambda^{-1})\lambda^3 - \frac{9}{4}\left(1 - \frac{5}{3}\Lambda^{-1}\right)\lambda^5 + \frac{5}{2}(1 - 2\Lambda^{-1})\lambda^6, \tag{4.4-27}$$

where

$$\Lambda = \left(1 + \frac{2\kappa^s}{3\mu a} + \frac{\hat{\mu}}{\mu}\right) \tag{4.4-28}$$

2. An Expanding Gas Bubble

As in Figure 4.4-3, a spherical gas bubble is quasistatically formed at the end of a narrow capillary tube in an unbounded, incompressible Newtonian liquid at rest at infinity. Gas flows into the bubble at an instantaneous volumetric flow rate Q (which could vary with time t), causing a sufficiently slow expansion of the gas bubble to permit neglecting inertial effects in the resulting motion of the liquid phase. Moreover, the bubble is assumed sufficiently small such that the continuous-phase hydrostatic pressure is essentially constant over the bubble surface, yet of a radius a much larger than the capillary tube radius r_c. This assures that the fluid motion is not sensibly affected by the tube's presence within the liquid and bubble phases; additionally, it ultimately assures that the hydrostatic pressure remains essentially constant over the bubble surface. As in the preceding example, the *a priori* assumption of a spherical shape for the bubble will be confirmed *a posteriori* [cf. Eq. (4.4-33)]. (See §§6.3 and 8.3 for applications of this problem.)

As a trial solution, we suppose the liquid-phase vector velocity field to be a purely radial quasistatic flow of the form $\mathbf{v} = \mathbf{i}_r v_r(r, t)$, with $\mathbf{i}_r$ a spherical polar unit vector, pointing outward from the center of the bubble.

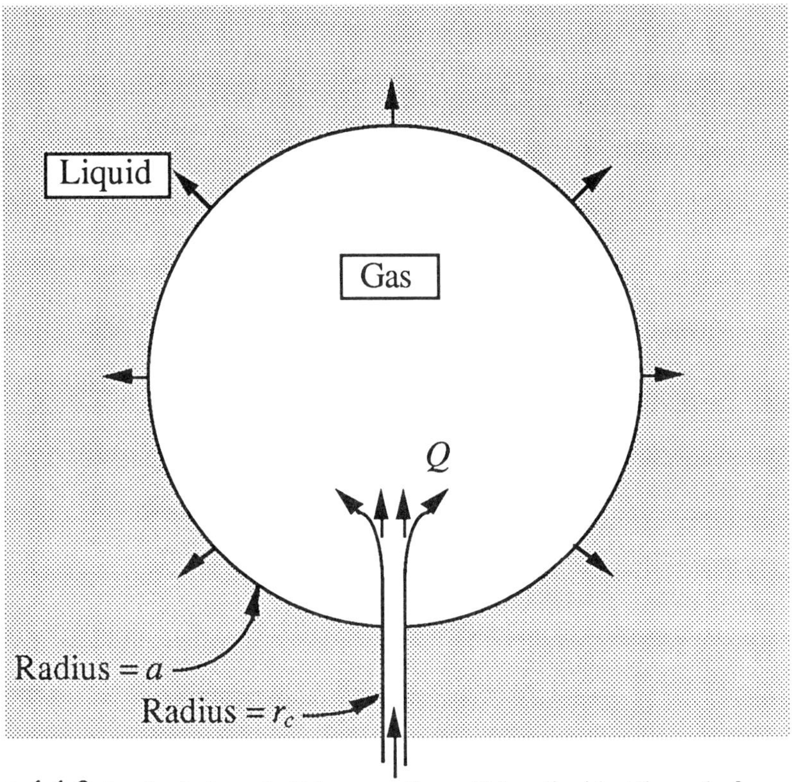

Figure 4.4-3 A spherical gas bubble expanding within a liquid at the end of a narrow capillary tube.

Upon employing the continuity equation in spherical coordinates (Table 4.1-3), it follows that

$$\frac{1}{r^2}\frac{\partial}{\partial r}(r^2 v_r) = 0,$$

or

$$r^2 v_r = f(t), \qquad (4.4\text{-}29)$$

where $f(t)$ will, in general, be a function of time if Q is, and conversely. With $V = 4\pi a^3/3$ the instantaneous volume of the bubble, the (instantaneous) volumetric flow rate $Q=dV/dt$ of gas into the bubble is, upon differentiation

$$Q = 4\pi a^2 \dot{a}, \qquad (4.4\text{-}30\,a)$$

where $\dot{a} = da/dt$. However, by definition of the interfacial velocity, $v_r^o = \dot{a}$, whence

$$Q = 4\pi a^2 v_r^o. \qquad (4.4\text{-}30\,b)$$

Of course, both a and v_r^o are both functions of time, even if Q is itself time independent. The latter equation may be used to evaluate the function appearing in Eq. (4.4-29), thereby yielding

$$v_r = \frac{Q}{4\pi r^2}. \tag{4.4-31}$$

Substitution of the above expression into the radial component of the Navier-Stokes equations in spherical coordinates (Table 4.1-3) shows that when both local and convective inertial terms are neglected, p=$p(t)$, independently of r. (See Question 4.18 for the case in which convective terms are present.)

Substitution of Eq. (4.4-31) into Eq. (4.2-18) yields

$$\mathbf{D}_s = \mathbf{I}_s \frac{Q}{4\pi a^3}$$

for the interfacial rate of deformation dyadic. In conjunction with Eqs. (4.2-15) and (4.2.17), the latter appropriate to a Newtonian interface, it follows that

$$\nabla_s \cdot \mathbf{P}^s \equiv 2H\bar{\sigma}\,\mathbf{n} = -\frac{2}{a}\left(\sigma + \frac{\kappa^s Q}{2\pi a^3}\right)\mathbf{n}. \tag{4.4-32 a}$$

with $H = -1/a$, $\mathbf{n} \equiv \mathbf{i}_r$, and

$$\bar{\sigma} = \sigma + \frac{\kappa^s Q}{2\pi a^3}. \tag{4.4-32 b}$$

Substitution of Eqs. (4.4-32) into the normal stress boundary condition (4.4-20a) (refer to Table 4.2-3) shows that the *a priori* assumption of a spherical droplet shape holds, provided that the equation

$$\tilde{p} = p + \frac{\mu Q}{\pi a^3} + \frac{2\bar{\sigma}}{a}. \tag{4.4-33 a}$$

is satisfied at all points on the bubble surface r=a. This equation relates the bubble- and liquid-phase pressures. Finally, elimination of Q via use of Eq. (4.4-30a) permits this pressure difference to be written in the more intrinsic form

$$\tilde{p} - p - \frac{2\sigma}{a} = 2\mu\,\Delta_s\left(1 + \frac{\kappa^s}{\mu a}\right) \tag{4.4-33 b}$$

wherein [cf. Eq. (3.4-20)]

$$\Delta_s = \frac{1}{A}\frac{dA}{dt} \equiv \frac{2\dot{a}}{a} \tag{4.4-33 c}$$

is the fractional rate of dilatation of the bubble surface, with A=$4\pi a^2$ the instantaneous interfacial area.

3. The Dilatational Viscosity of a Bubbly Liquid

An important application of interfacial rheology pertains to the rheology of emulsions and foams. (Also see example 6 of the present section.) In addition to their role in stabilizing such fluid-fluid dispersions (as discussed at length in Chapters 11 and 12), interfacial rheological properties influence the effective suspension-scale, bulk rheology of dispersions owing to the very large specific surfaces (interfacial area/droplet volume ratios) encountered in these systems. In particular, as is shown by the present example, the (potential) significance of interfacial rheological properties in relation to bulk-phase flows is determined by the following two factors: (i) the degree of surfactant adsorption on the interface; (ii) the magnitude of the specific surface. When surfactant adsorption and specific surface are large, the complete form of the interfacial stress boundary condition [e.g. Eq. (4.2-20) for the case of a Newtonian interface] needs to be applied, thereby including the effects of interfacial rheological properties—namely, interfacial viscosities and elasticities (the latter discussed in §§5.5 and 5.6). Whenever bubble or droplet dispersions meet these two criteria, the contribution of interfacial rheology to the effective bulk rheology of such suspensions is generally found to be qualitatively and quantitatively significant (see also Chapter 14).

As an illustration of the preceding facts, the effective dilatational viscosity of a dilute, surfactant-stabilized, bubbly liquid is calculated below upon using the results of the previous example.*

Consider a volume V of bubbly fluid comprising a dilute suspension ($\phi \equiv V_G/V \ll 1$, where V_G is the total gas-phase volume) of a large number n of radially dilating spherical gas bubbles uniformly dispersed in an incompressible Newtonian liquid continuum of viscosity μ_o.† Expansion of these compressible bubbles is achieved by reducing the pressure on the system as a whole, thereby giving rise to a microscale motion of the otherwise quiescent interstitial liquid. Viewed at the macroscale, the fluid motion of the effective continuum in such circumstances will be a spatially uniform dilatation or expansion $\langle \Delta \rangle$ of a compressible fluid. Thus,

$$\langle \Delta \rangle = \frac{1}{V}\frac{dV}{dt}, \qquad (4.4\text{-}34a)$$

or, equivalently,

---

* Whereas the dilatational viscosity κ of incompressible, three-dimensional Newtonian continua is irrelevant to the fluid mechanics of such systems [since the contribution $\kappa\, \nabla\cdot\mathbf{v}$ to the stress tensor vanishes in Eq. (4.1-15)] the compressibility of gas bubbles within effective, bubbly fluid continua nevertheless causes the latter to behave as compressible (Newtonian) fluids. This effective compressibility renders the effective dilatational viscosity of bubbly liquids and foams of practical interest in applications.

† This problem was first solved by Taylor (1954), who did not include the effects of surface viscosity in his analysis. Later, Prud'homme & Bird (1978) determined the effective dilatational viscosity of the bulk fluid for the case where the continuous phase was non-Newtonian, again ignoring possible interfacial rheological contributions. The present analysis incorporates results first obtained by Edwards (1987).

$$\langle \Delta \rangle = \frac{1}{V}\frac{dV_G}{dt},$$

since the liquid volume remains constant during the expansion.

In the subsequent calculation it proves useful, as in Figure 4.4-4, to subdivide the total bubbly fluid volume V into n identical cells, each containing, on average, a single dilating bubble, such that

$$V = \sum_{i=1}^{n} \tau_o\{i\} = n\tau_o, \qquad (4.4\text{-}34b)$$

where τ_o denotes the superficial volume of each unit cell. This gives,

$$\langle \Delta \rangle = \frac{4\pi a^2 \dot{a}}{\tau_o},$$

which, from Eq. (4.4-30a), yields

$$\langle \Delta \rangle = \frac{Q}{\tau_o}. \qquad (4.4\text{-}34c)$$

A macroscopic pressure tensor $\langle \mathbf{P} \rangle$ of the bubbly fluid may be defined by volume integration of the local pressure tensor $\mathbf{P}$ acting in both bubble and continuous phases, over the volume V, as

$$\langle \mathbf{P} \rangle = \frac{1}{V}\int_V \mathbf{P}\,dV. \qquad (4.4\text{-}35a)$$

As the mean pressure tensor is constant throughout the dispersion, this relation can be expressed in terms of a comparable integration over a unit cell upon noting Eq. (4.4-34b). Thus, Eq. (4.4-35a) may be written as

$$\langle \mathbf{P} \rangle = \frac{1}{V}\sum_{i-1}^{n} \int_{\tau_o\{i\}} \mathbf{P}\,d^3\mathbf{x}, \qquad (4.4\text{-}35b)$$

where $d^3\mathbf{x}$ denotes an element of volume. Owing to the homogeneity of the suspension, the volume integral

$$\int_{\tau_o\{i\}} \mathbf{P}\,d^3\mathbf{x}$$

over cell i will, on average, possess the same value for every i. Thus,

$$\sum_{i=1}^{n} \int_{\tau_o\{i\}} \mathbf{P}\,d^3\mathbf{x} = n\int_{\tau_o} \mathbf{P}\,d^3\mathbf{x}.$$

Substitution of this relation into (Eq. (4.4-35b) thereby yields

$$\langle \mathbf{P} \rangle = \frac{1}{\tau_o}\int_{\tau_o} \mathbf{P}\,d^3\mathbf{x}, \qquad (4.4\text{-}35c)$$

whence the calculation may now be effected by integration over a single unit cell (inclusive of both gas- and liquid-phase volumes),

$$\mathbf{P} = \nabla \cdot (\mathbf{P}\mathbf{x})^{\dagger} - \mathbf{x}(\nabla \cdot \mathbf{P}). \qquad (4.4\text{-}35\text{d})$$

The last term vanishes for quasistatic creeping flows since

$$\nabla \cdot \mathbf{P} = \mathbf{0}.$$

Thus, substitution of Eq. (4.4-35d) into Eq. (4.4-35c), followed by use of the divergence theorem (for a volume with a surface of discontinuity A; cf. Question 3.2), yields

$$\langle \mathbf{P} \rangle = \frac{1}{\tau_o} \oint_{\partial \tau_o} \mathbf{x}\, d\mathbf{S} \cdot \mathbf{P} - \frac{1}{\tau_o} \oint_{A} \mathbf{x}\, d\mathbf{S} \cdot \|\mathbf{P}\|, \qquad (4.4\text{-}35\text{e})$$

where A denotes the bubble surface and the vector $d\mathbf{S}$ represents an outwardly directed area element to $\partial \tau_o$.

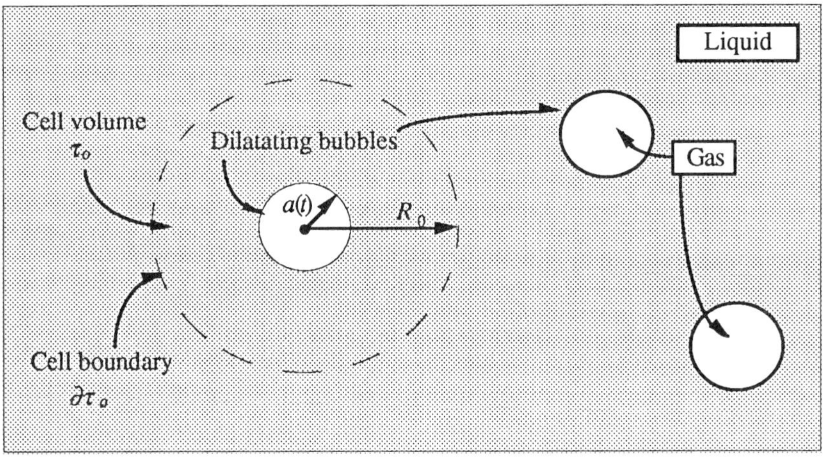

Figure 4.4-4 A dilute bubbly liquid composed of identical spherical gas bubbles. A spherical *cell* is depicted around one of the bubbles, by which the effective bulk rheological properties of the bubbly fluid as a whole may be deduced correctly to the terms of the first order in the volume fraction ϕ of gas bubbles.

Insofar as no external forces are exerted upon the neutrally buoyant dispersion,

$$\oint_{\partial \tau_o} d\mathbf{S} \cdot \mathbf{P} = \oint_{A} d\mathbf{S} \cdot \mathbf{P} = \mathbf{0},$$

(which employs the Stokes relation, $\nabla \cdot \mathbf{P} = \mathbf{0}$) it is immaterial from which origin the position vector $\mathbf{x}$ is measured, since if we replace $\mathbf{x}$ by $\mathbf{x}+\mathbf{c}$ where

c is a constant vector, the value of the integral is unaltered. For convenience we will therefore measure **x** from the center of the cell.

Owing to the symmetry and homogeneity, the global pressure tensor $\langle \mathbf{P} \rangle$ is necessarily isotropic as demonstrated below. Hence, we may write

$$\langle \mathbf{P} \rangle = -\mathbf{I}\langle \bar{p} \rangle, \qquad (4.4\text{-}36)$$

where the scalar $\langle \bar{p} \rangle$ is the mean pressure of the dispersion; namely,

$$\langle \bar{p} \rangle = \langle p \rangle - \langle \kappa \rangle \langle \Delta \rangle . \qquad (4.4\text{-}37)$$

[cf. Eq. (4.1-17)]. Here, $\langle p \rangle$ is the macroscopic thermodynamic pressure of the bubbly fluid and $\langle \kappa \rangle$ the macroscopic dilatational viscosity.

In the dilute limit, for which Eq. (4.4-35) applies, the precise shape of an *average* cell and distribution of bubbles is not a factor in the calculation of the global pressure tensor (hydrodynamic interactions between neighboring bubbles being negligible). Thus, for the purpose of calculating the macroscopic properties $\langle p \rangle$ and $\langle \kappa \rangle$ appearing in Eq. (4.4-37), a spherical cell of radius R_0 is chosen, containing a single bubble at its center (Figure 4.4-4). The volume of this unit cell is given by $\tau_o = 4\pi R_0^3/3$ and the bounding cell surface area by $\partial\tau_o = 4\pi R_0^2$. The vector $d\mathbf{S} = \mathbf{i}_r r^2 \sin\theta \, d\theta \, d\phi$ represents the outward directed area element on the bounding cell surface $\partial\tau_o$, and $\mathbf{x} = \mathbf{i}_r r$ is the position vector measured from the center of the cell.

Thus, Eq. (4.4-35e) becomes

$$\langle \mathbf{P} \rangle = \frac{1}{\tau_o} \oint_{\partial\tau_o} \mathbf{i}_r \mathbf{i}_r P_{rr} r^2 \sin\theta \, d\theta d\phi - \frac{1}{\tau_o} \oint_A \mathbf{i}_r \mathbf{i}_r \| P_{rr} \| r^2 \sin\theta \, d\theta d\phi , \qquad (4.4\text{-}38)$$

from which, upon employing Eqs. (4.4-31) and (4.4-33a), together with the expression for the pressure tensor **P** in spherical coordinates provided in Table 4.1-3, it follows that

$$\langle \mathbf{P} \rangle = \mathbf{I}\left\{ -3\tilde{p} + \frac{6\sigma}{a}(1+\phi) + \frac{4\mu\langle\Delta\rangle}{\phi}\left[1 - \phi + \frac{\kappa^s}{\mu a}(1+\phi)\right]\right\}. \qquad (4.4\text{-}39)$$

Here, $\phi \equiv \frac{4}{3}\pi a^3/\tau_o \ll 1$ is the volume fraction of bubbles.

Comparing Eqs. (4.4-36), (4.4-37) and (4.4-39), upon equating like-powers of Δ, it follows that to leading order in ϕ

$$\langle p \rangle = \tilde{p} - \frac{2\sigma}{a} \qquad (4.4\text{-}40)$$

and

$$\langle \kappa \rangle = \frac{4\mu}{3\phi}\left[1 + \frac{\kappa^s}{\mu a}\right]. \tag{4.4-41}$$

The above relations respectively provide expressions for the global fluid pressure and dilatational viscosity of the dilute bubbly fluid.

An explicit expression for the dependence of the macroscopic pressure (4.4-40) in terms of the radii of bubbles a, may be expressed, for example, in the case when an ideal gas fills the bubbles; whence

$$\langle p \rangle = \tilde{p}_o\left(\frac{a_o}{a}\right)^3 - \frac{2\sigma}{a}, \tag{4.4-42}$$

where $\tilde{p}_o$ is the bubble pressure at a particular (initial, say) bubble radius a_o. It is evident from this relation that the macroscopic thermodynamic pressure of a bubbly fluid continuum possesses a fundamentally different functional nature than that of familiar continua [cf. Eq. (4.1-13a)].

The magnitude of the interfacial viscosity contribution to the dilatational viscosity of the bubbly fluid, Eq. (4.4-41), is measured by a dimensionless group $(\kappa^s/\mu a)$, representing a particular form of the *Boussinesq number*. [Note its previous appearance also in Eqs. (4.4-24) and (4.4-32b).] This dimensionless group represents a particular form of the generic ratio

$$\mathrm{Bo} = \frac{\text{interfacial viscosity}}{\text{bulk viscosity} \times \text{length scale}}. \tag{4.4-43}$$

For Newtonian interfaces, the Boussinesq number appears either in terms of the interfacial dilatational viscosity,

$$\mathrm{Bo}_\kappa \equiv \frac{\kappa^s}{\mu a}, \tag{4.4-44}$$

or the interfacial shear viscosity,

$$\mathrm{Bo}_\mu \equiv \frac{\mu^s}{\mu a}. \tag{4.4-45}$$

The Boussinesq number measures the ratio of interfacial to bulk viscous effects. Thus, when Bo >1, interfacial viscosities strongly influence the nature of the bulk flow near or at the fluid interface.

The Boussinesq number Bo reflects the pair of criteria mentioned previously regarding the importance of interfacial to bulk rheology. In general, these numbers will be larger relative the larger are the surfactant adsorption (which has the potential effect of producing a large interfacial viscosity) and specific surface [corresponding to a small length scale a appearing in Eq. (4.4-43); e.g., a small curvature radius a].

4. The Deep-Channel Surface Shear Viscometer

The deep-channel surface shear viscometer (or viscous-traction canal viscometer) is the most common apparatus for measuring the shear rheological properties of surfactant-adsorbed interfaces. It may be used for both gas-liquid and liquid-liquid interfaces to measure the interfacial shear viscosity, nonlinear interfacial rheological properties, and viscoelastic interfacial properties. Here, we provide solutions for two examples illustrating the experimental modes by which this viscometer functions: (i) the steady rotational flow of a pair of immiscible Newtonian liquids possessing a common Newtonian interface; and (ii) the oscillatory flow of two immiscible Newtonian fluids whose common *non-Newtonian* interface displays Maxwell-Voight rheological constitutive behavior. These two solutions will subsequently prove useful in Chapters 7 and 9 (cf. §§7.1 and 9.1).

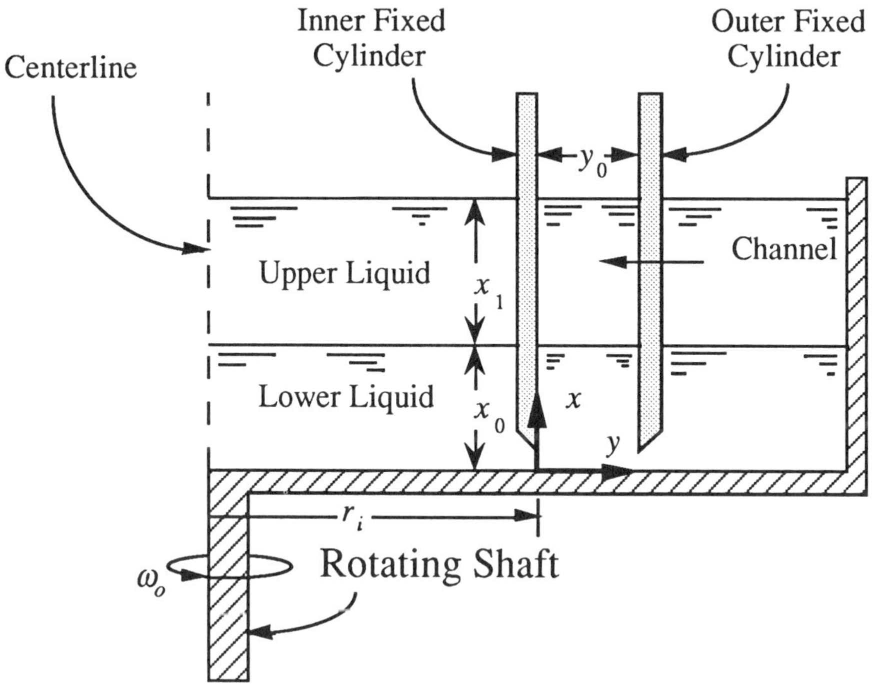

Figure 4.4-5 The deep-channel surface shear viscometer.

Case I: Steady Shear of a Newtonian Interface

Consider the apparatus depicted in Figure 4.4-5. Our analysis of this configuration will closely follow that of Wasan *et al.* (1971). The viscometer is composed of a flat-bottomed dish and two *stationary*, concentric circular cylinders. The cylinders are oriented vertically and placed near, though not touching, the bottom of the dish. The dish is rotated with angular speed ω_o, causing the fluids contained in the channel between the cylinders to rotate. The presence of the stationary inner and outer cylinders creates a shearing motion within the rotating fluid. On the assumption that $y_0/r_i \ll 1$, we suppose that the flow in between the fixed circular cylinder (i.e. in the 'channel') may be approximated by the comparable flow in a rectangular channel, at the same time neglecting the existence of the gap between the base of the channel walls and the bottom of the rotating dish. The first assumption is shown valid [cf. Eq. (7.1-3)] so long as the indicated inequality obtains, whereas the second assumption is valid for gaps much smaller than the height x_0 of the fluid-fluid interface above the bottom [cf. Eq. (7.1-4)]. We also assume the two interfaces to be planar, despite the existence of small curvatures near the channel walls due to wettability effects. Such curvature effects will be elucidated in §7.1.

Referring to Table 4.1-1, the Navier-Stokes equations may be expressed in the present case as

$$\frac{\partial^2 v}{\partial x^2} + \frac{\partial^2 v}{\partial y^2} = 0 \tag{4.4-46}$$

and

$$\frac{\partial^2 \tilde{v}}{\partial x^2} + \frac{\partial^2 \tilde{v}}{\partial y^2} = 0, \tag{4.4-47}$$

with v the z-component of the vector velocity field (with the z axis directed into the plane of the paper in Figure 4.4-4), and wherein the overbar identifies the lower phase. Inasmuch as $v \equiv v(x, y)$, the continuity equation in Table 4.1-1 is automatically satisfied.

The appropriate boundary conditions are

$$v = \tilde{v} = 0 \qquad \text{at } y = 0, \tag{4.4-48}$$

$$v = \tilde{v} = 0 \qquad \text{at } y = y_0, \tag{4.4-49}$$

$$\tilde{v} = (r_i + y)\omega_o \qquad \text{at } x = 0, \tag{4.4-50}$$

$$v = \tilde{v} \qquad \text{at } x = x_0, \tag{4.4-51}$$

$$\mu\frac{\partial v}{\partial x} = \mu^s \frac{\partial^2 v}{\partial y^2} \qquad \text{at } x = x_0 + x_1, \tag{4.4-52}$$

and

$$\tilde{\mu}\frac{\partial \tilde{v}}{\partial x} - \mu\frac{\partial v}{\partial x} = \tilde{\mu}^s \frac{\partial^2 v}{\partial y^2} \qquad \text{at } x = x_0. \tag{4.4-53}$$

The latter two conditions follow from Table 4.2-2 (upon paying attention to the change in the coordinates).

General solutions of Eqs. (4.4-46) and (4.4-47) may be obtained upon using separation of variables, yielding

$$v = \sum_{n=1}^{\infty}[a_n \cosh(\lambda_n x)\sin(\lambda_n y) + b_n \sinh(\lambda_n x)\sin(\lambda_n y)$$
$$+ c_n \cosh(\lambda_n x)\cos(\lambda_n y) + d_n \sinh(\lambda_n x)\cos(\lambda_n y)] \tag{4.4-54}$$

and

$$\tilde{v} = \sum_{n=1}^{\infty}[\tilde{a}_n \cosh(\tilde{\lambda}_n x)\sin(\tilde{\lambda}_n y) + \tilde{b}_n \sinh(\tilde{\lambda}_n x)\sin(\tilde{\lambda}_n y)$$
$$+ \tilde{c}_n \cosh(\tilde{\lambda}_n x)\cos(\tilde{\lambda}_n y) + \tilde{d}_n \sinh(\tilde{\lambda}_n x)\cos(\tilde{\lambda}_n y)], \tag{4.4-55}$$

where a_n, b_n, c_n, d_n and λ_n, together with the corresponding lower-phase (~) constants, are to be determined through application of the boundary conditions.

Condition (4.4-48) requires that

$$c_n = \tilde{c}_n = d_n = \tilde{d}_n = 0, \tag{4.4-56}$$

whereas Eq. (4.4-49) requires that

$$\lambda_n = \tilde{\lambda}_n = \frac{n\pi}{y_0} \tag{4.4-57}$$

in order to obtain nontrivial solutions. Condition (4.4-50) necessitates the equality

$$(r_i + y)\omega_o = \sum_{i=1}^{\infty} \tilde{a}_n \sin\left(\frac{n\pi y}{y_0}\right) \tag{4.4-58}$$

for all y in the range $0<y<y_0$. Integration, followed by use of the appropriate orthogonality properties of trigonometric functions, yields

$$\int_0^{y_0} (r_i + y)\omega_o \sin\left(\frac{m\pi y}{y_0}\right)dy = \begin{cases} 0 & (n \neq m) \\ \dfrac{\tilde{a}_n y_0}{2} & (n = m). \end{cases} \tag{4.4-59}$$

Upon performing the requisite integration on the left-hand side of the above we thereby obtain

$$\tilde{a}_n = \begin{cases} -\dfrac{2\omega_o y_0}{n\pi} & (n\ even) \\ \dfrac{4\omega_o(r_i + \frac{1}{2}y_0)}{n\pi} = \dfrac{4v_b}{n\pi} & (n\ odd), \end{cases} \tag{4.4-60}$$

where v_b is the midchannel velocity of the rotating dish.

Conditions (4.4-51)–(4.4-53) may be imposed simultaneously to determine the final three coefficients as

$$\tilde{b}_n = -\frac{\tilde{a}_n}{E_n}\{[\tilde{\mu}\sinh(n\pi D) + (\tilde{\mu}^s n\pi / y_0)\cosh n\pi D]$$
$$\times [\mu \cosh(n\pi D_1) + (\mu^s n\pi / y_0)\sinh(n\pi D_1)]$$
$$+ \mu\cosh(n\pi D)[\mu\sinh(n\pi D_1) + (\mu^s n\pi / y_0)\cosh(n\pi D_1)]\}, \tag{4.4-61}$$

$$a_n = \frac{\tilde{a}_n}{E_n}\tilde{\mu}\{m\cosh[n\pi(D + D_1)]$$
$$+ (\mu^s n\pi / y_0)\sinh[n\pi(D + D_1)]\} \tag{4.4-62}$$

and

$$b_n = -\frac{\tilde{a}_n}{E_n}\tilde{\mu}\{\mu\sinh[n\pi(D + D_1)]$$
$$+ (\mu^s n\pi / y_0)\cosh[n\pi(D + D_1)]\}, \tag{4.4-63}$$

where

$$E_n = [\tilde{\mu}\cosh n\pi D + (\tilde{\mu}^s n\pi / y_0)\sinh(n\pi D)]$$
$$\times [\mu\cosh n\pi D_1 + (\mu^s n\pi / y_0)\sinh(n\pi D_1)]$$
$$+ \mu\sinh(n\pi D)[\mu\sinh(n\pi D_1) + (\mu^s n\pi / y_0)\cosh(n\pi D_1)]. \tag{4.4-64}$$

Here $D=x_0/y_0$ and $D_1=x_1/y_0$.

Finally, then

$$v = \sum_{n=1}^{\infty}\tilde{a}_n \sin(n\pi y / y_0)\tilde{\mu}$$
$$\times \{\mu\cosh[n\pi(x - x_0 - x_1) / y_0]$$
$$- (\mu^s n\pi / y_0)\sinh[n\pi(x - x_0 - x_1) / y_0]\}$$
$$\div \{[\tilde{\mu}\cosh(n\pi D) + (\tilde{\mu}^s n\pi / y_0)\sinh(n\pi D)]$$
$$\times [\mu\cosh(n\pi D_1) + (\mu^s n\pi / y_0)\sinh(n\pi D_1)]$$
$$+ \mu\sinh n\pi D[\mu\sinh(n\pi D_1) + (\mu^s n\pi / y_0)\cosh(n\pi D_1)]\} \tag{4.4-65}$$

and

$$
\begin{aligned}
\tilde{v} = \sum_{n=1}^{\infty} \tilde{a}_n \sin(n\pi y / y_0) \\
&\times \{ \tilde{\mu}[\mu \cosh(n\pi D_1) + (\mu^s n\pi / y_0)\sinh(n\pi D_1)] \\
&\times \cosh[n\pi (x - x_0) / y_0] \\
&- \mu[\mu \sinh(n\pi D_1) + (\mu^s n\pi / y_0)\cosh(n\pi D_1)] \\
&\times \sinh[n\pi (x - x_0) / y_0] \\
&- (\tilde{\mu}_s n\pi / y_0)[\mu \cosh(n\pi D_1) + (\mu^s n\pi / y_0)\sinh(n\pi D_1)] \\
&\times \sinh[n\pi (x - x_0) / y_0]\} \\
&\div \{[\tilde{\mu} \cosh(n\pi D) + (\tilde{\mu}_s n\pi / y_0)\sinh n\pi D] \\
&\times [\mu \cosh(n\pi D_1) + (\mu^s n\pi / y_0)\sinh(n\pi D_1)] \\
&+ \mu \sinh(n\pi D)[\mu \sinh(n\pi D_1) + (\mu^s n\pi / y_0)\cosh(n\pi D_1)]\}. \quad (4.4\text{-}66)
\end{aligned}
$$

These expressions furnish the bulk-fluid velocity solutions as functions of both bulk-fluid and interfacial viscous and geometrical parameters.

The deep-channel approximation, which requires satisfaction of the inequalities $D>>1$ and $D_1>>1$[for $D>2/\pi$ and $D_1>2/\pi$ all but the first terms (i.e. n=1) in the above two series expansions may be neglected to within an error of less than 0.5% (Osborne 1968)], ultimately yields the following pair of relations for the interfacial shear viscosities at the respective gas-liquid and liquid-liquid interfaces:

$$
\frac{\mu^s \pi}{\mu y_0} = \frac{\tilde{v}_c / v_c}{\sinh(\pi D_1)} - 1 \qquad (4.4\text{-}67)
$$

and

$$
\frac{\tilde{\mu}^s \pi}{(\mu + \tilde{\mu}) y_0} = \frac{\tilde{v}_c^* - \tilde{v}_c}{\tilde{v}_c}. \qquad (4.4\text{-}68)
$$

Here, v_c and $\tilde{v}_c$ are the respective mid-channel velocities of the liquid-gas and liquid-liquid interfaces, whereas $\tilde{v}_c^*$ is the comparable velocity of the hypothetical surfactant-free liquid-liquid interface. Experimental measurement of these mid-channel velocities thereby allows the interfacial viscosities to be determined from Eqs. (4.4-67) and (4.4-68).

The following simple relation between the depth D of the liquid/liquid interface follows from inspection of the leading-order (n=1) terms of Eq. (4.4-66):

$$\frac{\tilde{v}_c}{v_b} \sim e^{-\pi / D}. \tag{4.4-69}$$

Insofar as the interfacial shear rate depends upon the height D of the liquid/liquid interface above the bottom dish, a semilog plot of $\tilde{v}_c / v_b$ versus D will yield a straight line of slope π provided that the assumptions of the analysis hold. Thus, in circumstances wherein the interfacial shear viscosity depends upon the rate of interfacial shear, a behavior deviating from Eq. (4.4-69) will occur. Other more direct methods of detecting nonlinear interfacial rheological behavior with the deep-channel surface viscometer are discussed in §9.1.

Case II. Oscillating Shear of a Maxwell-Voight Interface

Return to the schematic diagram of Figure 4.4-5, imagining now that the upper interface at $x=x_0+x_1$ is held stationary while simultaneously the bottom dish is oscillated with frequency ω, such that

$$v_b = \bar{v}_b \sin(\omega t). \tag{4.4-70}$$

The fluid interface, which in the absence of inertial effects in the bulk phase would oscillate *in phase* with the dish were the interface Newtonian, now exhibits a phase lead owing to shear viscoelastic behavior. Explicitly, the constitutive behavior of this interfacial viscoelasticity will be assumed to satisfy the Maxwell-Voight rheological model discussed in §4.3.

The subsequent analysis, which adheres closely to that of Mannheimer & Schechter (1970c), and in which the notation of the previous problem is maintained, will be performed in the complex plane. The complex time-dependent fluid velocity v^* may be expressed as

$$v^* = v \exp(i \omega t). \tag{4.4-71}$$

Using Eq. (4.4-51), the Navier-Stokes equations become [the problem statement closely parallels Eqs. (4.4.46)–(4.4-53)]

$$i\omega\rho v = \mu\left(\frac{\partial^2 v}{\partial x^2} + \frac{\partial^2 v}{\partial y^2}\right), \tag{4.4-72}$$

$$i\omega\tilde{\rho}\tilde{v} = \tilde{\mu}\left(\frac{\partial^2 \tilde{v}}{\partial x^2} + \frac{\partial^2 \tilde{v}}{\partial y^2}\right), \tag{4.4-73}$$

subject to the following boundary conditions:

$$v = \tilde{v} = 0 \qquad \text{at } y = 0,$$

$$v = \tilde{v} = 0 \qquad \text{at } y = y_0, \tag{4.4-75}$$

$$\tilde{v} = \tilde{v}_b \qquad \text{at } x = 0, \qquad (4.4\text{-}76)$$

$$v = \tilde{v} \qquad \text{at } x = x_0, \qquad (4.4\text{-}77)$$

$$v = 0 \qquad \text{at } x = x_0 + x_1 \qquad (4.4\text{-}78)$$

and

$$\tilde{\mu}\frac{\partial \tilde{v}}{dx} - \mu\frac{\partial v}{dx} = (\mu^s)^* \frac{\partial^2 v}{\partial y^2} \qquad \text{at } x = x_0. \qquad (4.4\text{-}79)$$

The final two conditions differ from Eqs. (4.4-52) and (4.4-53) by the now rigid nature of the upper surface and by the existence of a complex interfacial viscosity $(\mu^s)^*$, the latter being defined in Eq. (4.3-11). Equation (4.4-76) furnishes the oscillating bottom-dish condition.

The solution of the preceding system of equations proceeds similarly to the previous case. Thus, upon separating variables in Eqs. (4.4-72) and (4.4-73), and subsequently imposing the first two boundary conditions, Eqs. (4.4-74) and (4.4-75), one obtains

$$v = \sum_{n=1}^{\infty}\left[a_n \cosh\left(\frac{\phi_n x}{y_0}\right) + b_n \sinh\left(\frac{\phi_n x}{y_0}\right)\right]\sin\left[\frac{(2n-1)\pi y}{y_0}\right] \qquad (4.4\text{-}80)$$

and

$$\tilde{v} = \sum_{n=1}^{\infty}\left[\tilde{a}_n \cosh\left(\frac{\tilde{\phi}_n x}{y_0}\right) + \tilde{b}_n \sinh\left(\frac{\tilde{\phi}_n x}{y_0}\right)\right]\sin\left[\frac{(2n-1)\pi y}{y_0}\right], \qquad (4.4\text{-}81)$$

with complex eigenvalues

$$\phi_n = \sqrt{(2n-1)^2\pi^2 + i\frac{\omega\rho y_0^2}{\mu}} \qquad (4.4\text{-}82)$$

and

$$\tilde{\phi}_n = \sqrt{(2n-1)^2\pi^2 + i\frac{\omega\tilde{\rho} y_0^2}{\tilde{\mu}}}. \qquad (4.4\text{-}83)$$

Upon using Eq. (4.4-78), the upper-phase solution becomes

$$v = \sum_{n=1}^{\infty} b_n\left[\sinh\left(\frac{\phi_n x}{y_0}\right) - \tanh(\phi_n D_1)\cosh\left(\frac{\phi_n x}{y_0}\right)\right]\sin\left[\frac{(2n-1)\pi y}{y_0}\right],$$

$$(4.4\text{-}84)$$

whereas, with use of Eq. (4.4-76), we obtain

$$\tilde{v} = \sum_{n=1}^{\infty}\left[\frac{4\bar{v}_b}{(2n-1)\pi}\cosh\left(\frac{\tilde{\phi}_n x}{y_0}\right) + \tilde{b}_n \sinh\left(\frac{\tilde{\phi}_n x}{y_0}\right)\right]\sin\left[\frac{(2n-1)\pi y}{y_0}\right] \tag{4.4-85}$$

for the lower-phase solution. Satisfaction of Eq. (4.4-77) requires that

$$b_n = \frac{\dfrac{4\bar{v}_b}{(2n-1)\pi}\cosh(\tilde{\phi}_n D) + \tilde{b}_n \sinh(\tilde{\phi}_n D)}{\sinh(\phi_n D) - \tanh(\phi_n D_1)\cosh(\phi_n D)}. \tag{4.4-86}$$

The final constant is determined from Eq. (4.4-79) as

$$\begin{aligned}\tilde{b}_n = &- 4\bar{v}_b\{(2n-1)^2\pi^2(\mu^s)^*\cosh(\tilde{\phi}_n D) + \tilde{\phi}_n \sinh(\tilde{\phi}_n D)\\ &+ (\mu/\tilde{\mu})\phi_n \coth[\phi_n(D_1 - D)]\cosh(\tilde{\phi}_n D)\}\\ &\div (2n-1)\pi\{(2n-1)^2\pi^2(\mu^s)^*\sinh(\tilde{\phi}_n D) + \tilde{\phi}_n \cosh(\tilde{\phi}_n D)\\ &+ (\mu/\tilde{\mu})\phi_n \coth[\phi_n(D_1 - D)]\sinh(\tilde{\phi}_n D)\}.\end{aligned} \tag{4.4-87}$$

The complete velocity fields are now given by Eqs. (4.4-84) and (4.4-85). At the fluid interface $x = x_0$, the time-independent velocity field adopts the form

$$\begin{aligned}v^o = &\frac{4\bar{v}_b}{\pi}\sum_{n=1}^{\infty}\sin\frac{(2n-1)\pi y}{y_0}\\ &\div (2n-1)^2\{(2n-1)^2\pi^2(\mu^s)^* + \cosh(\tilde{\phi}_n D)\\ &+ (\mu/\tilde{\mu})\phi_n \coth[\phi_n(D_1 - D)][\sinh(\tilde{\phi}_n D)/\tilde{\phi}_n]\},\end{aligned} \tag{4.4-88}$$

which may be expressed in polar form as

$$\frac{v^o}{\bar{v}_b} = \left|\frac{v^o}{\bar{v}_b}\right|\exp(i\theta), \tag{4.4-89}$$

wherein

$$\left|\frac{v^o}{\bar{v}_b}\right| = \sqrt{\mathrm{Re}^2\left(\frac{v^o}{\bar{v}_b}\right) + \mathrm{Im}^2\left(\frac{v^o}{\bar{v}_b}\right)} \tag{4.4-90}$$

and

$$\theta = \tan^{-1}\left[\frac{\mathrm{Im}\left(\dfrac{v^o}{\bar{v}_b}\right)}{\mathrm{Re}\left(\dfrac{v^o}{\bar{v}_b}\right)}\right]. \tag{4.4-91}$$

As will be discussed in §9.2, measurements of the phase angle (4.4-91) and amplitude ratio (4.4-90) allow determination of the four material constants parameterizing the Maxwell-Voight model (4.3-11).

5. An Emulsion Droplet within a Homogeneous Shear Field

Velocity and pressure fields within and around a neutrally buoyant droplet whose interface exhibits Newtonian rheological behavior are determined below for the particular case wherein the emulsion droplet is placed within a homogeneous (undisturbed) shear field $\bar{\mathbf{S}}$ and for which the Reynolds (Re) and Capillary (Ca) numbers are small: namely,

$$\mathrm{Re} = \frac{4\rho a^2 \bar{S}}{\mu} << 1 \qquad (4.4\text{-}92)$$

and

$$\mathrm{Ca} = \frac{2\mu \bar{S} a}{\sigma} << 1. \qquad (4.4\text{-}93)$$

Here, μ and ρ respectively represent the viscosity and density of the continuous liquid and $\bar{S}$ the magnitude of the (undisturbed) shear field $\bar{\mathbf{S}}$. The latter restriction placed upon the Capillary number insures that interfacial tension forces dominate over bulk viscous forces. Thus, to leading order in Ca, the normal stress boundary condition (4.2-20a) reduces to Laplace's condition (2.1-3).* This is satisfied for a spherical droplet shape (here characterized by the radius a).

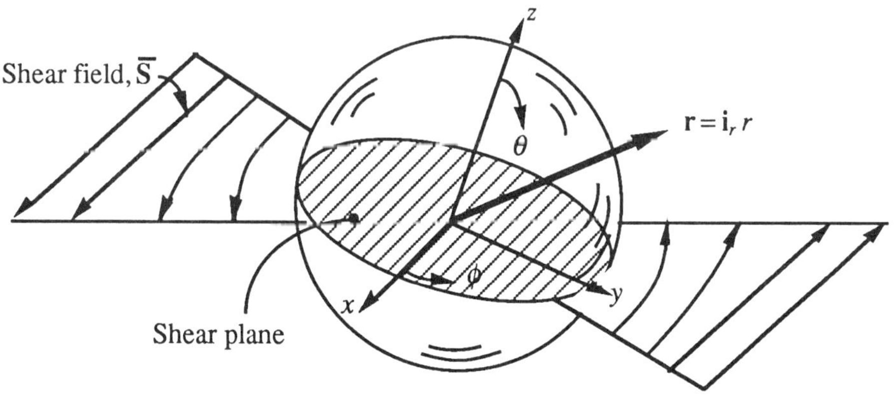

Figure 4.4-6 A spherical emulsion droplet suspended within a homogeneous shear field.

---

* In the usual circumstance that Bo = $o(1)$ [cf. Eqs. 4.4-43)–(4.4-45)], satisfaction of condition (4.4-93) furthermore assures that interfacial tension forces dominate over interfacial viscous forces in the normal stress condition (4.2-20a).

Imagine, as depicted in Figure 4.4-6, a neutrally buoyant spherical droplet of radius a suspended within a homogeneous shear field $\bar{\mathbf{S}} = (\mathbf{i}_x\mathbf{i}_y + \mathbf{i}_y\mathbf{i}_x)\bar{S}$. Subject to the low Reynolds number condition (4.4-92), the following creeping-flow form of the Navier-Stokes equations (4.1-23) obtains in both the continuous fluid ($a < r < \infty$) and droplet ($0 < r < a$) phases:

$$\mu\nabla^2\mathbf{v} = \nabla p \tag{4.4-94 a}$$

and

$$\tilde{\mu}\nabla^2\tilde{\mathbf{v}} = \nabla\tilde{p}, \tag{4.4-94 b}$$

together with the corresponding equations of continuity (4.1-24) in each phase, namely

$$\nabla\cdot\mathbf{v} = 0 \tag{4.4-95 a}$$

and

$$\nabla\cdot\tilde{\mathbf{v}} = 0. \tag{4.4-95 b}$$

Expressed in spherical coordinates, the velocity and pressure fields $(\mathbf{v}, \tilde{\mathbf{v}}, p, \tilde{p})$ satisfy the following boundary conditions:

$$v_r = \tilde{v}_r = 0 \qquad \text{at } r = a, \tag{4.4-96}$$

$$v_\theta = \tilde{v}_\theta \equiv v_\theta^o \qquad \text{at } r = a, \tag{4.4-97}$$

$$v_\phi = \tilde{v}_\phi \equiv v_\phi^o \qquad \text{at } r = a, \tag{4.4-98}$$

$$v_r = \tfrac{1}{2}\bar{S}r\sin^2\theta\sin 2\phi \qquad \text{as } r\to\infty, \tag{4.4-99}$$

$$v_\theta = \tfrac{1}{2}\bar{S}r\sin\theta\cos\theta\sin 2\phi \qquad \text{as } r\to\infty, \tag{4.4-100}$$

$$v_\phi = \bar{S}r\sin\theta\cos^2\phi \qquad \text{as } r\to\infty. \tag{4.4-101}$$

The tangential stress conditions at the Newtonian droplet interface follow from Table 4.2-3 as*

---

* The footnote preceding Eq. (4.1-13) regarding the neglect of interfacial tension gradients based upon the smallness of the surface Peclét number is applicable here as well.

$$\tilde{\mu}\left[r\frac{\partial}{\partial r}\left(\frac{\tilde{v}_\theta}{r}\right)+\frac{1}{r}\frac{\partial \tilde{v}_r}{\partial \theta}\right]-\mu\left[r\frac{\partial}{\partial r}\left(\frac{v_\theta}{r}\right)+\frac{1}{r}\frac{\partial v_r}{\partial \theta}\right]$$
$$=\frac{\kappa^s+\mu^s}{a}\frac{\partial}{\partial \theta}\left[\frac{1}{a\sin\theta}\left(\frac{\partial}{\partial \theta}(\sin\theta\, v_\theta^o)+\frac{\partial v_\phi^o}{\partial \phi}\right)\right]$$
$$+\mu^s\left\{\frac{2v_\theta^o}{a^2}+\frac{1}{a\sin\theta}\frac{\partial}{\partial \phi}\left[\frac{1}{a\sin\theta}\left(\frac{\partial v_\theta^o}{\partial \phi}-\frac{\partial}{\partial \theta}(\sin\theta\, v_\phi^o)\right)\right]\right\}$$
$$\text{at} \quad r=a \tag{4.4-102}$$

and

$$\tilde{\mu}\left[\frac{1}{r\sin\theta}\frac{\partial \tilde{v}_r}{\partial \phi}+r\frac{\partial}{\partial r}\left(\frac{\tilde{v}_\phi}{r}\right)\right]-\mu\left[\frac{1}{r\sin\theta}\frac{\partial v_r}{\partial \phi}+r\frac{\partial}{\partial r}\left(\frac{v_\phi}{r}\right)\right]$$
$$=\frac{\kappa^s+\mu^s}{a\sin\theta}\frac{\partial}{\partial \phi}\left[\frac{1}{a\sin\theta}\left(\frac{\partial}{\partial \theta}(\sin\theta\, v_\theta^o)+\frac{\partial v_\phi^o}{\partial \phi}\right)\right]$$
$$+\mu^s\left\{\frac{2v_\phi^o}{a^2}-\frac{1}{a}\frac{\partial}{\partial \theta}\left[\frac{1}{a\sin\theta}\left(\frac{\partial v_\theta^o}{\partial \phi}-\frac{\partial}{\partial \theta}(\sin\theta\, v_\phi^o)\right)\right]\right\}$$
$$\text{at} \quad r=a \tag{4.4-103}$$

In the above, (~) is used to denote droplet-phase variables.

Lamb (1945) provides the following general solution to Eqs. (4.4-94) and (4.4-95) in spherical coordinates:

$$\mathbf{v}=\mathbf{v}_\infty+\sum_{n=-\infty}^{n=\infty}\left[\nabla\times(\mathbf{r}\,\chi_n)+\nabla_n\Phi_n+\frac{(n+3)r^2}{2\mu(n+1)(2n+3)}\nabla\Psi_n-\frac{n\mathbf{r}\Psi_n}{\mu(n+1)(2n+3)}\right], \tag{4.4-104}$$

where $\mathbf{v}_\infty$ is the velocity field far from the droplet interface $r\to\infty$, $\mathbf{r}=\mathbf{i}_r r$ is a radial position vector measured from the center of the droplet, and χ_n, Ψ_n and Φ_n represent solid spherical harmonic functions. In addition,

$$p=\sum_{n=-\infty}^{n=\infty}\Psi_n \tag{4.4-105}$$

provides the corresponding solution for the pressure field.

The only permissible solid spherical harmonics satisfying the velocity conditions (4.4-99)–(4.4-101) are found to be (Wei *et al.* 1974):

$$\Psi_{-3}=A\frac{1}{r^3}\sin^2\theta\,\sin 2\phi \tag{4.4-106}$$

and

$$\Phi_{-3} = B\frac{1}{r^3}\sin^2\theta\sin 2\phi, \tag{4.4-107}$$

applicable to the continuous phase fields **v** and p, and

$$\tilde{\Psi}_2 = Cr^2\sin^2\theta\sin 2\phi, \tag{4.4-108}$$

$$\tilde{\Phi}_2 = Dr^2\sin^2\theta\sin 2\phi \tag{4.4-109}$$

and

$$\tilde{\chi}_1 = Er\cos\theta, \tag{4.4-110}$$

applicable to the droplet phase fields $\tilde{\mathbf{v}}$ and $\tilde{p}$.

Conditions (4.4-96)–(4.4-98) together with the tangential stress conditions (4.4-102) and (4.4-103) may be used to evaluate the coefficients A, B, C, D, and E. This ultimately yields [to leading order in Re and Ca; cf. Eqs. (4.4-92) and (4.4-93)]:

$$v_r \approx \tfrac{1}{2}a\bar{S}\left[\frac{r}{a} + F\left(\frac{a}{r}\right)^2 - (F+1)\left(\frac{a}{r}\right)^4\right]\sin^2\theta\sin 2\phi, \tag{4.4-111}$$

$$v_\theta \approx a\bar{S}\left[\frac{1}{2}\frac{r}{a} + \tfrac{1}{3}(F+1)\left(\frac{a}{r}\right)^4\right]\sin\theta\cos\theta\sin 2\phi, \tag{4.4-112}$$

$$v_\phi \approx a\bar{S}\left[\frac{r}{a}\sin\theta\cos^2\phi + \tfrac{1}{3}(F+1)\left(\frac{a}{r}\right)^4\sin\theta\cos 2\phi\right], \tag{4.4-113}$$

$$p \approx p_o + \mu\bar{S}\left(\frac{a}{r}\right)^3 F\ \sin^2\theta\sin 2\phi, \tag{4.4-114}$$

$$\tilde{v}_r \approx \tfrac{1}{2}a\bar{S}\left[\left(\frac{r}{a}\right)^3 - \left(\frac{r}{a}\right)\right]\left[F + \frac{5}{2}\right]\sin^2\theta\sin 2\phi, \tag{4.4-115}$$

$$\tilde{v}_\theta \approx \tfrac{1}{2}a\bar{S}\left[\tfrac{5}{3}\left(\frac{r}{a}\right)^3 - \left(\frac{r}{a}\right)\right]\left[F + \frac{5}{2}\right]\sin\theta\cos\theta\sin 2\phi, \tag{4.4-116}$$

$$\tilde{v}_\phi \approx a\bar{S}\left\{\tfrac{1}{2}\left[\tfrac{5}{3}\left(\frac{r}{a}\right)^3 - \left(\frac{r}{a}\right)\right]\left[F + \frac{5}{2}\right]\sin\theta\cos 2\phi + \tfrac{1}{2}\left(\frac{r}{a}\right)\sin\theta\right\}, \tag{4.4-117}$$

and

$$\tilde{p} \approx p_i + \tfrac{1}{2}a\bar{S}\left[7F + \frac{35}{2}\right]\left(\frac{r}{a}\right)^2\sin^2\theta\sin 2\phi. \tag{4.4-118}$$

Here, p_o and p_i are constants, related by the (limiting, low Capillary number form of the) normal stress boundary condition (Table 4.2-3):

$$p_i - p_o = \frac{2\sigma}{a},$$

which is the Laplace condition. In addition, we have defined

$$F \equiv \frac{1 + \dfrac{5\tilde{\mu}}{2\mu} + \left(\dfrac{2\mu^s + 3\kappa^s}{\mu a}\right)}{1 + \dfrac{\tilde{\mu}}{\mu} + \dfrac{2}{5}\left(\dfrac{2\mu^s + 3\kappa^s}{\mu a}\right)}. \tag{4.4-119}$$

In the limiting case wherein bulk viscous forces dominate over interfacial viscous forces, namely

$$\frac{2\mu^s + 3\kappa^s}{\mu a} << 1, \tag{4.4-120 a}$$

Eq. (4.4-119) reduces to the Taylor (1932) solution for a droplet with inviscid interface in a homogeneous shear field:

$$F = \frac{1 + \dfrac{5\tilde{\mu}}{2\mu}}{1 + \dfrac{\tilde{\mu}}{\mu}}. \tag{4.4-120 b}$$

When

$$\frac{2\mu^s + 3\kappa^s}{\mu a} >> 1, \tag{4.4-121 a}$$

corresponding to a highly viscous fluid interface, Eq. (4.4-119) reduces to

$$F = \frac{5}{2}, \tag{4.4-121 b}$$

which is the solution for a solid sphere in a shear field.

6. The Shear Viscosity and Viscoelasticity of a Dilute Emulsion

The preceding solution may be employed to determine the shear viscosity of a dilute emulsion, employing analogous methods to those used in example 3 [cf. Eq. (4.4-35)]. Thus, Oldroyd (1955) found the following limiting expression

$$\langle \mu \rangle = \mu(1 + F\phi) + O(\phi \text{ Ca}) + O(\phi^2) \tag{4.4-122}$$

for the shear viscosity of the dilute emulsion, to leading order in the dispersed-phase volume fraction ϕ and the Capillary number Ca [cf. Eq. (4.4-93)]. The coefficient F appearing in the above is that defined in Eq. (4.4-119). Explicitly:

$$\langle\mu\rangle = \mu\left(1+\phi\frac{1+\frac{5}{2}\frac{\tilde{\mu}}{\mu}+\left(\frac{2\mu^s+3\kappa^s}{\mu a}\right)}{1+\frac{\tilde{\mu}}{\mu}+\frac{2}{5}\left(\frac{2\mu^s+3\kappa^s}{\mu a}\right)}\right)+O(\phi\ \mathrm{Ca})+O(\phi^2). \tag{4.4-123}$$

In the limit of (4.4-120a), which applies to the case of a dilute suspension of droplets with inviscid interfaces,

$$\langle\mu\rangle = \mu\left(1+\phi\frac{1+\frac{5}{2}\frac{\tilde{\mu}}{\mu}}{1+\frac{\tilde{\mu}}{\mu}}\right)+O(\phi\ \mathrm{Ca})+O(\phi^2), \tag{4.4-124}$$

as first obtained by Taylor (1932). In the limit of (4.4-121a), which applies to the case of a dilute suspension of solid spheres,

$$\langle\mu\rangle = \mu\left(1+\frac{5}{2}\phi\right)+O(\phi^2), \tag{4.4-125}$$

as first obtained by Einstein (1906).

Oldroyd (1955) furthermore showed that shape deformation of droplet interfaces owing to unsteady shear flow results in a macroscopic *viscoelasticity* of the dilute emulsion. Based upon his postulate of a linear relation between global stress and rate of strain, he obtained

$$\left(1+\lambda_1\frac{D}{Dt}\right)\langle\tau\rangle = 2\langle\mu\rangle\left(1+\lambda_2\frac{D}{Dt}\right)\langle\mathbf{S}\rangle+O(\phi\ \mathrm{Ca})+O(\phi^2), \tag{4.4-126}$$

to leading order in the dispersed phase volume fraction ϕ and Capillary number Ca. Equation (4.4-126) provides a constitutive relation for the stress tensor of a dilute emulsion for which droplet interfaces possess a Newtonian rheological behavior. Here, $\langle\mu\rangle$ is as defined in Eq. (4.4-123) and the elastic constants appearing in (4.4-126) are:

$$\lambda_1 = \lambda+\phi\Lambda, \tag{4.4-127}$$

$$\lambda_2 = \lambda-\frac{3}{2}\phi\Lambda, \tag{4.4-128}$$

where

$$\lambda = \frac{\left[a(3\mu+2\tilde{\mu})(16\mu+19\tilde{\mu})+4\mu^s(12\mu+13\tilde{\mu}) + 2\kappa^s(32\mu+23\tilde{\mu})+32\mu^s\kappa^s/a\right]}{8\sigma\left[5(\mu+\tilde{\mu})+4\mu^s/a+6\kappa^s/a\right]} \tag{4.4-129}$$

and

$$\Lambda = \frac{a\mu[\alpha(16\mu + 19\tilde{\mu}) + 8(\mu^s + 3\kappa^s)]^2}{8\sigma[5\alpha(\mu + \tilde{\mu}) + 4\mu^s + 6\kappa^s]^2}. \tag{4.4-130}$$

Here, also

$$\frac{D}{Dt} = \frac{\partial}{\partial t} + \langle \mathbf{v} \rangle \cdot \nabla \tag{4.4-131}$$

represents the total time derivative following the macroscopic material fluid velocity $\langle \mathbf{v} \rangle$.

4.5 Summary

The interfacial stress boundary condition (4.2-9) has been developed in this chapter for fluid interfaces adsorbed by surfactant. Special attention has been given to the particular class of *Newtonian* interfaces, for which the explicit form of the interfacial stress condition is furnished in an invariant form by Eq. (4.2-20), or, in component form, by Tables 4.2-1–4.2-3. Non-Newtonian interfacial rheological behavior has been discussed in §4.3.

Principal interfacial rheological properties introduced by the Newtonian interfacial stress boundary condition (4.2-20) are the interfacial shear and dilatational viscosities μ^s and κ^s. These have been shown to influence hydrodynamic motion in the vicinity of a Newtonian interface, particularly for large Boussinesq number Bo [cf. Eq.(4.4-43)], which corresponds to the conditions of large surfactant adsorption and large specific surface. Thus, for example, the interfacial viscous effect in Eqs. (4.4-24), (4.4-26), (4.4-33b), (4.4-41) and (4.4-123) increases with increasing Boussinesq number.

Appendix 4.A

The Stress Balance at a Newtonian Interface: Derivation of Eq. (4.2-20)

Below it is demonstrated that for the case of a Newtonian interface, for which class of interfacial rheological behavior

$$\mathbf{P}^s = \mathbf{I}_s[\sigma + (\kappa^s - \mu^s)\nabla_s \cdot \mathbf{v}^o] + \mu^s[\nabla_s \mathbf{v}^o \cdot \mathbf{I}_s + \mathbf{I}_s \cdot \nabla_s \mathbf{v}^{o\dagger}] \tag{4. A-1}$$

the vector defined by the divergence operation

$$\nabla_s \cdot \mathbf{P}^s, \tag{4. A-2}$$

furnishes the terms on the right-hand side of Eq. (4.2-20) (minus the surface-excess force density vector $\mathbf{F}^s$).

Upon substitution of Eq. (4.A-1) into Eq. (4.A-2) there obtains

$$\nabla_s \cdot \mathbf{P}^s = \nabla_s \cdot \{\mathbf{I}_s[\sigma + (\kappa^s - \mu^s)\nabla_s \cdot \mathbf{v}^o]\} + \nabla_s \cdot \left\{\mu^s[\nabla_s \mathbf{v}^o \cdot \mathbf{I}_s + \mathbf{I}_s \cdot \nabla_s \mathbf{v}^{o\dagger}]\right\}. \qquad (4.\text{A-}3)$$

This may be simplified by employing Eq. (3.3-8), thereby yielding

$$\begin{aligned}\nabla_s \cdot \mathbf{P}^s &= \nabla_s \sigma + 2H\,\mathbf{n}[\sigma + (\kappa^s - \mu^s)\nabla_s \cdot \mathbf{v}^o] \\ &\quad + (\kappa^s - \mu^s)\nabla_s(\nabla_s \cdot \mathbf{v}^o) \\ &\quad + \mu^s\left[(\nabla_s^2 \mathbf{v}^o) \cdot \mathbf{I}_s + (\nabla_s \mathbf{v}^o)^\dagger : \nabla_s \mathbf{I}_s\right] \\ &\quad + 2\mu^s H\,\nabla_s \mathbf{v}^o \cdot \mathbf{n} + \mu^s \nabla_s \cdot (\nabla_s \mathbf{v}^{o\dagger}).\end{aligned} \qquad (4.\text{A-}4)$$

The following three additional identities are required (see Question 4.9):

$$\nabla_s \mathbf{I}_s \equiv \mathbf{b}\,\mathbf{n} + \mathbf{n}\mathbf{b};$$

$$\nabla_s \cdot (\nabla_s \mathbf{v}^{o\dagger}) \equiv \nabla_s(\nabla_s \cdot \mathbf{v}^o) - \mathbf{b} \cdot (\nabla_s \mathbf{v}^o) \cdot \mathbf{n} + \mathbf{b} : (\nabla_s \mathbf{v}^o)\mathbf{n}\,; \qquad (4.\text{A-}5)$$

$$\begin{aligned}\mathbf{n} \times \nabla_s[(\nabla_s \times \mathbf{v}^o) \cdot \mathbf{n}] &\equiv (\nabla_s^2 \mathbf{v}^o) \cdot \mathbf{I}_s - \nabla_s(\nabla_s \cdot \mathbf{v}^o) \\ &\quad + (2\mathbf{b} - 2H\,\mathbf{I}_s) \cdot (\nabla_s \mathbf{v}^o) \cdot \mathbf{n}.\end{aligned}$$

Substituting the first two relations of the above into Eq. (4.A-4) yields, after considerable manipulation,

$$\begin{aligned}\nabla_s \cdot \mathbf{P}^s &= \nabla_s \sigma + 2H\,\mathbf{n}[\sigma + (\kappa^s + \mu^s)\nabla_s \cdot \mathbf{v}^o] \\ &\quad + (\kappa^s + \mu^s)\nabla_s(\nabla_s \cdot \mathbf{v}^o) + 2H\,\nabla_s \mathbf{v}^o \cdot \mathbf{n} \\ &\quad + 2H\,\nabla_s \mathbf{v}^o \cdot \mathbf{n} + 2\mathbf{n}(\mathbf{b} - 2H\,\mathbf{I}_s) : (\nabla_s \mathbf{v}^o) - \nabla_s(\nabla_s \cdot \mathbf{v}^o)]\end{aligned} \qquad (4.\text{A-}6)$$

Substituting the final identity of Eq. (4.A-5) into the above equation yields

$$\begin{aligned}\nabla_s \cdot \mathbf{P}^s &= 2H\sigma\,\mathbf{n} + \nabla_s \sigma + (\kappa^s + \mu^s)\nabla_s \nabla_s \cdot \mathbf{v}^o \\ &\quad + 2\mu^s \mathbf{n}(\mathbf{b} - 2H\,\mathbf{I}_s) : \nabla_s \mathbf{v}^o + 2H\,\mathbf{n}(\kappa^s + \mu^s)\nabla_s \cdot \mathbf{v}^o \\ &\quad + \mu^s\{\mathbf{n} \times \nabla_s[(\nabla_s \times \mathbf{v}^o) \cdot \mathbf{n}] - 2(\mathbf{b} - 2H\,\mathbf{I}_s) \cdot (\nabla_s \mathbf{v}^o) \cdot \mathbf{n}\},\end{aligned} \qquad (4.\text{A-}7)$$

as in Eq. (4.2-20).

Appendix 4.B

The Stress Balance at a Newtonian Interface: Eq. (4.2-20) in Orthogonal Curvilinear Coordinates

A space vector **f** may generally be expressed in orthogonal curvilinear component form as

$$\mathbf{f} = \mathbf{i}_1 f_1 + \mathbf{i}_2 f_2 + \mathbf{i}_3 f_3. \qquad (4.\text{B-}1)$$

The vector **f** may [cf. Eq. (3.4-12)] be decomposed into vectors respectively normal and tangential to the surface whose unit tangent vectors are $(\mathbf{i}_1, \mathbf{i}_2)$; thus,

$$\mathbf{f} = \mathbf{f}_n + \mathbf{f}_s, \qquad (4.\text{B-}2)$$

where

$$\mathbf{f}_n \equiv \mathbf{n}\,\mathbf{n} \cdot \mathbf{f} = \mathbf{i}_3 f_3 \qquad (4.\text{B-}3)$$

and

$$\mathbf{f}_s \equiv \mathbf{I}_s \cdot \mathbf{f} = \mathbf{i}_1 f_1 + \mathbf{i}_2 f_2. \qquad (4.\text{B-}4)$$

Employing Eqs. (3.2-11), (B.1-13) and (4.B-4), provides

$$\nabla_s \cdot \mathbf{f}_s = h_1 h_2 \left[\frac{\partial}{\partial q^1}\left(\frac{f_1}{h_2}\right) + \frac{\partial}{\partial q^2}\left(\frac{f_2}{h_1}\right) \right]. \qquad (4.\text{B-}5)$$

Likewise, with Eq. (4.B-1), the surface divergence of the space vector **f** is given by

$$\begin{aligned} \nabla_s \cdot \mathbf{f} = {} & h_1 h_2 \left[\frac{\partial}{\partial q^1}\left(\frac{f_1}{h_2}\right) + \frac{\partial}{\partial q^2}\left(\frac{f_2}{h_1}\right) \right. \\ & \left. + f_3 \left[h_1 h_3 \frac{\partial}{\partial q^3}\left(\frac{1}{h_1}\right) + h_2 h_3 \frac{\partial}{\partial q^3}\left(\frac{1}{h_2}\right) \right] \right] \\ \equiv {} & \nabla_s \cdot \mathbf{f}_s - 2H\, \mathbf{f} \cdot \mathbf{n}. \end{aligned} \qquad (4.\text{B-}6)$$

Here, Eqs. (B.1-21), (B.1-22) and (3.2-13) have been used to show

$$2H = -\, h_1 h_3 \frac{\partial}{\partial q^3}\left(\frac{1}{h_1}\right) + h_2 h_3 \frac{\partial}{\partial q^3}\left(\frac{1}{h_2}\right). \qquad (4.\text{B-}7)$$

The surface gradient of the space vector **f** may be expressed by the dyadic (Question 4.10)

$$
\begin{aligned}
\nabla_s \mathbf{f} = \mathbf{i}_1\mathbf{i}_1 & \left[h_1 \frac{\partial}{\partial q^1} f_1 + h_1 h_2 f_2 \frac{\partial}{\partial q^2}\left(\frac{1}{h_1}\right) + h_1 h_3 f_3 \frac{\partial}{\partial q^3}\left(\frac{1}{h_1}\right) \right] \\
+ \mathbf{i}_2\mathbf{i}_2 & \left[h_2 \frac{\partial}{\partial q^2} f_2 + h_1 h_2 f_1 \frac{\partial}{\partial q^1}\left(\frac{1}{h_2}\right) + h_2 h_3 f_3 \frac{\partial}{\partial q^3}\left(\frac{1}{h_2}\right) \right] \\
+ \mathbf{i}_1\mathbf{i}_2 & \left[h_1 \frac{\partial}{\partial q^1} f_2 - h_2 h_1 f_1 \frac{\partial}{\partial q^2}\left(\frac{1}{h_1}\right) \right] \\
+ \mathbf{i}_2\mathbf{i}_1 & \left[h_2 \frac{\partial}{\partial q^2} f_1 - h_1 h_2 f_2 \frac{\partial}{\partial q^1}\left(\frac{1}{h_2}\right) \right] \\
+ \mathbf{i}_1\mathbf{i}_3 & \left[h_1 \frac{\partial}{\partial q^1} f_3 - h_1 h_3 f_1 \frac{\partial}{\partial q^3}\left(\frac{1}{h_1}\right) \right] \\
+ \mathbf{i}_2\mathbf{i}_3 & \left[h_2 \frac{\partial}{\partial q^2} f_3 - h_2 h_3 f_2 \frac{\partial}{\partial q^3}\left(\frac{1}{h_2}\right) \right]
\end{aligned} \tag{4. B- 8}
$$

whence it follows that

$$
\begin{aligned}
(\nabla_s \mathbf{f}) \cdot \mathbf{n} = \mathbf{i}_1 & \left[h_1 \frac{\partial}{\partial q^1} f_3 - h_1 h_3 f_1 \frac{\partial}{\partial q^3}\left(\frac{1}{h_1}\right) \right] \\
+ \mathbf{i}_2 & \left[h_2 \frac{\partial}{\partial q^2} f_3 - h_2 h_3 f_2 \frac{\partial}{\partial q^3}\left(\frac{1}{h_2}\right) \right].
\end{aligned} \tag{4. B- 9}
$$

Likewise, with some manipulation,

$$
(\nabla_s \times \mathbf{f}) \cdot \mathbf{n} = (\varepsilon \colon \nabla_s \mathbf{f}) \cdot \mathbf{n} = h_1 h_2 \left[\frac{\partial}{\partial q^1}\left(\frac{f_2}{h_2}\right) - \frac{\partial}{\partial q^2}\left(\frac{f_1}{h_1}\right) \right]. \tag{4. B- 10}
$$

The preceding relationships suffice to provide the desired decomposition of the surface divergence of the surface-excess pressure tensor for a Newtonian fluid interface in general orthogonal curvilinear coordinates: explicitly, the tangential components satisfy the relations

$$
\begin{aligned}
\mathbf{i}_1 \cdot \nabla_s \cdot \mathbf{P}^s = {} & h_1 \frac{\partial \sigma}{\partial q^1} + (\kappa^s + \mu^s) h_1 \frac{\partial}{\partial q^1} \left\{ h_1 h_2 \left[\frac{\partial}{\partial q^1}\left(\frac{v_1^o}{h_2}\right) + \frac{\partial}{\partial q^2}\left(\frac{v_2^o}{h_1}\right) \right] \right. \\
& \left. + v_3^o \left[h_1 h_3 \frac{\partial}{\partial q^3}\left(\frac{1}{h_1}\right) + h_2 h_3 \frac{\partial}{\partial q^3}\left(\frac{1}{h_2}\right) \right] \right\} \\
& - \mu^s h_2 \frac{\partial}{\partial q^2} \left\{ h_1 h_2 \left[\frac{\partial}{\partial q^1}\left(\frac{v_2^o}{h_2}\right) - \frac{\partial}{\partial q^2}\left(\frac{v_1^o}{h_1}\right) \right] \right\} \\
& - 2\mu^s h_2 h_3 \frac{\partial}{\partial q^3}\left(\frac{1}{h_2}\right) \left[h_1 \frac{\partial}{\partial q^1} v_3^o - h_1 h_3 v_1^o \frac{\partial}{\partial q^3}\left(\frac{1}{h_1}\right) \right]
\end{aligned} \tag{4. B- 11}
$$

and

$$\mathbf{i}_2 \cdot \nabla_s \cdot \mathbf{P}^s = h_2 \frac{\partial \sigma}{\partial q^2} + (\kappa^s + \mu^s) h_2 \frac{\partial}{\partial q^2} \left\{ h_1 h_2 \left[\frac{\partial}{\partial q^1}\left(\frac{v_1^o}{h_2}\right) + \frac{\partial}{\partial q^2}\left(\frac{v_2^o}{h_1}\right) \right] \right.$$
$$\left. + v_3^o \left[h_1 h_3 \frac{\partial}{\partial q^3}\left(\frac{1}{h_1}\right) + h_2 h_3 \frac{\partial}{\partial q^3}\left(\frac{1}{h_2}\right) \right] \right\}$$
$$+ \mu^s h_1 \frac{\partial}{\partial q^1} \left\{ h_1 h_2 \left[\frac{\partial}{\partial q^1}\left(\frac{v_2^o}{h_2}\right) - \frac{\partial}{\partial q^2}\left(\frac{v_1^o}{h_1}\right) \right] \right\}$$
$$- 2\mu^s h_1 h_3 \frac{\partial}{\partial q^3}\left(\frac{1}{h_1}\right) \left[h_2 \frac{\partial}{\partial q^2} v_3^o - h_2 h_3 v_2^o \frac{\partial}{\partial q^3}\left(\frac{1}{h_2}\right) \right].$$

(4. B-12)

The normal component is given by

$$\mathbf{i}_3 \cdot \nabla_s \cdot \mathbf{P}^s = - \left[h_1 h_3 \frac{\partial}{\partial q^3}\left(\frac{1}{h_1}\right) + h_2 h_3 \frac{\partial}{\partial q^3}\left(\frac{1}{h_2}\right) \right]$$
$$\times \left[\sigma + (\kappa^s + \mu^s) \left\{ h_1 h_2 \left[\frac{\partial}{\partial q^1}\left(\frac{v_1^o}{h_2}\right) + \frac{\partial}{\partial q^2}\left(\frac{v_2^o}{h_1}\right) \right] \right. \right.$$
$$\left. \left. + v_3^o \left[h_1 h_3 \frac{\partial}{\partial q^3}\left(\frac{1}{h_1}\right) + h_2 h_3 \frac{\partial}{\partial q^3}\left(\frac{1}{h_2}\right) \right] \right\} \right]$$
$$+ 2\mu^s h_1 h_3 \frac{\partial}{\partial q^3}\left(\frac{1}{h_1}\right) \left[h_2 \frac{\partial v_2^o}{\partial q^1} + v_1^o h_1 h_2 \frac{\partial}{\partial q^1}\left(\frac{1}{h_2}\right) \right.$$
$$\left. + v_3^o h_2 h_3 \frac{\partial}{\partial q^3}\left(\frac{1}{h_2}\right) \right]$$
$$+ 2\mu^s h_2 h_3 \frac{\partial}{\partial q^3}\left(\frac{1}{h_2}\right) \left[h_1 \frac{\partial v_1^o}{\partial q^1} + v_2^o h_1 h_2 \frac{\partial}{\partial q^2}\left(\frac{1}{h_1}\right) \right.$$
$$\left. + v_3^o h_1 h_3 \frac{\partial}{\partial q^3}\left(\frac{1}{h_1}\right) \right].$$

(4. B-13)

Appendix 4.C

Physical Interpretation of the 2-D Deformation Rate Tensor, $\mathbf{D}_s$

A physical interpretation of the interfacial rate of deformation tensor (4.2-16) is provided by considering the rate of physical displacement of surface points on a moving and deforming interface. According to Eq. (B.1-10a), a scalar displacement of surface points satisfies the expression

$$dl^2 = a_{\alpha\beta}\, dq^\alpha dq^\beta , \qquad (4.\ C\text{-}1)$$

relating physical displacement dl with a displacement dq^α of surface coordinates. Noting the definition of the surface metric $a_{\alpha\beta}$ provided in Eq. (3.2-3), together with Eqs. (3.2-2) and (3.4-8), permits

$$\frac{D_s}{Dt}\left(dl^2\right) = \frac{D_s}{Dt}\left(\frac{\partial \mathbf{x}_s}{\partial q^\alpha} \cdot \frac{\partial \mathbf{x}_s}{\partial q^\beta}\right) dq^\alpha dq^\beta. \tag{4.C-2}$$

Comparing Eqs. (3.4-5) and (3.4-6), this may be rewritten as

$$\frac{D_s}{Dt}\left(dl^2\right) = \left[\left(\frac{\partial \mathbf{v}^o}{\partial q^\alpha} \cdot \frac{\partial \mathbf{x}_s}{\partial q^\beta}\right) + \left(\frac{\partial \mathbf{x}_s}{\partial q^\alpha} \cdot \frac{\partial \mathbf{v}^o}{\partial q^\beta}\right)\right] dq^\alpha dq^\beta,$$

or, equivalently, upon use of the chain rule, the nesting convention for the double-dot product in Eq. (A.1-25), and the definition of the surface idemfactor provided in Question 3.14,

$$\begin{aligned}
\frac{D_s}{Dt}\left(dl^2\right) &= \left(\frac{\partial \mathbf{v}^o}{\partial q^\alpha} dq^\alpha \cdot \frac{\partial \mathbf{x}_s}{\partial q^\beta} dq^\beta\right) + \left(\frac{\partial \mathbf{x}_s}{\partial q^\alpha} dq^\alpha \cdot \frac{\partial \mathbf{v}^o}{\partial q^\beta} dq^\beta\right), \\
&= \left(\frac{\partial \mathbf{v}^o}{\partial \mathbf{x}_s} \cdot \frac{\partial \mathbf{x}_s}{\partial q^\alpha} dq^\alpha\right) \cdot d\mathbf{x}_s + d\mathbf{x}_s \cdot \left(\frac{\partial \mathbf{v}^o}{\partial \mathbf{x}_s} \cdot \frac{\partial \mathbf{x}_s}{\partial q^\beta} dq^\beta\right), \\
&= \frac{\partial \mathbf{v}^o}{\partial \mathbf{x}_s} \cdot \left(\frac{\partial \mathbf{x}_s}{\partial \mathbf{x}_s} \cdot d\mathbf{x}_s\right) \cdot d\mathbf{x}_s + d\mathbf{x}_s \cdot \frac{\partial \mathbf{v}^o}{\partial \mathbf{x}_s} \cdot \left(\frac{\partial \mathbf{x}_s}{\partial \mathbf{x}_s} \cdot d\mathbf{x}_s\right), \\
&= \frac{\partial \mathbf{v}^o}{\partial \mathbf{x}_s} \cdot \mathbf{I}_s \cdot d\mathbf{x}_s \cdot d\mathbf{x}_s + d\mathbf{x}_s \cdot \frac{\partial \mathbf{v}^o}{\partial \mathbf{x}_s} \cdot \mathbf{I}_s \cdot d\mathbf{x}_s, \\
&\equiv \left\{\left(\frac{\partial \mathbf{v}^o}{\partial \mathbf{x}_s}\right) \cdot \mathbf{I}_s + \mathbf{I}_s \cdot \left(\frac{\partial \mathbf{v}^o}{\partial \mathbf{x}_s}\right)^\dagger\right\} : d\mathbf{x}_s d\mathbf{x}_s.
\end{aligned} \tag{4.C-3}$$

This final relation may be rewritten with the definition of the surface rate of deformation tensor provided in Eq. (4.2-16) as

$$\frac{D_s}{Dt}\left(dl^2\right) = 2\mathbf{D}_s : d\mathbf{x}_s d\mathbf{x}_s, \tag{4.C-4}$$

or, alternatively [cf. Eq. (B.1-9)],

$$\frac{1}{(dl)} \frac{D_s}{Dt}(dl) = \mathbf{D}_s : \frac{d\mathbf{x}_s}{|d\mathbf{x}_s|} \frac{d\mathbf{x}_s}{|d\mathbf{x}_s|}. \tag{4.C-5}$$

Here, $d\mathbf{x}_s / |d\mathbf{x}_s|$ represents a unit surface tangent vector in the direction of $d\mathbf{x}_s$. Thus, $\mathbf{D}_s$, which constitutes the symmetric portion of the dyadic $(\nabla_s \mathbf{v}^o) \cdot \mathbf{I}_s$, represents the rate of relative displacement of surface points. Equation (4.C-5) is the two-dimensional analog to the comparable expression (see Question 4.11) for the three-dimensional rate of deformation tensor **D**, which latter tensor represents the rate of displacement of points in three-dimensional space.

Questions for Chapter 4

4.1 Which interfacial stress condition is responsible for defining the curvature of the interface?

4.2 Derive the three components of the Newtonian surface stress equation (4.2-20) for the case of a spherical fluid interface (Table 4.2-3), using the results of Appendix 4.B. Refer to Appendix B for the scale factors appropriate to the spherical fluid interface.

4.3 Derive the components of the Newtonian interfacial stress condition (4.2-20) for the case of the prolate spheroidal interface depicted in Figure 3.7-1, employing the results of Appendix 4.B. The prolate spheroidal coordinates are (η, ϕ, ξ) with the fluid interface corresponding to the $\xi=\xi_o$ coordinate surface, the latter being parameterized as $(q^1=\eta, q^2=\phi)$.

4.4 *i*. Derive the coefficients A, B, C, and E by substituting Eqs. (4.4-15)–(4.4-18) into conditions (4.4-11)–(4.4-13). *ii*. Show from (i) that Eq. (4.4-23) yields Eq. (4.4-24).

4.5 Derive the expression (4.4-25) for the settling velocity of a fluid sphere in the hollow of a solid spherical shell by employing analogous methods as those utilized in obtaining Eq. (4.4-24).

4.6 Verify Eq. (4.4-69) by maintaining only the n=1 terms in the series expansion of Eq. (4.4-66).

4.7 *i*. Refer to Appendix A and prove the identity

$$\boldsymbol{\varepsilon} : \boldsymbol{\varepsilon} = -\ 2\mathbf{I}.$$

ii. Establish the equivalence of Eq. (4.1-6) and its inverse (4.1-11) using the above identity.

4.8 Use the identity of Question 4.7 to show the equivalence between Eq. (4.2-6) and its inverse (4.2-13).

4.9 Derive the three identities of Eq. (4.A-5). Refer to §3.2, Appendix B, and Appendix 4.B.

4.10 Derive expression (4.B-8) by employing Eqs. (3.2-11), (B.1-13) and (4.B-1).

4.11 Beginning with the metrical relation [cf. Eq. (A.1-12)]

$$dl^2 = g_{ij}\, dq^i\, dq^j$$

derive an expression for the rate of displacement of spatial points [for which Eq. (4.C-5) represents the two-dimensional equivalent] in terms of the deformation rate dyadic **D**.

4.12 Derive the explicit form of the Newtonian surface stress boundary condition (4.2-20) for a radially expanding sphere: make reference to Appendix 4.C [cf. Eq. (4.4-32)].

4.13 Show that, when $I_s=0$, $\dot{\gamma} = \sqrt{II_s}$ physically represents the rate of shear of interfacial points. [Hint: Refer to Appendix 4.C, using also Eqs. (4.3-2) and (4.3-3).]

4.14 Show that the principal invariants I_s and II_s of the (symmetric) interfacial deformation-rate tensor $\mathbf{D}_s$ are related to the eigenvalues D_{s1} and D_{s2} of the latter by the expressions

$$I_s = D_{s1} + D_{s2}$$

and

$$II_s = D_{s1} D_{s2}.$$

4.15 Beginning with Eq. (4.2-18), demonstrate that the principal axes of deformation $(\mathbf{d}_1, \mathbf{d}_2)$, representing solutions to the characteristic equation

$$\mathbf{D}_s \cdot \mathbf{d}_\alpha = D_{s\alpha} \mathbf{d}_\alpha \qquad (\alpha = 1, 2)$$

lie entirely in the tangent plane of the interface.

4.16 A circular cylindrical interface translates with velocity $\mathbf{v}^o$ along the x axis (i.e. $\mathbf{v}^o = \mathbf{i}_x U$) as illustrated in the figure on the following page. Demonstrate by Eq. (4.2-2) that

$$\frac{D_s \rho^s}{Dt} = 0.$$

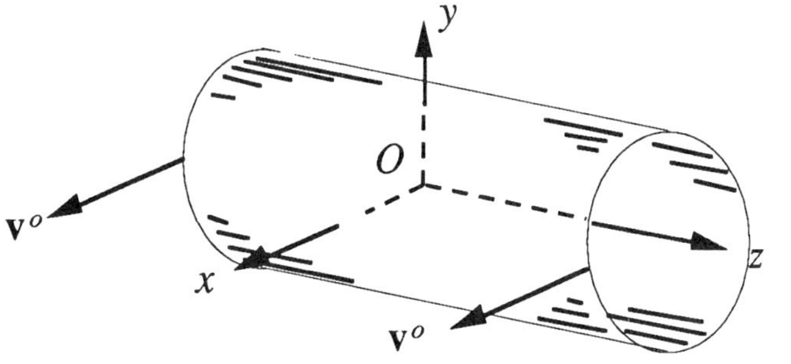

4.17 Equation (4.2-4) represents the two-dimensional analog to Eq. (4.1-4). The velocity $\mathbf{v}^o$ appearing in the acceleration term of Eq. (4.2-4) is, however, a three-dimensional vector. Upon letting,

$$\frac{D_s \mathbf{v}^o}{Dt} = \frac{D_s \mathbf{v}_s}{Dt} + \frac{D_s \mathbf{v}_n}{Dt}$$

according to the decomposition (3.4-12), it is clear that

$$\frac{D_s \mathbf{v}_n}{Dt}$$

represents a physically unique surface acceleration term [i.e. for which there is no three-dimensional analogy in Eq. (4.1-4)]. Show that:

$$\frac{D_s \mathbf{v}_n}{Dt} = \left(\frac{\partial^2 \mathbf{x}_s}{\partial t^2} \right)_{q^1, q^2} - \mathbf{v}_s \cdot \mathbf{b}(\mathbf{v}^o \cdot \mathbf{n}) + \mathbf{v}_s \cdot \left(\frac{D_s \mathbf{n}}{Dt} \right) \mathbf{n}.$$

(In the above, the first term represents a normal acceleration of the two-dimensional surface in three-dimensional space. The second term represents acceleration tangent to the interface owing to surface curvature. The third term represents acceleration normal to the interface owing to a surface bending motion.)

4.18 As a generalization of example 2 of §4.4, find an expression for the continuous-phase pressure field $p(r, t)$ that results when a spherical bubble expands radially [as in Eq. (4.4-31)], *including* inertial terms. [Use Eq. (4.4-33b) as a boundary condition upon the pressure field.]

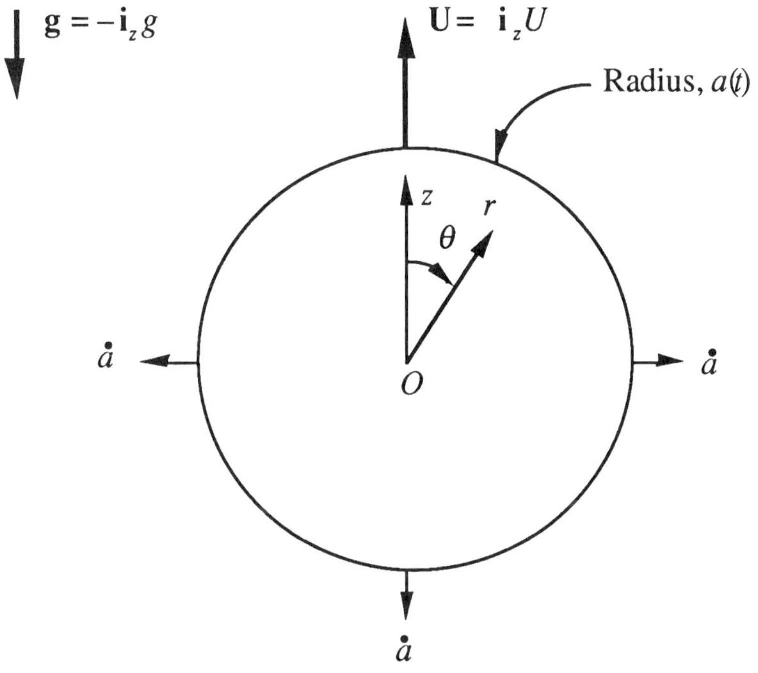

4.19 In the process of flotation, gas bubbles are introduced into an aqueous slurry of mineral particles. The hydrophobic particles adsorb to the surfaces of the gas bubbles during the ascent of the bubbles through the slurry. The bubbles rise to the top of the slurry where they form a froth phase in which the mineral content is high, thereby achieving separation of the mineral from the aqueous slurry. Clearly, the rate of ascent of the bubbles will be influenced by the interfacial rheological properties acquired by the bubble surfaces owing to the adsorption of the solid mineral particles. Assuming these properties to be those of the Newtonian rheological model (4.2-17), derive an expression for the ascent velocity U and hydrostatic pressure difference [i.e. $K - \tilde{K}$; cf. Eqs. (4.4-19)–(4.4-21)] upon assuming the following:

a. Interfacial tension gradients are negligibly small;

b. $Bo \equiv \kappa^s/\mu a = O(1)$;
c. Re <<1;
d. The (spherical) bubble *expands* during its ascent with a constant radial velocity $\dot{a}$.
(Hint: Use a similar development as in example 1 of §4.4; cf. also example 2.)

Additional Reading for Chapter 4

§4.2 Interfacial Conditions upon the Equations of Motion

Scriven, L.E. 1960 The dynamics of a fluid interface. *Chem. Engng Sci.* **12**, 98–108.

§4.3 Non-Newtonian Interfacial Rheological Behavior

Mannheimer, R.J. & Schechter, R.S. 1969 The theory of interfacial viscoelastic measurement by the viscous-traction method. *J. Colloid Interface Sci.* **32**, 225–241.

Pintar, A.J., Israel, A.B. & Wasan, D.T. 1971 Interfacial shear viscosity phenomena in solutions of macromolecules. *J. Colloid Interface Sci.* **37**, 52–66.

Gardner, J.W. & Schechter, R.S. 1976 Evaluation of surface rheological models. *Colloid Interface Sci.* **4**, 98–115.

Li, D. & Slattery, J.C. 1988 Measuring nonlinear surface stress-deformation behavior for aqueous solutions of dodecyl sodium sulfate and dodecyl alcohol. *J. Colloid Interface Sci.* **125**, 190–197.

5

"To make further progress in the study of oil films, it is important to know what is the cause of the spreading of the oil upon the water. The problem is greatly simplified by the knowledge that the films formed are one molecule thick. It is natural to assume that the force which causes the spreading is due to the attraction between the group of molecules of the oil and those of the water. From the chemical viewpoint . . . we should not regard this attraction as emanating from the molecule as a whole, but rather from certain atoms in the molecule . . .. The spreading of an oil upon water is thus due to the presence of an 'active group' in the molecule; that is, some group which has a marked affinity for water."

Irving Langmuir (1917)

CHAPTER 5

Interfacial Transport of Species

In the previous chapter the interfacial stress boundary condition was developed for fluid interfaces possessing both Newtonian and non-Newtonian characteristics. Subsequently, this boundary condition was used in the solution of several hydrodynamic problems involving surfactant-adsorbed fluid interfaces, all of which were assumed *a priori* to exhibit a homogeneous interfacial tension.

Often, however, bulk fluid motion near an interface has the effect of disturbing the homogeneity of the interfacial surfactant composition that would otherwise exist in the absence of flow. The ensuing interfacial tension gradients act in a manner such as to restore the interface to its homogeneous, equilibrium state by engendering flow in proximity to the interface. This phenomenon of interfacial tension gradient-induced flow constitutes the Marangoni effect (Figure 5.0-1). Solution of those hydrodynamic problems for which compositionally-induced Marangoni effects* are present generally requires simultaneous solution of the associated species conservation equation, as the latter is now coupled to the equations of motion through the dependence of interfacial tension [in the interfacial stress boundary condition (4.2-20)] upon surfactant concentration (as well as in the reciprocal appearance of the fluid velocity as a convective contribution in the surfactant transport problem).

In the present chapter the generic transport equations developed in Chapter 3, specifically Eqs. (3.5-9) and (3.6-16), are applied to the particular case of species (i.e. surfactant) transport. These equations, together with constitutive relations pertaining to the surfactant interphase

* The case of temperature-induced Marangoni flows (i.e. thermocapillarity), whereby the interfacial tension gradient-induced flow arises from the existence of interfacial temperature gradients, is addressed in Questions 5.4 through 5.6. Further information may be found in the additional reading citations pertaining to such thermocapillary flows at the conclusion of the chapter.

partitioning [cf. Eq. (5.2-1)] between the bulk phases, and the bulk/interface adsorption-isotherm [cf. Eqs. (5.3-1) and (5.3-2)], serve to define the surfactant transport problem. 'Surface equations of state', appropriate to Langmuir and Frumkin isotherms [cf. Eqs. (5.5-1) and (5.5-2)], provide the functional dependence of interfacial tension upon surfactant concentration. Subsequently, these equations of state (or, more precisely, *linearized* versions thereof, pertaining to small perturbations away from equilibrium conditions) are used jointly with the basic dynamical transport equations to examine two classical hydrodynamic problems serving to illustrate Marangoni flow phenomena.

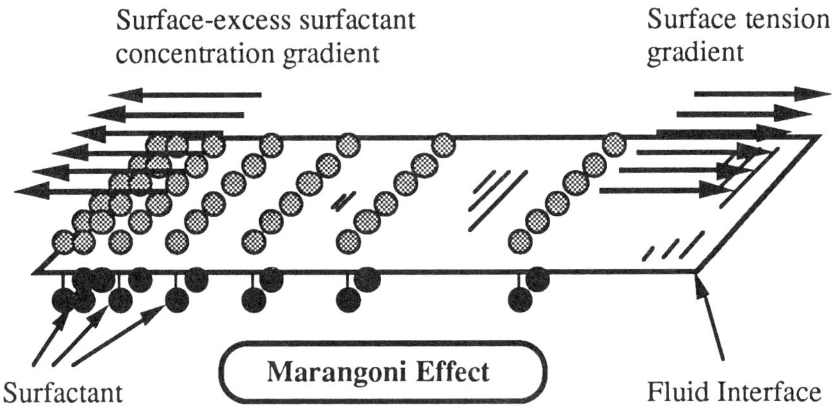

Figure 5.0-1 Since surfactant adsorption at a fluid interface generally lowers the interfacial tension, an inhomogeneous distribution of surfactant within the interface results in the creation of local interfacial tension gradients. The highest tension occurs in those regions where the surfactant density is lowest; thus, the tensile restoring force acts in an opposite direction to the surfactant density gradient. The ensuing fluid motion arising from this interfacial tension gradient, first observed by Marangoni (1871), is now known by his name.

5.1 The Equations of Species Transport in Continuous Three-Dimensional Media

The species transport equation follows from the generic equation (3.5-9), upon substitution of the physical variables identified in Table 3.5-1 for the species volumetric density [$\psi \equiv \rho_i$]. This yields

$$\frac{\partial \rho_i}{\partial t} + \nabla \cdot (\mathbf{v} \rho_i) + \nabla \cdot \mathbf{j}_i = R_i \,, \tag{5.1-1}$$

with ρ_i the mass of species i per unit volume, $\mathbf{j}_i$ the species flux density vector relative to the mass-average velocity $\mathbf{v}$, and R_i the volumetric rate of production of species i via chemical reaction.

Upon assuming the diffusive contribution to the species flux vector to be Fickian, we obtain*

$$\mathbf{j}_i \overset{\text{def}}{=} (\mathbf{v}_i - \mathbf{v})\rho_i \equiv - D_i \nabla \rho_i , \qquad (5.1\text{-}2)$$

with $\mathbf{v}_i$ and D_i respectively the velocity and molecular diffusivity of the ith species. Substitution of Eq. (5.1-2) into Eq. (5.1-1) furnishes the species transport equation,

$$\frac{\partial \rho_i}{\partial t} + \nabla \cdot (\mathbf{v}\,\rho_i) = \nabla \cdot (D_i \nabla \rho_i) + R_i \ . \qquad (5.1\text{-}3)$$

This relation may, in turn, be summed over all components to yield the overall continuity equation of mass conservation (4.1-1), upon utilizing Eq. (5.1-2) together with the following definition (Bird *et al.* 1960) of the mass-average velocity vector,

$$\mathbf{v} = \frac{\sum_{i=1}^{n} \rho_i \mathbf{v}_i}{\sum_{i=1}^{n} \rho_i} , \qquad (5.1\text{-}4)$$

appropriate to a mixture of n components.

In the case of a position-independent solute diffusivity D_i and an incompressible fluid, Eq. (5.1-3) simplifies to the form

$$\boxed{\frac{\partial \rho_i}{\partial t} + \mathbf{v} \cdot \nabla \rho_i = D_i \nabla^2 \rho_i + R_i \ .} \qquad (5.1\text{-}5)$$

This relation may be expressed equivalently in terms of the molar density $C_i = \rho_i / M_i$, (with M_i the molecular weight of the ith component) as

---

* More generally, the species flux vector $\mathbf{j}_i$ may be expressed as a sum of three contributions; explicitly,

$$\mathbf{j}_i = \mathbf{j}_i^D + \mathbf{j}_i^F + \mathbf{j}_i^T ,$$

where $\mathbf{j}_i^D$ represents the molecular diffusive flux [i.e. Eq. (5.1-2)], $\mathbf{j}_i^F$ denotes a flux arising by the action of an external force upon species i and $\mathbf{j}_i^T$ is the thermal diffusive flux (Bird *et al.* 1960). An example of forced diffusion will be encountered in §17.2 [cf. Eq. (17.2-1)].

Table 5.1-1
THE BULK-PHASE SPECIES BALANCE EQUATION (5.1-5)

Cartesian coordinates (x, y, z):

$$\frac{\partial \rho_i}{\partial t} + v_x \frac{\partial \rho_i}{\partial x} + v_y \frac{\partial \rho_i}{\partial y} + v_z \frac{\partial \rho_i}{\partial z} = D_i \left(\frac{\partial^2 \rho_i}{\partial x^2} + \frac{\partial^2 \rho_i}{\partial y^2} + \frac{\partial^2 \rho_i}{\partial z^2} \right) + R_i ;$$

Circular cylindrical coordinates $(R\ \phi, z)$:

$$\frac{\partial \rho_i}{\partial t} + v_R \frac{\partial \rho_i}{\partial R} + v_\phi \frac{1}{R} \frac{\partial \rho_i}{\partial \phi} + v_z \frac{\partial \rho_i}{\partial z}$$

$$= D_i \left[\frac{1}{R} \frac{\partial}{\partial R} \left(R \frac{\partial \rho_i}{\partial R} \right) + \frac{1}{R^2} \frac{\partial^2 \rho_i}{\partial \phi^2} + \frac{\partial^2 \rho_i}{\partial z^2} \right] + R_i ;$$

Spherical coordinates (r, θ, ϕ):

$$\frac{\partial \rho_i}{\partial t} + v_r \frac{\partial \rho_i}{\partial r} + v_\theta \frac{1}{r} \frac{\partial \rho_i}{\partial \theta} + v_\phi \frac{1}{r \sin \theta} \frac{\partial \rho_i}{\partial \phi} =$$

$$D_i \left[\frac{1}{r^2} \frac{\partial}{\partial r} \left(r^2 \frac{\partial \rho_i}{\partial r} \right) + \frac{1}{r^2 \sin \theta} \frac{\partial}{\partial \theta} \left(\sin \theta \frac{\partial \rho_i}{\partial \theta} \right) + \frac{1}{r^2 \sin^2 \theta} \frac{\partial^2 \rho_i}{\partial \phi^2} \right] + R_i .$$

THE FICKIAN SPECIES FLUX VECTOR, Eq. (5.1-2)

Cartesian coordinates:

$$j_{i\,x} = - D_i \frac{\partial \rho_i}{\partial x},$$

$$j_{i\,y} = - D_i \frac{\partial \rho_i}{\partial y},$$

$$j_{i\,z} = - D_i \frac{\partial \rho_i}{\partial z};$$

Circular cylindrical coordinates:

$$j_{i\,R} = - D_i \frac{\partial \rho_i}{\partial R},$$

$$j_{i\,\phi} = - D_i \frac{1}{R} \frac{\partial \rho_i}{\partial \phi},$$

$$j_{i\,z} = - D_i \frac{\partial \rho_i}{\partial z};$$

Spherical coordinates:

$$j_{i\,r} = - D_i \frac{\partial \rho_i}{\partial r},$$

$$j_{i\,\theta} = - D_i \frac{1}{R} \frac{\partial \rho_i}{\partial \theta},$$

$$j_{i\,\phi} = - D_i \frac{1}{r \sin \theta} \frac{\partial \rho_i}{\partial \phi}.$$

$$\frac{\partial C_i}{\partial t} + \mathbf{v} \cdot \nabla C_i = D_i \nabla^2 C_i + r_i , \qquad (5.1\text{-}6)$$

where r_i is the molar rate of production of i.

Component forms of Eq. (5.1-5) for planar, circular cylindrical, and spherical coordinate systems are provided in Table 5.1-1.

5.2 The Equations of Interfacial Species Transport

Macroscale boundary conditions imposed upon the fields appearing in the volumetric species transport equation (5.1-5) [or (5.1-6)] at a moving and deforming fluid interface must be specified for the solute concentration and flux fields.

In the present context of the species concentration field, the equivalent of the velocity continuity condition (3.6-17) across a fluid interface is the bulk interphase partition relation

$$\bar{\rho}_i(0+) = K_i \, \bar{\rho}_i(0-), \qquad (5.2\text{-}1)$$

where K_i is the interphase partition coefficient of species i.* This condition, which describes the *equilibrium* partitioning of surfactant between contiguous bulk phases, is, in the circumstance of diffusion-controlled surfactant transport (cf. §5.4), equally applicable under *dynamic* conditions too, as will be illustrated in §17.2.

The surface-excess species balance equation follows from the generic equation (3.6-16) when applied to the areal property of species mass density [$\psi^s \equiv \rho_i^s$]; explicitly,

$$\frac{\partial \rho_i^s}{\partial t} + \nabla_s \cdot (\mathbf{v}^o \rho_i^s) + \nabla_s \cdot (\mathbf{I}_s \cdot \mathbf{j}_i^s) = R_i^s + \mathbf{n} \cdot \left\| \bar{\mathbf{j}}_i \right\|, \qquad (5.2\text{-}2)$$

with ρ_i^s the surface-excess species mass density, $\mathbf{j}_i^s$ the surface-excess species diffusion flux vector, and R_i^s the surface-excess species production rate, all per unit area. This equation provides the ('jump') boundary condition imposed upon the normal component of the bulk-phase species flux vector $\mathbf{j}_i$ at a phase boundary.

Constitutive relations are required for $\mathbf{j}_i^s$ and R_i^s in (5.2-2). Attention in the following will be confined to circumstances for which the surface-

---

* The appearance of overbars on bulk-phase fields in this and the following two sections corresponds to the notation introduced in §3.6, wherein discontinuous variables were distinguished by an overbar. As in Chapter 4, for the sake of notational simplicity this notation is not carried over to applications (i.e. §5.6).

excess species (surfactant) flux vector may be expressed in the "surface Fickian" form

$$\mathbf{j}_i^s \overset{\text{def}}{=} (\mathbf{v}_i^o - \mathbf{v}^o)\rho_i^s \equiv - D_i^s \nabla_s \rho_i^s, \tag{5.2-3}$$

where D_i^s is the surface diffusivity of species i.* (See §17.3 for a microscale derivation of this relation.) The surface-excess species flux vector $\mathbf{j}_i^s$ is a lineal flux density vector defined wholly within the interface [cf. the footnote pertaining to the appearance of the surface-excess pressure tensor $\mathbf{P}^s$ in Eq. (4.2-3)], such that its surface projection

$$\mathbf{j}_i^s = \mathbf{I}_s \cdot \mathbf{j}_i^s \tag{5.2-4}$$

is equivalent to the flux vector itself. Hence, with use of Eq. (5.2-3), Eq. (5.2-2) may be expressed as

$$\frac{\partial \rho_i^s}{\partial t} + \nabla_s \cdot (\mathbf{v}^o \rho_i^s) - \nabla_s \cdot (D_i^s \nabla_s \rho_i^s) - R_i^s = \mathbf{n} \cdot \left\| \bar{\mathbf{j}}_i \right\|, \tag{5.2-5}$$

or, for the case of constant surface diffusivity D_i^s,

$$\boxed{\frac{\partial \rho_i^s}{\partial t} + \nabla_s \cdot (\mathbf{v}^o \rho_i^s) - D_i^s \nabla_s^2 \rho_i^s - R_i^s = \mathbf{n} \cdot \left\| \bar{\mathbf{j}}_i \right\|,} \tag{5.2-6}$$

where $\nabla_s^2 = \nabla_s \cdot \nabla_s$. In terms of the surface-excess molar species concentration,

$$\Gamma_i^s \overset{\text{def}}{=} \rho_i^s / M_i, \tag{5.2-7}$$

we obtain the molar form of the surface-excess species flux boundary condition

$$\frac{\partial \Gamma_i^s}{\partial t} + \nabla_s \cdot (\mathbf{v}^o \Gamma_i^s) - D_i^s \nabla_s^2 \Gamma_i^s - r_i^s = \mathbf{n} \cdot \left\| \bar{\mathbf{j}}_i \right\|, \tag{5.2-8}$$

with r_i^s the surface-excess molar areal production rate.

Component forms of Eq. (5.2-6) for planar, circular cylindrical and spherical interfaces (cf. Figures B.1-4 to B.1-6) are provided in Table 5.2-1.

---

* Other contributions to the surface-excess species flux vector $\mathbf{j}_i^s$ may arise in addition to the molecular diffusion contribution (5.2-3), these two-dimensional surface fluxes corresponding to the non-convective, three-dimensional fluxes $\mathbf{j}_i^F$ and $\mathbf{j}_i^T$ mentioned in the footnote pertaining to Eq. (5.1-2).

Table 5.2-1
THE SURFACE-EXCESS SPECIES BALANCE EQUATION (5.2-6)

Cartesian coordinates (for the surface z = const.):

$$\frac{\partial \rho_i^s}{\partial t} + \frac{\partial}{\partial x}(v_x^o \rho_i^s) + \frac{\partial}{\partial y}(v_y^o \rho_i^s) = D_i^s\left(\frac{\partial^2 \rho_i^s}{\partial x^2} + \frac{\partial^2 \rho_i^s}{\partial y^2}\right) + R_i^s + \left\| \bar{j}_{iz} \right\|;$$

Circular cylindrical coordinates [for the surface $R = R_0(t)$, say]:

$$\frac{\partial \rho_i^s}{\partial t} + \frac{v_r^o}{R_0}\rho_i^s + \frac{\partial}{\partial z}(v_z^o \rho_i^s) + \frac{1}{R_0}\frac{\partial}{\partial \phi}(v_\phi^o \rho_i^s)$$

$$= D_i^s\left(\frac{\partial^2 \rho_i^s}{\partial z^2} + \frac{1}{R_0^{\ 2}}\frac{\partial^2 \rho_i^s}{\partial \phi^2}\right) + R_i^s + \left\| \bar{j}_{iR} \right\|;$$

Spherical coordinates [for the surface $r = r_0(t)$, say]:

$$\frac{\partial \rho_i^s}{\partial t} + 2\frac{v_r^o}{r_0}\rho_i^s + \frac{1}{r_0 \sin\theta}\frac{\partial}{\partial \theta}(\sin\theta\, v_\theta^o \rho_i^s) + \frac{1}{r_0 \sin\theta}\frac{\partial}{\partial \phi}(v_\phi^o \rho_i^s)$$

$$= D_i^s\left[\frac{1}{r_0^{\ 2}\sin\theta}\frac{\partial}{\partial \theta}\left(\sin\theta\,\frac{\partial \rho_i^s}{\partial \theta}\right) + \frac{1}{r_0^{\ 2}\sin^2\theta}\frac{\partial^2 \rho_i^s}{\partial \phi^2}\right] + R_i^s + \left\| \bar{j}_{ir} \right\|.$$

5.3 Adsorption Kinetics

In addition to the interfacial boundary conditions (5.2-1) and (5.2-6) imposed upon the bulk-phase species transport processes, a kinetic adsorption relation is also required relating the instantaneous value of the bulk-phase species density $\bar{\rho}_i$ at the interface, namely $\bar{\rho}_i^o$ [which may be arbitrarily chosen to be either $\bar{\rho}_i(0+)$ or $\bar{\rho}_i(0-)$, the choice being strictly a matter of convenience], to the surface-excess species density ρ_i^s. For equilibrium conditions such a constitutive relation is known as the *adsorption isotherm.**

Equilibrium Adsorption

The functional constitutive form of the adsorption isotherm depends upon the physicochemical nature of the surfactant (e.g. nonionic, anionic, cationic), the presence or absence of an electrolyte, the thermodynamic ideality of the bulk phase or interface, etc., and has been the subject of extensive investigation [see the reviews of Lucassen-Reynders (1976, 1981)]. This section focuses upon the Langmuir (1918) and Frumkin (1925) isotherms. These two classical adsorption isotherms are commonly encountered in the literature, and possess relatively simple forms. Each presupposes the absence of an electrical double-layer at the interface, implying the adsorption of a *nonionic* surfactant; moreover, each should be considered strictly valid only for thermodynamically ideal bulk solutions (although not necessarily for a thermodynamically ideal interface).

The Frumkin isotherm possesses the form (Frumkin 1925)

$$\frac{\rho_i^s}{\rho_{i\infty}^s - \rho_i^s} \exp\left(-A\frac{\rho_i^s}{\rho_{i\infty}^s}\right) = K_a\bar{\rho}_i^o, \tag{5.3-1}$$

where K_a is the adsorption coefficient, A is a parameter which measures the degree of non-ideality of the interface, and $\rho_{i\infty}^s$ is the *surface-excess saturation density*, the latter corresponding to the maximum realizable surface-excess density of species i.The Langmuir isotherm, which is a special case of the above, further supposes the interface to be *ideal* (A=0), whence

* Specification of the appropriate (equilibrium or nonequilibrium) adsorption relations relative to both bulk phases contiguous to the interface implicitly specifies also the partition relation (5.2-1).

$$\frac{\rho_i^s}{\rho_{i\infty}^s - \rho_i^s} = K_a \bar{\rho}_i^o . \tag{5.3-2}$$

Nonequilibrium Adsorption

Under dynamical, nonequilibrium conditions, equilibrium adsorption isotherms of the type embodied in the preceding equations are generally invalid. In their stead, one rather seeks a kinetic rate expression for the normal component of the bulk-phase surfactant flux $\bar{\mathbf{j}}_i^o$ at the interface in terms of the local surfactant adsorption rate ϕ_i:

$$\mathbf{n} \cdot \bar{\mathbf{j}}_i^o = \phi_i . \tag{5.3-3}$$

The above relation reveals that surfactant transport from the bulk phase to the macroscale interface is a two-step process, whereby surfactant is first transported to the interface by bulk diffusion, thereafter 'adsorbing' to the interface [which latter adsorption step may be viewed as corresponding to a change of state (e.g. associated with orientational degrees of freedom) of the surfactant molecule rather than a physical displacement of surfactant in the direction normal to the interface]. The significance of the relative rates of these two steps (i.e. diffusion and adsorption) is discussed in the following section.

Borwanker & Wasan (1983) have proposed the following kinetic rate expression for Frumkin-like surfactant adsorption behavior in the equilibrium limit:

$$\phi_i = K_a \exp\left[\frac{A}{2}\left(\frac{\rho_i^s}{\rho_{i\infty}^s}\right)^2\right]\left[\bar{\rho}_i^o(\rho_{i\infty}^s - \rho_i^s) - \frac{\rho_i^s}{K_a}\exp\left(-A\frac{\rho_i^s}{\rho_{i\infty}^s}\right)\right]. \tag{5.3-4}$$

In the limit of thermodynamic ideality, with $A = 0$, the corresponding Langmuir-type kinetic expression obtains, namely

$$\phi_i = K_a\left[\bar{\rho}_i^o(\rho_{i\infty}^s - \rho_i^s) - \frac{\rho_i^s}{K_a}\right]. \tag{5.3-5}$$

For only small departures from the equilibrium state of adsorbed surfactant, a complete knowledge of adsorption kinetics, as in Eqs. (5.3-4) and (5.3-5), proves unnecessary; rather, the kinetic adsorption rate may be linearized about the equilibrium state, so as to obtain

$$\begin{aligned}\phi_i &= \phi_i^a - \phi_i^d \\ &= \left(\phi_i^a\right)_o - \left(\phi_i^d\right)_o + \left[\left(\frac{\partial \phi_i^a}{\partial \rho_i^s}\right)_o - \left(\frac{\partial \phi_i^d}{\partial \rho_i^s}\right)_o\right]\left(\rho_i^s - \rho_{i\,0}^s\right) + \dots \\ &\approx \alpha_i\left(\rho_{i0}^s - \rho_i^s\right),\end{aligned} \tag{5.3-6}$$

upon observing that the adsorption (a) and desorption (d) rates are necessarily identical at equilibrium. In such circumstances, the single adsorption parameter required is

$$\alpha_i \equiv \left(\frac{\partial \phi_i^d}{\partial \rho_i^s}\right)_o - \left(\frac{\partial \phi_i^a}{\partial \rho_i^s}\right)_o, \tag{5.3-7}$$

with subscript o representing the equilibrium state.

5.4 Adsorption- or Diffusion-Controlled Surfactant Transport To and From the Interface

Solution of the species (surfactant) transport equation (5.1-5) for a bulk fluid proximate to a fluid interface, requires, in principle, quantitative specification of the following three conditions: (i) the bulk-phase partitioning of surfactant across the fluid interface [i.e. Eq. (5.2-1)]; (ii) the normal bulk-phase species flux [i.e. Eq. (5.2-6)]; (iii) a kinetic adsorption relation [i.e. Eq. (5.3-3)]. An alternative to the above three conditions is *a priori* specification of the bulk-phase surfactant densities $\bar{\rho}_i(0+)$ and $\bar{\rho}_i(0-)$ at the interface. The existence of these alternative modes of specifying the species-specific interfacial boundary conditions is analogous to the linear momentum case of Chapter 4, where it was discussed following Eq. (4.2-20) that by specifying the normal velocity at an interface, the normal stress boundary condition (4.2-20a) is automatically satisfied.

The species flux boundary condition (5.2-6) may thus be circumvented if the bulk concentration fields $\bar{\rho}_i(0+)$ and $\bar{\rho}_i(0-)$ at the fluid interface are known at the outset.

In circumstances such that the normal component flux boundary condition proves unnecessary for solution of the bulk-phase species transport problem, the interfacial boundary condition reduces to a single equation [the surface-excess convection-diffusion equation (5.2-6)] embodying a single unknown—namely, the surface-excess species density ρ_i^s (assuming the interfacial velocity field to be known). Solution of this equation is necessary, for example, to establish the surface tension variation within the interface, whose gradients give rise to Marangoni flow.

Limiting-case solutions of Eq. (5.2-6) may be identified in the respective cases of: (i) *diffusion-controlled*, and (ii) *adsorption-controlled* adsorption kinetics. In the *diffusion-controlled* limit, surfactant is transported slowly by diffusion through the bulk phase to the interface, relative to which process the adsorption step of the overall surfactant mass transfer process appears to occur instantaneously. An equilibrium adsorption relation may therefore be assumed to obtain at the instantaneous conditions, and the known bulk-phase diffusion flux directly substituted into the right-hand side of Eq. (5.2-6), so as to eventually furnish ρ_i^s without having to devote attention to the details of the adsorption kinetics. In the *adsorption-controlled* limit, surfactant is transported rapidly to the interface via diffusion and/or convection, and the adsorption step becomes rate limiting. The surface-excess surfactant balance equation (5.2-6) may thus be regarded as divorced from the bulk-phase surfactant transport equation. Substitution of Eq. (5.3-3) into Eq. (5.2-6), jointly with an appropriate constitutive choice of kinetic rate expression, then provides a single equation for determining ρ_i^s.

5.5 The Surface Equation of State

The surface equation of state poses an equilibrium relation between interfacial tension and interfacial composition. Like the comparable equation of state for three-dimensional fluids, the surface equation of state (developed for equilibrium conditions) is usually assumed to be equally applicable to nonequilibrium conditions involving the *instantaneous* composition of the interface.

The Gibbs surface equation of state has been mentioned briefly in Chapter 2 [cf. Eq. (2.1-5)]. While the classical Gibbs equation is often useful for fluid interfaces possessing only a trace amount of surfactant, it is not generally applicable to concentrated monolayers.

Surface equations of state corresponding to the Frumkin and Langmir isotherms are respectively provided by the following expressions (Borwanker & Wasan 1983) for the '*surface pressure*':

$$\pi \overset{\text{def}}{=} \sigma_o - \sigma \equiv -\frac{RT}{M}\rho_{i\infty}^s\left[\ln\left(1-\frac{\rho_i^s}{\rho_{i\infty}^s}\right)+\frac{A}{2}\left(\frac{\rho_i^s}{\rho_{i\infty}^s}\right)^2\right] \tag{5.5-1}$$

and, with $A \to 0$,

$$\pi = \sigma_o - \sigma \equiv -\frac{RT}{M}\rho_{i\infty}^s \ln\left(1-\frac{\rho_i^s}{\rho_{i\infty}^s}\right), \tag{5.5-2}$$

where R is the gas constant and T the temperature.* Upon combining Eq. (5.5-2) with the Langmuir isotherm, Eq. (5.3-2), we obtain

---

* Observe that in the limit of a dilute surfactant monolayer, $\varepsilon \equiv \rho_i^s / \rho_{i\infty}^s << 1$, Eq. (5.5-2) reduces to

$$\pi = \sigma_o - \sigma = \frac{RT}{M}\rho^s_{i\infty} \ln\left(1 + K_a\bar{\rho}^o_i\right), \tag{5.5-3}$$

which is called the *Szyskowsky equation* (von Szyskowsky 1910). [Recall that the bulk-phase surfactant density $\bar{\rho}^o_i$ appearing in the above may be arbitrarily chosen to be either $\bar{\rho}_i(0+)$ or $\bar{\rho}_i(0-)$, the choice being strictly a matter of convenience since K^+_a and K^-_a will differ by the requisite amount necessary to make $K_a\bar{\rho}^o_i$ independent of the choice.]

Many, if not most, analytical investigations of Marangoni phenomena assume only small deviations from the equilibrium surfactant adsorption state. The theoretical simplification embodied in this limiting case stems from the fact that knowledge of the variation of interfacial tension with surfactant density over the *entire* range of interfacial compositions is now unnecessary as a consequence of the following arguments. Linearization of the interfacial tension about the equilibrium state yields

$$\begin{aligned}\sigma &= \sigma_o + \left(\frac{\partial\sigma}{\partial\rho^s_i}\right)_o (\rho^s_i - \rho^s_{io}) + \ldots \\ &\approx \sigma_o - \frac{1}{\rho^s_o} E^i_o(\rho^s_i - \rho^s_{io}),\end{aligned} \tag{5.5-4}$$

wherein

$$\boxed{E^i_o \overset{\text{def}}{=} -\left(\frac{\partial\sigma}{\partial \ln \rho^s_i}\right)_o} \tag{5.5-5}$$

is the Gibbs elasticity for the surfactant species i.* Here, partial differentiation is used to indicate differentiation at fixed values of the surface coordinates (q^1, q^2). In cases to which Eq. (5.5-4) applies, the interfacial tension gradient may be expressed in terms of the local surfactant density

---

$$\pi = \sigma_o - \sigma \equiv \frac{RT}{M}\rho^s_i + O(\varepsilon),$$

which is the Gibbs adsorption isotherm.

* An alternative definition of the Gibbs elasticity is (Rusanov & Krotov 1979)

$$E^i_o = \delta A\left(\frac{\partial\sigma}{\partial\delta A}\right)_o.$$

Here, δA is an infinitesimal area, as in Eq. (3.4-20), and $\partial\sigma$ is the differential increase in interfacial tension at the point (q^1, q^2) centered at δA owing to the differential change $\partial\delta A$. This definition may be demonstrated to be equivalent to that provided in Eq. (5.5-5) only for the special circumstances of an insoluble monolayer, and then only at large surface Peclét numbers (see Question 5.7).

ρ_i^s, with the Gibbs elasticity E_o^i appearing as the single material parameter of the interface; explicitly,

$$\nabla_s \sigma = -\frac{1}{\rho_{io}^s} E_o^i \nabla_s \rho_i^s . \tag{5.5-6}$$

This relation couples the interfacial stress condition (4.2-20) to the interfacial surfactant flux condition (5.2-6).

5.6 Examples

To illustrate the surfactant transport equations of the preceding sections, two hydrodynamic problems resulting in Marangoni flows will be considered. The first addresses an adsorption-controlled surfactant transport process, whereas the second pertains to a diffusion-controlled process.

1. Creeping Flow Past an Emulsion Droplet: Adsorption-Controlled Marangoni Flow

In the first example of §4.4, the problem of a spherical droplet translating though a quiescent, viscous fluid, and upon whose surface insoluble surfactant is adsorbed, was considered. Upon supposing satisfaction of the dual inequalities

$$\mathrm{Re} = \frac{2a\rho U}{\mu} << 1, \qquad \mathrm{Pe}_s = \frac{2aU}{D^s} << 1,$$

the interfacial dilatational viscosity κ^s was demonstrated to be the exclusive agency responsible for hindering interfacial motion of the fluid droplet [cf. Eq. (4.4-24)], causing the fluid interface to become increasingly rigid (with decreasing radius, in accordance with experimental findings).

Since publication of the original interfacial viscosity solution of Boussinesq (1913b), several studies (see the historical review of §1.1) have suggested that accompanying the translational motion of the fluid droplet, adsorbed surfactant is swept along the interface towards the rear of the droplet. The resulting inhomogeneity in surfactant concentration acts to create interfacial tension gradients along the droplet interface; in turn, these give rise to forces which tend to resist translational droplet motion by way of the Marangoni effect (see Figure 5.6-1).

This section offers an alternative solution to that of §4.4 for the sedimentation velocity U of a fluid sphere, appropriate to the conditions of small Reynolds number, *large* surface Peclet number, and small Boussinesq number:

$$\mathrm{Re} = \frac{2a\rho U}{\mu} << 1; \quad \mathrm{Pe}_s = \frac{2aU}{D^s} >> 1; \quad \mathrm{Bo}_\kappa = \frac{\kappa^s}{\mu a} << 1. \tag{5.6-1}$$

Physically, these conditions imply that: (i) inertial forces arising during the bulk fluid motion are negligible in comparison with comparable viscous forces (thereby allowing the *a priori* assumption of a spherical interface, as in

§4.4); (ii) surface convection dominates surface diffusion, thereby creating maximal conditions for the generation of interfacial tension gradients (since the effect of surface diffusion is to eliminate surfactant concentration gradients and concomitant interfacial tension gradients in the interface); (iii) bulk-phase viscous forces are much larger than interfacial viscous forces, whence interfacial rheological stresses arise only from the existence of interfacial tension gradients.

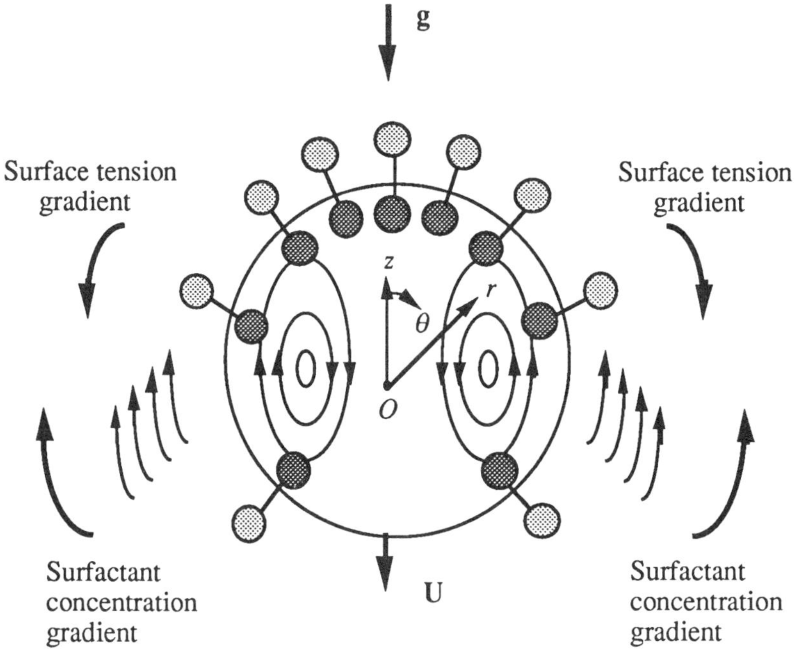

Figure 5.6-1 Surface tension gradients are created by the convective sweeping of adsorbed surfactant to the rear of the settling droplet. This phenomenon acts to diminish the sedimentation velocity of the droplet as a whole. (Surfactant molecules are disproportionately idealized in the figure.)

The present boundary value problem differs from that formulated in Eqs. (4.4-8)–(4.4-14) only in the respective forms adopted by the tangential, (4.4-13), and normal, (4.4-14), stress conditions. Thus, for the conditions embodied in Eqs. (5.6-1), the corresponding interfacial stress conditions are, respectively (see Table 4.2-3):

$$\tilde{\mu} a \frac{\partial}{\partial r}\left(\frac{\tilde{v}_\theta}{r}\right) - \mu a \frac{\partial}{\partial r}\left(\frac{v_\theta}{r}\right) = \frac{1}{a}\frac{\partial \sigma}{\partial \theta} \quad \text{at } r = a \qquad (5.6\text{-}2)$$

and

$$-\tilde{p} + 2\tilde{\mu}\frac{\partial \tilde{v}_r}{\partial r} + p - 2\mu\frac{\partial v_r}{\partial r} = -\frac{2\sigma}{a} \quad \text{at } r = a. \qquad (5.6\text{-}3)$$

As in Eq. (5.5-6), the existence of the interfacial tension gradient term in Eq. (5.6-2) couples the interfacial hydrodynamic problem to the interfacial surfactant transport problem. Following the classical Levich (1962) approach, it will be assumed in what follows that: (i) a single surfactant species is adsorbed from the continuous liquid at the interface; (ii) the surfactant transport process is adsorption-controlled (in the sense previously discussed in §5.4); (iii) only a small departure from the equilibrium surface-excess surfactant density state occurs, and this nonequilibrium state arises from the (purely) convective interfacial transport of surfactant towards the rear of the settling droplet. Thus, upon suppressing the index i that would have otherwise appeared as representing the (single) surfactant species, we write

$$\rho^s = \rho_o^s + \varepsilon\hat{\rho}^s, \tag{5.6-4a}$$

wherein

$$\varepsilon << 1, \tag{5.6-4b}$$

with the perturbation term $\hat{\rho}^s$ an algebraically-signed scalar of $O(1)$ with respect to ε. As in Eq. (5.5-6), the preceding assumption (5.6-4) furnishes the following explicit relation between the interfacial tension gradient at the droplet surface and the surfactant density gradient at the interface:

$$\frac{\partial \sigma}{\partial \theta} = -\frac{1}{\rho_o^s} E_o \frac{\partial \rho^s}{\partial \theta}. \tag{5.6-5}$$

According to Eq. (4.4-18),

$$v_\theta^o = U \sin\theta \left(-1 - \frac{E}{a^3} + \frac{C}{a}\right), \tag{5.6-6}$$

such that the surface-excess species transport equation (5.2-6), together with Eq. (5.3-3) (see Table 5.2-1), provides the expression

$$\frac{2\rho_o^s}{a}\left(-1 - \frac{E}{a^3} + \frac{C}{a}\right)U \cos\theta = \phi, \tag{5.6-7}$$

to leading order in ε. The assumptions (5.6-1) of large surface Peclét number and adsorption-controlled surfactant transport have been utilized in obtaining this relation.

According to the condition (5.6-4), Eqs. (5.3-6) and (5.3-7) may be used to supply the leading-order kinetic rate expression. Thus, with use of Eq. (5.6-7), it follows that

$$\frac{2\rho_o^s}{a}\left(-1 - \frac{E}{a^3} + \frac{C}{a}\right)U \cos\theta = \alpha\,(\rho_o^s - \rho^s), \tag{5.6-8}$$

which, together with Eq. (5.6-5), yields the expression

$$\frac{\partial \sigma}{\partial \theta} = -\frac{2E_o}{a\alpha}\left(-1 - \frac{E}{a^3} + \frac{C}{a}\right)U \sin\theta, \tag{5.6-9}$$

to leading order in ε.

The tangential and normal stress conditions, (5.6-2) and (5.6-3) respectively, may now be expressed explicitly as

$$\tilde{\mu} a \frac{\partial}{\partial r}\left(\frac{\tilde{v}_\theta}{r}\right) - \mu a \frac{\partial}{\partial r}\left(\frac{v_\theta}{r}\right) = -\frac{2E_o}{a^2\alpha}\left(-1 - \frac{E}{a^3} + \frac{C}{a}\right)\sin\theta \qquad \text{at } r = a \tag{5.6-10}$$

and

$$-\tilde{p} + 2\tilde{\mu}\frac{\partial \tilde{v}_r}{\partial r} + p - 2\mu\frac{\partial v_r}{\partial r} = -\frac{2\sigma_o}{a} \qquad \text{at } r = a, \tag{5.6-11}$$

each being valid to leading order in ε. Upon substituting the trial solutions (4.4-15)–(4.4-18) into the Eq. (5.6-10), together with the velocity conditions (4.4-11) and (4.4-12), the four unknown coefficients A, B, C and E appearing in these trial solutions may be evaluated (Question 5.8).

The *a priori* assumption of a spherical droplet shape may be verified *a posteriori* by showing that the velocity and pressure fields provided by Eqs. (4.4-15)–(4.4-20) satisfy the present form, (5.6-11), of the normal stress boundary condition. Substitution of Eqs. (4.4-15)–(4.4-20), (4.4-6) and (4.4-7) into Eq. (5.6-11) furnishes the following algebraic relation:

$$\tilde{K} - K - \frac{2\sigma_o}{a} + U\cos\theta\left[Ba\left(12\tilde{\mu} + 16\frac{E_o}{\alpha a}\right) - \frac{6\mu C}{a^2} - \frac{12\mu E}{a^4} + \frac{8AE_o}{\alpha a^2} + \frac{(\tilde{\rho} - \rho)ga}{U}\right] = 0. \tag{5.6-12}$$

(Note that the ρ's appearing here refer to the respective bulk-phase densities, rather than to areal surfactant concentrations.) As this equation must hold at all surface positions θ, the normal stress boundary condition may be satisfied identically for the assumed spherical droplet shape by separately requiring that

$$\tilde{K} - K - \frac{2\sigma_o}{a} = 0 \tag{5.6-13}$$

and

$$\frac{(\tilde{\rho} - \rho)ga}{U} - \mu\left[\frac{6C}{a^2} + \frac{12E}{a^4} - \frac{8AE_o}{\alpha\mu a^2} - Ba\left(12\frac{\tilde{\mu}}{\mu} + 16\frac{E_o}{\alpha\mu a}\right)\right] = 0. \tag{5.6-14}$$

The former equation represents the equilibrium Laplace condition (2.1-3) whereas the latter may be shown (see Question 5.9) to furnish a relation between the settling speed U of the droplet and the viscous/buoyancy terms (this speed being obtained alternatively by balancing the net hydrodynamic drag force with the buoyancy and gravitational forces acting upon the droplet). From Eq. (5.6-14), together with the explicit algebraic expressions obtained (Question 5.8) for the coefficients A, B, C and E, it follows that

$$U = \frac{2}{9}a^2\frac{(\tilde{\rho} - \rho)}{\mu}g\left[1 + \frac{1}{2}\left(1 + \frac{E_o}{\alpha a \mu} + \frac{3}{2}\frac{\tilde{\mu}}{\mu}\right)^{-1}\right]. \qquad (5.6\text{-}15)$$

It may be seen through comparison of Eq. (5.6-15) with Eq. (4.4-24) that the effect of interfacial tension gradients upon the settling velocity of the droplet is similar to that arising from the existence of an interfacial viscosity [bearing in mind that Eq. (5.6-15) was derived on the limiting basis of only small interfacial tension gradients, as well as adsorption-dominated surfactant transport].

Appearing in Eq. (5.6-15) is an important dimensionless group known as the *Marangoni* number. The latter may be defined generically by the ratio

$$\mathrm{Ma} = \frac{\text{interfacial dilatational elasticity}}{\text{surfactant adsorption parameter} \times \text{bulk viscosity}},$$

and appears explicitly in Eq. (5.6-15) in the form

$$\mathrm{Ma} = \frac{E_o}{\alpha a \mu}.$$

The Marangoni number represents the ratio of interfacial tension gradient forces to bulk viscosity forces. As may be observed by the form of the number, interfacial tension gradient effects are minimized either by small dilatational surface elasticity or fast surfactant adsorption kinetics.

Explicit values for the Gibbs elasticity E_o and adsorption parameter α may be obtained for specified adsorption kinetic models (see Question 5.11). For example, the Frumkin-type constitutive equation (5.3-1) predicts that

$$E_o = \frac{RT}{M}\rho_\infty^s\left\{\left(\frac{\rho_o^s / \rho_\infty^s}{1 - \rho_o^s / \rho_\infty^s}\right) - A\left(\frac{\rho_o^s}{\rho_\infty^s}\right)^2\right\}.$$

The Langmuir-type model, Eq. (5.3-2), which yields

$$E_o = \frac{RT}{M}\rho_\infty^s\left(\frac{\rho_o^s / \rho_\infty^s}{1 - \rho_o^s / \rho_\infty^s}\right)$$

for the Gibbs elasticity, is equally applicable to the above analysis. The kinetic factor α may be straightforwardly determined for each of these models (see Question 5.11).

2. Waves at a Fluid Interface: Diffusion-Controlled Marangoni Flow

A surfactant monolayer (or thin layer of oil) spread at a fluid interface has the effect of damping surface wave motion. This phenomenon owes to the fact that as the surfactant monolayer is compressed and expanded during the course of the wave motion, corresponding oscillations in the local surface-excess surfactant concentration occasion comparable oscillations in the local interfacial tension. As illustrated below, it is a combination of Marangoni and interfacial viscosity effects that results in damping of the wave motion.

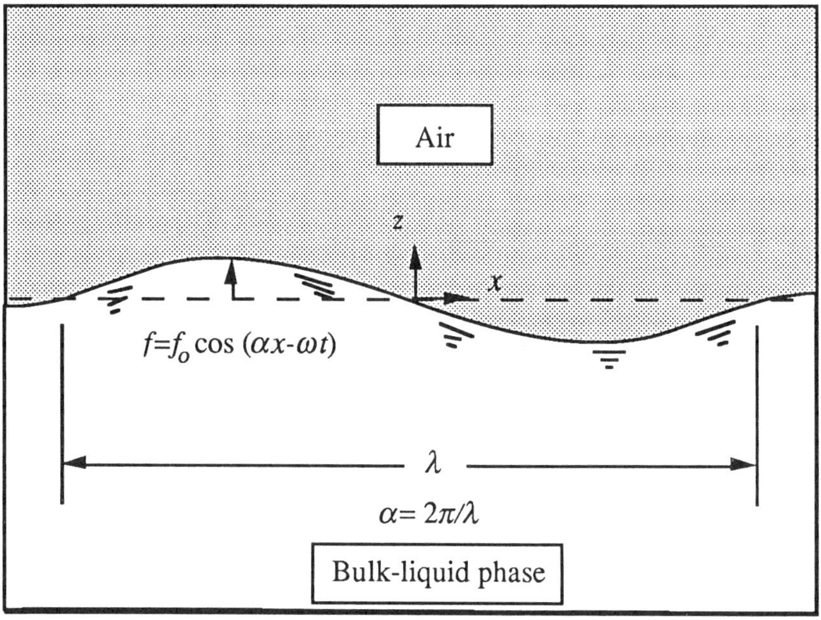

Figure 5.6-2 A capillary wave propagating at a liquid/air surface.

As depicted in Figure 5.6-2, imagine a capillary wave at a liquid/gas surface, propagating with an amplitude

$$f = f_o \exp(-\beta x)\cos(\alpha x - \omega t), \qquad (5.6\text{-}16)$$

where β is a damping coefficient, α the wave number (not to be confused with the α coefficient appearing in the preceding example), and ω the wave frequency. Following the classical approach (Hansen & Ahmad 1971) to investigating the ensuing capillary wave motion, we restrict attention to small-amplitude waves,

$$\alpha f_o << 1, \qquad (5.6\text{-}17)$$

while simultaneously supposing that the bulk-phase Reynolds and Peclét numbers are small, whereas the surface Peclét number is large:*

$$\mathrm{Re} = \frac{\rho f_o \omega}{\alpha \mu} << 1, \quad \mathrm{Pe} = \frac{f_o \omega}{\alpha D} << 1, \quad \mathrm{Pe}_s = \frac{f_o \omega}{\alpha D^s} >> 1. \qquad (5.6\text{-}18)$$

In these circumstances, the bulk hydrodynamic equations reduce to the unsteady Stokes equations (see Table 4.1-1), namely

$$\rho \frac{\partial v_x}{\partial t} = - \frac{\partial p}{\partial x} + \mu \left(\frac{\partial^2 v_x}{\partial x^2} + \frac{\partial^2 v_x}{\partial z^2} \right) \qquad (5.6\text{-}19)$$

and

$$\rho \frac{\partial v_z}{\partial t} = - \rho g - \frac{\partial p}{\partial z} + \mu \left(\frac{\partial^2 v_z}{\partial x^2} + \frac{\partial^2 v_z}{\partial z^2} \right), \qquad (5.6\text{-}20)$$

subject to the following boundary conditions:†

$$v_x = v_z = 0 \quad \text{as} \quad z \to -\infty, \qquad (5.6\text{-}21)$$

$$v_z = \frac{\partial f}{\partial t} \quad \text{at} \quad z = f, \qquad (5.6\text{-}22)$$

$$\mu \left(\frac{\partial v_x}{\partial z} + \frac{\partial v_z}{\partial x} \right) = \frac{\partial \sigma}{\partial x} + (\kappa^s + \mu^s) \frac{\partial^2 v_x}{\partial x^2} \quad \text{at} \quad z = f \qquad (5.6\text{-}23)$$

and

$$p - \tilde{p} - 2\mu \frac{\partial v_z}{\partial z} = - \sigma \frac{\partial^2 f}{\partial x^2} \quad \text{at} \quad z = f. \qquad (5.6\text{-}24)$$

Equation (5.6-23) which represents the tangential stress condition, follows directly from Table 4.2-1. The form of the normal stress condition (5.6-24) differs from that appearing in Table 4.2-1, owing to the existence of a

---

* The physical assumptions embodied in the inequalities (5.6-18) may be regarded as implying, in addition to (5.6-17), small wave frequency ω and small surfactant surface diffusivity D^s. The restrictions placed herein upon the bulk and surface Peclét numbers have been removed in the analysis of Hedge & Slattery (1971).

† The kinematical condition (5.6-22) may be derived by noting that the interface is material, whence it convects with the fluid motion. Thus, the surface equation (see Question 3.8)

$$F = f(x) - z = 0,$$

must satisfy

$$\begin{aligned} \frac{D_s F}{Dt} &= \frac{\partial F}{\partial t} + \mathbf{v} \cdot \nabla F x \\ &= \frac{\partial f}{\partial t} + v_x \frac{\partial f}{\partial x} - v_z \\ &= 0. \end{aligned}$$

Maintaining only the linear, leading-order terms in the above expression reproduces Eq. (5.6-22). The authors wish to thank Tal Hocherman for pointing out this generalization.

nonzero mean surface curvature term resulting from the wave perturbation to the surface (see Question 3.11). It should be noted that the precise surface shape [i.e. $f \equiv f(x, t)$] has not been specified *a priori* in this problem (as has been the case in all previous examples). Thus, the normal stress condition (5.6-24) must be imposed upon the velocity (v_x, v_z) and pressure (p) fields together with the other boundary conditions, so as to concurrently derive for the shape f of the interface simultaneously with the solutions for the fields v_x, v_z and p.

Solution of the hydrodynamic problem posed above is facilitated by introducing an imaginary wave number, k, defined as

$$k = \alpha + i\beta, \tag{5.6-25}$$

leading to the introduction of a complex wave amplitude

$$f^* = f_o^* \exp[i(kx - \omega t)], \tag{5.6-26}$$

whose real part furnishes Eq. (5.6-16). Owing to the convenience of the complex representation (5.6-26) [as opposed to (5.6-16)], a solution will be sought in the complex plane for the velocity and pressure fields v_x, v_z and p. As with (5.6-26), an asterisk will be postscripted to all subsequent complex variables, with an understanding that the corresponding *physical* variables are furnished by the real part of these expressions.

Owing to the linearity of the problem, each (complex) velocity component may be expressed as a linear combination of two independent (complex) scalar functions ϕ^* and φ^*, as

$$v_x^* = -\frac{\partial \phi^*}{\partial x} + \frac{\partial \varphi^*}{\partial z} \tag{5.6-27}$$

and

$$v_z^* = -\frac{\partial \phi^*}{\partial z} - \frac{\partial \varphi^*}{\partial x}. \tag{5.6-28}$$

Here, ϕ^* represents the *scalar* potential, obeying the equation

$$\nabla^2 \phi^* = 0, \tag{5.6-29}$$

whereas φ^* represents the y-component of a *vector* potential function Ψ^*, satisfying

$$\rho \frac{\partial \varphi^*}{\partial t} = \mu \nabla^2 \varphi^*. \tag{5.6-30}$$

For later reference, the expression for the velocity field may be rewritten in an invariant form. Explicitly, the real, physical scalar and vector potentials (ϕ, Ψ) appear in the following representations (see Question 5.2 for a physical interpretation of the respective $\mathbf{v}_1$ and $\mathbf{v}_2$ motions; see also §§6.2 and 8.1):

$$\mathbf{v}_1 = -\nabla\phi \tag{5.6-31}$$

and

$$\mathbf{v}_2 = -\nabla\times\Psi, \tag{5.6-32}$$

where $\mathbf{v}=\mathbf{v}_1+\mathbf{v}_2$ and $\Psi = \mathbf{i}_y\varphi$. Corresponding to Eqs. (5.6-29) and (5.6-30), are the respective forms

$$\rho\frac{\partial \mathbf{v}_1}{\partial t} = -\nabla p + \rho\mathbf{g}, \tag{5.6-33}$$

$$\rho\frac{\partial \mathbf{v}_2}{\partial t} = \mu\nabla^2\mathbf{v}_2. \tag{5.6-34}$$

Equations (5.6-29) and (5.6-30), satisfying Eqs. (5.6-21), (5.6-22) and (5.6-26), possess the respective solutions

$$\phi^* = A\exp(kz)\exp[i(kx - \omega t)] \tag{5.6-35}$$

and

$$\varphi^* = B\exp(mz)\exp[i(kx - \omega t)]. \tag{5.6-36}$$

Substitution of Eq. (5.6-36) into Eq. (5.6-30) yields

$$m = \sqrt{k^2 - \left(\frac{i\omega\rho}{\mu}\right)} \tag{5.6-37}$$

for the complex exponential factor m. Introduction of Eqs. (5.6-35) and (5.6-36) into Eqs. (5.6-27) and (5.6-28) provides the following expressions for the complex velocity fields:

$$v_x{}^* = [-ikA\exp(kz) + mB\exp(mz)]\exp[i(kx - \omega t)] \tag{5.6-38}$$

and

$$v_z{}^* = [-kA\exp(kz) - ikB\exp(mz)]\exp[i(kx - \omega t)]. \tag{5.6-39}$$

Furthermore, with use of Eqs. (5.6-19) and (5.6-20) we obtain

$$p^* = -i\rho\omega A\exp(kz)\exp[i(kx - \omega t)] - \rho gz, \tag{5.6-40}$$

for the complex pressure field, with $\mathbf{g} = -\mathbf{i}_z g$ the acceleration of gravity. Finally, upon substitution of Eq. (5.6-38) into Eq. (5.6-22) there follows for the complex wave amplitude,

$$f^* = \frac{k}{\omega}(-iA + B)\exp[i(kx - \omega t)]. \tag{5.6-41}$$

The real portions of Eqs. (5.6-38)–(5.6-41) provide the physical solutions for the principal variables of the propagating wave problem in terms of the two constants A and B, whose values must be determined by substituting these relations into conditions (5.6-23) and (5.6-24).

For this purpose explicit expressions for the surface tension and gradient thereof in terms of the surface-excess species density of adsorbed surfactant, are required. Unlike the previous example, where it was possible to deduce this dependence without an explicit solution of the bulk-phase species transfer problem, the full mass transfer problem cannot be avoided for the present case of rapid adsorption kinetics.

As in (5.6-4), we are concerned with the case of small deviations of order ε from the equilibrium state:

$$\rho^s = \rho_o^s + \varepsilon\hat{\rho}^s, \tag{5.6-42}$$

and

$$\rho = \rho_o + \varepsilon\hat{\rho}, \tag{5.6-43}$$

where ε is a small dimensionless parameter (and, as before, the ρ's appearing without an identifying index i refer to surfactant species densities). For the conditions described by (5.6-18), Tables 5.1-1 and 5.2-1 provide the following surfactant transport problem:

$$\frac{\partial\rho}{\partial t} = D\left(\frac{\partial^2\rho}{\partial x^2} + \frac{\partial^2\rho}{\partial z^2}\right), \tag{5.6-44}$$

subject to

$$\rho = \rho_o \qquad \text{at} \quad z \to -\infty \tag{5.6-45}$$

and

$$\frac{\partial\rho^s}{\partial t} + \rho_o^s\frac{\partial v_x}{\partial x} = -D\frac{\partial\rho}{\partial z} \qquad \text{at} \quad z = f. \tag{5.6-46}$$

An additional relation is required between the respective surface-excess and bulk-phase species densities. For diffusion-controlled surfactant transport (i.e. rapid adsorption kinetics), this relation is provided by an equilibrium adsorption isotherm. Owing to the conditions (5.6-42) and (5.6-43), the adsorption isotherm may be approximated by the linear relation

$$\rho \approx \rho_o + \left(\frac{\partial\rho}{\partial\rho^s}\right)_0 \varepsilon\hat{\rho}^s \quad \text{at } z = f. \tag{5.6-47}$$

Subject to *a posteriori* verification, we propose the following (complex) solutions to the preceding surfactant transfer problem, satisfying condition (5.6-45):

$$\rho^* = \rho_o + G\,\exp(nz)\exp[i(kx - \omega t)] \tag{5.6-48}$$

and

$$\rho^{s*} = \rho_o^s + H\,\exp[i(kx - \omega t)], \tag{5.6-49}$$

with

$$n = \sqrt{k^2 - \frac{i\omega}{D}} \,. \tag{5.6-50}$$

Upon substitution of the above into Eq. (5.6-47), it follows that

$$G = H\left(\frac{\partial \rho}{\partial \rho^s}\right)_o \,. \tag{5.6-51}$$

Moreover, upon substitution of Eqs. (5.6-48) and (5.6-49) into Eq. (5.6-46) one obtains

$$H = \rho_o^s \frac{k^2 A + ikmB}{i\omega - Dn\left(\dfrac{\partial \rho}{\partial \rho^s}\right)_o} \,. \tag{5.6-52}$$

From Eqs. (5.6-48) and (5.6-49), the following solutions obtain to leading order in ε:

$$\rho^* = \rho_o\left[1 + \left(\frac{\partial \rho}{\partial \rho^s}\right)_o \left(\frac{k^2 A + ikmB}{i\omega - Dn\left(\dfrac{\partial \rho}{\partial \rho^s}\right)_o}\right) \exp(nz)\right] \exp[i(kx - \omega t)] \tag{5.6-53}$$

and

$$\rho^{s*} = \rho_o^s\left[1 + \left(\frac{k^2 A + ikmB}{i\omega - Dn\left(\dfrac{\partial \rho}{\partial \rho^s}\right)_o}\right)\right] \exp[i(kx - \omega t)] \,. \tag{5.6-54}$$

Consistent with Eq. (5.6-42), a linear approximation to the interfacial tension is employed [cf. Eq. (5.5-4)]; together with Eq. (5.6-54) this provides the following leading-order expression for the complex interfacial tension gradient:

$$\frac{\partial \sigma^*}{\partial x} = \frac{\dfrac{k^2}{\omega} E_o(kA + imB)}{1 - \sqrt{\tau\left[\dfrac{2}{i\omega}\left(\dfrac{k^2}{i\omega D} - 1\right)\right]}} \exp[i(kx - \omega t)] \,, \tag{5.6-55}$$

with τ a diffusion parameter given by

$$\tau = \frac{D}{2}\left(\frac{\partial \rho}{\partial \rho^s}\right)_o^2 \,. \tag{5.6-56}$$

In many capillary wave theory applications, the wavelength of the propagating wave is sufficiently large such as to satisfy the inequality $k^2/\omega\, D \ll 1$ (Hansen & Ahmad 1971). In this case, Eq. (5.6-55) reduces to the expression

$$\frac{\partial \sigma^*}{\partial x} = -\frac{\frac{k^2}{\omega} E_o(kA + imB)}{1 - \sqrt{\tau \frac{2i}{\omega}}} \exp[i(kx - \omega t)]. \qquad (5.6\text{-}57)$$

[See Question 5.10 for the more general case where Eq. (5.6-55) applies.]

Substitute Eqs. (5.6-38)-(5.6-41) into the interfacial stress conditions (5.6-23) and (5.6-24), together with Eq. (5.6-57). Thereby, the following two algebraic equations are obtained in terms of the constants A and B:

$$A\left(-i\rho\omega^2 - 2\mu k^2\omega + i\sigma_o k^3 + i\rho g k\right) - B\left(2\omega\mu i k m + \sigma_o k^3 + \rho g k\right) = 0 \qquad (5.6\text{-}58)$$

and

$$A\left\{2\mu i k^2\omega - k^3[E^* - i\omega(\kappa^s + \mu^s)]\right\} - B\left\{\mu(m^2 + k^2)\omega + imk^2[E^* - i\omega(\kappa^s + \mu^s)]\right\} = 0, \quad (5.6\text{-}59)$$

where the dilatational modulus [cf. Eq. (4.3-8)] is here defined as

$$E^* \equiv \frac{E_o\left[1 + (\tau/\omega)^{1/2}\right]}{\left[1 + 2(\tau/\omega)^{1/2} + 2(\tau/\omega)\right]} - i\frac{E_o(\tau/\omega)^{1/2}}{\left[1 + 2(\tau/\omega)^{1/2} + 2(\tau/\omega)\right]}. \qquad (5.6\text{-}60)$$

In order that a solution of Eqs. (5.6-58) and (5.6-59) exist, the determinant of the coefficients must vanish. This gives

$$\begin{aligned}&\left(-i\rho\omega^2 - 2\mu k^2\omega + i\sigma k^3 + i\rho g k\right)\\ &\times\left\{i\mu\left(m^2 + k^2\right)\omega - mk^2[E^* - i\omega(\kappa^s + \mu^s)]\right\}\\ &+\left(2\omega\mu i k m + \sigma k^3 + \rho g k\right)\\ &\times\left\{2\mu k^2\omega + ik^3[E^* - i\omega(\kappa^s + \mu^s)]\right\} = 0,\end{aligned} \qquad (5.6\text{-}61)$$

which represents a dispersion relation between the imaginary wave number k and wave frequency ω.

Dispersion relations such as Eq. (5.6-61) are employed in analyses of capillary and longitudinal waves. The accompanying interfacial flows are often utilized in methods for measuring interfacial dilatational properties, as is further discussed in §§6.2 and 8.1

5.7 Summary

Inhomogeneities in surfactant concentration at a fluid interface give rise to interfacial tension gradients. These, in turn, either enhance or impede motion within the interface and adjacent bulk-fluid phases. The creation of interfacial tension gradients couples surfactant species and momentum transport problems, thereby requiring simultaneous consideration of the equations of interfacial species transfer.

In Chapter 5 the interfacial species flux boundary condition (5.2-6) has been developed for a Fickian fluid interface. This condition, together with a surfactant partitioning relation (5.2-1) and kinetic adsorption relation (5.3-3), constitutes the formulation of the species transport process within the bulk phases contiguous to the interface, by furnishing the interfacial boundary conditions imposed upon these bulk fields. For circumstances wherein the bulk-phase concentration at the fluid interface is known, solution of the surface-excess species balance equation (5.2-6) may be sought in the limits either of diffusion or adsorption-controlled surfactant mass transfer, as discussed in §5.4. Surface equations of state have been discussed in §5.5, and examples of adsorption-controlled and diffusion-controlled Marangoni problems presented in §5.6.

Questions for Chapter 5

5.1 If it is assumed that an equilibrium adsorption isotherm obtains at a dynamic fluid interface, which of the three conditions, (5.2-1), (5.2-6), or (5.3-3) would this isotherm replace in the species transfer problem statement? Would this assumption be appropriate to adsorption- or diffusion-controlled surfactant transport?

5.2 The two velocities defined by Eqs. (5.6-31) and (5.6-32) represent different types of wave motion: one corresponds to a *longitudinal* (or compressional) wave and the other to a *capillary* (or ripple) wave. *i.* Determine, by substituting Eqs. (5.6-31) and (5.6-32), into Eqs. (5.6-27) and (5.6-28), the identity of each velocity field; *ii.* Which type of wave is dissipative?; *iii.* Which type of wave would you expect to be most sensitive to surfactant adsorption/desorption phenomena?

5.3 Suppose that the surface-active species in the wave problem of §5.6 adsorbs from the gas phase. In this case the interfacial species flux boundary condition (5.6-46) will appear as

$$\frac{\partial \rho^s}{\partial t} + \rho_o^s \frac{\partial v_x}{\partial x} = - D \frac{\partial \rho}{\partial z} + K\,(\rho - \rho_o^g) \qquad \text{at} \quad z = f\,,$$

where K is the solute mass transfer coefficient from the gas phase, and ρ_o^g the mass density of solute in the gas phase. Derive an expression for the dispersion relation that generalizes Eq. (5.6-57) to include this effect.

5.4 A spherical gas bubble is animated within an otherwise quiescent fluid by the action of an (undisturbed) uniform temperature gradient **G** as depicted in the figure below.* As illustrated, the cause of the resulting bubble migration is the interfacial-tension gradient that develops along the bubble surface as a consequence of the temperature dependence of interfacial tension.

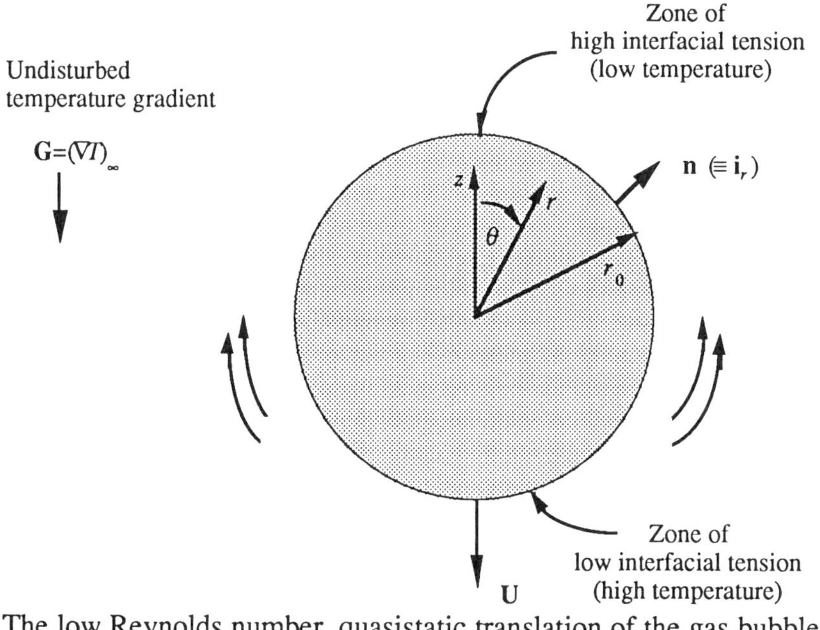

The low Reynolds number, quasistatic translation of the gas bubble is governed by the following bulk-phase equations defined in the domain $r_0 < r < \infty$:

$$\mu \nabla^2 \mathbf{v} = \nabla p\,, \tag{a}$$

$$\nabla \cdot \mathbf{v} = 0. \tag{b}$$

These are to be solved subject to the boundary conditions (relative to an origin fixed at the center of the translating droplet)

$$\mathbf{v} \to -\mathbf{U} \qquad \text{as} \quad r \to \infty, \tag{c}$$

* The original analysis of the thermocapillary problem outlined in Questions 5.4–5.6 owes to Hariri *et al.* (1990).

$$p \to p^{\infty} \quad \text{as} \quad r \to \infty, \tag{d}$$

$$\mathbf{n} \cdot \mathbf{v} = 0 \quad \text{at} \quad r = r_0, \tag{e}$$

wherein the latter condition follows from an *a priori* assumption of spherical bubble shape (to be confirmed *a posteriori* in Question 5.6). In the absence of surfactant there are no interfacial viscous effects; hence, upon assuming a linear relation between surface tension and temperature ($d\sigma/dT$ = constant)], the tangential stress boundary condition at the spherical bubble surface (4.2-20b) reduces to the following invariant expression [cf. Eq. (5.5-6)]:

$$\mathbf{n} \cdot \mathbf{P} \cdot \mathbf{I}_s = \frac{1}{\bar{T}_0} E_0^T \nabla_s T \quad \text{at} \quad r = r_0, \tag{f}$$

where $\bar{T}_0$ represents the mean temperature of the bubble surface [cf. Eq. (h)] and

$$E_0^T \equiv - \left(\frac{\partial \sigma}{\partial \ln T} \right)_0$$

represents the thermal analog of the Gibbs surface elasticity (5.5-5).

i. Show that for small Peclét numbers the bulk-phase temperature distribution satisfies the Laplace equation,

$$\nabla^2 T = 0. \tag{g}$$

Employing the boundary conditions

$$\mathbf{n} \cdot \nabla T = 0 \quad \text{at} \quad r = r_0$$

and

$$T \to \mathbf{r} \cdot \mathbf{G} \quad \text{as } r \to \infty,$$

determine the temperature distribution $T(r, \theta)$ in the bulk liquid phase, and hence show that the temperature along the spherical droplet surface is given by the expression:

$$T(r_0, \theta) = \bar{T}_0 + \frac{3}{2} r_0 \mathbf{n} \cdot \mathbf{G}, \tag{h}$$

where

$$\bar{T}_0 \overset{\text{def}}{=} \frac{1}{4\pi} \int_{\theta=0}^{\pi} \int_{\phi=0}^{2\pi} T(r_0, \theta) \sin\theta \, d\theta d\phi$$

is the mean temperature of the droplet surface.

ii. Show that Eqs. (a)–(h) are satisfied by the following vector-invariant solution for the velocity and pressure fields:

$$\mathbf{v} = \frac{E_0^T r_0^4}{4\mu r^3 \bar{T}_0} [3\mathbf{nn} - \mathbf{I}] \cdot \mathbf{G} - \mathbf{U}, \tag{i}$$

$$p = p^{\infty} \tag{j}$$

using the following expression for the migration velocity $\mathbf{U}$ of the bubble:

$$\mathbf{U} = \frac{E_o^T r_0}{2\mu \bar{T}_0}\mathbf{G}.$$

5.5 In the above, the bubble shape has been assumed spherical. As discussed in Chapter 4 [cf. Eq. (4.2-20a) and the discussion thereof] this is equivalent to assuming that the normal stress boundary condition

$$\mathbf{n} \cdot (\mathbf{P} - \tilde{\mathbf{P}}) \cdot \mathbf{n} = -2H\sigma$$

is identically satisfied (with $\tilde{\mathbf{P}} = -p_g\mathbf{I}$). Substitute Eqs. (i) and (j) from Question 5.4 into the preceding equation (noting the linear approximation to the thermal surface tension dependence), to obtain $H = -1/r_0$ for the dimensionless mean curvature (i.e. the solution for a sphere).

5.6 Show that the following condition is satisfied for the problem of Question 5.4:

$$\int_A \mathbf{n} \cdot \mathbf{P}\, dA + \int_A \nabla_s \sigma dA = \mathbf{0},$$

[i.e. that no net force is exerted upon the bubble (whose surface is A)].

5.7 Demonstrate by use of Eqs. (3.4-20) and (5.2-8) that the conditions under which the Gibbs elasticity E_o^i, defined through Eq. (5.5-5), is equivalent to $\partial\sigma/\partial \ln \delta A$ [as in Eq. (4.3-8)]. What type of surfactant monolayer does this imply?

5.8 Evaluate the coefficients A, B, C and E appearing in the velocity and pressure fields postulated in Eqs. (4.4-15)–(4.4-20) for the case in which the tangential stress condition (5.6-10) applies. (See also Question 4.4.)

5.9 Using the results obtained above, show that Eq. (5.6-14) leads to Eq. (5.6-15).

5.10 Employ Eq. (5.6-55) rather than the simplified expression (5.6-57) to obtain a more general form of the dispersion relation (5.6-61), appropriate to conditions wherein the inequality $k^2/\omega D << 1$ fails to apply.

5.11 Confirm the expressions for the Gibbs elasticity E_o obtained at the conclusion of example 1 in §5.6 for the Frumkin and Langmuir models. Find also the corresponding expressions for the adsorption parameter α, showing this latter parameter to be zero for the Langmuir model.

5.12 A radial surface wave

$$f = f_o e^{-\beta r} \exp(\alpha r - \omega t)$$

is assumed to propagate over an initially planar gas/liquid interface (i.e. $z = f$), as illustrated below.*

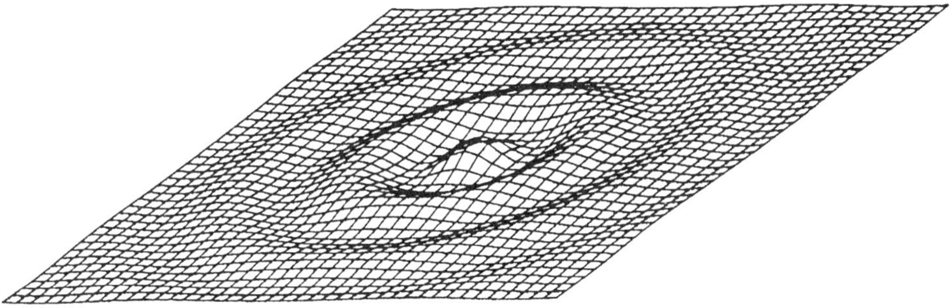

i. Derive expressions for the radial component of both the interfacial stress (4.2-20) and interfacial species flux (5.2-6) boundary conditions for a planar interface parameterized by the circular cylindrical coordinates (r, θ);

ii. Assuming the condition

$$\alpha f_o << 1,$$

derive a leading-order expression for the mean surface curvature $H = H(f) + O(\alpha f_o)$.

iii. Follow an identical procedure as in example 2 of §5.6 [including all the assumptions made there; i.e. Eq. (5.6-18)] to derive the dispersion relation appropriate to this case;

iv. What is the dispersion relation for a pure capillary wave?

v. What is the dispersion relation for a pure longitudinal wave?

vi. Which of the two waves (longitudinal or capillary) is more sensitive to interfacial rheological properties?

Additional Reading for Chapter 5

§5.1 The Equations of Species Transport in Continuous Three-Dimensional Media

Bird, R.B., Stewart, W.E. & Lightfoot, E.N. 1960 *Transport Phenomena*. New York: Wiley.

§5.2 The Equations of Interfacial Species Transport

Agrawal, M.L. & Neuman, R.D. 1988a Surface diffusion in monomolecular films. I. *J. Colloid Interface Sci.* **121**, 355–365.

---

* The authors wish to thank Dave Otis for supplying this figure.

Agrawal, M.L. & Neuman, R.D. 1988b Surface diffusion in monomolecular films. II. Experiment and theory. *J. Colloid Interface Sci.* **121**, 356–380.

§5.3 Adsorption Kinetics

Lucassen-Reynders, E.H. 1976 Adsorption of surfactant monolayers at gas/liquid and liquid/liquid interfaces. In *Progress in Surface and Membrane Science* (eds. D.A. Cadenhead and J.F. Danielli), pp. 253–359. New York: Marcel Dekker.

Lucassen-Reynders, E.H. 1981 Adsorption at fluid interfaces. In *Anionic Surfactants* (ed. E.H. Lucassen-Reynders), pp. 1–54. New York: Marcel Dekker.

§5.5 The Surface Equation of State

Adamson, A.W. 1982 *Physical Chemistry of Surfaces*, 4th ed. New York: Wiley.

§5.6 Examples

Levich, V.G. 1962 *Physicochemical Hydrodynamics*. Englewood Cliffs, New Jersey: Prentice Hall.

Haber, S. & Hetsroni, G. 1971 Hydrodynamics of a drop submerged in an unbounded arbitrary velocity field in the presence of surfactants. *Appl. Sci. Res.* **25**, 215–233.

Barton, K.D. & Subramanian, R.S. 1989 The migration of liquid drops in a vertical temperature gradient. *J. Colloid Interface Sci.* **133**, 214–222.

Feuillebois, F. 1989 Thermocapillary migration of two equal bubbles parallel to the line of centers. *J. Colloid Interface Sci.* **131**, 267–274.

Merritt, R.M. & Subramanian, R.S. 1989 Migration of a gas bubble normal to a plane horizontal surface in a vertical temperature gradient. *J. Colloid Interface Sci.* **131**, 514–525.

6

"The phænomena of capillary attraction in liquids are accounted for, according to the generally received theory of Dr. Young, by the existence of forces equivalent to a tension of the surface of the liquid, uniform in all directions, and independent of the form of the surface. The tensile force is not the same in different liquids. Thus it is found to be much less in alcohol than in water. This fact affords an explanation of several very curious motions observable under various circumstances, at the surfaces of alcoholic liquors. One part of these phænomena is, that if, in the middle of a glass of water, a small quantity of alcohol or strong spiritous liquor be gently introduced, a rapid rushing of the surface is found to occur outwards from the place where the spirit was introduced. . . . Another part of the phænomena is, that if the sides of the vessel be wet with water above the general level surface of the water, and if the spirit be introduced in sufficient quantity in the middle of the vessel, or if it be introduced near the side, the fluid is ever seen to ascend the inside of the glass until it accumulates in some places to such an extent, that its weight preponderates and it falls down again. The manner in which I observe these two parts of the phænomena is, that the more watery portions of the centre surface, having more tension than those which are more alcoholic, drag the latter briskly away, sometimes even so as to form a horizontal ring of liquid high up round the interior of the vessel, and thicker than by which the interior of the vessel was wet. Then the tendency is for the various parts of this ring or line to run together to those parts which happen to be most watery, and so there is no stable equilibrium, for the parts to which the various portions of the liquid aggregate themselves soon become too heavy to be sustained, and so they fall down."

W. Thompson (Lord Kelvin) (1855)

CHAPTER 6

Measurement of Dynamic Interfacial Tension and Dilatational Elasticity

Experimental methods currently employed for measuring dynamic interfacial properties are reviewed in this and the following three chapters. Throughout the perusal of these four chapters (6–9) it should be borne in mind that precise experimental values appear less commonly in the literature for interfacial rheological properties than for their bulk-phase counterparts. One important reason for this state of affairs is that interfacial transport theory has developed less rapidly in the past than has interest in its practical application. Thus, measurements have sometimes been performed without the benefit of a complete theoretical understanding of the dynamic interfacial phenomena under consideration, resulting in the partial accounting of the influence of bulk-phase viscous stresses at fluid interfaces (Langmuir 1936, Harkins & Meyers 1937, Davies 1957), confusion between the effects of interfacial-tension gradient and interfacial viscous phenomena (Derjaguin & Titievskaya 1957, Lucassen & van den Tempel 1972, Clint *et al.* 1981), as well as ambiguities between the interpretation of: (a) dynamic interfacial tension $\bar{\sigma}$ and (b) thermodynamic interfacial tension σ measured at a nonequilibrium surfactant adsorption (Clint *et al.* 1981, Defay & Hommelen 1959). Moreover, owing to the fact that interfacial rheological properties are extremely sensitive to the physical state of the interface, and particularly to trace amounts of surfactant impurities, the purity of surfactant and rigor invested toward maintaining experimental cleanliness may together be responsible for as much as an order of magnitude difference between measured values of interfacial rheological properties (Li & Slattery 1988). Temperature, vibrations, curvature of the interface and the existence of moving contact lines are additional factors that may influence measured values of dynamic interfacial properties.

As briefly discussed in Chapter 2 (and at length in Chapter 5), interfacial tension may assume a value in nonequilibrium conditions that significantly departs from its equilibrium value, σ_o. This may owe to spatial and/or temporal variations in the surface temperature (cf. Questions 5.4

through 5.6) or surface-excess surfactant density [cf. Eq. (5.5-6)], and/or the existence of a finite surface expansion rate (in which latter case the *dynamic* interfacial tension is indicated by the symbol, $\bar{\sigma}$) [cf. Eq. (4.2-19)]. Currently existing methods for measuring this 'elastic' surface behavior are the subject matter of the present chapter. These methods are classified herein as being either *direct* or *indirect* methods. *Direct* methods measure the dynamic interfacial tension, $\bar{\sigma}$, of an interface undergoing either steady, unsteady or pulse expansion. Examples of direct methods are provided in §§6.1, 6.3, 6.5, 6.6, 6.7 and 6.8. *Indirect* methods seek to measure the dilatational elasticity, E' [cf. Eq. (4.3-8)] of an interface undergoing a time-dependent surface expansion.* Examples of such indirect methods are provided in §§6.2, 6.4 and 6.7.

Techniques for measuring dynamic interfacial tension and dilatational elasticity date from the works of Rayleigh (1879), Thompson (Lord Kelvin) (1871), and Bohr (1909). These early techniques, in addition to several others, will be reviewed in the present chapter, beginning with a consideration of the oscillating jet method; this is followed by the consideration of surface wave, maximum bubble pressure, oscillating bubble, Langmuir trough, falling meniscus, and pulsed drop methods.

Experimental techniques for measuring *static* interfacial tension (e.g. spinning drop, du Nuoy, pendant or sessile drop tensiometers) are not discussed in this text. The reader may refer to standard textbooks of surface chemistry [e.g. Adamson (1982)] for a review of common methods.

6.1 Oscillating Jet Method

The oscillating jet technique for measuring dynamic interfacial tension is based upon the phenomenon of Rayleigh instability [Miller & Neogi (1985); see also §10.3]. Briefly, Rayleigh instability refers to the break-up of a fluid jet into individual droplets owing to the tendency of the interfacial tension to minimize interfacial area. Insofar as the radii of droplets formed through breakup depend directly upon the value of the interfacial tension, Rayleigh instability may furnish an indirect means for determining dynamic interfacial tension.

The standard analysis of the oscillating jet method (Bohr 1909) concerns the physical problem which is depicted in Figure 6.1-1. A liquid of viscosity μ is assumed to flow through a circular cylindrical capillary of radius a with constant (plug-flow) axial velocity U. Upon exiting the capillary into a gas medium, the position of the liquid jet surface (which surface is assumed to exhibit a homogeneous interfacial tension and

* The dilatational elasticity E' corresponds to the *Gibbs* elasticity E_o [cf. Eq. (5.5-5)] when the interfacial expansion rate is very small. When the surface expansion rate greatly exceeds the rate of surfactant adsorption, the dilatational elasticity is known as the *Marangoni* elasticity E_M.

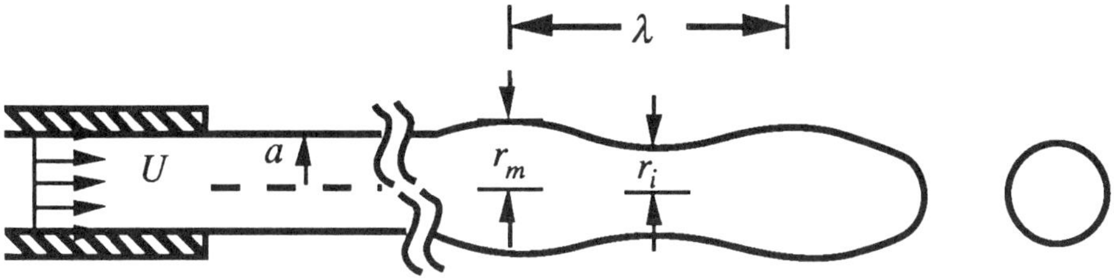

Figure 6.1-1 A fluid jet is ejected from a cylindrical nozzle, eventually breaking into small fluid droplets.

negligible interfacial viscosities) oscillates prior to breakup with wavelength λ and amplitude b; consequently, Bohr (1909) has shown that the dynamic interfacial tension may be determined by the relation

$$\bar{\sigma} = \frac{2\rho U \pi^2 a^3}{3}\left[\frac{1+\dfrac{37}{24}\dfrac{b^2}{a^2}}{\lambda^2+\dfrac{5}{24}\pi a^2}\right]\left[1+2\left(\frac{\mu\lambda}{\rho U \pi a^2}\right)^{3/2}+3\left(\frac{\mu\lambda}{\rho U \pi a^2}\right)^{2}\right]. \quad (6.1\text{-}1)$$

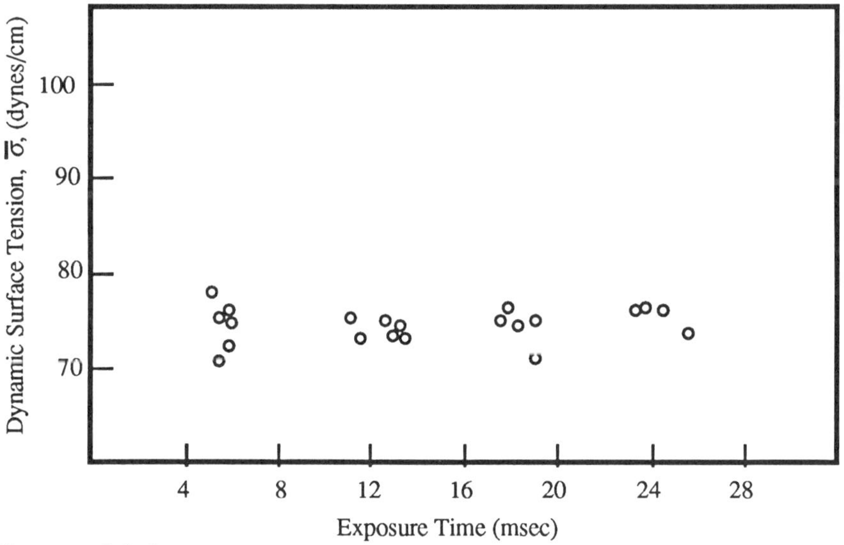

Figure 6.1-2 The dynamic surface tension of an aqueous/air surface vs time of exposure, as measured with the oscillating jet method at varying jet flow rates. Data taken from Caskey and Barlage (1971).

Here, the amplitude of the fluctuation is given by

$$b = \frac{r_m - r_i}{2}, \quad (6.1\text{-}2)$$

with r_m and r_i respectively denoting the maximum and minimum jet radii.

Owing to the numerous assumptions implicit in Eq. (6.1-1) [e.g. the assumption of plug flow through the orifice; see Caskey & Barlage (1971)], this technique has been found to be of limited value. This is evidenced by the scattered data appearing in Figure 6.1-1 (i.e. pure surface tension values should be independent of the time of surface exposure). Figure 6.1-3 shows curve-fitted data obtained by the oscillating jet method for a surfactant-adsorbed surface.

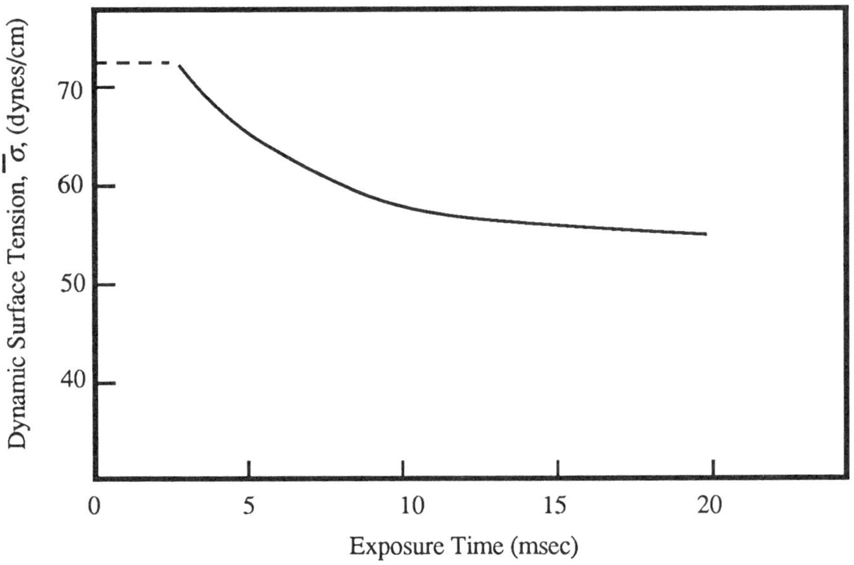

Figure 6.1-3 The dynamic surface tension of the surface formed between an aqueous solution of normal heptyl alcohol and air vs time of exposure, as determined by the oscillating jet method. Data taken from Defay & Hommelen (1958).

6.2 Surface Wave Methods

Surface wave techniques have been frequently employed to measure dynamic interfacial properties. As they generally do not involve a large departure of interfacial tension from its equilibrium value, they are primarily used for measuring dilatational elasticities of fluid interfaces (see also §8.1). Wave techniques are classified below as: (i) capillary wave methods; (ii) longitudinal wave methods; and (iii) light scattering methods.

Capillary Wave Method

As demonstrated in the second example of §5.6, the bulk-liquid motion that accompanies a propagating surface wave may be decomposed into a *transverse* (or $\mathbf{v}_1$) motion [as in Eq. (5.6-31)] and a *longitudinal* (or $\mathbf{v}_2$) motion [as in Eq. (5.6-32)], the sum of these two motions furnishing the composite wave motion ($\mathbf{v}_1 + \mathbf{v}_2 = \mathbf{v}$). While each of these wave motions may be expected to be generated to varying degree in all surface wave experiments, capillary (or 'ripple') techniques primarily generate only the transverse type. In capillary wave methods, as in all other surface wave techniques, a dispersion relation [of the type provided by Eq. (5.6-61)] is employed to relate measured wave properties (wave number α and damping coefficient β) to pertinent physical properties of the surfactant system, examples of such properties being the equilibrium interfacial tension, dilatational interfacial elasticity, and interfacial viscosities.

The pure transverse wave motion is governed by a dispersion relation that may be obtained from Eq. (5.6-58) (assuming satisfaction of the physical conditions upon which the analysis of §5.6 is based) upon setting B =0 [compare Eqs. (5.6-32) and (5.6-36)]; this yields

$$\rho\omega^2 - 2i\mu k^2\omega - \sigma k^3 - \rho g k = 0. \tag{6.2-1}$$

For undamped waves (μ =0, k real),

$$\rho\omega^2 = \sigma k^3 + \rho g k, \tag{6.2-2}$$

which latter relation is known as Kelvin's equation (Kelvin 1871). *Gravity* waves refer to transverse waves possessing 'long wavelength', i.e. for which $\sigma k^3 << \rho g k$, whereas *capillary* waves describe waves possessing 'short wavelength', for which $\sigma k^3 >> \rho g k$.

The dispersion relation (6.2-2) for capillary waves (i.e. $\sigma k^3 >> \rho g k$) propagating upon an inviscid liquid-gas surface yields

$$\alpha = \left(\frac{\rho\omega^2}{\sigma}\right)^{1/3} \tag{6.2-3}$$

(upon noting that k=α) which is known as the 'Kelvin relation' for a capillary wave. Equation (6.2-3) offers reasonable a estimate for the wave number of capillary waves in most capillary wave experiments.

Methods for propagating capillary ripples typically involve either a mechanical disturbance of the fluid interface or electrocapillary propagation, the latter representing the more attractive propagating mechanism, as it does not necessitate physical contact with the fluid surface (see Figure 6.2-1). The wave characteristics necessary for the evaluation of the dispersion relation are often determined by specular reflection of a laser beam from the fluid surface to a position-sensitive photodiode.

Since the transverse wave motion is independent of surface relaxation processes, the wave motion generated by capillary wave experiments has been found sensitive to dynamic surface properties only in the vicinity of maximum surface dilatational elasticity. In addition, as the transverse wave motion is essentially undamped [cf. Eq. (5.6-33)], to satisfy the conditions imposed by the linear wave theory, only low wave frequencies may be employed with mechanically induced capillary wave methods. Owing to the above, capillary wave experiments have been less frequently used (Brown 1940, Mann & Hansen 1963, Lucassen & Hansen 1966) for determining surface relaxational properties than have the longitudinal and light-scattering techniques described below.

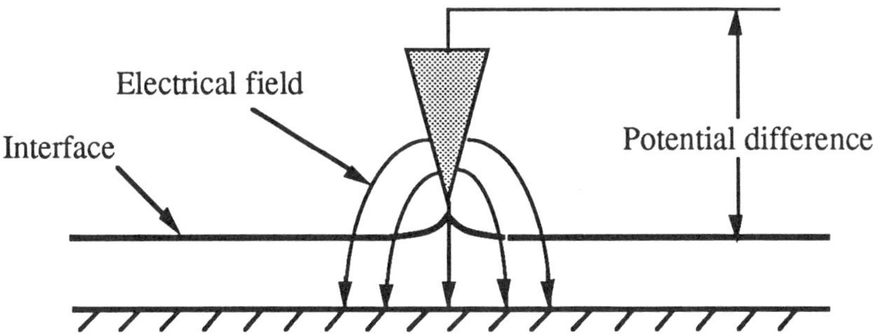

Figure 6.2-1 A capillary wave is created at a gas/liquid surface by electrocapillary action. For further details, see Sohl *et al.* (1978).

Longitudinal Wave Method

The existence of a longitudinal wave [$\mathbf{v}_2$ motion; cf. Eq. (5.6-32)] was first recognized by Lucassen (1968): it represents the second physical root of the dispersion relation (5.6-61). Since Lucassen's original work, propagation characteristics of longitudinal waves have been widely used for measuring dilatational elasticity values of adsorbed surfactant monolayers.

The pure longitudinal wave motion satisfies the simplified dispersion relation

$$i\mu\omega(m^2 + k^2) - mk^2[E^* - i\omega(\kappa^s + \mu^s)] = 0, \qquad (6.2\text{-}4)$$

as follows from Eq. (5.6-59), with A =0 [compare Eqs. (5.6-31) and (5.6-35)]. Hansen & Ahmad (1971) neglect the damping contributions of surface viscosity and the imaginary part of the dilatational modulus E^* [cf. Eq. (5.6-60)], to show that for high-frequency longitudinal waves [upon separating the real and imaginary contributions to Eq. (6.2-4)] there obtain the following relations for the wave number and damping coefficient [cf. Eqs. (4.3-8), (8.1-7) and (8.1-8)]:

$$\alpha = \frac{1}{2}\sqrt{(2+\sqrt{2})}\left(\frac{\rho\mu\omega^3}{E'^2}\right)^{1/4}, \qquad (6.2\text{-}5)$$

$$\beta = \frac{1}{2}\sqrt{(2-\sqrt{2})}\left(\frac{\rho\mu\omega^3}{E'^2}\right)^{1/4}, \qquad (6.2\text{-}6)$$

indicating that the longitudinal wave motion is more sensitive than the capillary wave motion to dynamic surface properties.

Longitudinal wave techniques involve the oscillation of a barrier within the plane of a fluid interface and the detection of longitudinal wave characteristics (α and β) by either a Wilhelmy plate or an electrocapillary method. If the frequency of oscillation of the longitudinal wave is large, such that the surface wave expands and contracts at a rate sufficiently exceeding rates of surfactant adsorption/desorption, the surfactant monolayer is rendered insoluble. In this case, the dilatational surface elasticity E' may be regarded as the Marangoni elasticity, E_M. Figure 6.2-2 displays Marangoni elasticity values for a mixed sodium dodecyl sulfate and high molecular weight poly-L-lysine monolayer. The lack of dependence of the Marangoni elasticity upon rate of surface expansion indicates that surface composition is independent of frequency of dilatation; this behavior may be viewed as a limiting characteristic of an insoluble monolayer (see also Figures 6.3-2 and 6.4-2).

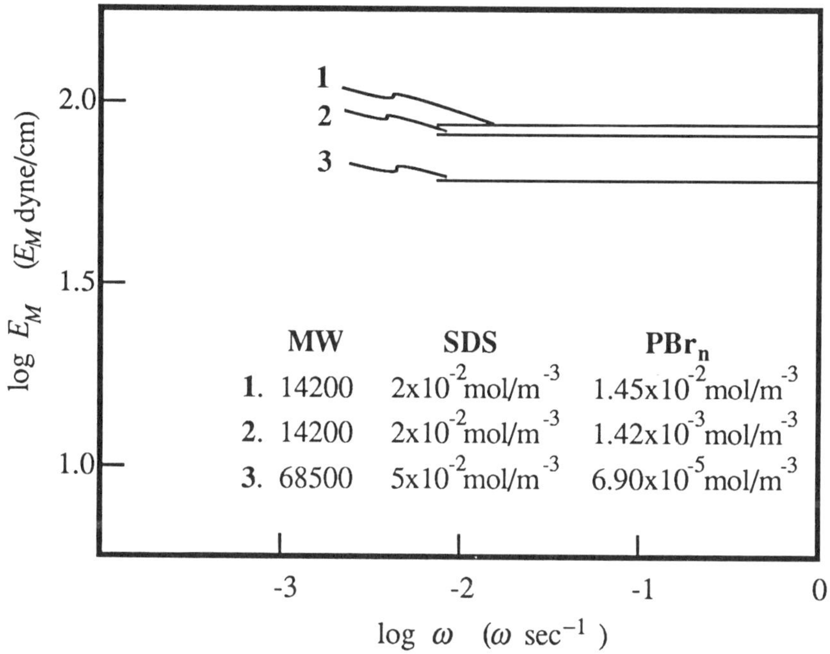

Figure 6.2-2 Marangoni elasticity values vs longitudinal wave frequency for poly-L-lysine/SDS aqueous mixtures. Data taken from Lucassen *et al.* (1978).

If the frequency of wave oscillation is sufficiently small, the wave disturbance will cause only a small departure from the equilibrium state of the interface, in which case the dilatational surface elasticity E' is regarded as being equivalent to the Gibbs elasticity, E_o.* In this latter case, wave propagation characteristics are most sensitive to relaxation properties of the interface; thus, as illustrated in Figure 6.2-3, measured Gibbs elasticity values depend strongly upon frequency of wave propagation.

---

* As discussed in the footnote following Eq. (5.5-5), this equivalence is only strictly appropriate for the case of an insoluble monolayer.

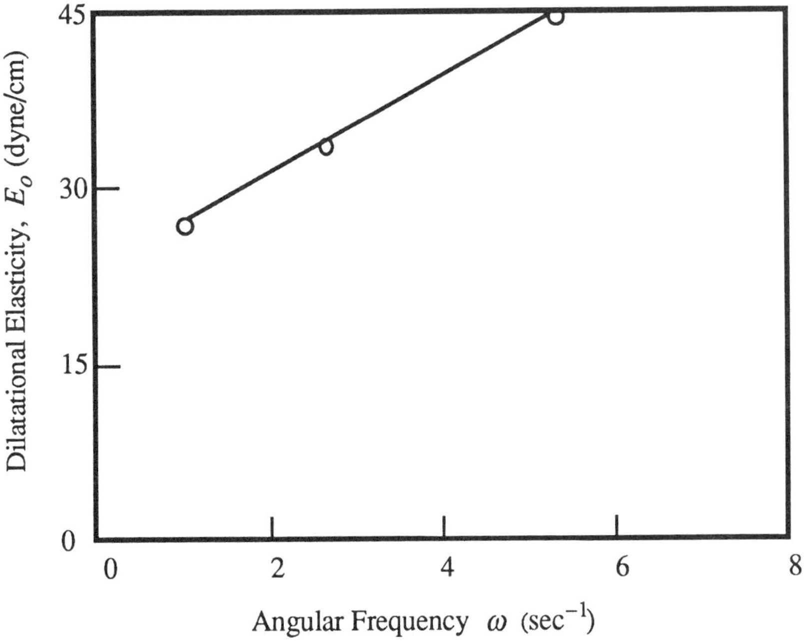

Figure 6.2-3 Gibbs elasticity values vs longitudinal wave frequency for aqueous solution of decanoic acid. Data taken from Snik *et al.* (1983).

Light Scattering Method

Very small amplitude (Å-scale), thermally-induced capillary waves may be observed to propagate at fluid interfaces in equilibrium. Upon viewing such (microscale) waves as propagating over a two-dimensional (macroscale)* surface, which surface possesses the macroscale properties of interfacial tension, etc., these ever-present waves may be theoretically described by classical capillary wave theory, thereby affording a method for the determination of dynamic interfacial properties by measurements of the thermal wave characteristics. One of the unique features of light-scattering techniques relative to mechanically generated wave techniques concerns the fact that light scattering provides a 'nonperturbative' technique, performed in macroscopic equilibrium conditions. Surfactant adsorption/desorption is normally ignored in theoretical analyses of light scattering, presumably owing to the fact that the fluctuation time scale is very small in comparison to time scales of adsorption/desorption; the mathematical description of the thermal capillary wave motion is thus appreciably simplified. In addition, the accuracy of measurement possible with this method combined with the ability to apply linear surface wave theory to thermal capillary waves of high frequency (owing to the existence of very small wave amplitudes), render

* See §§3.1 and 15.1 for a discussion of the macro- and microscale views of interfaces.

light scattering techniques attractive for the purpose of measuring dynamic surface properties.

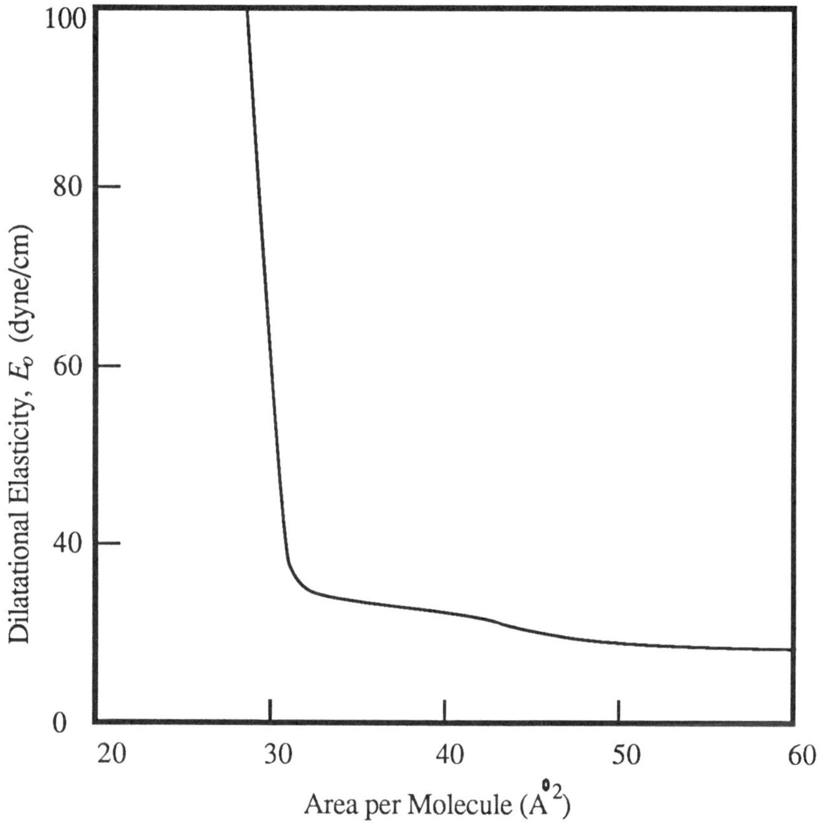

Figure 6.2-4 Gibbs elasticity values vs surface area per molecule for an aqueous solution of pentadecanoic acid. Data taken from Hard & Neuman (1981).

In the method of light scattering, correlated capillary wave characteristics are studied via reflected light [either randomly scattered light (Langevin & Griesmar 1980) or directed laser light (Byrne & Earnshaw 1979, Hard & Lofgren 1977, Hard & Neuman 1981)] to determine the Gibbs dilatational elasticity and interfacial viscosities in conditions of macroscopic equilibrium. Illustrative data are shown in Figure 6.2-4.

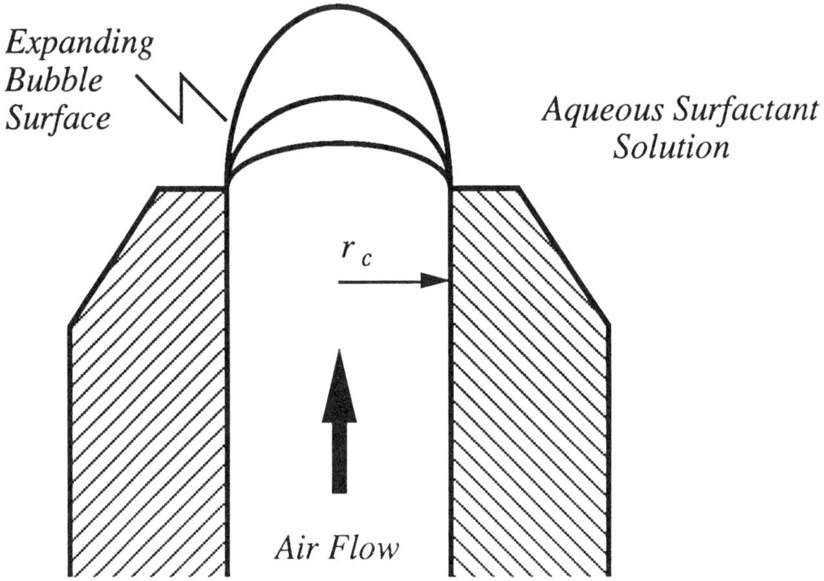

Figure 6.3-1 The principle of the maximum bubble pressure method is depicted. Air is injected through a narrow capillary into a surfactant solution. The maximum pressure differential between the air pressure within the capillary and the hydrostatic liquid pressure occurs when the bubble is hemispherical. This maximum bubble pressure may be used to directly infer the dynamic surface tension.

6.3 Maximum Bubble Pressure Method

The maximum bubble pressure method (MBPM) has been widely used as a direct method for measuring dynamic surface tension, $\bar{\sigma}$, vs rate of surface expansion (Bendure 1971, Mysels 1986, Garrett & Ward 1988, Kao *et al.* 1991a). In this technique, a bubble is formed at the tip of a capillary immersed within a surfactant solution (Figure 6.3-1) by injecting a gaseous stream through the circular cylindrical tube. The bubble grows until it reaches a hemispherical shape, evidenced by a maximum pressure within the capillary, after which the bubble rapidly expands and detaches from the capillary. This is subsequently followed by the creation of a fresh bubble at the capillary tip. This process of bubble formation is repeated at a frequency ω which may be directly related to the rate of surface expansion at the instant the bubble reaches its hemispherical shape.* By recording the maximum

* The viability of relating gas flow rate through the capillary to the rate of surface expansion at the moment of attainment of the maximum bubble pressure has been questioned by Garrett & Ward (1988), who provide visual data to suggest that the flow of gas through the capillary is at a minimum at the moment of the maximum bubble pressure. This follows from the fact that the pressure difference between the air pump and the capillary tip is minimum at the maximum bubble pressure. Nevertheless, the pressure

bubble pressure at a given frequency of bubble generation, and employing the normal stress boundary condition (4.2-20a) [which assumes the form of a dynamic Laplace condition, Eq. (4.4-33)] for a spherical surface, namely,

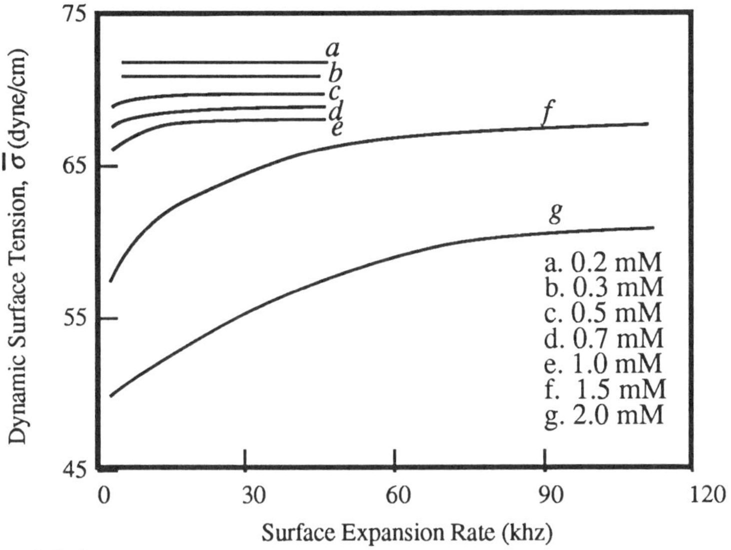

Figure 6.3-2 Dynamic surface tension vs rate of surface expansion for an aqueous solution of octanoic acid. Data is taken from Kao *et al.* (1991a).

$$p_i - p_o = \frac{2\bar{\sigma}}{r_c}, \tag{6.3-1}$$

the dynamic surface tension may be determined. Here, r_c is the capillary radius and p_i and p_o internal and external bubble pressures respectively. Bulk-phase viscous effects in Eq. (6.3-1) are negligible in comparison to capillary effects for standard capillary dimensions (Kao *et al.* 1991a).

The MBPM represents an effective means for measuring the dynamic surface tension of a gas-liquid surface. For liquid-liquid interfaces, experimental difficulties are encountered pertaining to the accurate appraisal of viscous pressure losses in the narrow-bore capillary.

Results for dynamic surface tension vs rate of surface expansion for an aqueous-air surface with adsorbed octanoic acid are shown in Figure 6.3-2.

---

difference between the pump and capillary tip generally varies only a small amount relative to the mean pressure drop in the capillary from the maximum to the minimum bubble pressure: this should not create a sizeable fluctuation in the flow of gas. On the other hand, the visual observations of Garrett & Ward (1988) may indicate an additional resistance for a period of time prior to the maximum bubble pressure owing to a moving contact line, as in their pictures the meniscus falls within the capillary tube during this period. In any event, the gas flow rate through the capillary is certainly smaller than the measured gas flow rate at the moment of the maximum bubble pressure, though by what amount remains unclear.

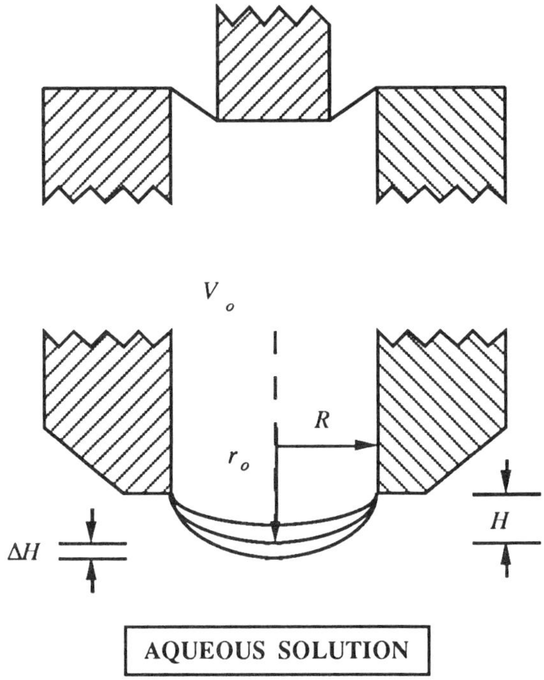

Figure 6.4-1 A schematic illustration of the oscillating bubble method. A bubble cap is oscillated at the tip of a capillary immersed within a surfactant solution. The pressure drop across the bubble surface is measured to deduce the dilatational elasticity.

6.4 Oscillating Bubble Method

The oscillating bubble method, like the MBPM, involves the rapid expansion and contraction of a bubble surface whose shape is near to a hemisphere. Unlike the MBPM, the induced motion is oscillatory about a mean radial position, hence, as with surface wave methods, the oscillating bubble method is most useful for measuring the surface dilatational elasticity. The oscillating bubble method may be preferable (relative to surface wave methods) for the measurement of surface dilatational elasticity when the particular case of interest concerns very large surface expansion rates.

Figure 6.4-1 illustrates the basic features of the oscillating bubble method. A bubble hemisphere is oscillated about a radius r_o with amplitude ΔH_o. The surface dilatational elasticity E' may be calculated from experimental data by the relation (Wantke *et al.* 1976)

$$E' = \frac{d \ln r}{d \ln A}\left[\frac{C(\omega)\kappa p_o r_o^2}{2V_o \Delta H}\frac{dH}{dr}\Delta a - \Delta\sigma_o\right]$$

$$+ \frac{d \ln r}{d \ln A}\left[\frac{4\pi\omega}{F}\right]^{1/2}\frac{RTr_o}{\Delta H}\frac{dH}{dr}(\xi_1 + \eta_1). \qquad (6.4\text{-}1)$$

Here, A and H respectively denote the surface area and height of the bubble cap, V_o is the mean gas volume between the membrane pump and bubble surface, $\kappa = c_p/c_v$ is an adiabatic constant, p_o is the mean pressure inside the bubble, $C(\omega)$ is a frequency-dependent apparatus variable, Δa represents a fitted constant, $\Delta\sigma_o$ is the difference between solution and solvent equilibrium surface tensions, R, T and D are, respectively, the gas constant, absolute temperature and diffusion coefficient and ξ_1 and η_1 represent adsorption kinetic factors.

Surface dilatational elasticity values obtained by the oscillating bubble method are shown in Figure 6.4-2. In the limit of high oscillation frequency, the dilatational elasticity, E', corresponds to the Marangoni elasticity, E_M.

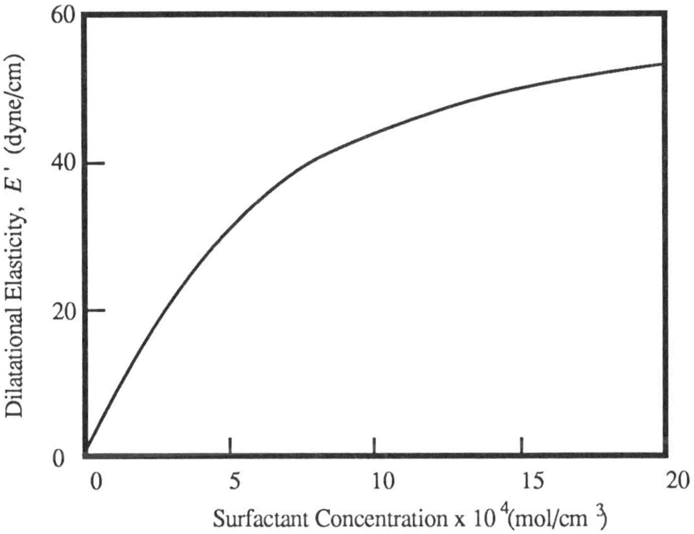

Figure 6.4-2 Surface dilatational elasticity values vs surfactant concentration measured by the oscillating bubble method for an aqueous system of octanoic acid. Values taken from Malysa *et al.* (1981).

6.5 Langmuir Trough Method

The Langmuir trough method represents one of the oldest techniques for measuring the surface tension of a newly created surfactant monolayer at a gas-liquid surface (Langmuir 1917). A sketch of the basic apparatus is provided in Figure 6.5-1. In this technique, a surfactant monolayer is spread upon a gas-liquid surface. A lineal barrier is employed to sweep this monolayer to one compartment of the trough, leaving a clean solvent interface in the other compartment: a surface pressure force is observed to act on the barrier. This surface pressure may be measured as a function of the position of the barrier to determine a relationship between area per molecule and surface pressure.

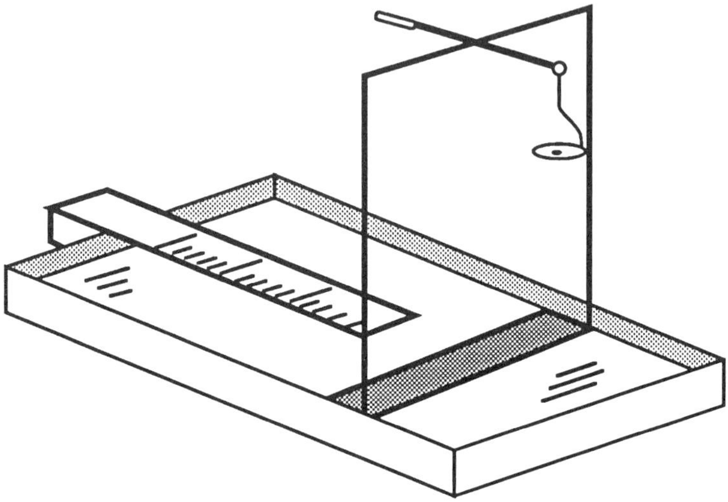

Figure 6.5-1 The Langmuir balance.

For the measurement of dynamic surface tension, the barrier may be oscillated at a known frequency, and the time-dependent surface pressure measured during the course of the oscillation.

6.6 Falling Meniscus Method

The falling meniscus method, like the pulsed drop method discussed in the following section, is designed to measure the time-dependent interfacial tension that results when a surfactant-adsorbed interface is deformed and then gradually returns to its equilibrium state. These techniques are *not* methods for determining the dynamic interfacial tension, $\bar{\sigma}$ [cf. Eq. (4.2-19)]; rather, the relaxation of the thermodynamic surface tension, σ, to its equilibrium value, σ_o, owing to the existence of a nonequilibrium surfactant adsorption is measured. The information gathered from such measurements may be useful in the investigation of the kinetics of surfactant adsorption.

The principle of the falling meniscus method is illustrated in Figure 6.6-1. A capillary tube with a small circular hole bored at the top is gradually immersed into a surfactant solution until the liquid fills the capillary. Ultimately, the liquid breaks through the hole at the top of the capillary. The height of the capillary tip, h_o, above the pool of liquid is noted the instant the liquid breaks through the capillary hole; this value may be used to infer the Laplace pressure difference across the hemispherical fluid interface. The surface tension may be obtained from known constants by the formula

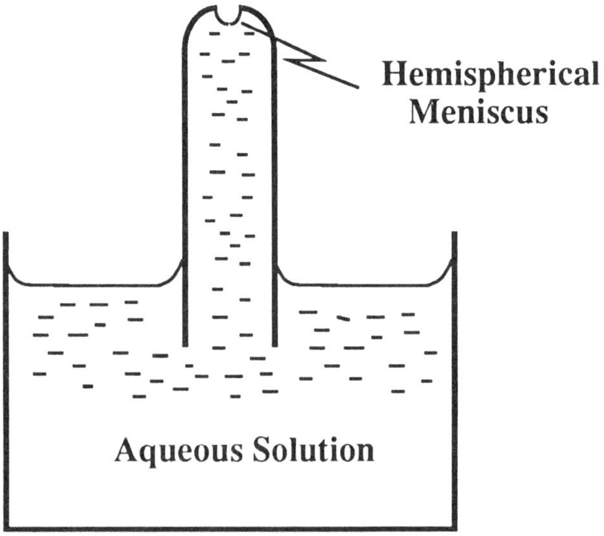

Figure 6.6-1 The falling meniscus method.

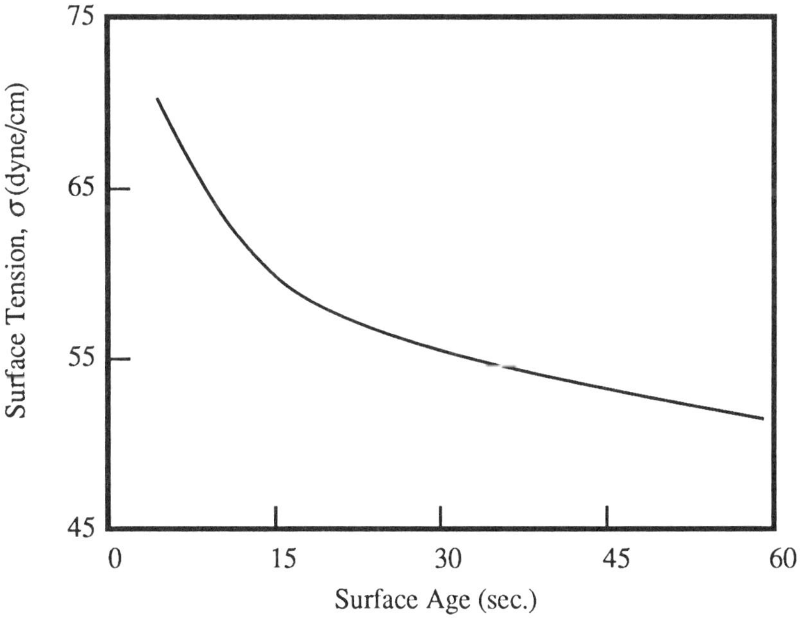

Figure 6.6-2 Surface tension vs surface age for 0.012g/l aqueous solution of normal dodecyl alcohol, by the falling meniscus method. Data taken from Defay & Hommelen (1959).

$$\sigma = \frac{\rho g r_c}{2}\left(h_o - \frac{2r_c}{3}\right), \qquad (6.6\text{-}1)$$

where r_c is the radius of the capillary hole and g the gravitational constant.

Time-dependent effects may be determined by placing the capillary tube at a height exceeding that which is suggested by the static surface tension value at the particular bulk surfactant concentration. The 'aging' time is reported to be the time required for the meniscus to break, at the particular surface tension. Figure 6.6-2 shows surface tension data obtained by the falling meniscus method as a function of surface age, for a normal dodecyl alcohol solution.

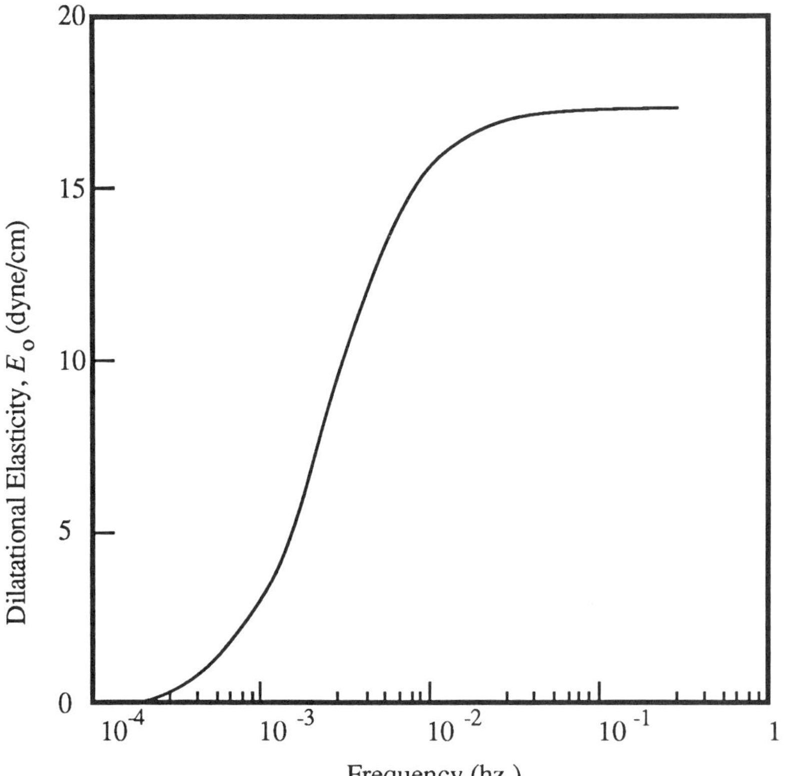

Figure 6.7-1 Gibbs dilatational elasticity vs frequency for 10 ppm stearic acid in aqueous solution of n-decane at pH=2.5, obtained by the pulsed drop method. Data taken from Clint *et al.* (1981).

6.7 Pulsed Drop Method

The pulsed drop method involves the rapid expansion of a droplet at the end of a capillary from a radius r_1 to r_2, and the monitoring of the pressure decay within the droplet as a function of time from the moment of expansion. As in the falling meniscus method, the interfacial tension may be measured as a function of time, attributing the continual lowering of the tension to adsorption relaxation.

An advantage of the pulsed drop method relative to the falling meniscus method is that a known surface deformation is the cause of the surface relaxation; hence, the Gibbs dilatational elasticity may be determined from values of the time-dependent interfacial tension. Elasticity data obtained by this method as a function of frequency (inverse time of aging) are illustrated in Figure 6.7-1.

An alternative technique to the pulsed drop method, which involves the *gradual* expansion of a droplet, has recently been proposed by Nagarajan *et al.* (1991).

6.8 Summary

Several techniques have been discussed in Chapter 6 for measuring the dynamic interfacial tension and/or the dilatational elasticity of surfactant-adsorbed fluid interfaces.

The oscillating jet method allows the determination of dynamic surface tension under the conditions of fluid jet break-up. This technique suffers limitations owing to the fact that it is an indirect method of measurement. Conditions of jet breakup, however, occur widely in engineering processes, and the oscillating jet method may prove to be useful for the examination of dynamic interfacial properties in these conditions.

Surface wave techniques have been classified as; (i) capillary, (ii) longitudinal, and (iii) light-scattering methods. Each of these methods may be employed for measuring the dilatational elasticity of a fluid interface. The capillary wave method is least commonly used owing to its limited sensitivity to surface properties and the restricted applicability of linear surface wave theory to low-frequency waves. The longitudinal wave technique is applicable both to low and high frequency waves, and is a sensitive method for measuring the dilatational elasticity of the surface. Light-scattering techniques differ from the other wave techniques insofar as they do not involve a mechanical interfacial disturbance; hence, the dilatational elasticity values obtained apply to conditions of zero surface deformation.

The maximum bubble pressure method is a practical, relatively simple means of measuring dynamic surface tension over a wide surface expansion rate range. The conditions of surface expansion in this technique are similar to those of the oscillating bubble method, though the latter is utilized for determining dilatational surface elasticities.

The Langmuir trough method is a standard technique for measuring dynamic and static surface tensions as a function of surface compositional changes.

The falling meniscus and pulsed drop methods are both designed to measure the decay of surface tension to its equilibrium value from a nonequilibrium value created by a surface deformation; they are most useful for examining the kinetics of surfactant adsorption.

Questions for Chapter 6

6.1 *a.* What effect does increasing the surface tension have upon the size of droplets created by the break-up of a fluid jet? *b.* Would you expect interfacial-tension gradients to cause an increase or decrease in the radii of droplets created by jet breakup? Why?

6.2 *a.* How is insoluble monolayer behavior achieved for a (soluble) surfactant-adsorbed fluid interface in a dynamic state? *b.* How may insoluble monolayer behavior be deduced from measurements of the surface dilatational elasticity?

6.3 Using Eq. (4.4-33), determine the magnitudes of the capillary pressure (i.e. the pressure drop owing to the existence of a finite surface tension) and the dynamic bulk-viscous pressure drop, upon choosing: capillary radius, r_c = 100 micron, radius of detached bubble = 1 mm, frequency of bubble creation, $\omega = 5$ hz, $\mu = 0.01$ g/cm-s, $\sigma = 40$ dyne/cm.

Additional Reading for Chapter 6

§6.1 Oscillating Jet Method

Caskey, J.A. & Barlage, W.B. 1971 An improved technique for determining dynamic surface tension of water and surfactant solutions. *J. Colloid Interface Sci.* **35**, 46–52.

Defay, R. & Hommelen, J.R. 1958 I. Measurement of dynamic surface tensions of aqueous solutions by the oscillating jet method. *J. Colloid Sci.* **13**, 553–564.

§6.2 Surface Wave Methods

Lucassen, J. & Hansen, R.S. 1966 Damping of waves on monolayer-covered surfaces. *J. Colloid Sci.* **22**, 32–44.

Ting, L., Wasan, D.T., Miyano, K. & Xu, Q. 1984 Longitudinal surface waves for the study of dynamic properties of surfactant systems II. Gas-liquid surface. *J. Colloid Interface Sci.* **102**, 248–258.

Hard, S. & Neumann, R. 1981 Laser light scattering measurements of viscoelastic monomolecular films. *J. Colloid Interface Sci.* **82**, 315–334.

§6.3 Maximum Bubble Pressure Method

Mysels, K.J. 1986 Improvements in the maximum bubble-pressure method of measuring surface tension. *Langmuir J.* **2**, 428–432.

§6.4 Oscillating Bubble Method

Lunkenheimer, K., Hartenstein, C., Miller, R. & Wantke, K.D. 1984 Investigations of the method of the radially oscillating bubble. *Colloids and Surfaces* **8**, 271–288.

§6.6 Langmuir Trough Method
Somasundaran, P., Danitz, M. & Mysels, K.J. 1974 A new apparatus for measurements of dynamic interfacial properties. *J. Colloid Interface Sci.* **48**, 410–416.

§6.6 Falling Meniscus Method
Defay, R. & Hommelen, J.R. 1959 II. Measurement of dynamic surface tensions of aqueous solutions by the falling meniscus method. *J. Colloid Sci.* **14**, 401–410.

§6.7 Pulsed Drop Method
Clint, J.H., Neustadter, E.L. & Jones, T.J. 1981 Dynamic interfacial phenomena related to EOR. *Dev. Pet. Sci.* **13**, 135–148.

7

"The superficial layer of liquids has a proper viscosity, independent of the viscosity of the interior of the mass. In some liquids this superficial viscosity is greater than the internal viscosity, and often much greater, as in water and, especially, in solution of saponine; in other liquids, on the contrary, it is less than the internal viscosity, and often much less, as in oil of turpentine, alcohol, etc."

J. Plateau (1869)

CHAPTER 7

Measurement of Interfacial Shear Viscosity

Numerous experimental techniques have been employed for the purpose of measuring the interfacial shear viscosity μ^s of Newtonian interfaces. Most of these techniques seek to avoid the existence of interfacial-tension gradients by utilizing viscous-traction driven flows, as the existence of Marangoni stresses in a surface viscometric flow may appreciably complicate the absolute detection of the interfacial viscous stress (cf. the introductory comments to Chapter 8).

Many of the earliest techniques for measuring interfacial shear viscosity were found to be plagued by the presence of interfacial-tension gradients, from the very first experiments of Plateau (1869). Marangoni (1871) soon recognized that the oscillation of a needle within a surfactant-adsorbed fluid interface, as was proposed by Plateau for the purpose of deducing the surface viscosity, should occasion interfacial-tension gradients owing to compression of the surface; nevertheless, oscillating needle experiments continued well into the current century (Cumper & Alexander 1950, Tschoegl & Alexander 1960), attracting further criticism (Derjaguin & Titievskaya 1957) along the lines of Marangoni (1871).

Modern experimental techniques succeed in avoiding the presence of interfacial-tension gradients by generating surface flows that are absent of a dilatational motion. Such surface viscometers are typically designed to examine interfacial velocity profiles (*indirect* methods) or interfacial torsional stress values (*direct* methods). Among all surface viscometers, the *indirect* deep-channel surface viscometer may be regarded as the most effective, owing to its relative sensitivity, 'exactness' of theoretical description, and simplicity of technique and analysis; it is also the most widely-used device. The deep-channel method possesses a simple design as well as an exact, analytical flow description, and allows sensitive measurements of the interfacial shear viscosity. One of the primary drawbacks of the technique pertains to the necessity of placing a small particle "in" the fluid interface to track the interfacial midchannel velocity. This drawback is avoided by most other classes of surface viscometers, which latter techniques are characteristically employed as torsional (i.e. *direct*) devices.

Most of the 'modern' surface viscometers reviewed in this chapter have been developed during the early period of interfacial rheology (see §1.1) when an accurate accounting of bulk viscous traction forces at fluid interfaces was not undertaken. The early (and later corrected) theoretical descriptions of these devices, though sometimes still employed, are thus often seriously in error. A common assumption in the early viscometric theories is that by measuring the difference between torque or surface velocity values for 'clean' and surfactant-adsorbed interfaces, the value of the surface viscosity may be directly deduced. This proposition is of course erroneous owing to the intimate coupling of bulk and interfacial viscous effects, as elucidated in Chapters 4 and 5.

The result of the preceding state of affairs, as has been discussed at length by Mannheimer & Burton (1970), is that the early theories have generally tended to over-predict interfacial shear viscosity values. Even 'relative' values, frequently the aim of early interfacial viscosity measurements, should such values be deduced by employing the original, approximate theories, can lead to qualitatively incorrect predictions (Mannheimer & Burton 1970).

In recent years, revised analyses of nearly all of the early viscometers still in use have been performed. The equations presented in this chapter properly adhere to the theoretical foundations outlined in Chapter 4.

Three basic types of surface viscometers for measuring interfacial shear viscosity are reviewed in the following; in particular, canal viscometers are considered in §7.1, disk viscometers in §7.2, and knife-edge viscometers in §7.3.

7.1 Canal Surface Viscometers: The Deep-Channel Surface Viscometer

The classical canal surface viscometer, first proposed by Dervichian & Joly (1939) and Harkins & Meyers (1937), utilized surface pressure-driven interfacial flows (cf. §2.4). This early technique proved impractical for measuring interfacial shear viscosity owing to the difficulty involved in unambiguously distinguishing between Marangoni and interfacial viscous effects. To avoid this complication, Davies (1957) later proposed a variation of the surface canal viscometer (now known as the *double knife-edge* viscometer; cf. §7.3), employing viscous traction forces. However, the induced surface flow of the Davies canal viscometer suffers the disadvantage of being relatively insensitive to the interfacial viscous stress, as further discussed in §7.3. The modern canal surface viscometer discussed in this section is known as the *deep-channel* surface viscometer. It was first proposed by Mannheimer & Schechter (1970a) as a modification of the earlier device of Davies (1957).

The design of the deep-channel surface viscometer consists of two concentric brass cylinders lowered into a pool of liquid contained within a brass dish, to a depth at which the brass cylinders nearly touch the bottom of the dish. During operation of the device, the dish is rotated with a known angular velocity. The midchannel (or 'centerline') surface motion of the interface within the channel formed by the concentric cylinders is monitored by a small (100-200 micron) tale or teflon particle placed within the fluid interface. To insure that the particle follows the interfacial motion, essentially unhindered by bulk-liquid viscous drag forces, several particle sizes may be used to demonstrate the lack of dependence of the measured surface velocity upon particle size. Typically, for surfactant-adsorbed interfaces, surface viscous forces dominate bulk viscous forces (i.e. Bo $\equiv \mu^s/\mu a >>1$) for spherical particle radii a in the 100 micron range.

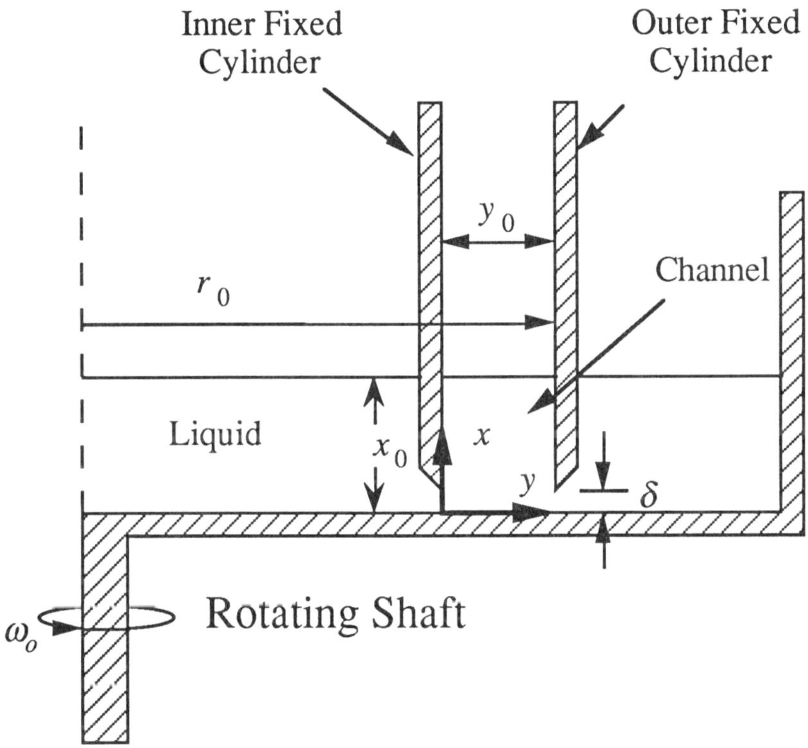

Figure 7.1-1 The deep-channel surface viscometer.

From the experimentally measured centerline surface velocity, the interfacial shear viscosity, μ^s, may be determined by a relation of the type [cf. Eq. (4.4-68)]

$$\frac{\mu^s \pi}{\mu y_0} = \frac{v_c^*}{v_c} - 1, \qquad (7.1\text{-}1)$$

which equation is appropriate to the single liquid system depicted in Figure 7.1-1, satisfying the 'deep-channel' restriction ($x_0/y_0>>1$). (The theoretical analysis of the deep-channel surface viscometer for the *liquid-liquid* system of Figure 4.4-5 is provided in example 4 of §4.4.) Here, v_C denotes the centerline surface velocity in the presence of surfactant, and v_c^* the corresponding value in the absence of surfactant.

Equation (7.1-1) may be in error for several design-related causes, as discussed by Pintar *et al.* (1971). First, the assumption of a 'linear canal' may not hold if the separation width of the concentric cylinders, y_0, possesses a value near to the radius of the outer-most cylinder r_0. Explicitly, the centerline surface velocity for a flat interface, in the absence of surface viscosity, may be expressed according to the linear canal theory of §4.4 by the formula

$$v_c^* = \frac{4v_b}{\pi \cosh(\pi D)}, \tag{7.1-2}$$

where v_b is the centerline velocity of the bottom dish. By comparison, a first-order approximation accounting for the 'circular' characteristic may be expressed as

$$v_c^* \approx \frac{4v_b}{\pi \cosh(\pi D)}\left[1+\frac{3}{32}\left(\frac{y_0}{r_0}\right)^2\right]. \tag{7.1-3}$$

A second possible cause for discrepancy between the classical analysis and experiment is the neglect of the gap width, δ, between the rotating dish and channel walls (see Figure 7.1-1). Pintar *et al.* (1971) have demonstrated that the influence of this gap upon the surface centerline velocity (once again, in the absence of surfactant) is provided to leading order in the dimensionless parameter (δ/x_0) by the expression

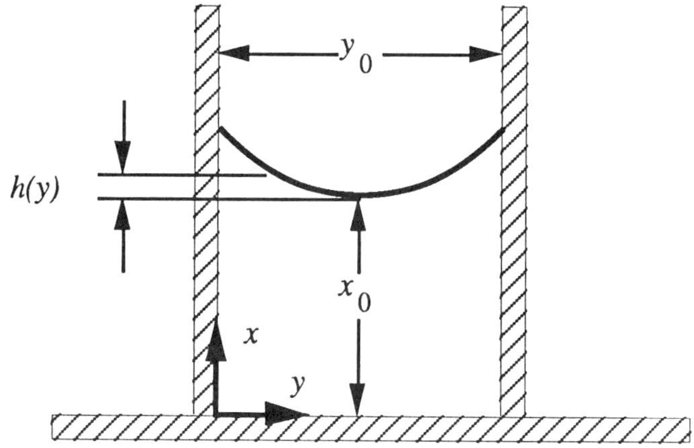

Figure 7.1-2 A curved fluid surface in the deep-channel viscometer, as occurs on account of a three-phase contact angle at the channel walls which is other than 90°.

$$v_c^* \approx \frac{4v_b}{\pi \cosh(\pi D)}\left[1+\left(\frac{\pi\delta}{2x_0}\right)^2 \frac{\cosh(\pi D)}{6} \sum_{n=0}^{\infty} \frac{(-1)^n (2n+1)}{\cosh[(2n+1)\pi/4D]}\right]. \tag{7.1-4}$$

The preceding corrections are relatively straightforward to avoid in practice.

Of potentially more serious consequence to measured values of surface viscosity is any deviation away from the assumed flat shape of the fluid interface, as may occur should the contact angle between the interface and channel walls possess a value other than 90°. An example of this behavior is illustrated in Figure 7.1-2. In such cases the centerline surface velocity of a surfactant-free interface may be expressed to leading order by the relation [see Pintar *et al.* (1972) for higher order corrections]

$$v_c^* \approx \frac{4v_b}{\pi \cosh(\pi D)}[1-\pi\bar{h}_0], \tag{7.1-5}$$

where

$$\bar{h}_0 = (1/y_o)\int_0^{y_0} h(y)dy . \tag{7.1-6}$$

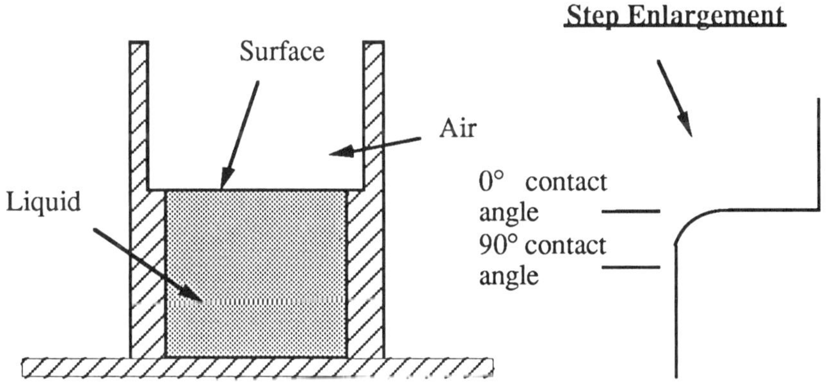

Figure 7.1-3 A step may be added in the deep-channel apparatus to insure a flat liquid surface.

Curvature of the fluid surface may be avoided by coating the channel walls with an oil film, say, thereby effecting a 90° contact angle; alternatively, as suggested (for gas-liquid systems) by Mannheimer & Schechter (1970a), the contact angle may be avoided by adding a notch in the channel walls at the level of the fluid-air surface (Figure 7.1-3). Ideally, the top of the step is made nonwetting to the liquid phase. The level of thc liquid is then raised to the level of the step. In certain cases the liquid tends to jump over the step rather than remain flat at the edge of the step. In this event it proves advantageous to lower the level of the liquid appropriately (depending upon the contact angle which the liquid seeks with the channel wall) to a level on the curved step surface at which the surface will remain flat.

The deep-channel surface viscometer is one of the most sensitive surface viscometers, characteristically able to detect interfacial shear viscosities as low as 10^{-4} sp. Figure 7.1-4 shows interfacial shear viscosity values of a crude oil system measured by the deep-channel surface viscometer as a function of *age*, or time of pre-equilibration.

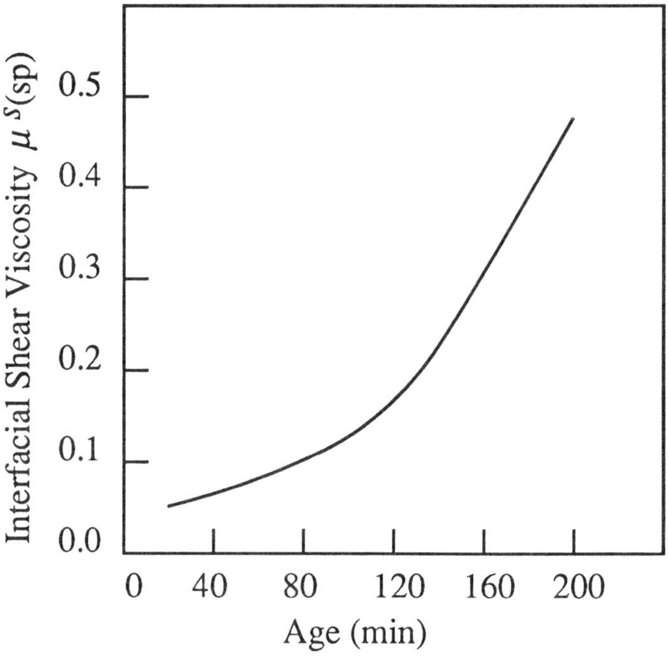

Figure 7.1-4 Interfacial shear viscosity vs pre-equilibration time for 1% NaCl Gach Saran Crude. Data taken from Wasan & Mohan (1977).

7.2 Disk Surface Viscometers

Many surface viscometers exist in the literature which utilize torsional stress measurements upon a rotating disc within or near to a liquid interface to determine the interfacial shear viscosity (Goodrich & Chatterjee 1970, Briley *et al.* 1976, Shail 1978, Oh & Slattery 1978, Davis & O'Neill 1979, Shail & Gooden 1981). This type viscometer is moderately sensitive (disk viscometers are typically sensitive to interfacial shear viscosity values in the range $\mu^s \geq 10^{-2}$ sp). In addition, such viscometers generally require numerical evaluation of measured viscometric flow parameters. Nevertheless, torsional measurements avoid the difficulties of tracking the interfacial velocity—the primary drawback of the deep-channel surface viscometer. Torsional devices are, therefore, practically useful for the measurement of interfacial shear viscosities of highly viscous surfactant monolayers (Jiang *et al.* 1983).

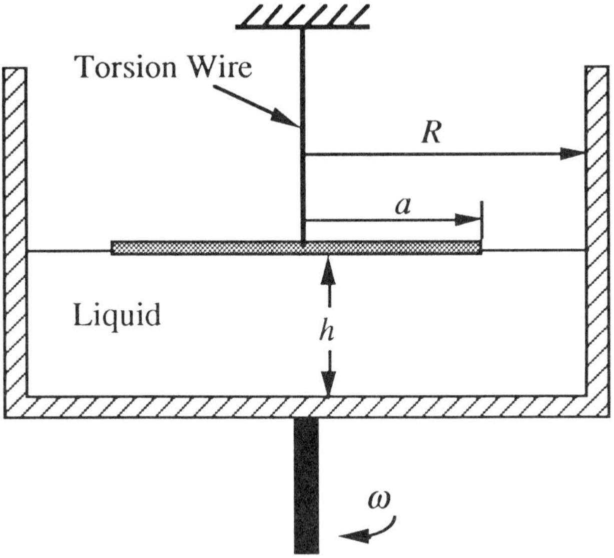

Figure 7.2-1 The disk surface viscometer.

The classical disk surface viscometer apparatus is depicted in Figure 7.2-1. A thin, flat, circular cylindrical disk is rotated within the plane of an interface with angular velocity ω. A torque is exerted upon the disk by both the surfactant film and the viscous liquid. According to the theoretical model, the torque exerted may be decomposed into a lineal surface traction torque along the rim of the disk (through which torque interfacial shear viscosity, μ^s, imparts a direct influence), and a tractional torque owing to intimacy of the bulk-phase liquid along the base of the disk. Standard analyses do not account for contact angle effects nor the effects of a finite disk thickness.

In the limit of zero Boussinesq number ($\mathrm{Bo} \equiv \mu^s/\mu R \to 0$), with also $a/h \to 0$, $a/R \to 0$, the torque exerted upon the rotating disk is provided to leading order by the expression (Goodrich & Chatterjee 1970)

$$\text{Disk torque} = (16/3)a^3\mu\,\omega. \tag{7.2-1}$$

In the limit of large Boussinesq number ($\mathrm{Bo} \to \infty$), with $a/h \to 0$ and $a/R \to 0$, the following relation holds:

$$\text{Disk torque} = (8/3)a^3\mu\,\omega + 4\pi a^2\mu^s\omega. \tag{7.2-2}$$

General torque expressions are provided by Briley *et al.* (1976) and Shail (1978), the latter reference additionally providing the next-order terms in the above expressions. Theoretical results for liquid-liquid systems are provided by Oh & Slattery (1978).

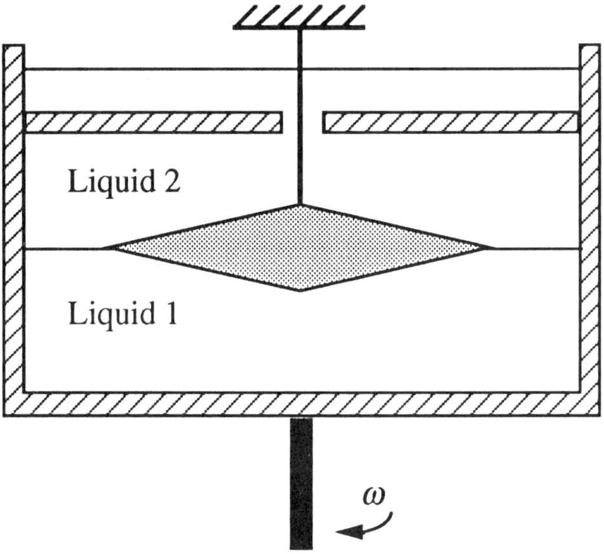

Figure 7.2-2 The biconical interfacial viscometer.

A second type of disk surface viscometer (Shail 1979, Shail & Gooden 1981) employs a thin disk rotating *below* the fluid interface. While this technique may exhibit an advantage relative to the preceding design by avoiding the delicate placement of the disc "in" the interface (in addition to the fact that it eliminates possible contact angle complications) the sensitivity of this configuration to interfacial shear viscosity is reduced by the fact that the disc is placed in the bulk fluid at a finite distance from the interface. A similar analysis has been carried out by Davis & O'neill (1979) for a rotating sphere beneath a surfactant-adsorbed fluid interface.

A variation of the disk viscometer is the biconical bob interfacial viscometer, which latter viscometer is depicted in Figure 7.2-2. The theoretical analysis (Oh & Slattery 1978) for this device is quite similar to that of the disc viscometer, and the technique may be regarded as possessing similar advantages and limitations.

7.3 Knife-Edge Surface Viscometers

The classical knife-edge surface viscometer of Brown *et al.* (1953), depicted in Figure 7.3-1, consists of a knife-edge bob suspended from a torsional wire such that the circular knife just touches the interface of a surfactant solution contained within a cylindrical cup. The cup is rotated and the torsional stress on the bob measured to determine the interfacial shear viscosity. Since the development of this technique, double knife-edge

(Davies 1957, Lifshutz *et al.* 1971), blunt knife-edge (Goodrich & Allen 1972, Briley *et al.* 1976), and rotating wall knife-edge (Goodrich *et al.* 1975, Poskanzer & Goodrich 1975a) viscometers have been developed upon similar principles of measurement, as discussed below. With the exception of the rotating wall knife-edge viscometer, these techniques are generally of moderate or low sensitivity (as with the disc viscometers, the range of sensitivity may be expected to be on the order of $\mu^s \geq 10^{-2}$ sp), are generally not amenable to an analytical theoretical description, are sensitive to the wetting characteristics of the knife-interface contact line, and are primarily advantageous for avoiding the placement of a tracer particle "in" the fluid interface to track the interfacial velocity.

Mannheimer & Burton (1970) offer the following expression for the torque upon the torsional wire in the single knife-edge experimental apparatus of Figure 7.3-1:

$$\text{Knife Torque} = \frac{\pi a^2 R \mu \omega}{\left[\sum_{i=1}^{\infty} \frac{\left[J_1(a\xi_i / R) \right]^2}{\left[J_o(\xi_i) \right]^2 \left[\frac{\mu^s}{\mu R} (\xi_i)^2 + \xi_i \coth(h\xi_i / R) \right]} \right]}, \tag{7.3-1}$$

where $J_1\,(a\xi_i/R)$ and $J_0\,(\xi_i)$ are, respectively, the first- and zeroth-order Bessel functions of the first kind, and ξ_i is the root of $J_1\,(\xi_i)=0$.

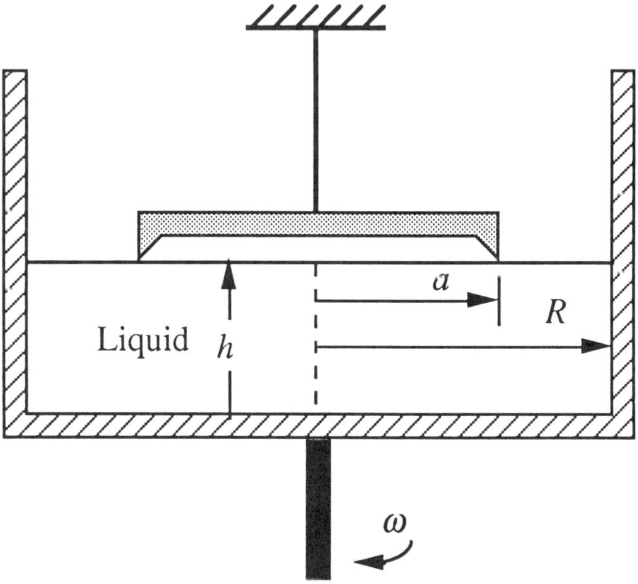

Figure 7.3-1 The single knife-edge surface viscometer.

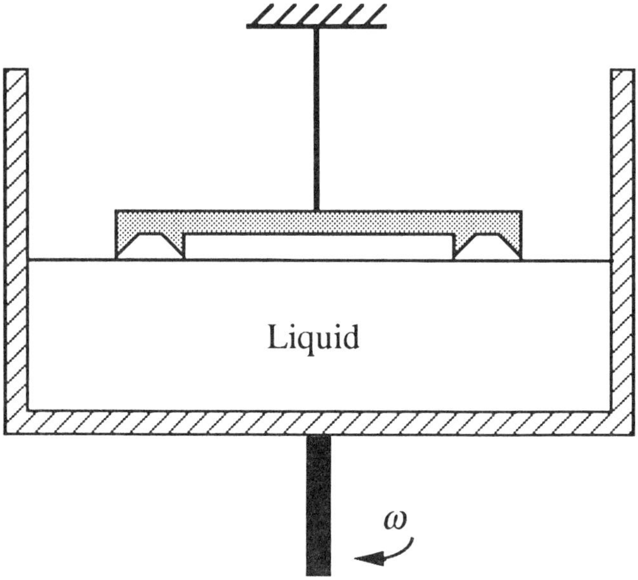

Figure 7.3-2 Double knife-edge surface viscometer.

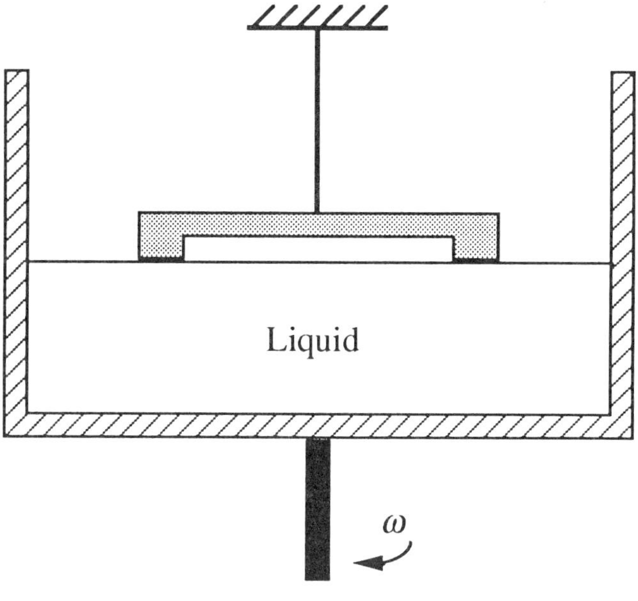

Figure 7.3-3 Blunt knife-edge surface viscometer.

The double knife-edge and blunt knife-edge surface viscometers, depicted in Figures 7.3-2 and 7.3-3 respectively, are similar in design and practicality as the single knife-edge viscometer earlier discussed. The theoretical description of the respective viscometric flows have been provided by Lifshutz *et al.* (1971) and Briley *et al.* (1976).

With each of the previous knife-edge viscometers it is crucially important that the knife make contact with, but not break, the fluid interface. For this reason the knife-edge may be made nonwetting, though the problem of knife placement is still apparently cumbersome (Mannheimer & Burton 1970).

One of the most promising knife-edge surface viscometers, first introduced by Goodrich *et al.* (1975), is known as the rotating wall knife-edge surface viscometer. The method is capable of achieving a high sensitivity ($10^{-5} \leq \mu^s \leq 0.1$ sp) and does not suffer from the difficulties of placing a knife "at" the interface, yet without breaking the plane of the interface. As with the deep-channel surface viscometer, a small particle is placed within the surface to discern the rotational speed of the fluid interface; measurement of this rotational speed permits the determination of the interfacial shear viscosity.

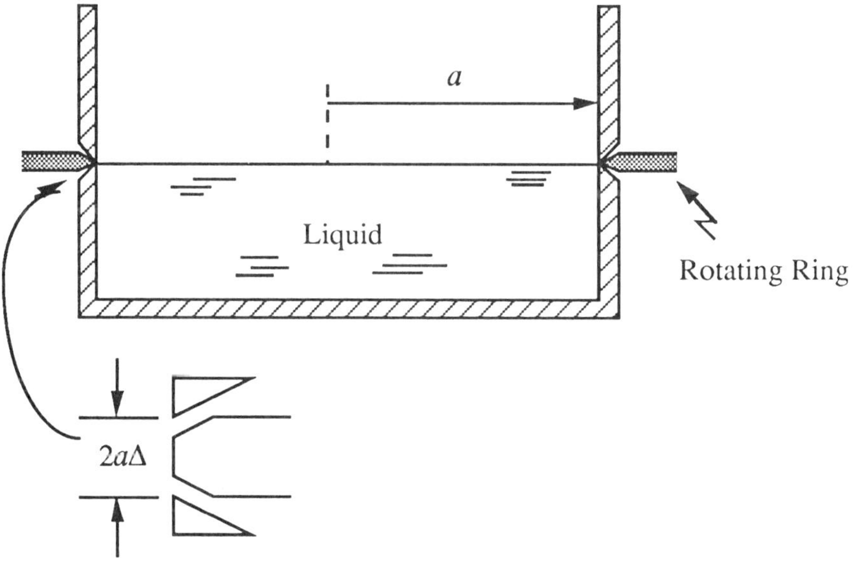

Figure 7.3-4 The rotating wall knife-edge surface viscometer.

The basic experimental design of the rotating wall knife-edge surface viscometer is depicted in Figure 7.3-4. Liquid is contained within a cylindrical dish, at a depth which is at least equivalent to the radius of the dish. The upper surface of the liquid contacts the walls of the container along a knife-edge ring that has been inserted into the wall and is flush with it. Rotation of the ring creates a viscometric flow in the vicinity of the interface. A small talc or teflon dust particle is placed within the interface, in the center portion of the surface, and the period of rotation is monitored. From the ratio of the displacement velocity to the known angular velocity of the rotating ring, the interfacial shear viscosity may be determined.

The theoretical analysis of Goodrich *et al.* (1975) affords a simple analytical expression for very small interfacial shear viscosity and $\Delta \rightarrow 0$ (cf. Figure 7.3-4); explicitly

$$\mu^s \approx 0.5631(\Omega / \omega)\mu a, \tag{7.3-2}$$

where (Ω/ω) is the ratio of particle angular rotation to the angular rotation of the machine-driven ring. A least-squares polynomial fit of the general integral expression (Goodrich *et al.* 1975) relating angular rotation of the tale particle Ω and the surface shear viscosity μ^s, applicable to arbitrarily large interfacial shear viscosity for the range of dimensionless rotation speeds $0 \leq \Omega/\omega \leq 0.7$, furnishes the relation

$$\mu^s = \mu a\left[0.5631\left(\frac{\Omega}{\omega}\right) + 1.1189\left(\frac{\Omega}{\omega}\right)^3 - 0.6254\left(\frac{\Omega}{\omega}\right)^5 + 3.4489\left(\frac{\Omega}{\omega}\right)^7\right], \tag{7.3-3}$$

which is valid to within an accuracy of 0.3%.

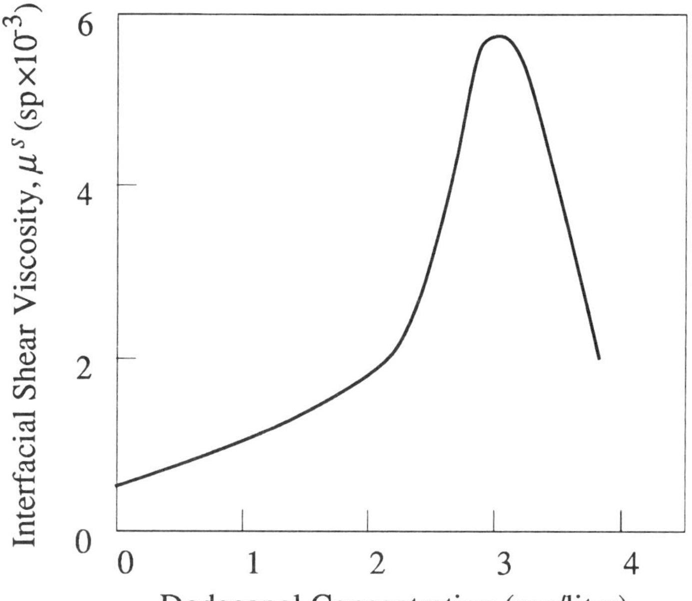

Figure 7.3-5 Interfacial shear viscosity vs dodecanol concentration for an aqueous solution of 1.75 g/l sodium dodecyl sulfate. Data taken from Poskanzer & Goodrich (1975)

Figure 7.3-5 shows interfacial shear viscosity data measured by the rotating knife-edge surface viscometer for a sodium dodecyl sulfate solution as a function of increasing dodecanol concentration.

Another knife-edge surface viscometer has been proposed by Shahin (1986), in which a thin rod is placed within the planar surface of a solution contained within a trough. A magnetic field results in the motion of the rod within the interface, and small particles are placed within the wake of the moving rod to track the interfacial shear field. By establishing a relation between the force exerted upon the rod and the shear field created, interfacial shear viscosity (as well as non-Newtonian interfacial rheological properties) may potentially be detected. Shahin (1986) cites a sensitivity to interfacial shear viscosity of 10^{-5} sp, though an inconclusive amount of data has been reported.

7.4 Summary

Several surface viscometers have been reviewed in Chapter 7 which may be employed for the purpose of measuring the interfacial shear viscosity, μ^s; these include canal, disc, and knife-edge methods.

The deep-channel surface (canal) viscometer is the standard experimental device for measuring interfacial shear viscosity, owing to its sensitivity ($\mu^s \geq 10^{-4}$ sp), simplicity, and exact analytical description. The primary drawback of the technique is the necessity of placing a small tracer particle within the interfacial flow field for tracking the centerline surface velocity. This may be particularly cumbersome with heavy oil systems, for which systems the particle may require several hours or more to execute a complete revolution, and with liquid-liquid systems for which the placement of the particle at the interface may be difficult.

Disk viscometers are generally less sensitive devices for measuring interfacial shear viscosity than are canal viscometers (typically $\mu^s \geq 10^{-2}$ sp) and exhibit a more complex flow field. The primary advantage of disk techniques is their ability to directly measure torsional stress. Nevertheless, placement of the disc "at" the interface and avoidance of contact angle anomalies are problematic issues. Disk viscometers appear most useful for the measurement of μ^s at highly viscous surfactant interfaces.

Knife-edge viscometers are similar to disc viscometers in most respects. A notable exception is the rotating wall knife-edge viscometer, which viscometer (similar to the deep-channel viscometer) entails the measurement of the velocity of a tracer particle within the fluid interface and is quite sensitive to interfacial shear viscosity ($\mu^s \geq 10^{-5}$ sp). This device appears to be the most promising current surface viscometer, together with the standard deep-channel device.

Questions for Chapter 7

7.1 Name three reasons why a tracer particle that has been placed in the surface along the centerline between the channel walls in the deep-channel surface viscometer, might be attracted toward the channel walls, rather than follow the centerline motion (a phenomenon often observed in practice).

7.2 *a*. Could a "step", such as depicted in Figure 7.1-3, be used to eliminate the curvature of a fluid-fluid interface at the channel walls in the deep-channel surface viscometer? *b*. What would be an alternative method for eliminating interface curvature at the walls?

7.3 Owing to the difficulty of placing a solid particle at the interface between two fluids in the deep-channel surface viscometer, small bubbles are sometimes created at the interface, and the motion of these bubbles is tracked to determine the centerline surface velocity. How could you determine whether the motion of the bubble truly reflects the surface motion: *a*. experimentally? *b*. theoretically?

Additional Reading for Chapter 7

§7.1 Canal Surface Viscometers: The Deep-channel Surface Viscometer

Mannheimer, R.J. & Schechter, R.S. 1970 An improved apparatus and analysis for surface rheological measurements. *J. Colloid Interface Sci.* **32**, 195–211.

Joly, M. 1964 Surface viscosity. *Recent Progr. Surface Sci.* **1**, 1–50.

§7.2 Disk Surface Viscometers

Oh, S.G. & Slattery, J.C. 1978 Disk and biconical interfacial viscometers. *J. Colloid Interface Sci.* **67**, 516–525.

Shail, R. 1978 The torque on a rotating disk in the surface of a liquid with an adsorbed film. *J. Colloid Interface Sci.* **12**, 59–76.

§7.3 Knife-edge Surface Viscometers

Lifshutz, N., Hedge, M.G. & Slattery, J.C. 1971 Knife-edge surface viscometers. *J. Colloid Interface Sci.* **37**, 73–79.

Goodrich, F.C., Allen, L.H., & Poskanzer, A. 1975 A new surface viscometer of high sensitivity I. Theory. *J. Colloid Interface Sci.* **52**, 201–220.

8

"I thank you for the remarks of your learned friend at Carlisle. I had when a youth, read and smiled at Pliny's account of a practice among the seamen of his time, to still the waves in a storm by pouring oil onto the sea . . .. But the stilling of a tempest by throwing vinegar into the air had escaped me. I think with your friend, that it has been of late too much the mode to slight the learning of the Ancients. The Learned too, are apt to slight too much the Knowledge of the Vulger. The cooling by evaporation was long an instance of the former. The Art of smoothing of waves with oil is an instance of both.

". . . In 1757, being at sea in a fleet of 96 sail bound against Loiusbourg, I observed the waves of two of the ships to be remarkably smooth, while all the others were ruffed by the wind, which blew fresh. Being puzzled by this differing appearance, I at last pointed it out to our captain, and asked him the meaning of it. 'The cooks,' says he, 'have I suppose, been just emptying their greasy water thro the scupper, which has greased the sides of those ships a little,' and this answer he gave to me with an air of some little contempt, as to a person ignorant of what every body else knew. In my own mind I at first slighted his solution . . .. But recollecting what I had formerly read in Pliny, I resolved to make some experiment of the Effect of Oil on Water when I should have opportunity.

". . . At length being at Chapham, where there is, on the common, a large pond, which I observed one day very rough with wind, I fetched a cruet of oil, and dropt a little of it on the water. I saw it spread itself with surprising swiftness upon the surface, but the effect of smoothing the waves was not produced; for I had applied it first on the Leeward side of the pond where the waves were largest, and the wind drove my back upon the shore. I went to the windward side, where they began to form; and there the oil, tho no more than a teaspoonful, produced an instant calm, over a space several yards square . . ."

Benjamin Franklin (1773)

CHAPTER 8

Measurement of Interfacial Dilatational Viscosity

The experimental detection of the interfacial dilatational viscosity κ^s of a Newtonian interface poses greater difficulties than the detection of the interfacial shear viscosity μ^s. This fact may be attributed to the coupling that arises in all interfacial dilatational flows between interfacial dilatational viscous and compositional elastic† effects. Measured values of κ^s for Newtonian interfaces are consequently distinguished in the literature as representing either *apparent* or *absolute* values, respectively corresponding to whether the values implicitly include, or do not include, a compositional elastic contribution.

Absolute measurements of interfacial dilatational viscosity generally require the application of an explicit surfactant transport model, in addition to an appropriate surface equation of state, the development of which has been outlined in §§5.2–5.5. Knowledge of a viable surfactant transport model permits the theoretical prediction of the dependence of interfacial tension upon surface expansion rate, thereby affording the possibility of theoretically distinguishing between compositionally induced and viscous interfacial stresses. Measurements of κ^s performed without the aid of a realistic surfactant transport description generally furnish apparent values, as the experimentalist is unable in this case to distinguish between interfacial viscous and compositional Marangoni contributions. Nevertheless, as will be demonstrated in §8.1, Marangoni and interfacial dilatational viscous effects may in certain circumstances be distinguished even *without* an explicit surfactant transport model, particularly when the surfactant-adsorbed interface expands or contracts as an insoluble monolayer (cf. §2.2 as well as Figure 6.2-2 and the discussion thereof for a description of 'insoluble

† In this chapter, the dilatational modulus, E^* [cf. Eq. (4.3-8)], is referred to as the *compositional elasticity*. Both names appear in the literature.

monolayer behavior'). Especially effective methods for measuring absolute values of the interfacial dilatational viscosity render the interfacial surfactant layer insoluble by imposing upon the interface a rate of surface dilatation that sufficiently exceeds the rate of surfactant transport to or from contiguous bulk phases.

In this chapter, techniques for measuring absolute and apparent interfacial dilatational viscosities (or some combination of κ^s and μ^s) are grouped into the following three classes: (i) surface wave techniques (§8.1); (ii) techniques involving the deformation of a bubble or droplet (§8.2); and (iii) the maximum bubble pressure method (§8.3).

8.1. Surface Wave Methods

Surface wave methods are generally employed for the purpose of measuring *absolute* interfacial dilatational viscosities. This usually requires an accurate theoretical description of the surfactant transport, Marangoni process, as further discussed below.

Return to the capillary wave analysis of §5.6 and observe, with the aid of Eq. (5.6-23), that the tangential (i.e. x-) component of the bulk-phase pressure tensor acting normal to the propagating wave surface may be expressed according to the linear surface wave theory of that section by the relation

$$P^o_{xz} = \frac{\partial \sigma}{\partial x} + (\kappa^s + \mu^s)\frac{\partial^2 v^o_x}{\partial x^2}\,. \tag{8.1-1}$$

Here, as elsewhere in the text (cf. Table 2.3-1), the superscript o is employed to indicate the evaluation of a bulk-phase field quantity at the interface. Following the complex variable notation introduced in §5.6 [cf. Eq. (5.6-26) and the discussion thereof], the interfacial-tension gradient appearing in Eq. (8.1-1) may be expressed in complex form, with the aid of Eqs. (5.6-57) and (5.6-60), as

$$\frac{\partial \sigma^*}{\partial x} = - E^*\left(\frac{k^3}{\omega} A + i\frac{k^2}{\omega} mB\right)\exp[\,i(kx - \omega t)]\,. \tag{8.1-2}$$

Furthermore, upon employing Eq. (5.6-38), there follows from the above,

$$P^o_{xz}{}^* \approx -\,[\,E^* - i\omega(\kappa^s + \mu^s)]\left(\frac{k^3}{\omega} A + i\frac{k^2}{\omega} mB\right)\exp[\,i(kx - \omega t)]\,. \tag{8.1-3}$$

The (complex) amplitude ξ^* of the compressional (x-component of the) surface wave is related to the x-component of the surface velocity by

$$v^{o*}_x = \frac{\partial \xi^*}{\partial t} \equiv - i\omega\xi^*, \tag{8.1-4}$$

whence, by Eq. (8.1-1),

$$P^o_{xz}{}^* = [E^* - i\omega(\kappa^s + \mu^s)]\frac{\partial^2 \xi^*}{\partial x^2}. \tag{8.1-5}$$

Finally, according to the decomposition (4.3-8),

$$P^o_{xz}{}^* = [E' - i\omega(\kappa^s + \mu^s - E''/\omega)]\frac{\partial^2 \xi^*}{\partial x^2}, \tag{8.1-6}$$

where, by Eq. (5.6-60), the dilatational elasticity E' satisfies the relation

$$E' \equiv \frac{E_o\left[1 + (\tau/\omega)^{1/2}\right]}{\left[1 + 2(\tau/\omega)^{1/2} + 2(\tau/\omega)\right]} \tag{8.1-7}$$

and the relaxational elasticity E'' obeys the relation

$$E'' \equiv -\frac{E_o(\tau/\omega)^{1/2}}{\left[1 + 2(\tau/\omega)^{1/2} + 2(\tau/\omega)\right]}. \tag{8.1-8}$$

In the above, the parameters E_o and τ are, respectively, the Gibbs elasticity (5.5-5) and 'diffusion parameter' (5.6-56).

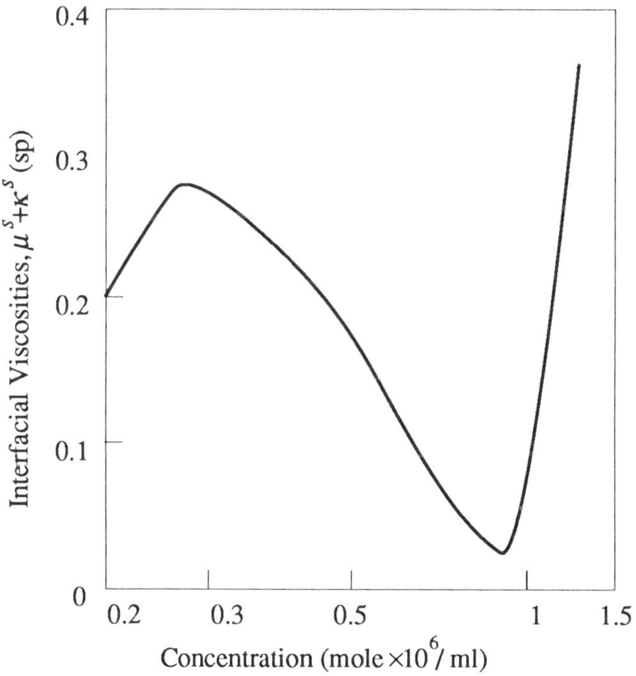

Figure 8.1.1 Combined values of the shear and dilatational interfacial viscosities for the interface between aqueous solutions of sodium dodecyl sulfate and decane, obtained by the longitudinal wave method. Data taken from Ting *et al.* (1985).

The tangential bulk-phase stress component evaluated at the interface thus combines an *elastic* (interfacial-tension gradient) effect, E', and an apparent *viscous* effect, $(\kappa^s + \mu^s) + E''/\omega$. Whereas both the viscous $(\kappa^s + \mu^s)$ and interfacial tension gradient (E''/ω) terms contribute to a phase shift in the propagating wave, only the term involving the group $(\kappa^s + \mu^s)$ is mechanically dissipative, and therefore truly 'viscous'. According to the representation (8.1-8), obtained by assuming diffusion-controlled surfactant transfer to the interface and small perturbations away from the equilibrium surfactant concentration, the relaxational term E''/ω may be considered known in terms of the measured parameters E_o and τ . Having thus modeled the surfactant transport process, the surface viscosity sum $(\kappa^s + \mu^s)$ may be determined experimentally together with E_o and τ .

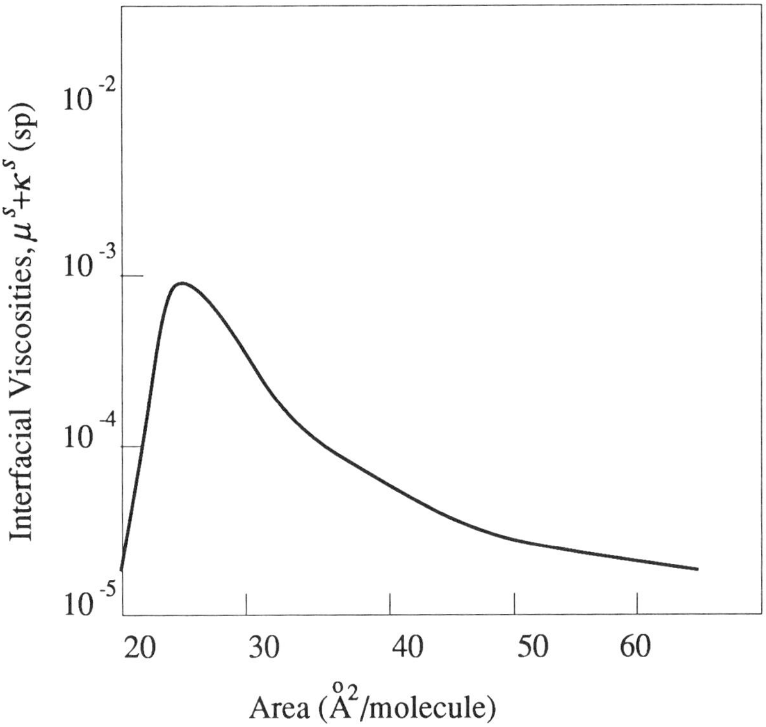

Figure 8.1.2 Combined values from light scattering of the shear and dilatational interfacial viscosities for the surface of an arachidic acid solution. Data taken from Byrne *et al.* (1979).

It is further evident from Eq. (8.1-8) that $E''/\omega \to 0$ in the limit of an insoluble monolayer (either $\tau \to 0$ or $\omega \to \infty$), revealing that under these physical conditions [only the interfacial viscosities $(\kappa^s + \mu^s)$ contributing to a surface wave phase shift] an explicit surfactant transport model is not required. Most longitudinal wave measurements of interfacial dilatational viscosity have been performed, however, for soluble monolayers in the low wave frequency regime (Lucassen & Hansen 1966, 1967, Ting *et al.* 1984, 1985), thereby requiring an explicit surfactant transport model, as in Eqs. (8.1-7) and (8.1-8).

Figure 8.1-1 displays values of combined (absolute) dilatational and shear interfacial viscosities measured by the longitudinal wave technique with low wave frequencies, i.e. on the order of 1 sec^{-1}.

In the light scattering capillary wave method (Langevin & Griesmar 1980, Byrne & Earnshaw 1979, Hard & Lofgren 1977, Hard & Neuman 1981) dynamic interfacial properties are measured at sufficiently high thermal wave frequencies so as to regard the interfacial-tension gradient as contributing a purely elastic effect, i.e. $E''/\omega \to 0$. Interfacial viscosities measured by the light scattering method are shown in Figure 8.1-2.

8.2 Droplet Deformational Methods

Several methods exist for measuring a combination of shear and dilatational interfacial viscosities through the rotation, translation, or deformation of bubbles and droplets. Each of these techniques involves an interfacial flow that is accompanied by interfacial-tension gradients. Thus, these methods appear most practically useful for measuring *apparent* values of the dilatational interfacial viscosity.

Agrawal & Wasan (1979) have suggested that the translational velocity of bubbles or droplets in a quiescent viscous liquid might be measured for the purpose of determining an apparent dilatational viscosity (see the first examples of §§4.4 and 5.6). Employing a theoretical model which acknowledges the existence of both interfacial viscous and Marangoni stresses, they obtained the following expression for the settling velocity of a fluid sphere whose surface is adsorbed by surfactant [cf. Eqs. (4.4-24) and (5.6-15)]:

$$\mathbf{U} = \frac{2}{3}a^2\frac{(\rho - \tilde{\rho})}{\mu}\mathbf{g}\left(\frac{\frac{2}{3a}\kappa^s_{app} + \mu + \tilde{\mu}}{\frac{2}{a}\kappa^s_{app} + 2\mu + 3\tilde{\mu}}\right), \tag{8.2-1}$$

with κ^s_{app} an apparent interfacial dilatational viscosity, given in terms of the surfactant transport model of §5.6 by the relation

$$\kappa^s_{app} = \kappa^s + E_o / \alpha . \tag{8.2-2}$$

Other possible expressions arising by different surfactant kinetic behaviors are reported by Agrawal & Wasan (1979).

Though relatively simple to implement, droplet settling measurements are not generally used for deducing surface dilatational viscosity values owing to the relative insensitivity of $\mathbf{U}$ upon κ^s_{app}.

Wei *et al.* (1974) have suggested that measurements of the circulation velocity of a tracer particle in the equatorial plane of a spherical droplet rotating within a shear field (Figure 8.2-1) might provide a means for deducing a combination of shear and dilatational interfacial viscosities for the case of small capillary number [see example 5 of §4.4]. In such circumstances, as demonstrated in Question 8.3, the following expression is obtained, relating interfacial shear and dilatational viscosities to the period of circulation T (the notation of §4.4 is maintained):

$$\frac{3\kappa^s + 2\mu^s}{\mu a} = \frac{5}{2}\left[T\left(T - \frac{4\pi}{\bar{S}}\right)^{-1/2} - \left(1 + \frac{\hat{\mu}}{\mu}\right)\right]. \qquad (8.2\text{-}3)$$

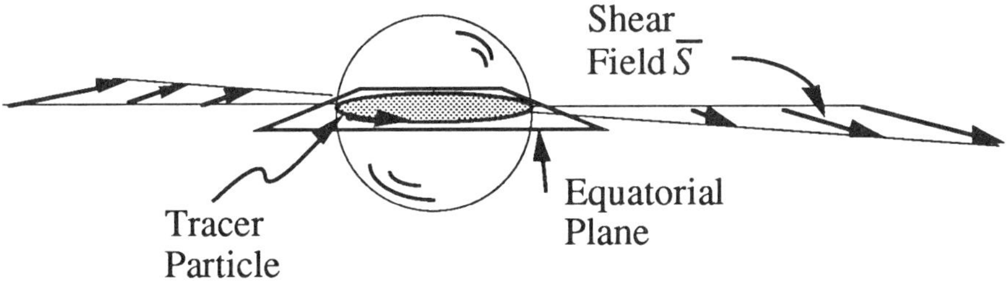

Figure 8.2-1 A spherical droplet is depicted rotating in an (undisturbed, homogeneous) shear field. A tracer particle placed at the droplet interface in the equatorial plane is shown; its velocity of revolution may be measured to determine the circulation period, T.

Flummerfelt (1980) has shown that the above result is equally appropriate to the more common circumstance wherein interfacial-tension gradients act along the droplet interface, allowing for the fact that the dilatational surface viscosity κ^s appearing in Eq. (8.2-3) represents an *apparent* value [i.e. of the form (8.2-2)], combining the effects of interfacial viscosity and compositional elasticity.

Employing Eq. (8.2-3), Wei *et al.* (1974) have re-analyzed the droplet circulation data of Rumsheidt and Mason (1961). They deduced the existence of significant interfacial viscosity values, evidently owing to impurities adsorbed to the interface between water and silicone oil. Circulation experiments have also been performed by Phillips *et al.* (1980) for pure (surfactant-free) and surfactant systems.

The study of droplet shape deformation in a shear field provides another method for measuring apparent interfacial dilatational viscosities.

Taylor (1932, 1934) first predicted that an initially spherical droplet placed within a shear field deforms into an elipsoidal shape with its major axis tending to orient in the direction of flow. The theoretical results of Taylor showed that to leading order in the capillary number, upon neglecting interfacial-tension gradient and interfacial viscosity effects, a deformation parameter,

$$D \equiv \frac{r_{max} - r_{min}}{r_{max} + r_{min}}, \tag{8.2-4}$$

defined in terms of the maximum and minimum curvature radii (see Figure 8.2-2) satisfies the relation

$$D = \mathrm{Ca}\,\frac{19\,\mathrm{N}_\mu + 16}{16\,\mathrm{N}_\mu + 16}. \tag{8.2-5}$$

Here, the capillary number is defined as

$$\mathrm{Ca} = \frac{\mu S a}{\sigma} \tag{8.2-6}$$

and the bulk viscosity ratio as

$$\mathrm{N}_\mu = \frac{\hat{\mu}}{\mu}. \tag{8.2-7}$$

A (^) is used to denote droplet phase quantities. The angle of orientation, α, of the major axis of the spheroid from the plane of shear is given by

$$\alpha = \frac{\pi}{4}. \tag{8.2-8}$$

These leading-order deformation and orientation results have since been extended by several researchers (Chaffey & Brenner 1967, Cox 1969, Frankel & Acrivos 1970, Barthes & Acrivos 1973, Choi & Schowalter 1975). Cox (1969) has shown that to leading order in a dimensionless deformation parameter δ, where $r = r_0 + f(\theta, \phi)\delta$ and r_0 is the radius of the initially spherical droplet, the following generalizations of the preceding results obtain:

$$D = \mathrm{Ca}\frac{\dfrac{19\,\mathrm{N}_\mu + 16}{16\,\mathrm{N}_\mu + 16}}{\sqrt{\left[1 + \left(\dfrac{19\,\mathrm{CaN}_\mu}{20}\right)^2\right]}}, \tag{8.2-9}$$

and

$$\alpha = \frac{\pi}{4} + \frac{1}{2}\tan^{-1}\left(\frac{19\,\mathrm{CaN}_\mu}{20}\right). \tag{8.2-10}$$

The applicability of relations such as (8.2-9) and (8.2-10) to actual droplet deformations may be verified by measuring both D and α. Thus, upon defining

$$F = \frac{19D}{20}\left[1 + \frac{1}{\tan^2\left(2\alpha - \frac{\pi}{2}\right)}\right]^{1/2} \tag{8.2-11}$$

which parameter is dependent only upon the viscosity ratio, namely,

$$F = \frac{(19\,\mathrm{N}_\mu + 16)}{(16\,\mathrm{N}_\mu + 16)} \frac{1}{\mathrm{N}_\mu}, \tag{8.2-12}$$

the validity of the theoretical description may be confirmed through bulk viscosity measurements alone, thereby avoiding the explicit need for measured interfacial tension values.

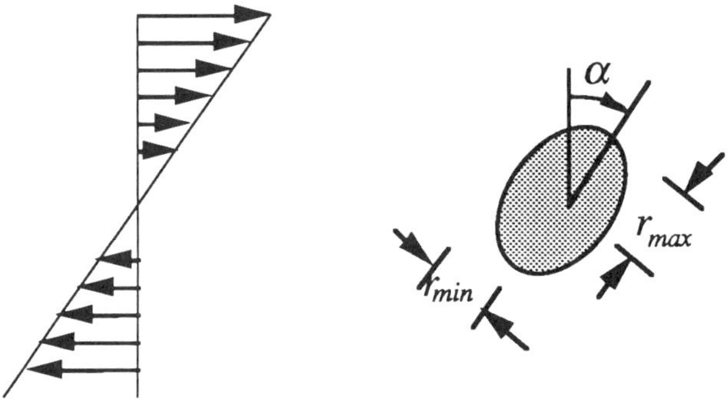

Figure 8.2-2 An elipsoidal liquid droplet in an (undisturbed) homogeneous shear field.

Flummerfelt (1980) has generalized the Cox (1969) theory by including interfacial rheological effects, thus furnishing a method for measuring apparent interfacial dilatational viscosities through the study of droplet deformation in a shear field. Flummerfelt (1980) found that to the same order of approximation as in the Cox theory, including the combined effects of interfacial-tension gradients and interfacial viscosities, there obtain the expressions;

$$D = \frac{5}{16}\,\mathrm{Ca}\,\frac{\dfrac{24(\mathrm{Bo}_\kappa + \mathrm{Ma}) + 8\,\mathrm{Bo}_\mu + 19\,\mathrm{N}_\mu + 16}{6(\mathrm{Bo}_\kappa + \mathrm{Ma}) + 4\,\mathrm{Bo}_\mu\, 5\,\mathrm{N}_\mu + 5}}{\sqrt{1 + \left(\dfrac{19\lambda R\ \mathrm{CaN}_\mu}{20}\right)^2}}, \tag{8.2-13}$$

and

$$\alpha = \frac{\pi}{4} + \frac{1}{2}\tan^{-1}\left(\frac{19\lambda R\ \mathrm{CaN}_\mu}{20}\right), \tag{8.2-14}$$

where the following dimensionless groups have been defined:

$$\mathrm{Bo}_\mu \equiv \frac{\mu^s}{a\mu}, \qquad \mathrm{Bo}_\kappa \equiv \frac{\kappa^s}{a\mu}, \tag{8.2-15}$$

represent the relevant Boussinesq numbers of the interfacial flow;

$$\mathrm{Ma} \equiv \frac{E_o \mu a}{mK_a/(1+mK_a/k) + \hat{m}\hat{K}_a(1+\hat{m}\hat{K}_a/\hat{k})}, \tag{8.2-16}$$

furnishes the relevant Marangoni number of the interfacial flow; and

$$\lambda \equiv \mathrm{N}_\mu + \frac{6}{5}(\mathrm{Bo}_\kappa + \mathrm{Ma}) + \frac{4}{5}\,\mathrm{Bo}_\mu, \tag{8.2-17}$$

and

$$\lambda R \equiv \frac{1}{19\lambda}\Big[19\,\mathrm{N}_\mu^2 + 23\,\mathrm{N}_\mu(\mathrm{Bo}_\kappa + \mathrm{Ma}) + 26\,\mathrm{N}_\mu\,\mathrm{Bo}_\mu + 16(\mathrm{Bo}_\kappa + \mathrm{Ma})\,\mathrm{Bo}_\mu\Big] \tag{8.2-18}$$

are dimensionless groups combining bulk and interfacial resistive effects. As in Eq. (8.1-6), terms involving κ^s and E_o appear coupled in the above expressions.

The surfactant transport model employed in obtaining the explicit relation (8.2-16) for the Marangoni number applies to both cases of diffusion- and adsorption-controlled surfactant transport. In particular, the species flux vectors at each side of the droplet interface [cf. Eq. (5.3-3)] are assumed to possess the following forms (omitting overbars):

$$\mathbf{n}\cdot\mathbf{j}^o = \frac{k}{(1+k/mK_a)}(\rho^s - \rho_o^s) \tag{8.2-19}$$

and

$$\mathbf{n}\cdot\hat{\mathbf{j}}^o = \frac{\hat{k}}{\left(1+\hat{k}/\hat{m}\hat{K}_a\right)}(\rho^s - \rho_o^s). \tag{8.2-20}$$

Here, $\mathbf{n}$ is the unit surface normal, ρ^s the surface-excess species density of surfactant (subscripts corresponding to species are also omitted, as in §5.6) with ρ_o^s its equilibrium value, k and $\hat{k}$ are adsorption/desorption rate constants in the continuous and droplet phases respectively, m and $\hat{m}$ mass transfer coefficients in continuous and droplet phases respectively, and K and $\hat{K}$ linear equilibrium adsorption coefficients relative to the continuous and droplet phases, respectively. (Thus, the case of an adsorption-controlled surfactant transfer process may be obtained in the limit $k/mK \ll 1$, and the case of a diffusion-controlled process in the limit $k/mK \gg 1$.)

The scalar F, defined in Eq. (8.2-11), may be expressed with the aid of Eqs. (8.2-13) and (8.2-14), by the relation

$$F = \frac{5}{16}\,\frac{24(\mathrm{Bo}_\kappa + \mathrm{Ma}) + 8\,\mathrm{Bo}_\mu + 19\,\mathrm{N}_\mu + 16}{6(\mathrm{Bo}_\kappa + \mathrm{Ma}) + 4\,\mathrm{Bo}_\mu + 5\,\mathrm{N}_\mu + 5}\,\frac{1}{\lambda R}. \tag{8.2-21}$$

Upon comparison of Eqs. (8.2-11) and (8.2-21), it may be seen that by performing measurements of D, α, a, and N_μ, explicit values of the interfacial groups Bo_μ and $Bo_\kappa + Ma$ may be obtained. The interfacial shear viscosity μ^s may be independently measured by any of the techniques discussed in Chapter 7 to separately determine values for the apparent interfacial dilatational viscosity group $Bo_\kappa + Ma$.

Such measurements have been carried out by Phillips *et al.* (1980) upon several pure and surfactant systems. By attributing the difference between experimental values of F and the Cox formula (8.2-12) to interfacial viscous effects, Eq. (8.2-21) was employed to determine both the shear and apparent dilatational viscosities of the interface. [For this purpose, various radii of droplets were investigated and a nonlinear least-squares regression subsequently performed on Eq. (8.2-21).] Results were compared with circulation experiments, using Eq. (8.2-3), and order-of-magnitude agreement was found.

$$\Omega = \Omega_o \qquad \text{Spinning Drop Tensiometer}$$

$$\Omega = \Omega_o + A \cos (\omega t) \qquad \text{Spinning Drop Viscometer}$$

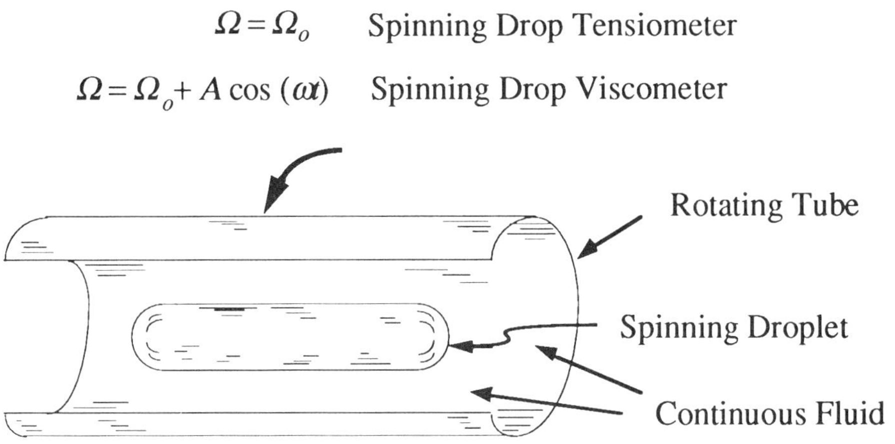

Figure 8.2-3 The spinning drop method. A droplet is elongated within a spinning tube rotating either with a constant angular velocity (spinning drop tensiometer) or an oscillating angular velocity (spinning drop viscometer).

Droplet deformation is also the principle of the spinning drop tensiometer, which latter device furnishes one of the standard methods for measuring ultra-low interfacial tensions. The basic configuration of the spinning drop tensiometer is depicted in Figure 8.2-3. A circular cylindrical tube containing a droplet suspended within a continuous fluid is rotated with constant angular velocity Ω about the axis of the tube. The surrounding fluid and elongated droplet are assumed to rotate as rigid bodies; hence, by measuring appropriate geometrical parameters, the interfacial tension may be determined.

Slattery *et al.* (1980) have proposed a method whereby the spinning drop tensiometer might be used to measure interfacial viscosities. In this technique, the tube undergoes small oscillations in the rotational velocity. By measuring the amplitude of droplet radius oscillations (in the region where the droplet is reasonably approximated as being cylindrical), as well as the phase lag, both interfacial shear and dilatational viscosities may, in principle, be determined.

There exist several drawbacks to this technique, as it has been proposed by Slattery and coworkers. Primary among these is the need for *a priori* knowledge of the dilatational elasticity of the fluid interface. To avoid this issue, Slattery *et al.* (1980) suggest that dilute surfactant systems should be examined for which the Gibbs elasticity is small, although this would severely limit the usefulness of the technique. In addition, the amplitude of the oscillating droplet radius is apparently too small for practical measurement when liquid-liquid systems are used; the technique is thus further restricted to gas-liquid systems.

8.3 The Maximum Bubble Pressure Method

Many of the drawbacks of the preceding surface wave and droplet deformational methods for measuring interfacial dilatational viscosity may be overcome by the maximum bubble pressure method (MBPM), the latter technique having first been employed as a surface viscometer by Kao *et al.* (1991a, 1991b). Advantages of this technique include: (i) the coupling of compositional elasticity and interfacial viscosity is avoided by achieving high rates of surface expansion; (ii) it is sensitive to the interfacial dilatational viscosity of low- as well as highly-concentrated surfactant systems; and (iii) the (absolute) interfacial dilatational viscosity is measured alone, rather than in combination with the interfacial shear viscosity. The chief limitation of the technique appears to be the need for very high surface expansion rates, possibly larger than encountered in practice, for which rates the fluid interface may exhibit nonlinear rheological behavior. Uncertainties may also arise regarding the accuracy of experimentally determining the rate of surface expansion (see the first footnote of §6.3).

The MBPM has been previously described in §6.3, the analysis of the surface motion near the maximum bubble pressure having been provided by the second example of §4.4. Equation (4.4-33) is reproduced below:

$$\tilde{p} = p + \frac{\mu Q}{\pi a^3} + \frac{2}{a}\left[\sigma + \frac{\kappa^s Q}{2\pi a^3}\right]. \tag{8.3-1}$$

According to this equation, the pressure drop across the bubble surface at the instant when the bubble possesses a hemispherical shape (corresponding to the maximum bubble pressure) owes to a combination of bulk viscous, surface tension, and surface dilatational viscosity effects. The bulk viscous term in Eq. (8.3-1) is negligible for typical experimental conditions of the MBPM; this may be verified [as in the work of Kao *et al.* (1991a)] by demonstrating the independence of the pressure drop with gas flow rate Q for pure deionized water. The term in brackets represents the dynamic surface tension [cf. Eqs. (4.2-19) and (6.3-1)]. It may be measured by the MBPM as a function of frequency of bubble generation, which in this technique is directly related to the flow rate Q, or the rate of surface expansion by $\nabla_s \cdot \mathbf{v}^o = Q/\pi r_c^3$.

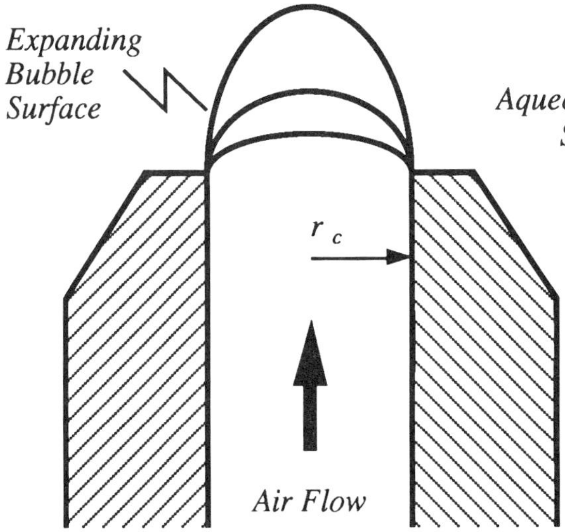

Figure 8.3-1 The maximum bubble pressure method.

Generally, the surface tension σ will vary according to the rate of surface expansion, with the consequence that the surface viscosity and surface tension contributions to the mean surface tension will be inseparable without the aid of a plausible surfactant transfer model [see Joos & Rillaerts (1981)]. However, at high frequencies of bubble generation (on the order of 10–30 hz, say), the bubble surface will expand far more rapidly than the rate of surfactant adsorption/desorption, such that the surface will expand essentially as an insoluble monolayer. In this high frequency range, the dynamic surface tension may thus be measured versus frequency of bubble creation in order to determine the *absolute* surface dilatational viscosity. As the fluid surface will likely exhibit nonlinear stress-deformation behavior at these high expansion rates, the frequency range used to determine κ^s should be relatively small such that the interfacial dilatational viscosity is essentially constant throughout this range.

The data of Figure 6.3-2 illustrate the limiting (insoluble monolayer) behavior of the dynamic surface tension versus surface expansion rate for high rates of expansion. The slope of these curves in the high frequency limit provides a direct measure of the surface dilatational viscosity [cf. Eq. (8.3-1)]. The values obtained by Kao and coworkers are shown in Figure 8.3-2, where comparisons are also provided with data taken from the longitudinal wave method. The magnitudes of surface dilatational viscosity obtained by the MBPM and the longitudinal wave method may be seen to differ considerably. This may owe to a rate-dependence of interfacial viscosity, as the surface expansion rate in the MBPM lies in the khz range. [By contrast, longitudinal wave methods typically subject the interface to an expansion rate ($\approx f_o \alpha \omega$) on the order of 0.01 hz.] Moreover, an underestimate of gas flow rate may contribute to smaller surface viscosity values in the MBPM (cf. the first footnote of §6.3).

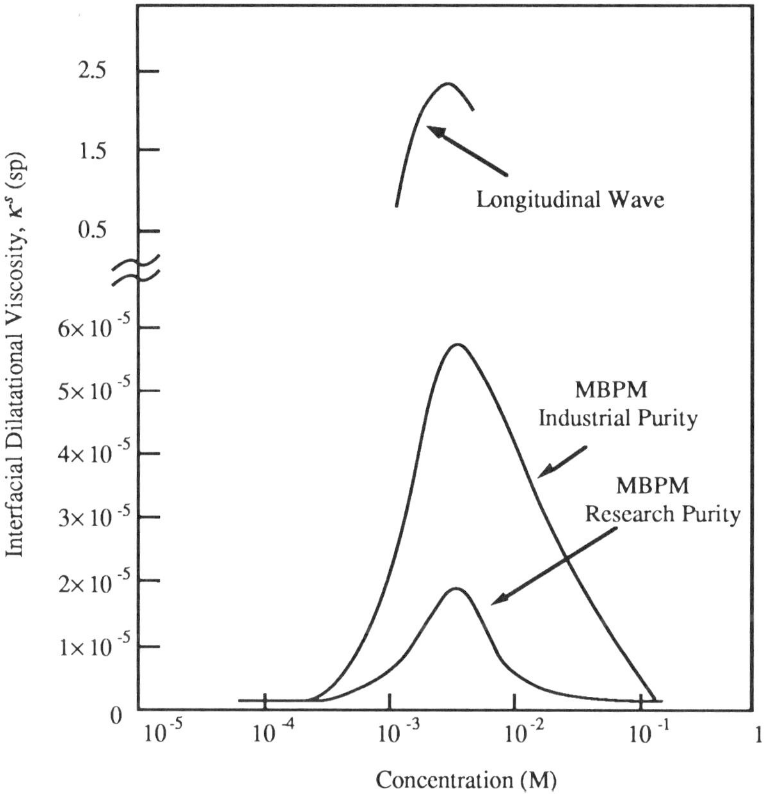

Figure 8.3-2 Surface dilatational viscosity values as measured by the MBPM (Kao *et al.* 1991a) and by the longitudinal wave method (Djabbarah & Wasan 1982) versus sodium dodecyl sulfate (SDS) concentration.

Application of the MBPM to liquid-liquid systems poses further difficulties, owing to viscous pressure losses in the capillary. Preliminary measurements have been made by Kao *et al.* (1991b).

8.4 Summary

Several methods have been reviewed in this chapter for the purpose of measuring the interfacial dilatational viscosity of both gas-liquid and liquid-liquid interfaces. In many of these, interfacial viscous and compositional elastic effects (the latter owing to the expansion rate dependence of the interfacial tension) are not distinguished; together, they are measured as *apparent* interfacial dilatational viscosities. Other techniques allow an *absolute* determination of interfacial dilatational viscosity.

Of the three classes of measurement techniques presented, only the surface wave (§8.1) and MBPM (§8.3) methods are presently capable of absolute interfacial viscosity measurements. One advantage of the wave methods is that sufficiently small surface expansion rates are employed to reasonably guarantee that the surface remains Newtonian. Since surfactant adsorption/desorption accompanies longitudinal wave motion, an explicit surfactant transfer model is required to determine the interfacial viscosities, whereas with the light scattering method, an equilibrium surfactant concentration is maintained. Current surface wave methods are capable of measuring only the sum of interfacial shear and dilatational viscosity. An advantage of the MBPM is that the interfacial dilatational viscosity may be directly measured (there being no contribution of surface shear viscosity to the interfacial flow) and at sufficiently high rates of surface expansion such that a surfactant adsorption model is not required. This technique is quite straightforward in both theory and practice, though it possesses the disadvantage of requiring very high rates of surface expansion, for which interfacial viscosities will likely display a deformation rate dependence.

Circulation measurements and deformation studies of a droplet placed within a homogeneous shear field may yield apparent estimates of the interfacial dilatational viscosity, including both viscous and compositional elastic effects. Such methods have been discussed in §8.3.

Questions for Chapter 8

8.1 Why do both shear and dilatational interfacial viscosities appear in the stress relation (8.1-5)?

8.2 Could the surface wave problem of Question 5.12 be employed for the purpose of measuring the interfacial dilatational viscosity? Would the interfacial viscous stress include a combination of interfacial shear and dilatational viscous effects as in the wave problem of §5.6? Explain why this occurs.

8.3 Derive Eq. (8.2-3) by employing Eq. (4.4-113).

Additional Reading for Chapter 8

§8.1 Surface Wave Methods

Ting, L., Wasan, D.T., Miyano, K. & Xu, Q. 1984 Longitudinal surface waves for the study of dynamic properties of surfactant systems II. Gas-liquid surface. *J. Colloid Interface Sci.* **102**, 248–258.

Ting, L., Wasan, D.T., & Miyano, K. 1985 Longitudinal surface waves for the study of dynamic properties of surfactant systems

III. Liquid-liquid interface. *J. Colloid Interface Sci.* **107**, 345–354.

§8.2 Drop Deformational Methods

Flummerfelt, R.W. 1980 Effects of dynamic interfacial properties on drop deformation and orientation in shear and extensional flow fields. *J. Colloid Interface Sci.* **76**, 330–349.

Slattery, J.C., Chen, J.D., Thomas, C.P., & Fleming, P.D. 1980 Spinning drop interfacial viscometer. *J. Colloid Interface Sci.* **73**, 483–499.

§8.3 The Maximum Bubble Pressure Method

Kao, R.L., Edwards, D.A., Wasan, D.T. & Chen, E. 1991a Measurements of the interfacial dilatational viscosity at high rates of interfacial expansion using the maximum bubble pressure method. I. Gas-liquid surface. (Submitted).

9

"It is a matter of common observation that a strongly alcoholic liquid, such as port wine, exposed to the air in an open glass vessel, creeps up the sides of the vessel in the form of a fine film which terminates in a thickened marginal 'roll', and that this 'roll' is studded with drops which grow in size until they detach themselves and return to the mother liquid in the form of 'tears'."

Max Loewenthal (1931)

CHAPTER 9

Measurement of Non-Newtonian Interfacial Rheological Properties

Fluid interfaces adsorbed by polymers, proteins or other large surfactant molecules often display non-Newtonian interfacial rheological behavior, as do most surfactant-adsorbed fluid interfaces in the presence of sufficiently large rates of deformation. Interfacial rheological models appropriate to such non-Newtonian behavior have been previously discussed in §4.3.

In the current chapter, adaptation of experimental devices introduced in Chapter 7 for the purpose of measuring non-Newtonian properties is discussed for each of the following three categories of interfacial behavior†; (i) nonlinear stress-deformation behavior (§9.1); (ii) viscoelastic behavior (§9.2); and (iii) viscoplastic behavior (§9.3). For each of these rheology classes, the deep-channel surface viscometer is most commonly employed for measurement purposes, though knife-edge, disc and biconical bob viscometers are also used.

9.1 Nonlinear Interfacial Rheological Behavior

Current methods for measuring nonlinear interfacial rheological properties owe largely to the work of Slattery and coworkers (Hedge & Slattery 1971, Jiang *et al.* 1983, Wei & Slattery 1976, Li & Slattery 1988). These techniques generally employ adaptations of the classical Newtonian analyses for deep-channel, knife-edge, disk and biconical bob viscometers.

---

† In Chapter 8 it was discussed that surfactant monolayers may display non-Newtonian behavior at high surface expansion rates. The methods and experimental findings discussed there represent the current extent of research into the non-Newtonian dilatational properties of fluid interfaces. Most techniques for measuring non-Newtonian interfacial rheological behavior (including all the techniques discussed in this chapter) focus upon interfacial shear properties.

As analyses of flow in torsional devices for interfaces exhibiting nonlinear rheological behavior have been limited to highly viscous fluid interfaces, the most widely-used technique is presently the deep-channel surface viscometer.

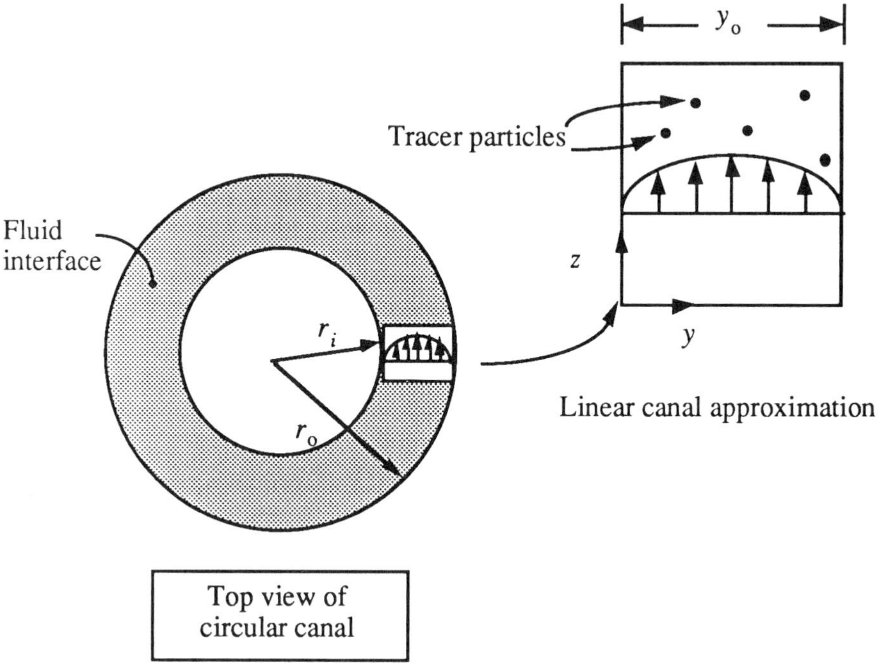

Figure 9.1-1 Small tracer particles are placed on the fluid interface to measure the interfacial velocity profile in the deep channel method for measuring nonlinear interfacial rheological behavior.

In the deep-channel method, several small tracer particles are placed on the fluid interface at different radial positions, as depicted in Figure 9.1-1; angular velocities $f(y)$ are determined from measurements of the period of revolution.

From this information, the deformation rate tensor of the interface [cf. Eq. (4.2-18)] may be experimentally determined by the relation

$$\mathbf{D}_s = \frac{1}{2}(\mathbf{i}_x\mathbf{i}_y + \mathbf{i}_y\mathbf{i}_x)\frac{df}{dy}. \qquad (9.1\text{-}1)$$

Hedge & Slattery (1971) furthermore show that the surface-excess stress tensor for a gas-liquid surface [see Wei & Slattery (1976) for the case of a liquid-liquid interface] adopts the form

$$\tau_s = (\mathbf{i}_x \mathbf{i}_y + \mathbf{i}_y \mathbf{i}_x)$$
$$\times \left\{ 2\mu \sum_{n=1}^{\infty} \left[f_n \coth(n\pi D) - \frac{1-(-1)^n}{n\pi \sinh(n\pi D)} \right] \cos\left(\frac{n\pi \bar{y}}{y_0} \right) \right.$$
$$\left. - 2\mu \sum_{n=1}^{\infty} \left[f_n \coth(n\pi D) - \frac{1-(-1)^n}{n\pi \sinh(n\pi D)} \right] \cos\left(\frac{n\pi y}{y_0} \right) \right\}, \quad (9.1\text{-}2)$$

where μ is the liquid-phase viscosity,

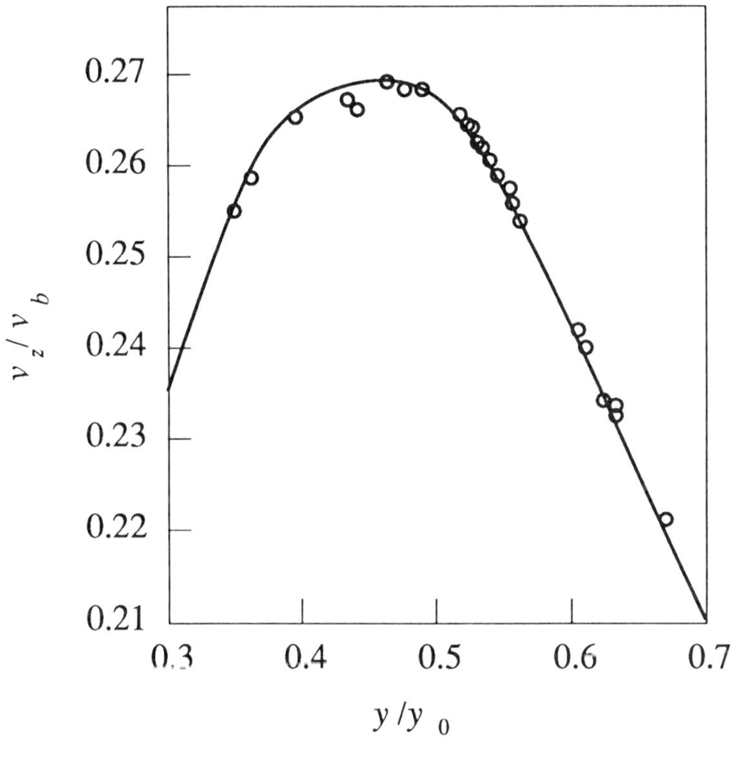

Figure 9.1-2 Interfacial velocity profile (nondimensionalized by the midchannel velocity v_b of the rotating dish) for the interface between a 6% aqueous solution of potassium oleate and air (μ^s=1.74×10^{-4}). The solid line indicates the predicted behavior for a Newtonian interface. Data taken from Wei & Slattery (1976).

$$f_n \equiv \int_0^{y_0} f(y) \sin\left(\frac{n\pi y}{y_0} \right) dy , \quad (9.1\text{-}3)$$

represents a weighted surface velocity average,

$$D \equiv \frac{x_0}{y_0}, \quad (9.1\text{-}4)$$

(where, as demanded by the deep-channel approximation, D>>1) and $\bar{y}$ is a dimensional value of the y-coordinate, defined such that

$$\frac{df}{dy} = 0 \qquad \text{at} \quad y = \bar{y}. \tag{9.1-5}$$

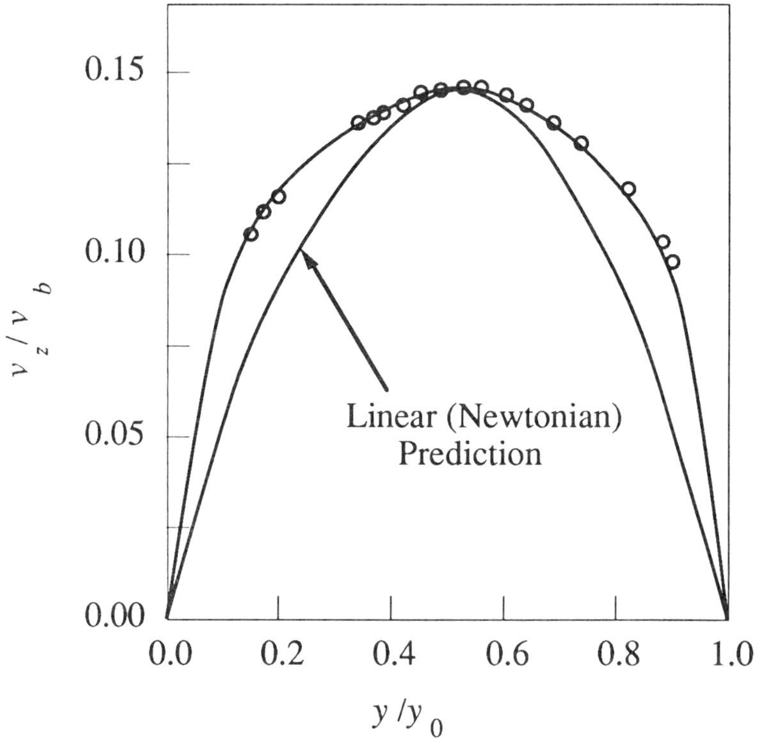

Figure 9.1-3 Interfacial velocity profile for n-octadecanol monolayer (20.5 Å^2/molecule) over distilled water. The line through the data represents a least-squares fit. The second solid line indicates the predicted behavior for a Newtonian interface (μ^s=4.70×10^{-3} sp). Data taken from Wei & Slattery (1976).

By measuring several discrete values of the interfacial velocity $f(y)$, it is possible to calculate both the surface deformation rate tensor (9.1-1) and the surface-excess stress tensor (9.1-2). Assuming the interfacial constitutive relationship [cf. Eq. (4.2-17)]

$$\boldsymbol{\tau}^s = 2\mu^s(\dot{\gamma})\mathbf{D}^s \tag{9.1-6}$$

($\dot{\gamma} = \left|\frac{df}{dy}\right|$ denoting the rate of surface shear), the apparent interfacial shear viscosity $\mu^s(\dot{\gamma})$ may be determined.

Interfacial velocity profiles are depicted in Figures 9.1-2 and 9.1-3 for surfactant interfaces displaying Newtonian and non-Newtonian rheological behaviors.

The dependence of the apparent interfacial shear viscosity upon rate of shear deformation is illustrated in Figure 9.1-4.

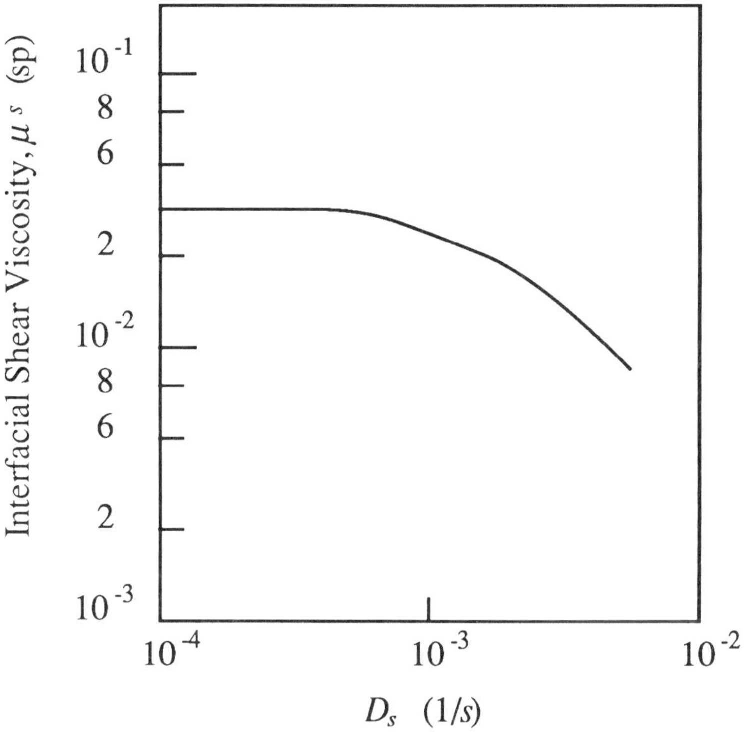

Figure 9.1-4 Apparent interfacial shear viscosity versus the magnitude of the shear deformation rate for n-octadecanol monolayer (20.5 $\mathring{A}^2$/molecule) over distilled water. Data taken from Wei & Slattery (1976).

In these measurements, a tendency of particles to move tangentially across fluid streamlines was attributed to the existence of small interfacial-tension gradients. Particle-particle hydrodynamic interactions may also produce this behavior, should particles not be sufficiently spaced.

Pintar *et al.* (1971) have provided an explicit model of nonlinear shear stress behavior, as discussed in Chapter 4 [cf. Eq. (4.3-6)], constituting the surface analog of the Powell-Erying model. They found that for such interfacial stress behavior, the centerline surface velocity may be expressed in the deep-channel limit (D>>1) as

$$v_c = v_c^* \frac{\alpha\beta^3}{6(\mu_0^s + \alpha\beta)} \left(\frac{4v_b\sqrt{(\pi D/2) - 1/16}}{y_0 \cosh(\pi D)} \right)^2, \qquad (9.1\text{-}7)$$

with α, β and μ_0^s the coefficients that appear in the constitutive relation (4.3-6). These parameters may be experimentally determined by varying the shear rate [cf. Eq. (4.3-3)], namely,

$$\dot{\gamma} = \sqrt{\mathrm{II}_s} = \frac{4v_b\sqrt{\pi D / 2}}{y_0 \cosh(\pi D)}, \tag{9.1-8}$$

rather than placing several particles in the interface, as was the procedure in the previous method.

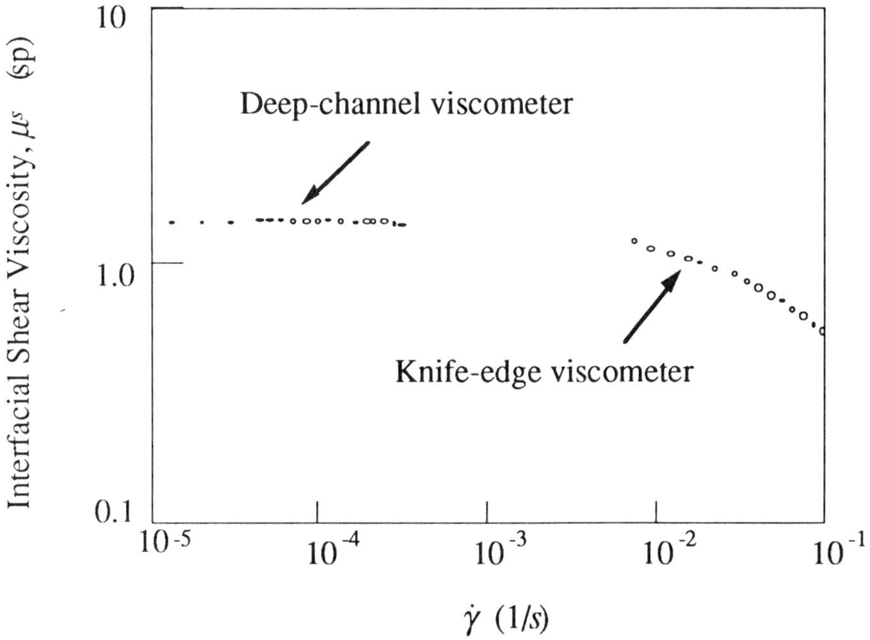

Figure 9.1-5 Apparent interfacial shear viscosity versus the surface shear rate for mixed sodium dodecyl sulfate and sodium dodecyl alcohol (weight ratio 0.031 SDOH/SDS) monolayer (c=3.47×10^{-3} M) over distilled water. Data taken from Li & Slattery (1988).

Torsional surface viscometers may also be employed to measure nonlinear interfacial rheological behavior. Jiang *et al.* (1983) have shown that the apparent interfacial shear viscosity may be determined from independent measurements of the torsional stress, τ_z (using knife edge, disc, or biconical bob surface viscometers) by employing two torsional viscometers (A and B, say) with differing dish-to-bob ratios. Thus, with R the dish radius and a the bob radius,

$$s_A \equiv \frac{R}{a_A} \tag{9.1-9}$$

$$s_B \equiv \frac{R}{a_B}, \tag{9.1-10}$$

Jiang *et al.* (1983) have demonstrated that the formula

$$\mu^s = \frac{\tau_z}{4\pi R^2}\left(\frac{s_A^2 - s_B^2}{\omega_A - \omega_B}\right) \qquad \text{as} \quad s_A - s_B \to 0 \tag{9.1-11}$$

obtains, wherein the angular velocities of the respective dishes (ω_A and ω_B) are chosen such that the identical torque τ_z is measured for both viscometers (A and B).

Equation (9.1-11), which is valid only for highly viscous fluid interfaces ($\mathrm{Bo} \equiv \mu^s/\mu R \to \infty$), may be employed to interpret torsional viscometer measurements in order to obtain values of the apparent interfacial shear viscosity as a function of the rate of surface shear, as in Figure 9.1-5. The data appearing in the figure reveal shear-thinning behavior of a surfactant-adsorbed interface that is Newtonian for sufficiently small rates of interfacial shear. (The data also demonstrate the consistency of surface viscosity values obtained by alternative measurement techniques.)

9.2 Viscoelastic Interfacial Rheological Behavior

Torsional viscometers have previously been used to investigate viscoelastic interfacial rheological behavior (Tachibana & Inokuchi 1958, Criddle & Meader 1955, Motomura & Matamura 1963, Biswas & Haydon 1963). However, these devices often suffer disadvantages owing to inertial effects of the bob; in addition, current theoretical analyses are constrained to highly-viscous interfacial layers (Davies & Rideal 1963, Joly 1964, Ewers & Sack 1954). A relatively successful experimental technique developed by Mannheimer & Schechter (1970c) for investigating viscoelastic interfacial rheological behavior that employs the deep-channel surface viscometer is discussed below.

When used for the purpose of measuring viscoelastic properties, the deep-channel surface viscometer is operated in an oscillatory mode, in which mode of operation the floor of the viscometer is oscillated sinusoidally. Simultaneous measurements of the phase angle between the surface motion and the oscillating motion of the bottom dish, and the 'surface-to-floor' amplitude ratio, may permit determination of the viscoelastic properties of the fluid interface, presuming knowledge of an appropriate rheological model. In many cases the viscoelastic behavior of the surface will not satisfy a known surface viscoelastic model, in which case direct data of phase angle and amplitude ratio may still provide valuable information.

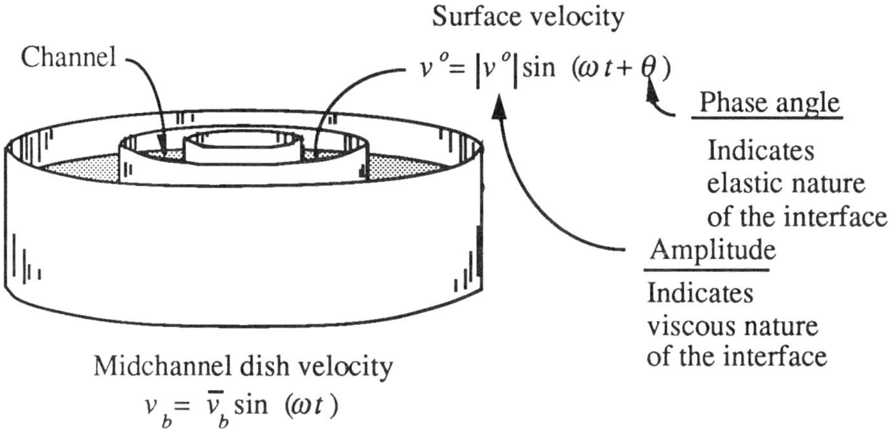

Figure 9.2-1 The deep-channel method for measuring viscoelastic interfacial rheological behavior.

The basic theoretical analysis of this method has been previously outlined in example 4 (Case II) of §4.4. As follows from Eq. (4.4-89), the surface-to-floor velocity ratio may be expressed as

$$v^o / \bar{v}_b = |v^o / \bar{v}_b| [\cos(\omega t + \theta) + i \sin(\omega t + \theta)] . \tag{9.2-1}$$

Since the floor oscillates in a sinusoidal motion [cf. Eq. (4.4-70)], only the imaginary part of this relation represents the actual physical velocity; explicitly

$$v^o / \bar{v}_b = |v^o / \bar{v}_b| \sin(\omega t + \theta) . \tag{9.2-2}$$

The maximum, time-independent, centerline surface-to-floor amplitude ratio may be defined as

$$\begin{aligned} A_{\max} &\stackrel{\text{def}}{=} \int_0^{\tau/2} |v^o / \bar{v}_b|_{y_0/2} \sin(\omega t + \theta)\, dt \\ &\equiv |\bar{v}_c^o / \bar{v}_b| , \end{aligned} \tag{9.2-3}$$

where $\tau = 2\pi / \omega$.

In the deep-channel method, direct measurements of the phase angle θ and the amplitude ratio $A_{\max}$ are performed. These values may be used to deduce viscoelastic surface properties, as is illustrated below for the special case of the two-parameter Maxwell surface viscoelasticity model. The Maxwell model assumes the applicability of the following simplification of Eq. (4.3-11) (i.e. for which the Voight time constant τ_o, [cf. Eq. (4.3-12)] becomes vanishingly small):

$$(\mu^s)^* = \frac{\mu_s''}{1 + i\omega t_o}. \tag{9.2-4}$$

Here, μ_s'' is the dynamic Maxwell shear viscosity and t_o a characteristic relaxation time.

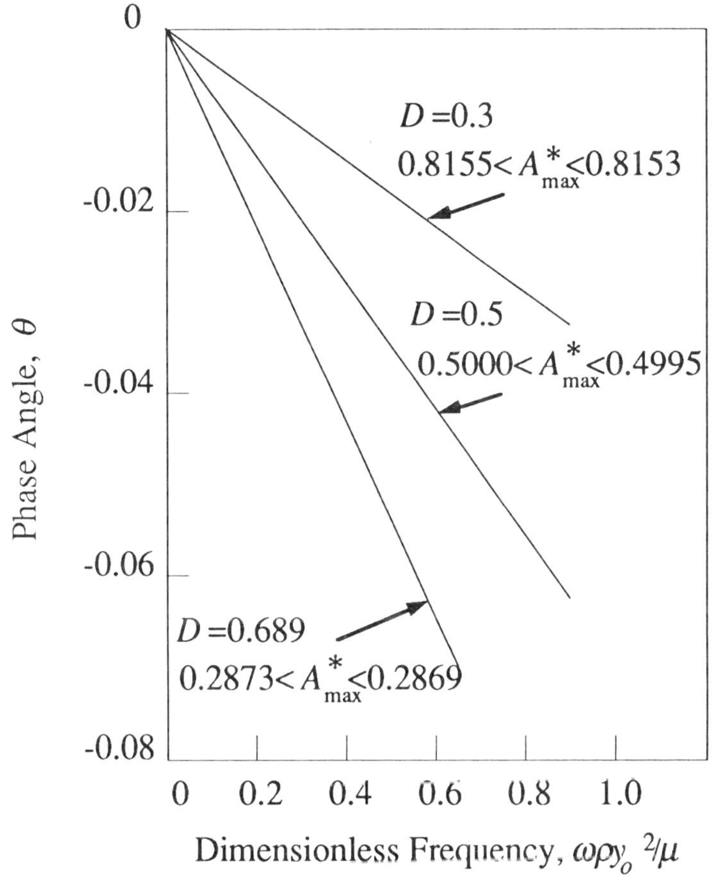

Figure 9.2-2 Phase angle vs dimensionless oscillation frequency (i.e. Reynolds number) for a clean fluid surface with negligible surface viscosity and elasticity. Theoretical values taken from Mannheimer & Schechter (1970).

According to Eq. (4.4-88), the (complex) amplitude of the time-independent, centerline surface velocity [cf. Eq. (4.4-71)] may be expressed to leading order in the deep-channel approximation $D >> 1$ as

$$v_c^o \approx \frac{4}{\pi}\left[\frac{\pi^2(\mu^s)^*}{\phi_1}\sinh(\phi_1 D) + \cosh(\phi_1 D)\right]^{-1}, \tag{9.2-5}$$

which relation assumes evaluation at a gas/liquid surface. It follows from Eqs. (4.4-72) and (4.4-82) that the eigenvalue ϕ_1 tends to a value of π when inertial effects in the bulk phase are negligible (i.e. $\mathrm{Re} = \omega \rho y_0^2 / \mu << 1$). Hence, application of Eqs. (9.2-3) and (9.2-4) furnishes the following relations:

$$\mu_s'' \approx \frac{\left[\left(1 - A^*_{\max} \sec \theta\right)^2 + \tan^2 \theta\right] \coth(\pi D)}{\pi\left(1 - A^*_{\max} \sec \theta\right) A^*_{\max} \sec \theta} \tag{9.2-6}$$

for the dynamic surface shear viscosity, and

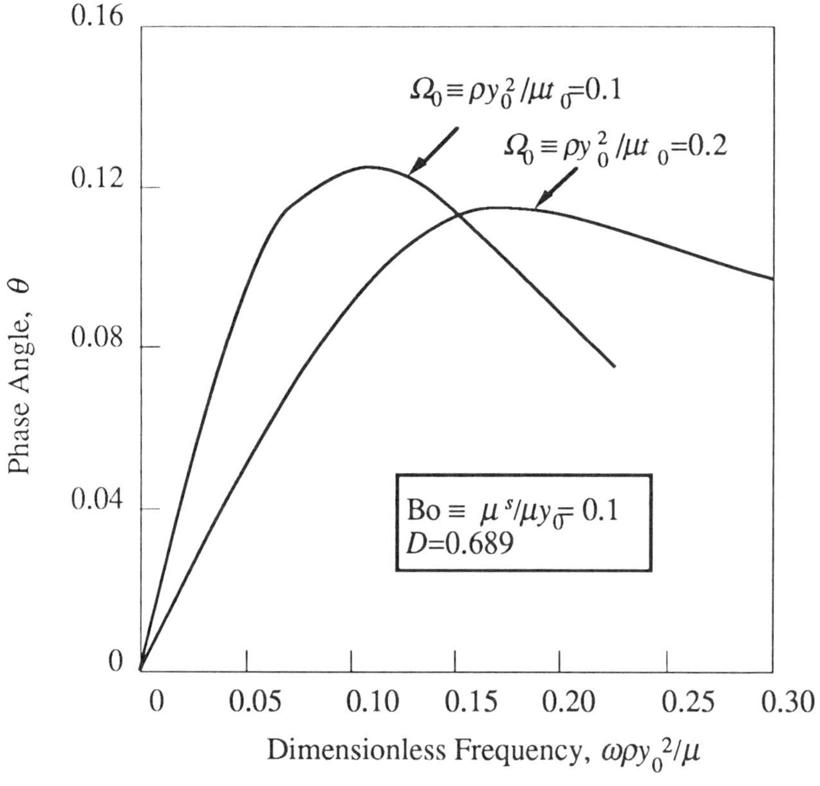

Figure 9.2-3 Phase angle vs dimensionless oscillation frequency for a Maxwell surface. Theoretical values taken from Mannheimer & Schechter (1970c).

$$t_o \approx \frac{\tan \theta}{\omega\left(1 - A^*_{\max} \sec \theta\right)} \tag{9.2-7}$$

for the characteristic relaxation constant. In the above, $A^*_{\max}$ is the surface-to-floor amplitude ratio for a clean fluid interface, neglecting bulk-fluid inertial effects.

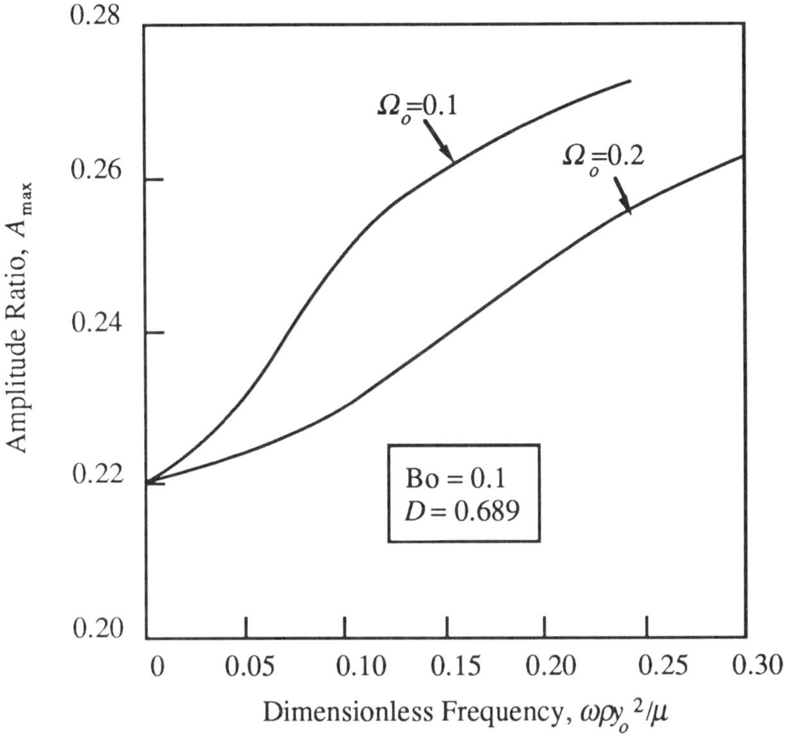

Figure 9.2-4 Amplitude ratio vs dimensionless oscillation frequency for a Maxwell surface. Theoretical values taken from Mannheimer & Schechter (1970).

To determine whether it is possible to neglect bulk-liquid inertial effects [as in Eqs. (9.2-6) and (9.2-7)], measurements of phase angle for the pure (surfactant-free) system should be performed. The expected behavior is illustrated in Figure 9.2-2, where the dimensionless phase angle is plotted versus dimensionless frequency of oscillation for various values of the depth/width ratio D [using Eq. (4.4-91) with $\mu_s''=0$]. In particular, the phase lag observed in this figure, which owes strictly to inertial effects of the bulk-phase fluid, should be much smaller than the phase lead produced by surface elasticity effects. Since the absolute value of the phase angle diminishes with smaller values of D, inertial effects may be minimized by adjusting this parameter (assuming satisfaction of the deep-channel assumption D>>1). The insensitivity of the phase angle to the amplitude ratio and the linear dependence of phase angle upon frequency, both valid only for pure systems, may also be used to verify the accuracy of experimental measurements.

Figure 9.2-3 displays the predicted phase angle behavior for a Maxwell surface fluid, for two values of the dimensionless relaxation frequency, $\Omega_o=y_o^2/\nu t_o$, employing Eq. (4.4-91). The phase lead results from the tendency of the surface—owing to its shear elastic properties—to stop, and continue its motion ahead of the floor surface. The maxima occur in the vicinity of the relaxation frequency, beyond which frequency the surface is seen to be increasingly less elastic. When the frequency of oscillation is much larger than the relaxation frequency, the surface behaves as a clean fluid surface, i.e. exhibiting negligible surface elasticity and viscosity, as predicted by Eq. (9.2-4). The influence of the dynamic interfacial shear viscosity is reflected in the amplitude ratio A_{max}, which ratio varies from a low to high value depending upon the fluidity of the interface. As illustrated in Figure 9.2-4, at very low frequencies, the amplitude ratio converges to a value corresponding to a purely viscous fluid interface, which value may be confirmed by standard surface shear viscosity measurements (see Chapter 7). At high oscillation frequencies, the amplitude ratio approaches a value corresponding to a clean fluid surface, entirely absent of surface viscous effects.

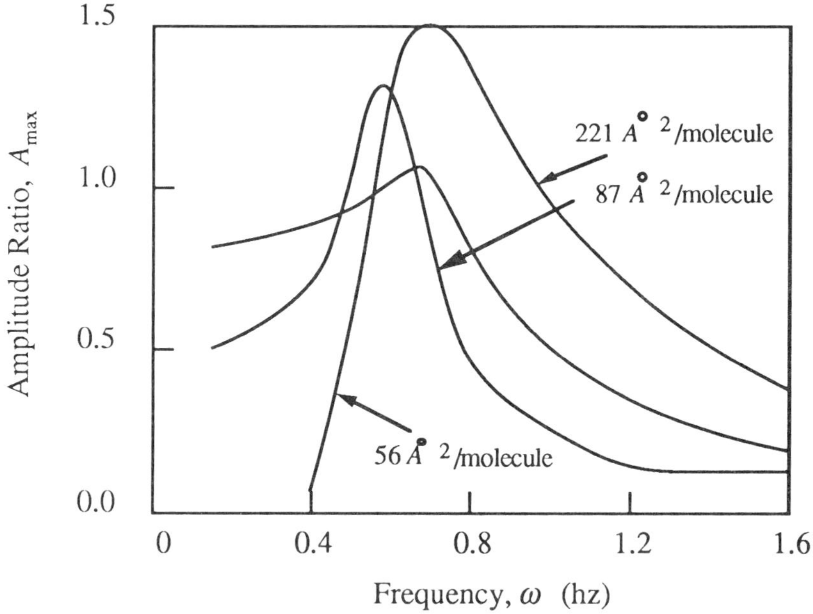

Figure 9.2-5 Amplitude ratio vs dimensionless oscillation frequency for dipalmitoyllecithin (DPL) films. Data taken from Kott *et al.* (1974).

The results shown in Figures 9.2-3 and 9.2-4 represent theoretical values based upon Maxwellian surface viscoelastic behavior. Often, however, the interface will display a more complex viscoelastic nature, in which case direct data for phase angle and amplitude ratio will provide the most practically worthwhile information. For example, in Figure 9.2-5, experimental values of the amplitude ratio are shown for various pulmonary lung surfactant systems. The depicted behavior differs considerably from that predicted in Figure 9.2-4 for the Maxwell surface, and no surface viscoelastic model has yet been proposed for describing this behavior. Nevertheless, the relaxation effect and the existence of a frequency range in which the floor oscillation is actually amplified at the surface (i.e. for $A_{max}>1$) provide examples of potentially significant rheological effects that may be directly gleaned from the data.

9.3 Viscoplastic Interfacial Rheological Behavior

Plastic-like interfacial rheological behavior (for which class of non-Newtonian interfacial behavior the 'apparent' interfacial shear viscosity increases with decreasing shear rate, and the surface film may become rigid below a critical yield stress value) has been observed to exist at the interfaces of several nonaqueous systems (Mcbain & Robinson 1949, Criddle & Meader 1955, Mannheimer 1969). Such interfacial rheological behavior differs fundamentally from any of the rheological behaviors previously discussed, and has been modeled by Mannheimer & Schechter (1967, 1970b) in terms of the Bingham plastic surface model [cf. Eq. (4.3-7)].

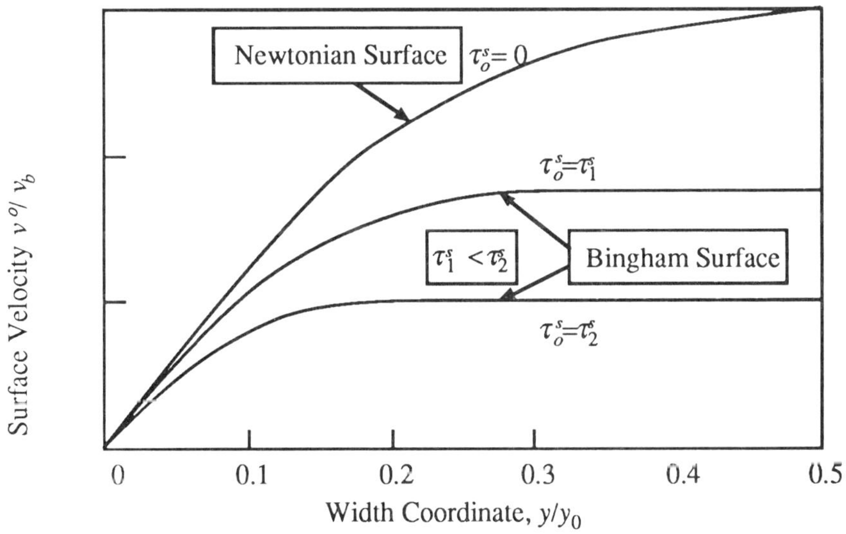

Figure 9.3-1 Hypothetical surface velocity profiles for Newtonian and Bingham fluids flowing in the linear canal of the deep-channel surface viscometer.

The interfacial response of a Bingham surface fluid will appear similar to that depicted in Figure 9.3-1 for the case of the deep-channel surface viscometer. The central portion of the surface film is depicted as remaining rigid, since in this region the interfacial shear stress falls below the critical yield stress value, τ_o^s. The remainder of the surface film near the walls where the velocity gradient is greatest flows as a Newtonian interface.

The general solution to the equation relating the surface centerline velocity to physical and geometrical parameters appears in the form of a tedious series solution, consequently only in limiting cases may an explicit analytical relation between centerline surface velocity, apparent surface shear viscosity and yield stress value be obtained. In the limit of zero surface yield stress, the solution is simply the standard Newtonian result, [e.g. Eq. (7.1-1) for a gas-liquid surface]. In the limit of a completely rigid film, one obtains by the deep-channel approximation (Mannheimer & Schechter 1967, 1970b)

$$\tau_o^s = \frac{4 v_b^{\min} \mu}{\pi \sinh(\pi D)},$$

where $v_b^{\min}$ denotes the minimum (midchannel, or 'centerline') floor speed required to effect surface motion. Measurements of the surface centerline velocity, $v^*_{c\,\min}$ (appropriate to a surfactant-free system rotating at this same minimum floor speed), may be used in the following simple equation to furnish values of the surface yield stress:

$$\tau_o^s = v^*_{c_{\min}} \mu. \qquad (9.3\text{-}1)$$

A *general* solution (the above being limited to a nearly rigid film) is graphically illustrated in Figure 9.3-2 for the case of a gas-liquid surface, at various values of the Boussinesq number, $\mathrm{Bo} \equiv \mu^s / \mu y_o$. One observes that in the limit of zero surface yield, the surface centerline velocity is that of a Newtonian fluid surface, whereas in the limit of a rigid surface, Eq. (9.3-1) is obeyed independently of the interfacial shear viscosity value.

Figure 9.3-2 may be used in at least two different ways for interpreting centerline velocity data of Bingham-like fluid surfaces.* The yield stress of the surfactant film may be determined by measuring the floor speed necessary to cause interfacial motion (or the maximum floor speed before the surface will flow), employing Eq. (9.2-1). Thus, by measuring the surface centerline velocity for clean and surfactant-adsorbed interfaces at larger rotational floor speeds, the value of the yield stress may be used to determine the appropriate interfacial shear viscosity. Otherwise, measurements of centerline surface velocities for several rotational floor speeds may be used to interpolate appropriate values of μ^s and τ_o^s from Figure 9.3-2.

---

* Figure 9.3-2 is only a reproduction of results displayed by Mannheimer & Schechter (1970b). For actual experimental measurements, a similar plot should be constructed by numerical solution of the general centerline velocity formulas found in their original paper.

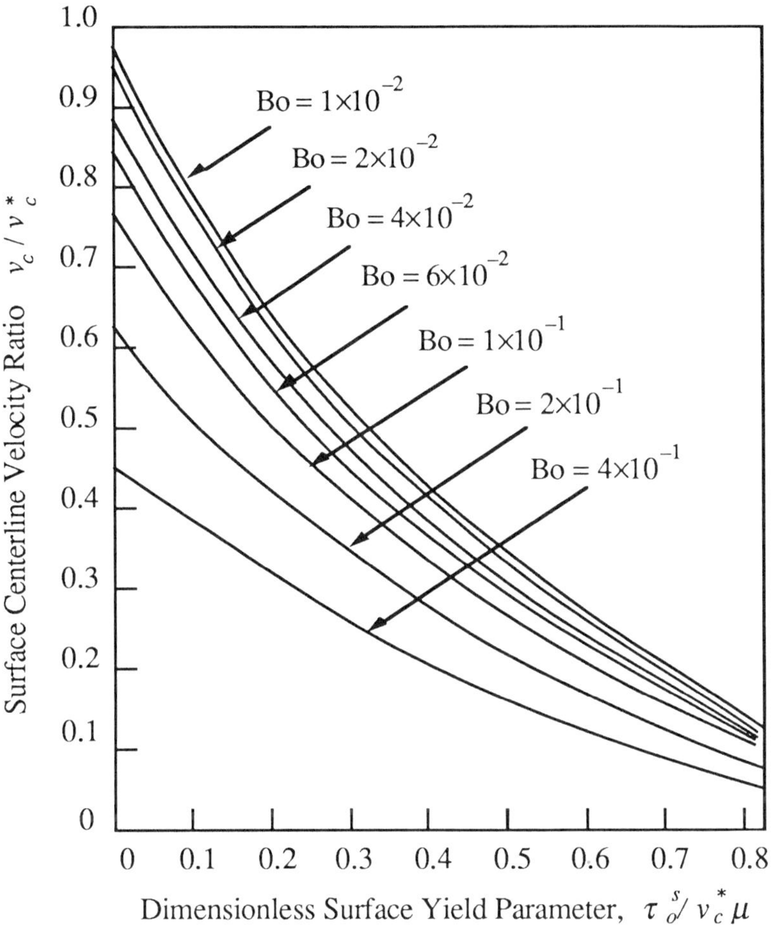

Figure 9.3-2 Theoretical surface centerline velocity ratio versus dimensionless surface yield parameter for various values of the Boussinesq number. Adapted with permission from Mannheimer & Schechter, *J. Colloid Interface Sci.* **32**, p.212 (1970) © Academic Press Inc.

Experimental measurements have also been performed by Mannheimer & Schechter (1970b) for blends of calcium sulfonate in mineral oil. The effect of humidity and temperature was studied upon the surface rheological properties of a nitrogen gas-mineral oil surface; these properties were found to obey the Bingham surface model. It was found that the apparent surface shear viscosity decreases over a long period of continual exposure to moist nitrogen (evidenced by a more rapid surface centerline velocity), and that, with increasing temperature, the apparent surface shear viscosity actually increases. This anomalous behavior was attributed to the effect of moisture absorption to the surfactant film, which adsorption was demonstrated to diminish the rigidity of the monolayer. Thus, at constant relative humidity, the effect of increasing temperature was found to diminish the apparent surface shear viscosity, as expected.

Data are shown in Figure 9.3-3 for controlled (0%) humidity experiments at different temperatures. Upon employing Figure 9.3-2 in an iterative fashion, the surface rheological behavior was satisfactorily described by the Bingham surface model. Both the surface yield and surface viscosity values are seen in this figure to diminish with increasing temperature.

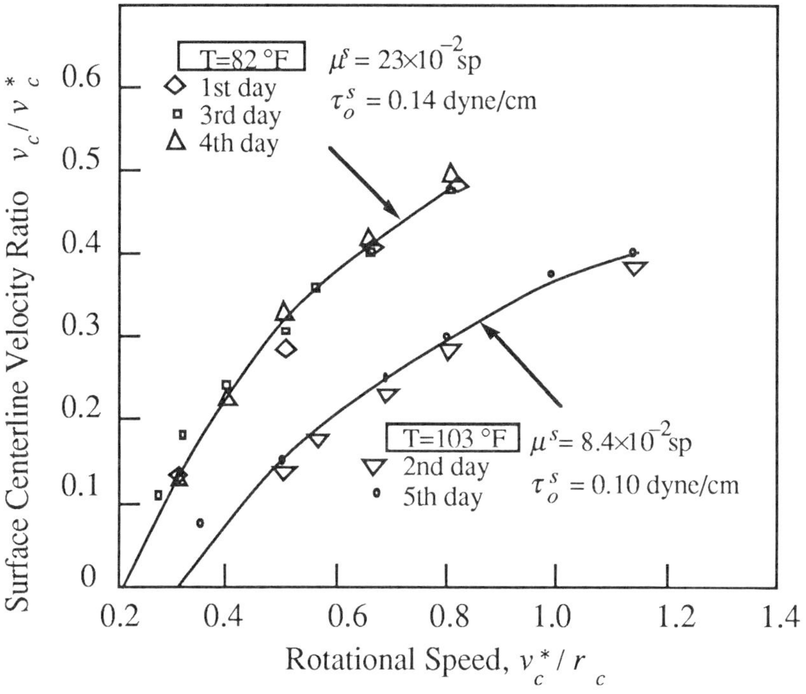

Figure 9.3-3 Experimental values of the surface centerline velocity ratio versus rotational clean surface speed for two temperatures at 0% humidity. Adapted with permission from Mannheimer & Schechter, *J. Colloid Interface Sci.* **32**, p.212 (1970) © Academic Press Inc.

9.4 Summary

Fluid interfaces adsorbed by surfactant may exhibit rheological behavior considerably more complex than the classical Newtonian surface behavior discussed in Chapters 7 and 8. The deep-channel surface viscometer may be employed to examine the shear-rate dependence of the apparent surface shear viscosity, either by placing several small tracer particles within the fluid surface and measuring the speeds of rotation, or by varying the liquid depth (say), and measuring the dependence of the centerline velocity upon the varying shear rate. The former method has been used successfully to examine nonlinear interfacial rheological behavior, and the values obtained have been demonstrated to be consistent with nonlinear rheological measurements taken from torsional viscometers. The latter type surface viscometer is, however, presently limited to highly viscous surfactant monolayers.

Shear elastic behavior of surfactant-adsorbed interfaces may also be measured by the deep-channel surface viscometer. Explicit values for shear elasticities are available currently only for very simple Maxwell or Maxwell-Voight interfaces, though directly measured values of phase angle and amplitude ratio may furnish meaningful information.

An interfacial yield stress has also been observed at fluid interfaces. As the yield stress and shear viscosity of the surface may exhibit an opposite response to moisture adsorption, surfactant adsorption and temperature, the two values may yield independent information of the rigidity of the fluid interface, which may be of value for understanding phenomena which involve interfacial or thin film stability.

Questions for Chapter 9

9.1 Given the following data for interfacial velocities f in the deep-channel device as a function of dimensionless distance from the channel walls y/y_0:

f (cm/s)	y/y_0
0.000	0.0
0.016	0.2
0.026	0.3
0.029	0.4
0.030	0.5
0.029	0.6
0.026	0.7
0.016	0.8
0.000	0.0

i. Use the trapezoidal rule to determine f_1, f_2, f_3 and f_4 from Eq. (9.1-3). *ii*. Determine an approximation to the surface stress defined by Eq. (9.1-2) at the channel wall $y = 0$, using the first four f values determined in (i), with $D = 2$ and $\mu = 1$ cP. *iii*. What is the percentage contribution of each term ($n = 1, 2, 3, 4$) in the series solution, Eq. (9.1-2), to the total surface stress approximation determined in (ii)?

Additional Reading for Chapter 9

§9.1 Nonlinear Interfacial Rheological Behavior

Jiang, T.S., Chen, J.D. & Slattery, J.C. 1983 Nonlinear interfacial stress-deformation behavior measured with several interfacial viscometers. *J. Colloid Interface Sci.* **96**, 7–19.

Li, D. & Slattery, J.C. 1988 Measuring nonlinear surface stress-deformation behavior for aqueous solutions of dodecyl sodium sulfate and dodecyl alcohol. *J. Colloid Interface Sci.* **125**, 190–197.

§9.2 Viscoelastic Interfacial Rheological Behavior

Mannheimer, R.J. & Schechter, R.S. 1970 The theory of interfacial viscoelastic measurement by the viscous-traction method. *J. Colloid Interface Sci.* **32**, 225–241.

Gardner, J.W., Addison, J.V. & Schechter, R.S. 1978 A constitutive equation for a viscoelastic interface. *AIChE J.* **24**, 400–405.

§9.3 Viscoplastic Interfacial Rheological Behavior

Mannheimer, R.J. & Schechter, R.S. 1970 Shear-dependent surface rheological measurements of foam stabilizers in nonaqueous liquids. *J. Colloid Interface Sci.* **32**, 212–224.

10

"May it even be said, most of the still unexplained phenomena of Acoustics are connected with the instability of jets of fluid. For this instability there are two causes; the first is operative in the case of jets of heavy liquids, e.g. water, projected into air (whose relative density is negligible), and has been investigated by Plateau . . .. It consists in the operation of the capillary force, whose effect is to render the infinite cylinder an unstable form of equilibrium, and to favour its disintegration into detached masses whose aggregate surface is less than the cylinder."

Lord Rayleigh (1878)

"My attention was first drawn to the subject . . . last spring, by Mr. H.F. Newall, of the Rugby School Natural-History Society, who showed me the mark made by drops of water and mercury falling on a smoked glass plate, the lampblack being swept away in concentric circles and radial striæ. The patterns thus left were generally symmetrical and beautiful, and varied with the height of fall of the drop. I have since sought to investigate the cause of these appearances in Professor Helmholtz's laboratory in Berlin."

A. Worthington (1876)

CHAPTER 10

Interfacial Stability

Fluid interfacial instabilities, such as arise in the break-up of cresting waves, the "splash" created by a blunt object falling into a liquid pool and in the break-up of a fluid jet, are commonly encountered in both natural and engineering processes. A review of these phenomena and of their stability analyses is provided by Miller (1978) [see also Miller and Neogi (1985)]; Chandrasekhar (1961) may be consulted for a standard treatment of the more general aspects of hydrodynamic stability.

In the remaining chapters of Part I (Chapters 10-14) attention will be focused upon practical applications of the basic interfacial rheological theory developed in Chapters 3–5. Emphasis in these chapters will be primarily devoted to the underlying physics of the various interfacial phenomena considered and [where it appears necessary to demonstrate application of the basic theory or (particularly in Chapters 11, 12 and 14) to illustrate a new theoretical concept (e.g. 'disjoining pressure')] to problem formulation and solution; in certain cases, problem formulation will be only briefly discussed or not discussed at all (e.g. the subject of interfacial turbulence in §10.4); in such cases the reader will be directed to the original literature for further details. Finally, in all subsequent examples involving deformation of an interface, our attention will consistently be restricted to the special case of 'small' interfacial deformations, the mathematical analyses of which may reasonably assume a linearization of the deformation about an equilibrium or steady-state nonequilibrium surface shape.

We begin our consideration of interfacial stability in the present chapter with a discussion of linear stability analysis in §10.1. This is followed by a brief discussion of the stability of superposed fluids in §10.2. Next, in §10.3, the Rayleigh instability of a fluid jet is considered: emphasis is made upon the influence of interfacial-tension gradients caused by an adsorbing/desorbing solute species. Interfacial turbulence, caused by placing two nonequilibrated fluids in mutual contact is treated in §10.4. The effect of both interfacial-tension gradients and interfacial viscosities is considered in this problem. Following this, a superficial, qualitative discussion of Bénard instability is provided in §10.5.

10.1 Normal Mode Interfacial Stability Analysis

The stability of stationary fluid flows may be theoretically examined by means of a linear *normal mode* analysis. Such stability analyses involve small, dynamic fluctuations of physical variables of the stationary flow. The linearized equations of motion are subsequently solved to determine the evolution of these fluctuations, thereby revealing whether the applied disturbances will grow or decay with time. The amplitude of the perturbations is expressed in terms of a Fourier series of 'normal modes', the stability of the interface being investigated with respect to all such normal modes.

Linear interfacial stability analysis characteristically involves a normal perturbation of interfacial position, of the form (for a planar surface)

$$f(x, y, t) = \int_{-\infty}^{\infty} dk_x \int_{-\infty}^{\infty} dk_y\, A_k \exp[i(k_x x + k_y y)] e^{\beta t}, \quad (10.1\text{-}1)$$

with

$$k^2 \equiv k_x^2 + k_y^2, \quad (10.1\text{-}2)$$

where $f(x,y,t)$ is the amplitude of the normal displacement of the interface (satisfying $fk<<1$), β the frequency of the perturbation (generally a complex number), k the wave number, x and y Cartesian coordinates within the plane of the interface and $\exp[i(k_x x+k_y y)]$ the set of normal modes of wave number k. For a circular cylindrical jet, the corresponding surface perturbation is

$$f(z, \phi, t) = \int_{-\infty}^{\infty} dk_z \int_{-\infty}^{\infty} dk_\phi\, A_k \exp[i(k_z z + k_\phi \phi)] e^{\phi t}, \quad (10.1\text{-}3)$$

with z and ϕ the interfacial coordinates (cf. Figure B.1-5). The wave number k_ϕ may be regarded as defining the number of ϕ-planes of symmetry.

The primary objective of linear stability analysis is the determination of the growth constant β, obtained by solving the linearized equations of motion together with appropriate boundary conditions. The interface is observed to be 'stable' to small fluctuations when the real part of β is negative; it is termed 'unstable' when β is positive. Marginal stability is defined by the condition that $\beta = 0$. Once a solution for the growth constant has been obtained, rates of disturbance growth and decay may also be investigated.

10.2 The Stability of Superposed Fluids

Normal mode stability analysis is demonstrated in this section by considering two superposed fluid phases A and B, possessing the respective densities ρ_A and ρ_B, with $\rho_A > \rho_B$, say. By placing the heavier fluid A above the lighter fluid B ('above' in the sense that a gravitational field points 'downward' in the direction of fluid B out of fluid A), the superposed configuration will generally be unstable to any mechanical disturbance of the interface from its equilibrium planar shape. This may be confirmed through the linear stability analysis referred to in the preceding section by imposing an interfacial disturbance of the form [cf. Eq. (10.1-1)]

$$f = f_o \exp[\, i(kx + \omega t)] , \tag{10.2-1}$$

at the planar fluid interface $z = f$ (cf. Figure 5.6-2). The perturbation f is expected to grow with time if fluid A is placed above fluid B; this will be reflected by a positive, real value for the disturbance frequency, $\beta = i\omega$.

The dispersion relation which governs the (undamped) wave disturbance f at a liquid-liquid interface may be obtained by generalizing Eq. (6.2-1) to the case of a liquid-liquid system (with $\mu = 0$, k real); explicitly,

$$\left[(\rho_B + \rho_A)\omega^2 - \sigma k^3 - (\rho_B - \rho_A)gk\,\right]\rho_B\omega^2 = 0, \tag{10.2-2}$$

with g the gravitational constant and σ the interfacial tension. Equation (10.2-2) provides a relation between the complex wave frequency ω and the wave number k.

From Eq. (10.2-2) it follows that the growth coefficient β is provided by

$$\boxed{\beta = i\omega = \sqrt{\frac{(\rho_A - \rho_B)gk \; - k^3\sigma}{(\rho_A + \rho_B)}}\;.} \tag{10.2-3}$$

Two opposing effects may be distinguished in this relation: (i) whereas the buoyancy force (with $\rho_A > \rho_B$) causes the wave amplitude to grow, rendering the system unstable, the effect of interfacial tension is to diminish the growth constant β, thereby suppressing the instability: (ii) short wavelength 'capillary' wave disturbances (corresponding to large wave number k) enhance the stabilizing effect of interfacial tension; long wavelength 'gravity' waves minimize the interfacial tension effect, whence the density disparity leads to interfacial instability.

The existence of surfactant at the fluid interface gives rise to interfacial-tension gradients: these may either stabilize or destabilize the fluid interface, possibly depending upon the direction of transfer of a surface tension-lowering solute which transfers across the interface (cf. §§10.3 and 10.4). Interfacial viscosities also arise with surfactant adsorption: these stabilize the fluid interface by damping interfacial disturbances (cf. §10.4).

10.3 Rayleigh Instability

When liquid is ejected from an orifice, the exiting liquid jet eventually disintegrates into small liquid rivulets, as illustrated in Figure 10.3-1: such an interfacial instability is known as *Rayleigh* instability. The size of droplets which are formed by the break-up of a liquid jet (as observed in numerous engineering applications, such as fuel injection, ink-jet printing and gas absorption or stripping in spray towers) is strongly dependent upon the magnitude of the interfacial tension, as interfacial tension tends to minimize interfacial area [cf. Eq. (6.1-1)]. The effect of surfactant adsorption, which results in the lowering of the interfacial tension, is hence to produce a more stable liquid jet, resulting in a relative growth of droplet size. Interfacial-tension gradients (in the *absence* of solute mass transfer across the fluid interface) also stabilize the liquid jet, though this latter dynamic effect is relatively minor.

The interfacial hydrodynamic stress that poses the greatest stabilizing/destabilizing influence upon Rayleigh instability results from a net transfer of solute material across the liquid interface. As shown below, should the liquid be a solution possessing a volatile species (which species lowers interfacial tension), upon its exit through a nozzle into an air or other gaseous phase, the solute will transfer across the interface and into the air, being accompanied by interfacial-tension gradients which act to stabilize the jet. On the other hand, surface tension-reducing material might absorb from the air into an ejecting liquid jet, resulting in a transfer of solute material across the interface—though in this case in the direction of the liquid phase. In this latter case, the interfacial-tension gradients accompanying solute transfer across the interface are observed to *enhance* Rayleigh instability. These conclusions may be deduced from a solution of the problem formulated below.

Consider a cylindrical liquid jet that is ejected from an orifice into a gas phase, as depicted in Figure 10.3-1 [the analogous problem of a liquid jet ejected into an immiscible liquid has been addressed by Burkholder and Berg (1974b)]. Imagine this jet to be of an infinite length, initially cylindrical (possessing radius a), traveling in the axial direction with a uniform [and sufficiently large (Keller *et al.* 1973)] velocity U, and subject to small dynamic fluctuations in the radial location of the interface; these latter fluctuations will be assumed to be of the form [cf. Eq. (10.1-3)]

$$f = f_o \exp\left(ik_z z + ik_\phi \phi + \beta t \right) . \tag{10.3-1}$$

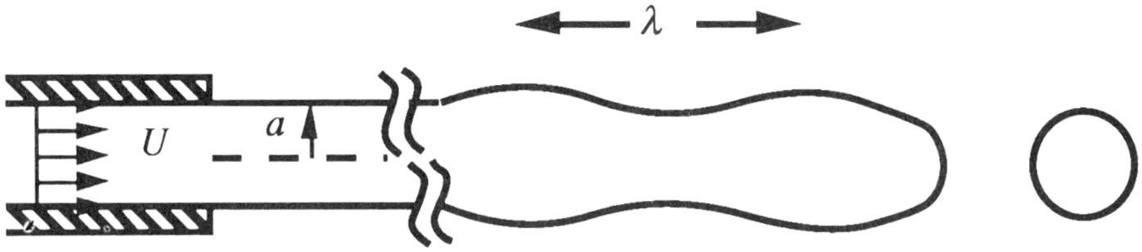

Figure 10.3-1 A jet of liquid is ejected from a circular cylindrical nozzle into a gas phase, eventually breaking up into small liquid droplets.

The equations of motion that are assumed to govern the propagation of this small disturbance ($f_o k << 1$) follow upon linearizing the Navier-Stokes equations in circular cylindrical coordinates (cf. Table 4.1-2). By symmetry considerations, only the R and z components must be considered; explicitly

$$\rho \frac{\partial v_R}{\partial t} = -\frac{\partial p}{\partial R} + \mu\left[\frac{\partial}{\partial R}\left(\frac{1}{R}\frac{\partial}{\partial R} R v_R\right) + \frac{1}{R^2}\frac{\partial^2 v_R}{\partial \phi^2} - \frac{2}{R^2}\frac{\partial v_\phi}{\partial \phi} + \frac{\partial^2 v_R}{\partial z^2}\right] \tag{10.3-2}$$

and

$$\rho \frac{\partial v_z}{\partial t} = -\frac{\partial p}{\partial z} + \mu\left[\frac{\partial}{\partial R}\left(R \frac{\partial}{\partial R} v_z\right) + \frac{1}{R^2}\frac{\partial^2 v_z}{\partial \phi^2} + \frac{\partial^2 v_z}{\partial z^2}\right]. \tag{10.3-3}$$

Here, the system of coordinates (R, ϕ, z) are chosen as translating along the axis of the jet with velocity U. The incompressibility condition in this moving system of coordinates is provided by

$$\frac{1}{R}\frac{\partial}{\partial R}(R v_R) + \frac{1}{R}\frac{\partial v_\phi}{\partial \phi} + \frac{\partial v_z}{\partial z} = 0. \tag{10.3-4}$$

The pertinent boundary conditions to the above are the continuity of normal velocity condition

$$v_R = \frac{\partial f}{\partial t} \quad \text{at} \quad R = a, \tag{10.3-5}$$

the normal interfacial stress condition (cf. Table 4.2-2)

$$\hat{p} - p + 2\mu \frac{\partial v_R}{\partial R} - \sigma\left(\frac{1}{R} + \frac{1}{R^2}\frac{\partial^2 f}{\partial \phi^2} + \frac{\partial^2 f}{\partial z^2}\right) = 0 \quad \text{at } R = a, \tag{10.3-6}$$

and the tangential interfacial stress condition

$$\mu\left(\frac{\partial v_z}{\partial R} + \frac{\partial v_R}{\partial z}\right) = \frac{\partial \sigma}{\partial z} \quad \text{at } R = a. \tag{10.3-7}$$

[Note in Eq. (10.3-6) that the mean surface curvature, H, has been expressed in terms of the fluctuation f by linearizing the surface shape about

$R = a$ (cf. Question 3.12).] Interfacial viscosities have not been included in Eqs. (10.3-6) and (10.3-7), i.e. it is assumed that Bo $\equiv \mu^s/\mu a \ll 1$.

Consistent with the assumption of small radial disturbance ($f_o k \ll 1$), the interfacial tension may be assumed to satisfy a linearized surface equation of state [cf. Eq. (5.5-4)]. Thus, the latter two boundary conditions may be approximated respectively as

$$\hat{p} - p + 2\mu\frac{\partial v_R}{\partial R} - \left[\sigma_o - \frac{1}{\rho^s_{1o}}E^1_o\left(\rho^s_1 - \rho^s_{1o}\right)\right]\left(\frac{1}{a} + \frac{1}{a^2}\frac{\partial^2 f}{\partial \phi^2} + \frac{\partial^2 f}{\partial z^2}\right) = 0, \tag{10.3-8}$$

and

$$\mu\left(\frac{\partial v_z}{\partial R} + \frac{\partial v_R}{\partial z}\right) = -\frac{1}{\rho^s_{1o}}E^1_o\frac{\partial \rho^s_1}{\partial z}. \tag{10.3-9}$$

Here, "1" denotes the solute species and ρ^s_{1o} the equilibrium value of the surface-excess solute density ρ^s_1.

Complete specification of the preceding boundary value problem requires simultaneous consideration of the coupled solute transfer problem (as in each of the examples of §5.6).

Burkholder & Berg (1974a) have numerically solved a similarly formulated problem by assuming the existence of a constant species density gradient $(\partial\rho_1/\partial R)_o$ across the liquid jet, as well as a constant species density ρ_{1o} in the surroundings ($R \to \infty$). To account for the absence of chemical equilibrium between the jet and its gaseous surroundings [which disequilibrium results in a solute flux from (to) the gas phase, across the interface, into (out of) the liquid] Burkholder & Berg (1974a) introduce a mass transfer coefficient K into the surface-excess species balance equation (as in Question 5.3). The linearized solute conservation equation may thus be expressed as (cf. Table 5.1-1)

$$\frac{\partial \rho_1}{\partial t} + v_R\left(\frac{\partial \rho_1}{\partial R}\right)_o = D\left[\frac{1}{R}\frac{\partial}{\partial R}\left(R\frac{\partial \rho_1}{\partial R}\right) + \frac{\partial^2 \rho_1}{\partial z^2}\right]. \tag{10.3-10}$$

The surface-excess species balance equation becomes (cf. Table 5.2-1)

$$\frac{\partial \rho^s_1}{\partial t} + \rho^s_{1o}\left(\frac{v_R}{R} + \frac{\partial v_z}{\partial z}\right) = D^s\left[\frac{1}{R}\frac{\partial}{\partial R}\left(R\frac{\partial \rho^s_1}{\partial R}\right) + \frac{\partial^2 \rho^s_1}{\partial z^2}\right] - D\left(\frac{\partial \rho_1}{\partial R}\right)_o + K\left(\rho_1 - \rho_{1o}\right) \quad \text{at } R = a. \tag{10.3-11}$$

The problem statement provided by Eqs. (10.3-1) through (10.3-11) may be solved to determine the characteristic dispersion relation, by which equation the stability of the liquid jet (as interpreted through the behavior of the growth constant β) may be examined as a function of interfacial and bulk-phase properties. The essential conclusions of the analysis of Burkholder & Berg (1974a) are shown in Figures 10.3-2 and 10.3-3.

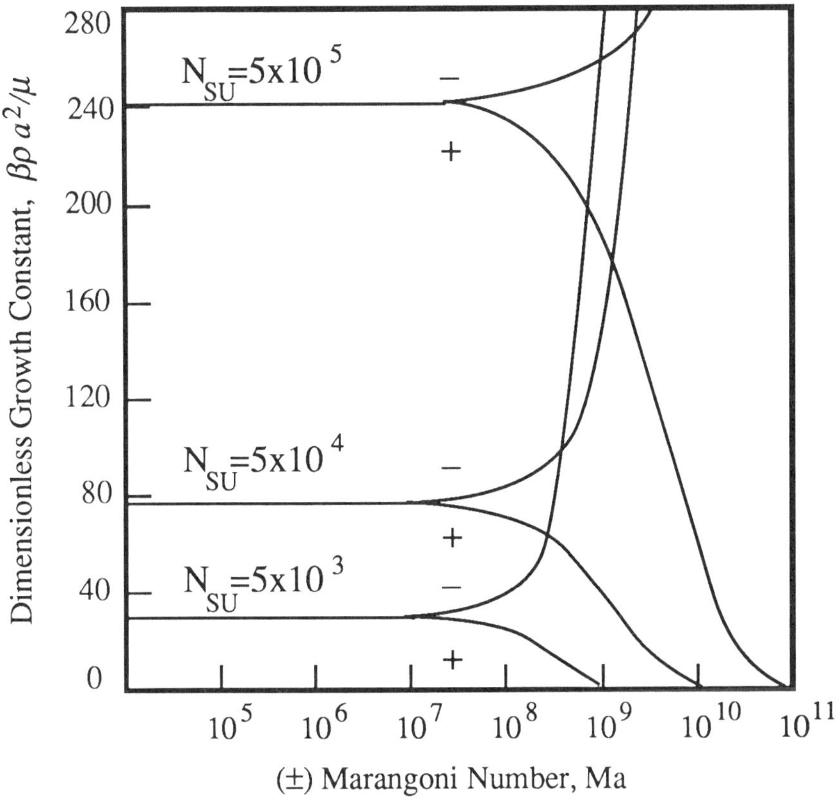

Figure 10.3-2 The dimensionless growth constant is plotted versus the Marangoni number at various values of the dimensionless interfacial tension (Suratman number). The theoretical values are taken from Burkholder & Berg (1974a).

In Figure 10.3-2, the dimensionless growth constant is plotted versus the Marangoni number,

$$\mathrm{Ma} \equiv \frac{-\dfrac{1}{\rho_o^s} E_o^1 \left(\dfrac{\partial \rho_1^s}{\partial \rho_1}\right)_o \left(\dfrac{\partial \rho_1}{\partial r}\right)_o}{\mu D},$$

for various values of the Suratman number,

$$N_{SU} \equiv \frac{\sigma_o \rho a}{\mu}.$$

When the Marangoni number is positive, a flux of surfactant *out* of the liquid jet and *into* the surroundings is indicated. A negative value of the Marangoni number indicates a flux of solute *into* the jet. (These conclusions assume that the solute causes a decrease in surface tension with surface concentration.)

Transfer of solute *out* of the liquid jet is observed to be stabilizing (i.e. the growth constant diminishes) resulting in larger droplets upon breakup. Transfer from the surroundings *into* the jet is correspondingly destabilizing. Decreasing interfacial tension, as accompanies surfactant adsorption, is clearly stabilizing.

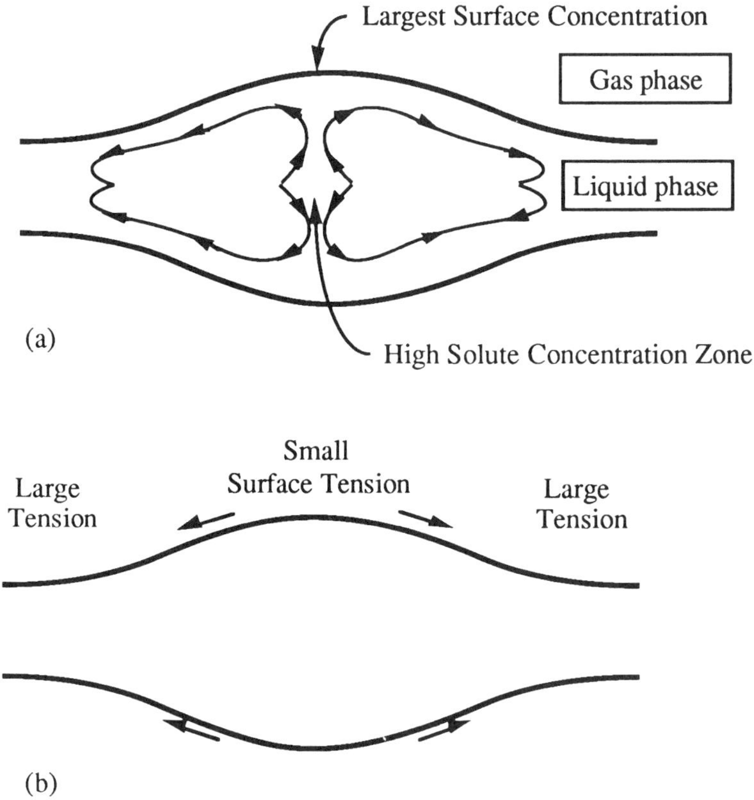

Figure 10.3-3 Fluid pathlines generated by the wave disturbance are depicted in (a). The indicated zones of high surface concentration of solute pertain to the case of mass transfer of solute *out* of the fluid jet. In (b), the direction of the surface tension gradient is indicated (also for the case of solute transfer out of the fluid jet).

The cause for the observed behavior is illustrated in Figure 10.3-3. The interfacial disturbance (10.3-1) leads to a flow pattern within the jet (the pattern is depicted as viewed by one traveling with the ejecting liquid jet, at a velocity U), as shown in Figure 10.3-3a. If the solute is transferring *out* of the jet, then the maximum solute concentration is within the core of the jet (i.e. in the vicinity of $R = 0$). As may be seen from the depicted fluid pathlines, the region of highest solute concentration will be convected to the surface near the crest of the wave. This means that the highest surface concentration of (interfacial tension-lowering) solute will be at the crest of the wave—which will thus be the zone of lowest interfacial tension. The effect of the interfacial-tension gradient will be as indicated in Figure 10.3-3b; i.e. liquid near the interface will be pulled in the direction of the wave troughs, thereby suppressing the growth of the wave disturbance. If solute absorbs from the surroundings, the jet core region will be the zone of lowest solute concentration, hence the opposite (destabilizing) effect will occur.

10.4 Interfacial Turbulence

Interfacial turbulence refers to the spontaneous interfacial mixing generated during the course of interphase species transfer. It is observed when two unequilibrated liquids (at least one of which contains a solute that is soluble in both phases) are joined in contact. The varied dynamic phenomena of interfacial turbulence range from a subtle twitching of the interface, to the 'kicking' of a pendant drop in one of the phases or the formation of long, narrow streams of one phase reaching into the next, breaking finally into small emulsion droplets [e.g. Sherwood & Wei (1957), Orell (1961) and Thiessen (1966)]. Interfacial turbulence may also markedly increase the rate of solute mass transfer across an interface, which may prove useful to processes ranging from extraction to emulsification and detergency.

In the classical theoretical study of Scriven & Sternling (1959), the origins of interfacial turbulence were associated with local interfacial fluctuations in solute surface concentration. These were shown to occasion local interfacial-tension gradients, which gradients lead to local interfacial agitation and to the ultimate observance of gross interfacial motion. As may be gathered from the interfacial turbulence study of Hennenberg *et al.* (1977) (the results of which are summarized below), interfacial turbulence is enhanced by; (i) solute transfer from the phase of highest viscosity and lowest diffusivity; (ii) small viscosities and diffusivities; and (iii) large interfacial-tension gradients and small interfacial viscosities.

The theoretical analysis of interfacial turbulence originating at an initially planar phase interface possesses many similarities to the stability analysis of superposed fluids (cf. §10.2);* hence, we will forego the details of such an analysis here, limiting our attention to the basic influence of interfacial rheological parameters in the process of interfacial turbulence generation.

Hennenberg *et al.* (1977) consider a fluctuation of pressure and velocity fields about their hydrostatic values [whereas solute interphase transfer is assumed to exist in this hydrostatic state, this (purely diffusive) transport of solute will not give rise to any mass-average, convective motion] in the bi-liquid system depicted in Figure 10.4-1; these fluctuations are assumed to be of the generic form

---

* Indeed, one method for theoretically investigating interfacial turbulence is to employ the capillary wave theory of §5.6, including transport between bulk phases of a solute that lowers interfacial tension, as is the procedure of Gouda and Joos (1975).

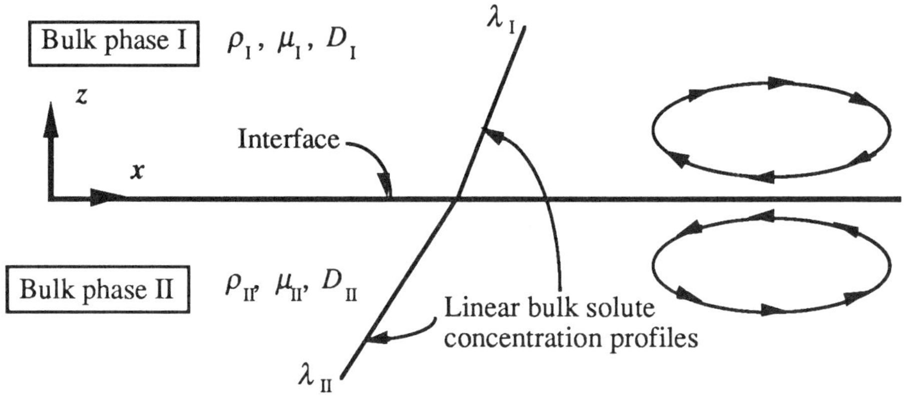

Figure 10.4-1 The interfacial turbulence model of Hennenberg *et al.* (1977) is depicted for two unequilibrated liquid phases, through which an interfacial tension-lowering solute diffuses. Small interfacial fluctuations give rise to local solute inhomogeneities near the interface; these produce the concomitant interfacial-tension gradients which generate interfacial turbulence.

$$f(x, y, z) = f_o g(z) h_x(x) h_y(y) \exp(\beta t). \qquad (10.4\text{-}1)$$

Upon assuming: (i) small fluctuations

$$f_o k << 1, \qquad (10.4\text{-}2)$$

with k representing a characteristic disturbance wave number; (ii) linear bulk concentration gradients in each liquid phase [molar variables are used; cf. Eqs. (5.1-6) and (5.2-8)],

$$\left(\frac{\partial C}{\partial z}\right)_{I} = \lambda_I, \qquad (10.4\text{-}3)$$

$$\left(\frac{\partial C}{\partial z}\right)_{II} = \lambda_{II}; \qquad (10.4\text{-}4)$$

(iii) local equilibrium at the interface (i.e. adsorption-controlled species transport; cf. §5.4), satisfying the linear adsorption relations

$$\Gamma^s = K_a^I C_I = K_a^I C_I; \qquad (10.4\text{-}5)$$

(iv) the validity of both the Newtonian interfacial stress (4.2-20) and the Fickian interfacial flux (5.2-8) boundary conditions; as well as (v) the linear approximations (5.5-4) and (5.5-5), Hennenberg *et al.* (1977) obtain the following leading-order analytical solution for the growth constant β in the vicinity of marginal stability (the latter defined by the condition $\beta = 0$):

$$\boxed{\beta \approx -\frac{\mu_{\mathrm{I}}+\mu_{\mathrm{II}}}{k\sigma_o}\left[\frac{\left(1-g\dfrac{\rho_{\mathrm{I}}-\rho_{\mathrm{II}}}{k^2\sigma_o}\right)\Pi_1}{(\Pi_2+\Pi_3)\left(1-g\dfrac{\rho_{\mathrm{I}}-\rho_{\mathrm{II}}}{k^2\sigma_o}\right)+2\,\Pi_1}\right]\cdot} \quad (10.4\text{-}6)$$

Here, σ_o is the equilibrium interfacial tension, $(\rho_{\mathrm{I}}, \rho_{\mathrm{II}})$ the bulk-phase mass densities, $(\mu_{\mathrm{I}}, \mu_{\mathrm{II}})$ the bulk-phase viscosities, g the acceleration of gravity, and:

$$\Pi_1 = \frac{1}{\sigma_o}\left\{[2(\mu_{\mathrm{I}}+\mu_{\mathrm{II}})+k(\kappa^s+\mu^s)]\left(kD^s+\frac{D_{\mathrm{I}}}{K_a^{\mathrm{I}}}+\frac{D_{\mathrm{II}}}{K_a^{\mathrm{II}}}\right)\right.$$
$$\left.+E_o-\frac{E_o\lambda_{\mathrm{II}}}{4\Gamma_o^s k^2}\left(1-\frac{D_{\mathrm{II}}}{D_{\mathrm{I}}}\right)\right\}, \quad (10.4\text{-}7)$$

$$\Pi_2 = \left(1+\frac{1}{2kK_{\mathrm{II}}}+\frac{1}{2kK_{\mathrm{I}}}\right)\left(2+\frac{\kappa^s+\mu^s}{\mu_{\mathrm{I}}+\mu_{\mathrm{II}}}k\right)$$
$$+\frac{1}{\mu_{\mathrm{I}}+\mu_{\mathrm{II}}}\left(kD^s+\frac{D_{\mathrm{I}}}{K_a^{\mathrm{I}}}+\frac{D_{\mathrm{II}}}{K_a^{\mathrm{II}}}\right)\left(\Gamma_o^s+\frac{\rho_{\mathrm{I}}+\rho_{\mathrm{II}}}{2k}\right)$$
$$+\frac{E_o\lambda_{\mathrm{II}}}{8\Gamma_o^s k^3[\mu_{\mathrm{I}}+\mu_{\mathrm{II}}]}\left(1-\frac{D_{\mathrm{II}}}{D_{\mathrm{I}}}\right)\left(\frac{1}{D_{\mathrm{I}}}+\frac{1}{D_{\mathrm{II}}}\right) \quad (10.4\text{-}8)$$

and

$$\Pi_3 = \frac{3g\lambda_{\mathrm{II}}\left(1+\dfrac{D_{\mathrm{II}}}{D_{\mathrm{I}}}\right)\left(2+\dfrac{\kappa^s+\mu^s}{\mu_{\mathrm{I}}+\mu_{\mathrm{II}}}k\right)}{4k^3\sigma_o\left(1-\dfrac{g(\rho_{\mathrm{I}}-\rho_{\mathrm{II}})}{k^2\sigma_o}\right)}. \quad (10.4\text{-}9)$$

In the above, E_o is the Gibbs elasticity, Γ_o^s the equilibrium surface-excess solute concentration, $(D_{\mathrm{I}}, D_{\mathrm{II}})$ the bulk-phase diffusivities and $(\mu^s+\kappa^s)$ the sum of the interfacial shear and dilatational viscosities.

The influence of the relevant physical parameters upon interfacial turbulence may be understood by considering the relation (10.4-6) for the growth coefficient in the vicinity of marginal stability. Upon recalling that stability occurs for cases in which $\beta < 0$ (as in this case disturbances damp), whereas instability occurs for $\beta > 0$, Eq. (10.4-6) may be seen to suggest the existence of two general cases for which instability may arise: (i) the upper phase density ρ_{I} is appreciably greater than the lower phase density ρ_{II} (this is the problem of superposed fluids considered in §10.2, although here the stability condition is modified owing to the adsorption of surfactant at the interface): (ii) there exists a substantial Gibbs elasticity, E_o, resulting in the generation of interfacial-tension gradients which act to drive bulk-phase flow. This second case pertains to the general circumstances of interfacial turbulence.

When interfacial-tension gradients are present at the interface between two unequilibrated liquids ($E_o \neq 0$) owing to the interphase mass transfer of an interfacial tension-lowering solute, Eqs. (10.4-6) and (10.4-7) indicate that (among other factors) interfacial turbulence is enhanced by: (i) mass transfer of solute from the phase of lowest diffusivity (i.e. $D_I > D_{II}$); (ii) a relatively large concentration gradient in the phase from which transfer occurs (i.e. $\lambda_{II} / \Gamma_o^s k^2 >> 1$); (iii) relatively small interfacial tension [i.e. $g(\rho_I - \rho_{II}) / k^2 \sigma_o >> 1$].

Interfacial turbulence is *suppressed* by: (i) large bulk and interfacial viscosities and (ii) large diffusivities, either of which may lead to the inequality $k^2(\mu_I + \mu_{II})(D_I + D_{II}) / \sigma_o >> 1$; $k^3(\mu^s + \kappa^s) D^s / \sigma_o >> 1$.

10.5 Bénard Instability

When paint dries, periodic, cellular patterns often form within the thin films of the drying paint: these patterns are known as *Bénard cells*, after the pioneering experimental investigations of Bénard (1901). In his studies, Bénard observed that cellular (hexagonal) circulation patterns form within a thin film of liquid heated from below. Although for many years this cellular instability was attributed to buoyancy-driven convection, following Rayleigh (1916), the phenomenon is now understood to result from gradients of interfacial tension (see §1.1 for pertinent historical comments). (As previously illustrated in Questions 5.4–5-6, interfacial-tension gradients result when temperature varies tangentially within the interface. This occurs owing to the temperature dependence of interfacial tension.)

Depicted in Figure 10.5-1 is a liquid layer resting upon a solid surface, which surface is heated relative to the film temperature. After a sufficient time, a steady-state is achieved wherein a temperature gradient acts across the thickness of the liquid film (Figure 10.5-1a). A mechanical disturbance or thermal fluctuation results in a small disturbance in velocity and temperature fields in the vicinity of the interface. Inhomogeneities in surface temperature cause a variation of interfacial tension (Pearson 1958, Scriven & Sternling 1964). Interfacial-tension gradients cause flow in the interface and contiguous liquid layer, carrying liquid from areas of low interfacial tension to those of high tension (Figure 10.5-1b). Circulatory convection patterns are thereby created within the film, with liquid rising toward (depressed) areas of low tension and away from (raised) areas of high tension (Figure 10.5-1c). Thus, Bénard convection cells are produced.

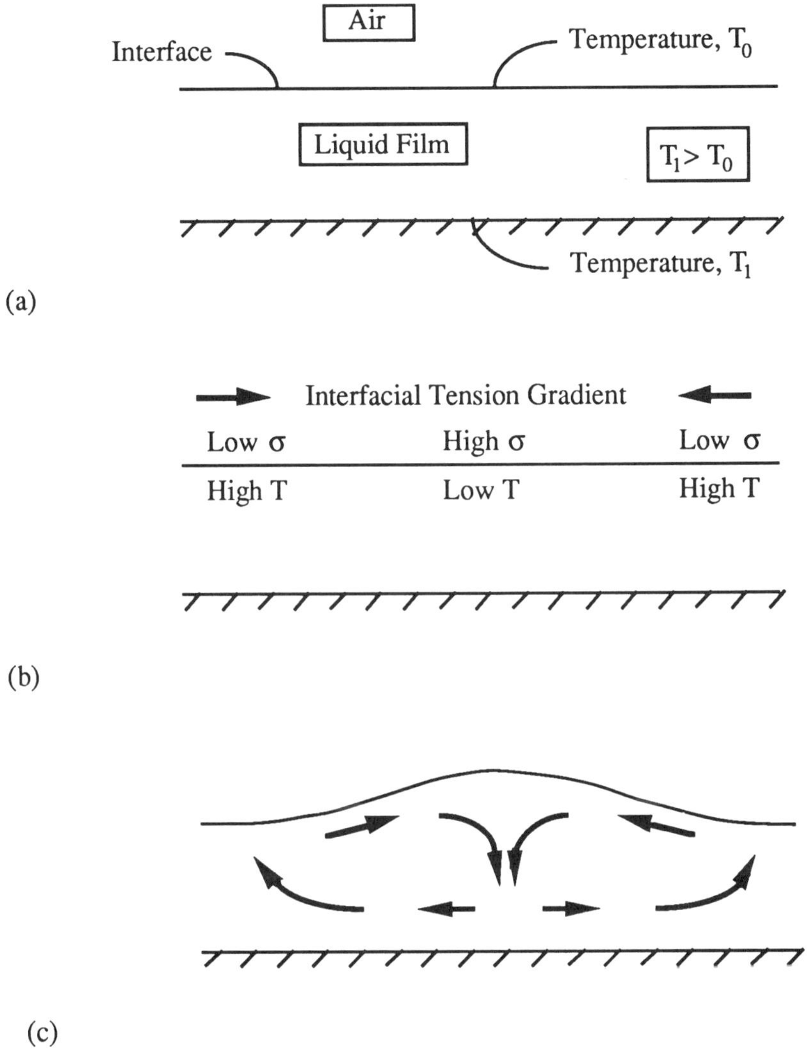

Figure 10.5-1 The onset of Bénard instability. (a) A thin liquid film rests upon a heated solid surface. (b) Thermal fluctuations occur within the film, resulting in temperature inhomogeneities along the interface which lead to interfacial tension gradients. (c) The interfacial tension gradients drive flow in the bulk liquid phase, creating cellular convection patterns.

Note that the rise of liquid toward centers of depression within the film, rather than towards centers of high elevation (as occurs in cases of buoyancy-driven convection) provides an experimentally verifiable distinction between the above Marangoni-driven instability and buoyancy-driven instabilities.

10.6 Summary

Interfacial instabilities play an important role in each of the applications of interfacial rheology considered in the following chapters of Part I. In the present chapter it has been observed that intrinsically unstable interfacial systems [as occur with very thin liquid films (Chapters 11 and 12), or gas bubbles dispersed in a continuous liquid (Chapters 13 and 14)] are often rendered stable to mechanical or thermal disturbances by the adsorption (or interphase transfer) of surface tension-lowering solute. This stabilization process has been observed with superposed fluids in §10.4 (see also §10.2) and the break-up of jets of liquid (§10.3). It is primarily associated with the development of interfacial-tension gradients which act to resist the formation of instabilities; interfacial viscosities have this same effect [cf. Eq. (10.4-7)]. On the other hand, interfacial rheological stresses, may, in certain cases, be the *source* of interfacial instabilities. This has been observed with the phenomena of interfacial turbulence (§10.4) as well as Bénard instability (§10.5). In these examples, interfacial-tension gradients generate destabilizing interfacial motion.

Questions for Chapter 10

10.1 State whether the following conditions stabilize or destabilize a fluid interface:
i. Large dilatational elasticity owing to the adsorption of surfactant;
ii. Large interfacial viscosities;
iii. Rapid surfactant adsorption;
iv. Large interfacial tension;
v. Large interfacial (surfactant) diffusivities;
vi. Transfer of a surface-tension lowering solute into a fluid jet from a surrounding gas phase.
vii. High bulk-phase viscosities.

Additional Reading for Chapter 10

§10.1 Normal Mode Interfacial Stability Analysis

Miller, C.A. 1978 Stability of interfaces. In *Surface and Colloid Science*, Vol. 10 (ed. E. Matijevic), pp. 227–293. New York: Plenum.

§10.2 The Stability of Superposed Fluids

Miller, C.A. & Neogi, P. 1985 *Interfacial Phenomena*. Surfactant Science Series, Vol. 17. New York: Marcel Dekker.

§10.3 Rayleigh Instability

Entov, V.M. & Yarin, A.L. 1984 The dynaimcs of thin liquid jets in air. *J. Fluid Mech.* **140**, 91–111.

Allen, R.F. 1988 The mechanics of splash. *J. Colloid Interface Sci.* **134**, 309–316.

Hajiloo, A., Ramamohan, T.R. & Slattery, J.C. 1987 Effect of interfacial viscosity on the stability of a liquid thread. *J. Colloid Interface Sci.* **117**, 384–393.

§10.4 Interfacial Turbulence

Sternling, C.V. & Scriven, L.E. 1959 Interfacial turbulence: Hydrodynamic instability and the Marangoni effect *AIChE J.* **5**, 514–523.

§10.5 Bénard Instability

Dijkstra, H.A. 1990 The coupling of Marangoni and capillary instabilities in an annular thread of liquid. *J. Colloid Interface Sci.* **136**, 151–159.

11

"A film which is truly fluid in its interior is in general subject to a continual diminution of thickness by the internal current, due to gravity and the suction at its edge. Sooner or later, the interior will somewhere cease to have the properties of matter in mass. The film will then probably become unstable . . ., the thinnest parts tending to thin . . . very much as if there were an attraction between the surfaces of the film, insensible at greater distances, but becoming sensible when the thickness of the film is sufficiently reduced . . .. In a film of soap-water, however, the rupture does not take place, and the processes which go on can be watched. It is apparent even to a very superficial observation that a film of which the tint is approaching black exhibits a remarkable instability . . ..

"That which is most difficult to account for in the formation of the black spot is the arrest of the process by which the film grows thinner. It seems most natural to account for this, if possible, by passive resistance to motion due to a very viscous or gelatinous condition of the film. For it does not seem likely, that the film, after becoming unstable by the flux of matter from its interior, would become stable (without the support of such resistance) by a continuance of the same process."

J. Willard Gibbs (1906)

CHAPTER 11

Thin Liquid Film Hydrodynamics

Thin liquid films may be formed by the surfaces of two approaching droplets or by the spreading of liquid upon a solid surface; they have been the focus of scientific interest since Hooke's report (1672) to the Royal Society regarding 'holes' within stable soap films [later understood by Newton (1704) and Gibbs (1957) to be film regions sufficiently thin to prevent the interference of light rays reflected from upper and lower film surfaces]. In addition to their well-known optical properties, thin liquid films are studied for their importance to processes ranging from foam and emulsion coalescence (see §§14.1 and 14.2), to the stability of liquid films on solid surfaces (Whitaker 1964, Smith 1970, Yih & Seagrave 1978), and detergency (Adamson 1982). They are useful as probes for detecting long-range molecular forces (Derjaguin & Titiyevskaya 1953, Derjaguin 1955, Derjaguin & Gutop 1965, Lyklema & Mysels 1965, Nikolov *et al.* 1989), as well as the magnetism of gases (Faraday 1851), and are studied as models of biological cell membranes (Porter *et al.* 1973, Nicolson 1974, Willingham & Pastan 1975, Pasternak 1976).

Theoretical studies of thin liquid films may be grouped into investigations of geometrical and optical properties (Newton 1704, Courant & Robbins 1941), thermodynamics (Derjaguin & Kussakov 1937, Frumkin 1938, Frenkel 1955, Rusanov 1967, Scheludko 1967, Toshev & Ivanov 1975), hydrodynamics (Sonntag & Strenge 1972, Hartland 1967, Levich 1962, Lucassen *et al.* 1970, Woods & Burrill 1972, Johannes & Whitaker 1965, Ivanov & Dimitrov 1974, Radoev *et al.* 1974, Traykov *et al.* 1977, Barber & Hartland 1976, Zapryanov *et al.* 1983, Malhotra & Wasan 1987), and stability (see the references cited in §12.1). In the present chapter, attention is devoted primarily to the role played by interfacial rheological stresses in the hydrodynamics of thin films. This subject is addressed following a brief discussion of geometrical and basic hydrostatic relations for liquid films, throughout adhering to the simplifying context of a plane-parallel thin liquid film.

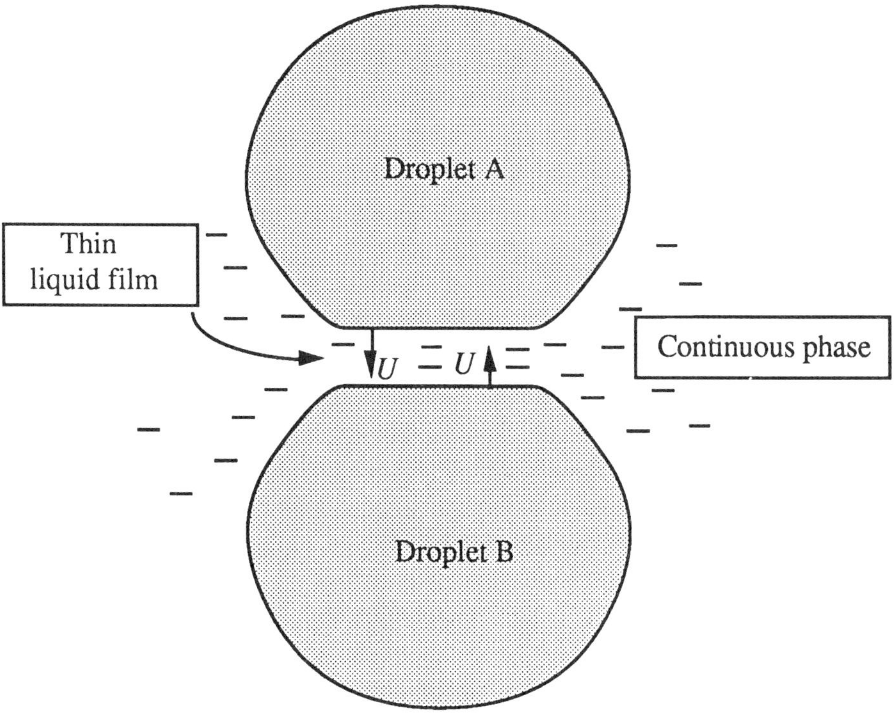

Figure 11.0-1 The creation of a thin liquid film between two approaching liquid droplets.

The hydrodynamics of thin film drainage may be strongly influenced by interfacial-tension gradients and interfacial viscous stresses. These interfacial stresses tend generally to resist film thinning, rendering the surfaces of the thin film *immobile*, or solid-like. Interfacial-tension gradients are especially effective in resisting film drainage, as surfactant swept to the meniscus, or *Plateau border*, by the efflux of fluid from within the draining film establishes a surface force opposing the direction of film fluid flow (cf. Figure 11.3-3; Figure 5.6-1 depicts a related phenomenon). As the specific surface (interfacial area-to-volume ratio) of the thin film system is relatively large, the effect of the tension gradient may result in a significant slowing of film drainage.

In Figure 11.0-1 two identical droplets are shown approaching along their lines of centers (ultimately; but see Figure 11.1-2) forming a plane-parallel film of continuous liquid that proceeds to thin under the influence of either external or internal (e.g. long-range dispersion) forces. The potential (theoretical) influence of interfacial rheological properties upon the hydrodynamics of such a thinning film is illustrated in Figure 11.0-2. In this figure, values for the approach velocity U are shown versus film thickness for each of two limiting cases: (i) the upper curve depicts the case wherein film surfaces are assumed completely mobile (i.e. there exists no interfacial resistance to drainage); (ii) the lower curve provides theoretical predictions for the limiting case wherein film surfaces are assumed completely immobile (i.e. the surfaces are rigid). In the figure, the approach velocity U is divided by the immobile drainage velocity U_{Re} [called the *Reynolds velocity* (Reynolds 1886)]: this forms a nondimensional group called the *interfacial mobility*.

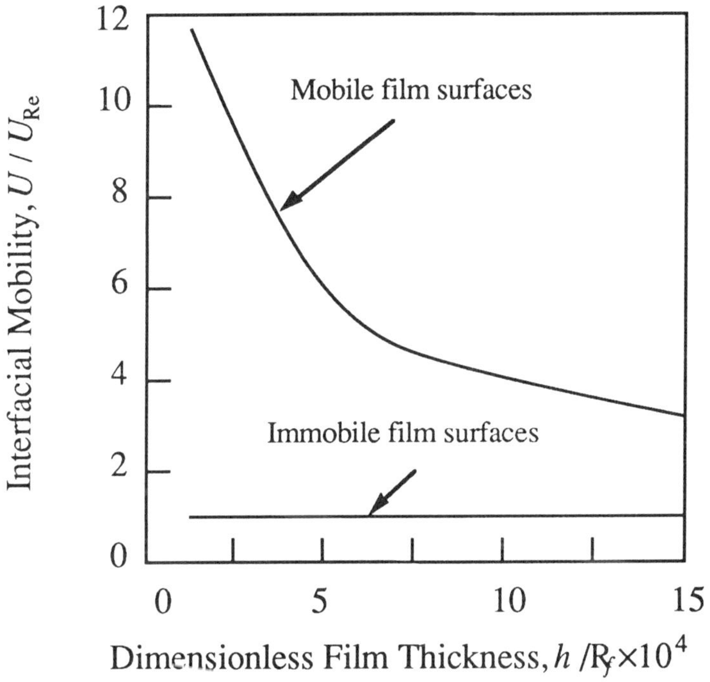

Figure 11.0-2 Theoretical values for the dimensionless drainage velocity, or *interfacial mobility* versus dimensionless thin film thickness. Theoretical values taken from Malhotra & Wasan (1987).

The range of influence of interfacial rheological properties upon drainage rates may be regarded as bound between the upper and lower thinning velocity curves of Figure 11.0-2: this dynamic interfacial effect represents a potential several-fold reduction of the drainage velocity. The overall influence of interfacial rheological properties may vary owing to several indirect factors; these include film thickness, the radius of the thin film, surfactant structuring within the film and adsorption kinetics. In §11.3 the drainage problem described briefly above, arising by the symmetrical approach of two droplets, is considered in detail as a straightforward illustration of interfacial rheological effects in the hydrodynamics of thin liquid films.

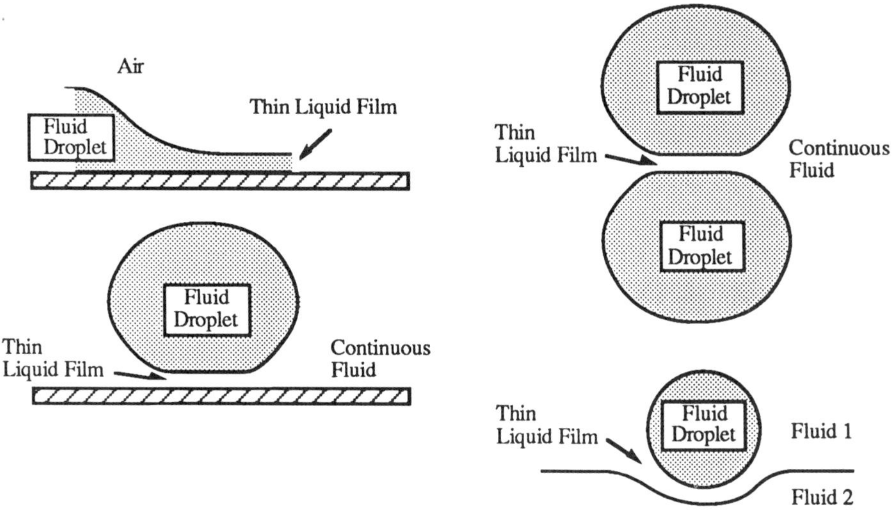

Figure 11.1-1 Thin liquid films formed between liquid, solid, and gas phases.

11.1 Geometrical Aspects of Thin Liquid Films

A thin liquid film, whether formed upon a solid surface or by liquid trapped between two (or more) droplets dispersed within a continuous liquid (Figure 11.1-1), may undergo several stages of thinning owing to the influence of gravitational, convective, and/or suction forces, before arriving at a stable, or meta-stable film configuration. Figure 11.1-2 illustrates the configurational stages through which a liquid film formed between two fluid droplets may pass before achieving a stable, plane parallel, *black* film. This plane parallel geometry is actually observed well before the formation of the black film (see Figure 11.1-2c). Typically, the planar configuration is observed at thicknesses less than 200 nm (whereas, in contrast, black films generally occur at thicknesses less than 10 nm).

The formation of an equilibrium plane-parallel liquid film owes to the action of long-range dispersion (i.e. van der Waal) and electrostatic forces; these intermolecular attractions and/or repulsions arise on account of the close proximity of the pair of film surfaces. Structural forces are also sometimes observed in a thinning liquid film, although presently their origin and theoretical description is less understood than are the dispersion and electrostatic forces. Structural forces appear to arise by ordered micellar layering within the thinning film, as depicted in Figure 11.1-3 (Nikolov & Wasan 1989).

Most theoretical thin film drainage studies have focused upon the plane parallel film configuration. Intermolecular film forces, such as those mentioned above, are generally accounted for explicitly in these analyses through their contribution to the net 'external' force acting upon the droplet [cf. Eq. (11.3-30)], in addition to their indirect influence [cf. Eqs. (11.2-7)-

(11.2-11)] upon film geometrical parameters, such as the film contact angle θ and equilibrium film thickness h (see Figure 11.1-4).

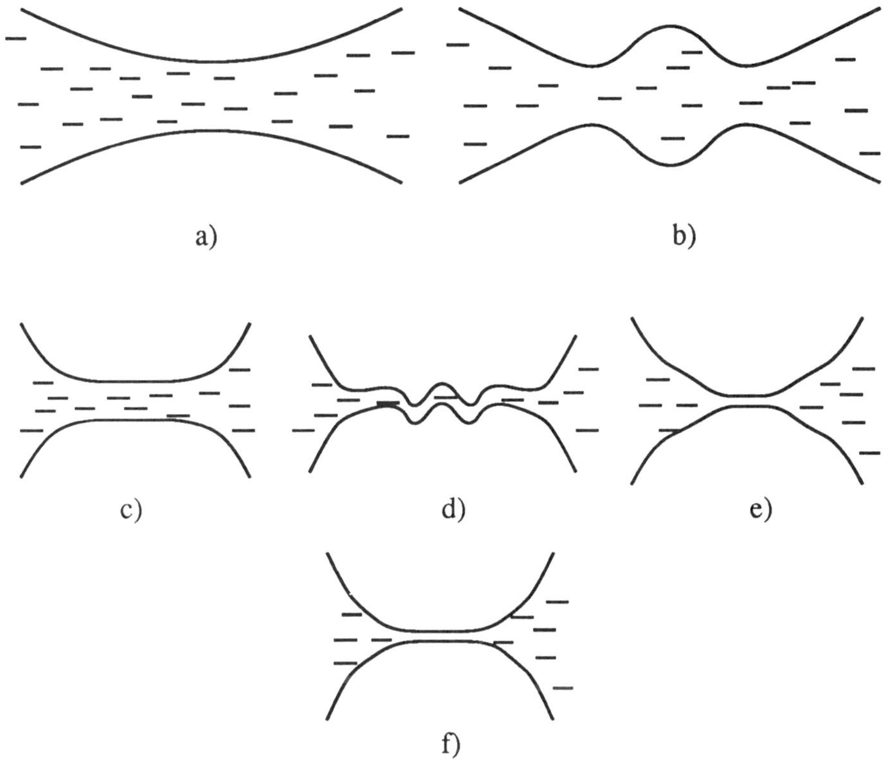

Figure 11.1-2 Main stages in the formation and evolution of a thin liquid film between two bubbles or drops: a) mutual approach of slightly deformed bubble surfaces; b) formation of a dimple; c) formation of nearly plane-parallel film; d) thermal or mechanical fluctuations may cause instability; e) formation of a black film; f) black film grows to equilibrium radius (Ivanov & Dimitrov 1989).

Contact angles at the meniscus regions of plane-parallel films may be defined in various ways. The convention adopted in this chapter is to define the film meniscus contact angle, θ [in addition, there may arise a contact angle within the center of the film, corresponding to the transition to a thinner (perhaps black) film, θ_r] by extrapolating a tangent (in the cross-sectional plane) from the meniscus surface to the extended plane of the film surface. An alternative convention is to extrapolate the meniscus tangent to the center of the film (or to a solid surface).

Whereas only a single contact angle is required to locally define the macroscopic geometry of the three-phase contact line region encircling the film, should the film be formed of continuous liquid between two *differing* fluid phases (as illustrated in Figure 11.1-5c), two contact angles will be needed.

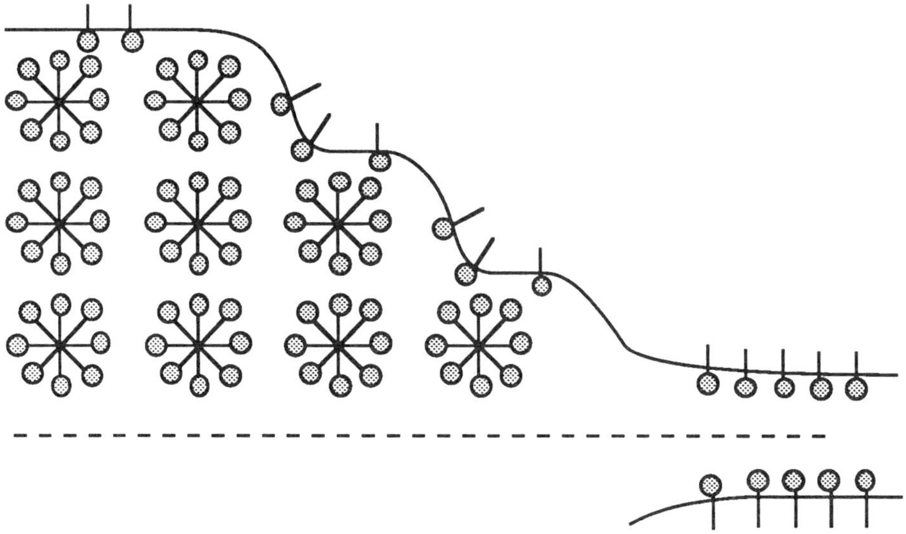

Figure 11.1-3 Micellar structuring within a thinning liquid film may lead to step-wise thickness changes, as layers of micelles form within the film, preventing a continuous transition from thick to thin film.

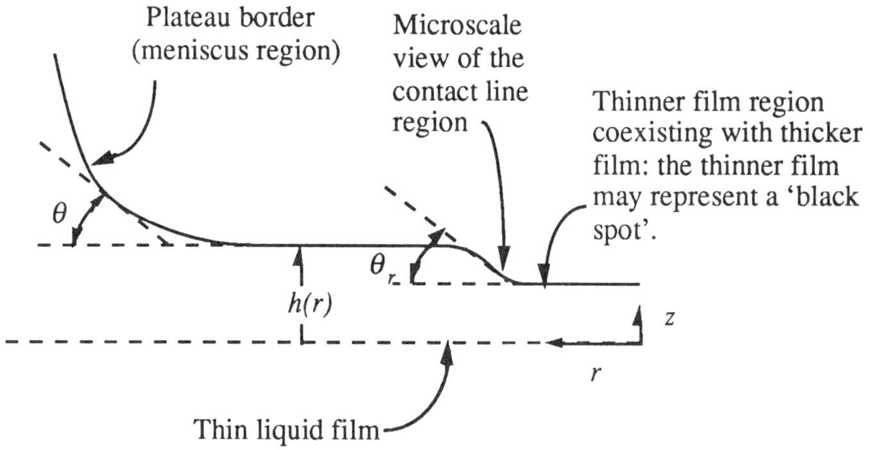

Figure 11.1-4 Contact angle and film thickness for a thin liquid film formed between two fluid droplets (or, equivalently, at a solid surface).

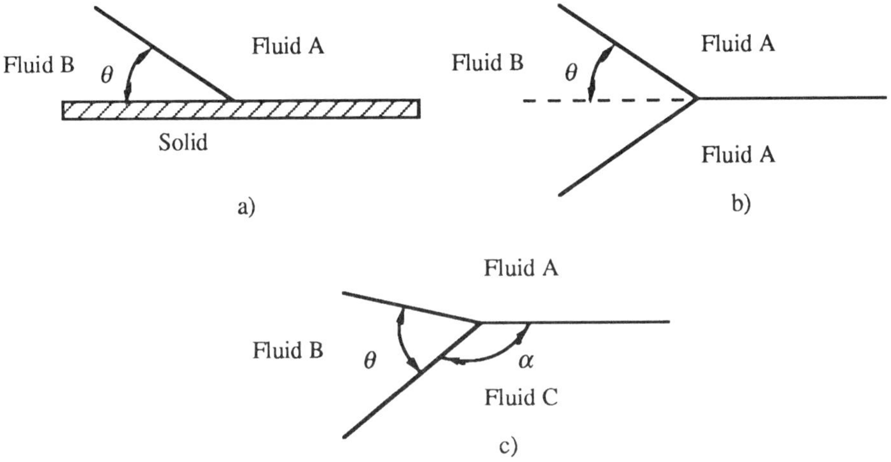

Figure 11.1-5 Macroscopic contact angles at three-phase contact lines.

11.2 Hydrostatics of Thin Liquid Films: The Disjoining Pressure Model

Derjaguin & Kusakov (1937) first observed that when a liquid film is formed between two flat plates, such that the film liquid is allowed to exist in hydrostatic equilibrium with the surrounding liquid, an external force directed perpendicular to the plates is required to balance a force Π of internal origin, thereby maintaining a constant separation distance: this latter, internal film force is known as the *disjoining pressure* (Derjaguin 1955). It may be viewed as the difference between the (apparent) homogeneous pressure within the film p_f' [the prime is used here to distinguish this apparent pressure from the 'true' pressure in the film; cf. Eq. (11.2-5)] and the hydrostatic pressure in the bulk-liquid phase, p_l; explicitly

$$\Pi = p_f' - p_l . \tag{11.2-1}$$

The disjoining pressure, Π, arises as the macroscopic consequence of intermolecular van der Waal, electrostatic, and/or structural forces [cf. Eqs. (12.1-7)–(12.1-11)] acting between the respective surfaces of the thin film; it is, of course, sensible only when film surfaces achieve a separation that is of molecular dimensions. Since Derjaguin & Kusakov (1937), disjoining pressure has also been observed to act between two liquid surfaces in close proximity, whence it directly influences the nature of the hydrostatic film geometry by providing a direct contribution to the normal interfacial stress boundary condition $\left\|\bar{\mathbf{P}}_{nn}\right\|$ [cf. Eq. (4.2-20a)].

Consider the thin liquid film configuration depicted in Figure 11.2-1. Assuming the absence of surface-excess forces, a continuity of hydrostatic bulk-phase pressures must exist across the planar fluid surface as required by the Laplace condition (2.1-3) [representing the hydrostatic limit of Eq. (4.2-20a)]; thus

$$p_f' = p_d. \tag{11.2-2}$$

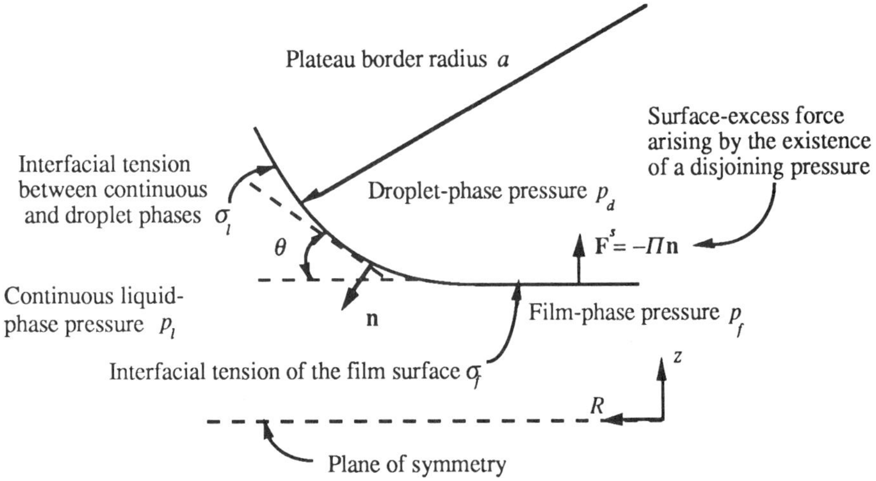

Figure 11.2-1 The disjoining pressure model of a thin liquid film formed by the surfaces of two symmetrical fluid droplets in hydrostatic equilibrium.

Here, p_d denotes the droplet-phase pressure. Outside the film, in the meniscus region, where there exists the continuous liquid-phase pressure p_l, the Laplace condition (2.1-3) requires that

$$p_d - p_l = 2\sigma_l / a, \tag{11.2-3}$$

where a is the radius of the liquid droplet and σ_l the interfacial tension existing at the interface between the droplet and liquid phases. Comparing the above two relations with Eq. (11.2-1), it follows as a consequence that

$$\Pi = 2\sigma_l / a, \tag{11.2-4}$$

whence the disjoining pressure is sometimes referred to as the *capillary pressure* (through association with the term on the right-hand side of the above expression).

It is evident from Eq. (11.2-1) that the hypothetical pressure p_f' cannot, in fact, exist within the thin liquid film (barring its coexistence with internal body forces of the type discussed in §12.1) without there resulting flow from the film into the Plateau border regions. Thus, in the disjoining pressure model of thin liquid films, a homogeneous pressure p_f is defined within the thin liquid film as

$$p_f \stackrel{\text{def}}{=} p_f' - \Pi. \tag{11.2-5}$$

It possesses the necessary quality that in equilibrium,

$$p_f = p_1,$$

in addition to possessing a discontinuity across the planar film surfaces, namely,

$$p_f - p_d = \Pi. \tag{11.2-6}$$

The disjoining pressure, Π, thus arises in the normal surface stress condition $\|\mathbf{P}_{nn}\|$ as a normal surface-excess force, as expressed by the relation

$$\boxed{\mathbf{F}^s = -\,\Pi\,(h)\mathbf{n}.} \tag{11.2-7}$$

Here, **n** represents the unit outward-directed normal to the liquid droplet interface and the negative sign corresponds to the fact that the disjoining pressure is taken to be positive when acting to 'disjoin', or separate, the film surfaces. The surface force defined in Eq. (11.2-7) depends upon the distance of separation, h, of the thin film surfaces. It may be computed (neglecting structural forces) in a relatively straightforward manner using DLVO theory [Derjaguin & Landau (1941), Verwey & Overbeek (1948); cf. also Eqs. (12.1-6)–(12.1-11)].*

Basic hydrostatic conditions appropriate to the disjoining pressure model may be stated briefly as follows [see Toshev & Ivanov (1975) for a thermodynamic derivation of these relations]. In the absence of flow, the hydrostatic pressure within the film p_f is equivalent to the pressure in the liquid phase, p_l, and suffers a jump discontinuity across the film surfaces, as governed by the hydrostatic normal stress boundary condition (4.2-20a):

$$-\mathbf{n}\cdot\|\mathbf{P}\| = 2H\sigma_f\,\mathbf{n} + \mathbf{F}^s, \tag{11.2-8}$$

where the jump condition on the left-hand side is provided by [cf. Eqs. (3.6-12b) and (4.1-13)]

$$\mathbf{n}\cdot\|\mathbf{P}\| = \mathbf{n}\,(p_d - p_f). \tag{11.2-9}$$

The surface-excess force, $\mathbf{F}^s$, is given by Eq. (11.2-7). The film surface tension σ_f may be related to the interfacial tension σ_l acting along the interface in the Plateau border and bulk-liquid regions in terms of the macroscopic contact angle θ as

$$\sigma_f = \sigma_l \cos\,\theta. \tag{11.2-10}$$

---

* A body-force approach is sometimes preferred to the disjoining pressure model, particularly for the description of asymmetrical or nonplanar film surfaces (such as the surfaces of the film depicted in Figure 11.1-4d). The body-force model, which is considered in §12.1, is used primarily for investigations of the stability of thin films.

A *film tension* (regarded as the tension acting upon the thin film when viewed macroscopically as a single membrane) may be defined as (Toshev & Ivanov 1975)

$$\gamma_f = 2\sigma_f + \Pi h. \tag{11.2-11}$$

The equilibrium *spreading coefficient* $S^f_{\alpha\beta}$ characterizing the ability of a film formed between two (generally dissimilar) phases α and β to spontaneously spread may be represented as (Adamson 1982)

$$S^f_{\alpha\beta} \equiv \gamma_f - (\sigma^\alpha_l + \sigma^\beta_l). \tag{11.2-12}$$

This quantity has the sense of being positive for a spreading film and negative for a nonspreading film.

In the particular circumstance of an equilibrium film possessing two different thicknesses (e.g. Figure 11.1-4) the interfacial tensions of the film may suffer a discontinuity across the macroscopic contact line which separates the two film regions owing to the existence of a 'line tension', the latter property being discussed at length in Chapter 18.

11.3 Hydrodynamics of Thin Liquid Films

Experiments have shown (Scheludko 1967, Burrill & Woods 1973, Manev *et al.* 1984) that two droplets whose radii are less than 1 mm which approach along their lines of center will form a nearly plane-parallel film at distances of separation less than 200 nm. At distances less than 100 nm, long-range van der Waal forces begin to attract the approaching surfaces. These forces are soon balanced by repulsive electrostatic and structural forces. The nature of the film thinning process in this plane-parallel configuration will often be a strong determinant as to whether the droplets or bubbles coalesce; indeed, should the contact time between droplets prove insufficient for the thinning film to reach a stage of instability, the slow thinning stage may be responsible for the stabilization of the dispersed system.

Many investigations (Hartland 1967, Levich 1962, Lucassen *et al.* 1970, Woods & Burrill 1972, Johannes & Whitaker 1965, Ivanov & Dimitrov 1974, Radoev *et al.* 1974, Traykov & Ivanov 1972, Barber & Hartland 1976, Zapryanov *et al.* 1983, Malhotra & Wasan 1987) have been made of thinning, plane-parallel films in recent years. Most of these seek a generalization to the classic Reynolds (1886) solution for the thinning velocity between two plane-parallel plates, namely (see Question 11.2),

$$U_{\mathrm{Re}} = -\frac{dh}{dt} = \frac{8h^3}{3\mu R^2}\Delta P\ , \tag{11.3-1}$$

where R is the film radius, μ the liquid viscosity in the film, and ΔP the pressure drop imposed across each planar film surface to effect thinning of the film. In the following, a mathematical formulation is provided for a draining foam/emulsion film, the primary intent of which is to illustrate the effects of interfacial rheological properties. The model accounts for: (i) flow in both film and droplet phases; (ii) the partitioning of surfactant; (iii) interfacial viscosities and their gradients; (iv) interfacial tension and its gradient and (v) surfactant diffusion and adsorption. The model employs much of the theoretical material developed in Chapters 4 and 5, offering a fairly comprehensive representation of the influence of dynamic interfacial properties upon film drainage; nevertheless, several assumptions remain whereby the quantitative applicability of the model to experimental circumstances proves limited. These appear in the form of *ad hoc* boundary conditions imposed at the periphery of the film [e.g. the assumption of no radial flow in the droplet phase at the periphery of the film; cf. Eq. (11.3-13)] as well as the assumption of planar film surfaces. Generally speaking, the model provides no real assessment of the role which drainage forces play in the vicinity of the Plateau borders. For these reasons the quantitative agreement between model and experiment has been observed to diminish with decreasing film thickness (Malhotra & Wasan 1987); nevertheless, the basic qualitative conclusions pertaining to the influence of interfacial rheological properties upon the thin film drainage process are observed in practice. These latter conclusions will be discussed toward the conclusion of this section through a parametric study of the solution to the boundary value problem developed below.

Consider the symmetrical film geometry depicted in Figure 11.3-1. The quasistatic Stokes equations are assumed for both droplet and film phases. In the film phase, the lubrication approximation

$$h / R_f << 1,$$

(with R_f the film radius) permits the following lubrication equations (Table 4.1-2):

$$\frac{\partial p}{\partial R} = \mu \frac{\partial^2 v_R}{\partial z^2}, \qquad (11.3\text{-}2)$$

$$\frac{\partial p}{\partial z} = 0, \qquad (11.3\text{-}3)$$

in addition to which there is the continuity condition

$$\nabla_R v_R + \frac{\partial v_z}{\partial z} = 0. \qquad (11.3\text{-}4)$$

In the above we have defined the spatial gradient,

$$\nabla_R = \frac{1}{R}\frac{\partial}{\partial R} R. \qquad (11.3\text{-}5)$$

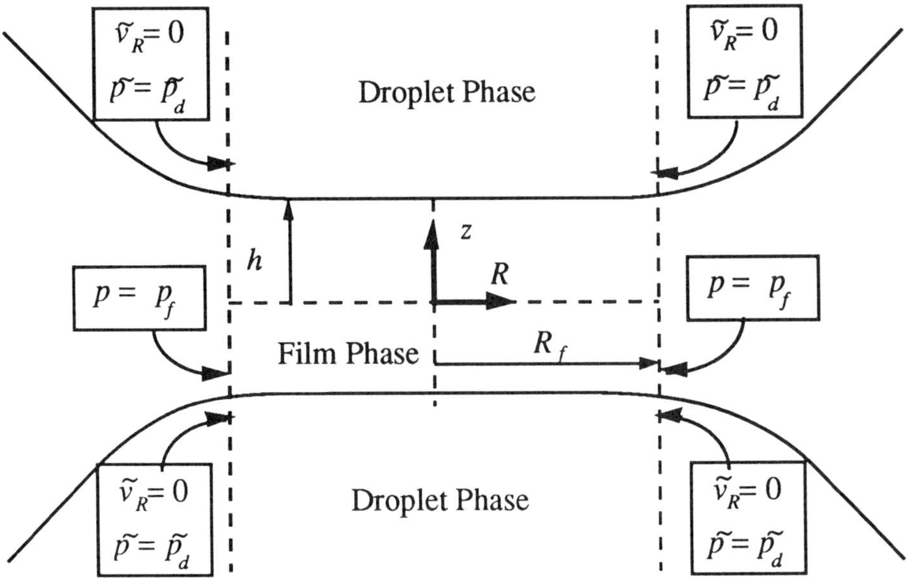

Figure 11.3-1 Hydrodynamic boundary conditions for the problem of a thin film draining between two identical fluid droplets approaching under the action of capillary suction forces along their lines of centers.

The low-Reynolds number, Stokes form of the Navier-Stokes equations are assumed to apply in the droplet phase; thus, from Table 4.1-2 there obtains

$$\frac{\partial \tilde{p}}{\partial R} = \tilde{\mu}\left(\nabla^2 \tilde{v}_R - \frac{\tilde{v}_R}{R^2}\right), \tag{11.3-6}$$

$$\frac{\partial \tilde{p}}{\partial z} = \tilde{\mu}\nabla^2 \tilde{v}_z, \tag{11.3-7}$$

together with the continuity equation

$$\nabla_R \tilde{v}_R + \frac{\partial \tilde{v}_z}{\partial z} = 0, \tag{11.3-8}$$

where

$$\nabla^2 = \frac{\partial^2}{\partial R^2} + \frac{1}{R}\frac{\partial}{\partial R} + \frac{\partial^2}{\partial z^2}. \tag{11.3-9}$$

In these relations, the symbol (~) has been used to designate droplet-phase quantities

As mentioned previously, we assume *a priori* that film surfaces remain parallel throughout the drainage process [at least in the vicinity of the lubrication zone of the film, where Eqs. (11.3-2)–(11.3-4) apply]. Hence, the above equations are subject to the following velocity continuity conditions at the upper film surface:

$$v_R^o = \tilde{v}_R^o \equiv v_s(r) \qquad \text{at} \quad z = h, \tag{11.3-10}$$

$$v_z^o = \tilde{v}_z^o \equiv - U, \qquad \text{at} \quad z = h, \tag{11.3-11}$$

with $U = -dh/dt$ the drainage velocity of the film. In addition, the tangential stress condition (4.2-20b) adopts the form

$$\mu \frac{\partial v_R}{\partial z} - \tilde{\mu}\frac{\partial \tilde{v}_R}{\partial z} = (\kappa^s + \mu^s)\nabla_R\left(\frac{\partial v_s}{\partial R}\right)$$
$$- \frac{E_o^1}{\rho_{1o}^s}\frac{\partial \rho_1^s}{\partial R} \qquad \text{at} \quad z = h, \tag{11.3-12}$$

where the interfacial-tension gradient has been approximated as in Eq. (5.5-6); namely,

$$\sigma \approx \sigma_o - \frac{1}{\rho_{1o}^s} E_o^1(\rho_1^s - \rho_{1o}^s),$$

which applies to the case of small departure from surface compositional equilibrium,

$$\rho_1^s = \rho_{1o}^s + \varepsilon \hat{\rho}_1^s; \qquad \varepsilon << 1.$$

In the above, the subscript o denotes an equilibrium value and E_o^1 represents the Gibbs elasticity; cf. Eq. (5.5-5). At the periphery of the thin film ($R = R_f$), the film pressure p is assumed to equate with the hydrostatic film pressure p_f defined by Eq. (11.2-5) [satisfying, in equilibrium, Eq. (11.2-6)]. In addition, liquid within the droplet phase is assumed to circulate above the planar film surfaces such that the condition of no radial flow at $R = R_f$ is met. Thus,

$$\tilde{v}_R = 0 \qquad \text{at} \quad R = R_f, \tag{11.3-13}$$

$$p = p_f \qquad \text{at} \quad R = R_f, \tag{11.3-14}$$

Symmetry provides the condition,

$$\frac{\partial v_r}{\partial z} = 0 \qquad \text{at} \quad z = 0, \tag{11.3-15}$$

in addition to which there is the necessary condition,

$$\tilde{v}_R = v_R = 0 \qquad \text{at} \quad R = 0. \tag{11.3-16}$$

Finally, equilibrium prevails far away into the droplet phase:

$$\tilde{v}_R = 0 \qquad \text{at} \quad z \to \infty, \tag{11.3-17}$$

$$\tilde{v}_z = - U \qquad \text{at} \quad z \to \infty, \tag{11.3-18}$$

$$\tilde{p} = p_d \qquad \text{at} \quad z \to \infty. \tag{11.3-19}$$

The quasistatic, low-Peclét form of the convective diffusion equation is assumed to apply in the film and droplet phases. Thus, as in Table 5.1-1,

$$\nabla^2 \rho_1 = 0, \tag{11.3-20}$$

$$\nabla^2 \tilde{\rho}_1 = 0. \tag{11.3-21}$$

Adsorption kinetic fluxes of surfactant to the film surfaces, of the forms [cf. Eq. (5.3-3)]

$$-D\frac{\partial \rho_1}{\partial z} = K\left(\rho_1 - \frac{\rho_1^s}{K_a}\right) \quad \text{at} \quad z = h, \tag{11.3-22}$$

$$-\tilde{D}\frac{\partial \tilde{\rho}_1}{\partial z} = \tilde{K}\left(\tilde{\rho}_1 - \frac{\rho_1^s}{\tilde{K}_a}\right) \quad \text{at} \quad z = h, \tag{11.3-23}$$

are assumed to hold in each phase. By appropriate choices of the constants ($\tilde{K}$, $\tilde{K}_a$, K, K_a), these relations may be reduced to either Langmuir (5.3-5) or Frumkin (5.3-4) type relations. The quasistatic surface-excess species balance equation follows from Eq. (5.2-6) as

$$\rho_{1o}^s \nabla_R v_s - D^s \frac{1}{R}\frac{\partial}{\partial R}\left(R\frac{\partial \rho_1^s}{\partial R}\right) = -D\frac{\partial \rho_1}{\partial z} + \tilde{D}\frac{\partial \tilde{\rho}_1}{\partial z}. \tag{11.3-24}$$

In addition, consistent with conditions (11.3-13) and (11.3-14), equilibrium values of the surfactant density are assumed at the periphery of the film; explicitly,

$$\rho_1 = \rho_{1o} \quad \text{at} \quad R = R_f, \tag{11.3-25}$$

$$\tilde{\rho}_1 = \tilde{\rho}_{1o} \quad \text{at} \quad R = R_f. \tag{11.3-26}$$

By symmetry

$$\frac{\partial \rho_1}{\partial z} = 0 \quad \text{at} \quad z = 0. \tag{11.3-27}$$

Equilibrium prevails far away from the film surface and into the droplet phase; hence

$$\tilde{\rho}_1 = \tilde{\rho}_{1o} \quad \text{as} \quad z \to \infty. \tag{11.3-28}$$

The final condition necessary for the solution of the above boundary value problem is the balance of external and hydrodynamic forces acting upon each droplet. Thus,

$$\mathbf{F}^e + \mathbf{F}^h = \mathbf{0}, \tag{11.3-29}$$

with

$$\mathbf{F}^e \equiv -F\,\mathbf{i}_z, \tag{11.3-30}$$

$$\mathbf{F}^{h} \equiv - \oint_{\partial V} dA\,\mathbf{n} \cdot \mathbf{P}\,, \tag{11.3-31}$$

where $\mathbf{n}$ is the outward-drawn unit normal to the fluid droplet, $\mathbf{i}_z$ is the unit tangent vector to the z-coordinate (the assumed net direction of the applied external force, whose magnitude is F), $\mathbf{P}$ is the pressure tensor in the liquid phase surrounding the droplet [obtained by solving the preceding problem, cf. Eqs. (4.1-13) and (4.1-15)], and ∂V denotes the closed surface bounding the droplet. Since the continuous fluid surrounding the droplet is quiescent with the exception of the very thin layer of film fluid, Eq. (11.3-31) may be approximated as

$$\mathbf{F}^{h} \approx - \int_{A} dA\,\mathbf{n} \cdot \mathbf{P}\,, \tag{11.3-32}$$

with A the area of the planar thin film surface (i.e. $A = \pi R_f^2$).

In the remainder of this section, a parametric study of the numerical solution to the above boundary value problem is reviewed. For this purpose the following dimensionless quantities are defined:

$$\bar{h} \equiv \frac{h}{R_f}\,, \quad \mathrm{N}_{\mu} \equiv \frac{\tilde{\mu}}{\mu}\,, \quad \mathrm{Bo} \equiv \frac{\mu^s + \kappa^s}{\mu R_f}\,, \quad \mathrm{E_s} \equiv E_o^1 \frac{R_f}{\mu D}\,; \tag{11.3-33}$$

$$\bar{\mathrm{K}} \equiv \frac{K R_f}{D}\,, \quad \bar{\tilde{\mathrm{K}}} \equiv \frac{\tilde{K} R_f}{D}\,, \quad \bar{\mathrm{K}}_{\mathrm{a}} \equiv \frac{\rho_{1o} K_a}{\rho_{1o}^s}\,, \quad \bar{\tilde{\mathrm{K}}}_{\mathrm{a}} \equiv \frac{\tilde{\rho}_{1o} \tilde{K}_a}{\tilde{\rho}_{1o}^s}\,; \tag{11.3-34}$$

$$\mathrm{A_d} \equiv \frac{\rho_{1o}^s}{R''\,\rho_{1o}}\,, \quad \mathrm{N_s^D} \equiv \frac{D^s}{D}\,, \quad \mathrm{S_c} \equiv \frac{\mu}{\rho D}\,. \tag{11.3-35}$$

The following values will be assumed for the purpose of the parametric study:

Parameter	N_{μ}	$\mathrm{N_s^D}$	$\mathrm{A_d}$	F	$\mathrm{S_c}$	$\bar{h}_i$	$\bar{h}_f$	$\bar{\mathrm{K}}$	$\bar{\tilde{\mathrm{K}}}_{\mathrm{a}}$	$\bar{\mathrm{K}}_{\mathrm{a}}$	$\bar{\tilde{\mathrm{K}}}$
Magnitude	5	1	10^{-1}	10^{-1}	10^{3}	2×10^{-3}	1.5×10^{-4}	1	1	0.5	0.5

In Figure 11.3-2, the effect of interfacial-tension gradients upon interfacial mobility is illustrated in terms of the dimensionless elasticity number $\mathrm{E_s}$. The interfacial-tension gradient in the thinning film is created by the efflux of liquid from the film and the sweeping of surfactant along the film surfaces to the Plateau borders, as depicted in Figure 11.3-3. This creates an interfacial-tension gradient that opposes film drainage, creating "immobile" film surfaces in accordance with the Reynolds solution (11.3-1). The Marangoni effect is one of the primary stabilization factors in droplet coalescence.

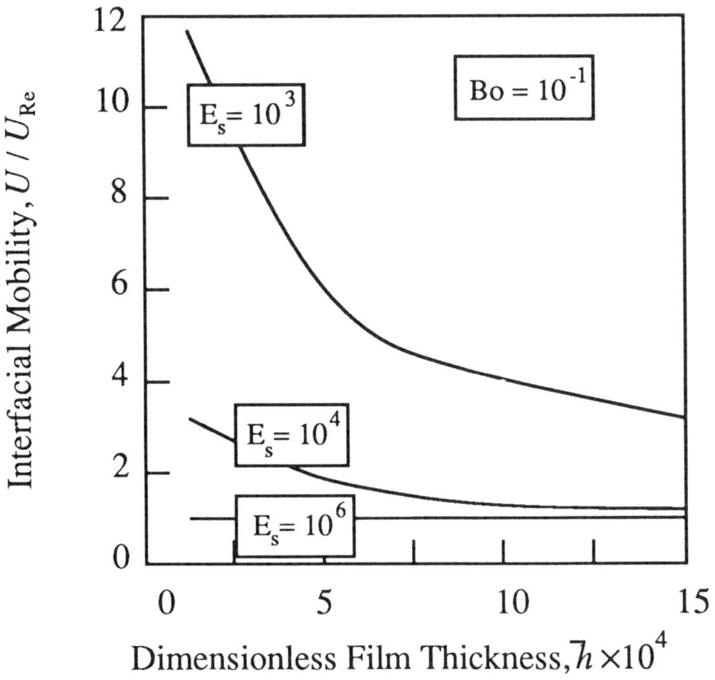

Figure 11.3-2 Interfacial mobility, or dimensionless drainage velocity, versus dimensionless film thickness, at three values of the dimensionless interfacial elasticity. Numerical values taken from Malhotra & Wasan (1987).

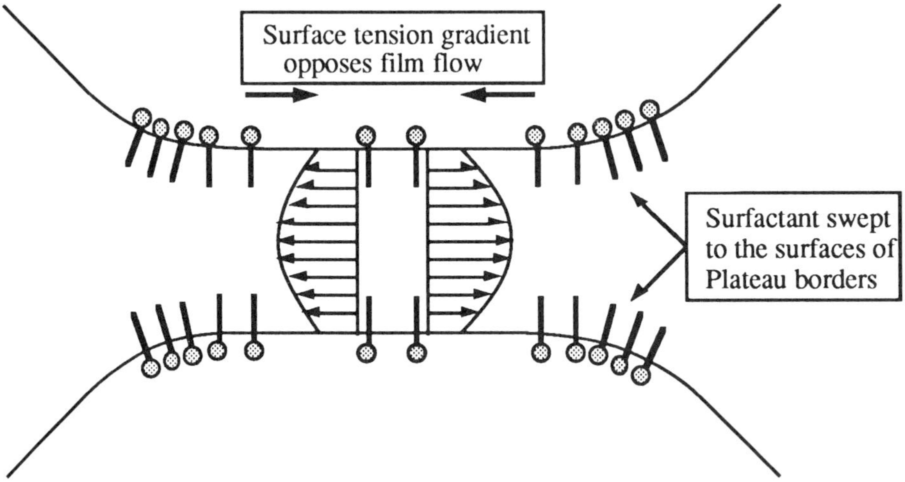

Figure 11.3-3 Marangoni effect in the thin film drainage process. Surfactant is swept to the Plateau borders by flow in the film and droplet phases, thereby creating surface concentration gradients which engender surface tension gradients.

Yet, as shown in Figure 11.3-4, for sufficiently high values of interfacial viscosity, Marangoni stresses may be secondary in importance to interfacial viscous stresses, and the film surfaces rendered immobile by virtue of their extremely viscous nature. This, however, may be expected to occur only in rare circumstances, as with the exception of very small radii droplets (generally 1 mm or less), such major influences of interfacial viscosity are not encountered. As may be seen from the definitions provided by Eq. (11.3-33), the Boussinesq number will typically fall within the range of 1-100 for films of radius on the order of 1 cm, whereas values of the dimensionless elasticity number may easily be on the order of 10^6. Hence, in most cases the Marangoni effect will impart the more significant influence upon film drainage.

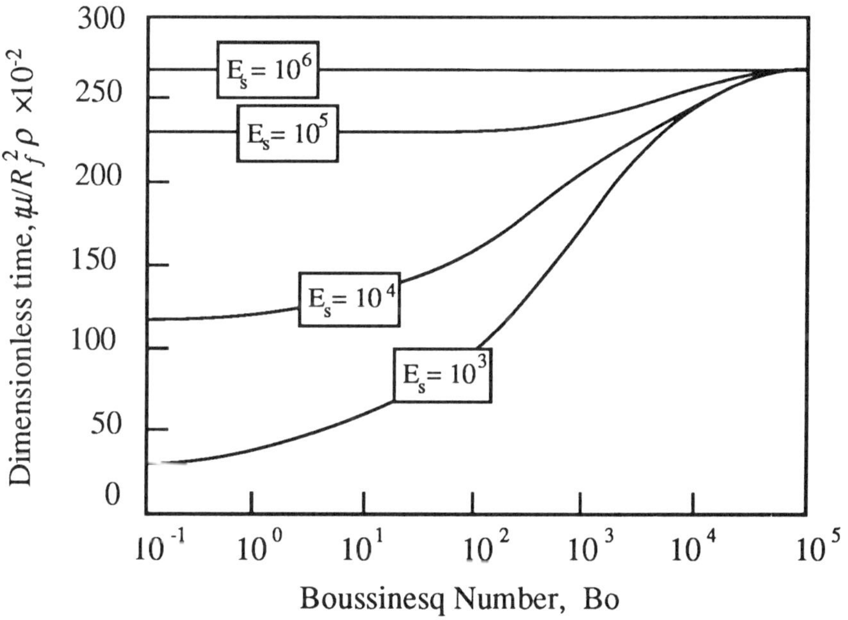

Figure 11.3-4 Dimensionless drainage time for the film to drain from a dimensionless thickness $\bar{h}_i$ to the thickness $\bar{h}_f$, versus Boussinesq number, at various values of the dimensionless interfacial elasticity. Numerical values taken from Malhotra & Wasan (1987).

Two factors that may significantly influence the magnitude of interfacial-tension gradients in the thin film are surface diffusion and surfactant adsorption. In Figure 11.3-5, drainage time is plotted versus interfacial viscosity, varying the surface diffusivity number. As the number increases, corresponding to increased surface diffusion, the drainage time diminishes, which is evidence of the fact that the film surfaces have become more *mobile*. The cause for this effect may be understood by reference to

Figure 11.3-3. If a large surface diffusion occurs in the film surfaces, the surfactant may diffuse back into the film, diminishing the surface concentration gradient, thereby cancelling the Marangoni effect.

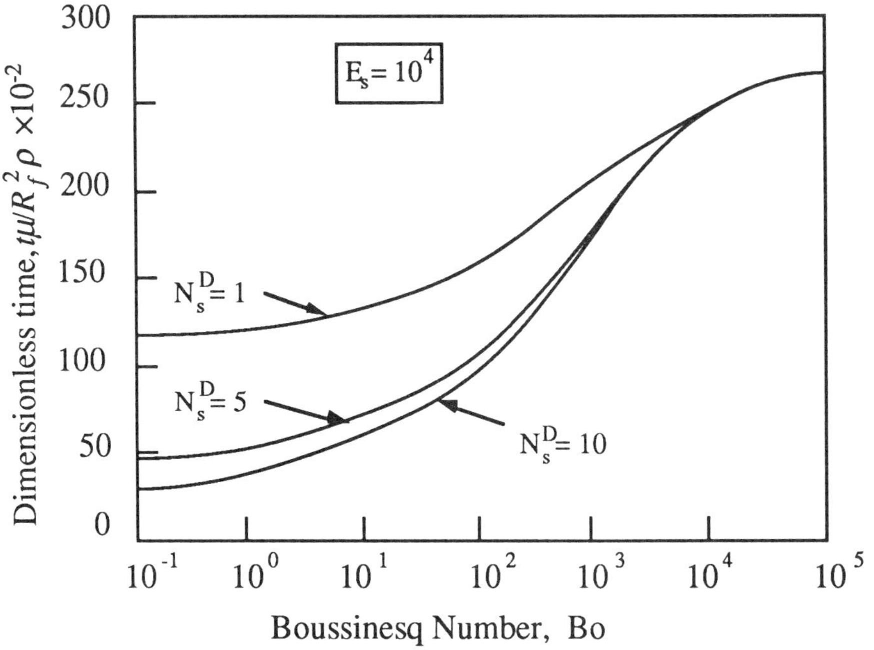

Figure 11.3-5 Dimensionless drainage time for the film to drain from a dimensionless thickness $\bar{h}_i$ to the thickness $\bar{h}_f$, versus Boussinesq number at various values of the surface diffusivity number. Numerical values taken from Malhotra & Wasan (1987).

A similar effect may be achieved by increasing the rate of surfactant adsorption to the film surfaces *from* the *dispersed phase* (the influence of adsorption from the *film* phase is negligible, as the film-phase surfactant transfer process is diffusion-controlled). The adsorption effect is illustrated in Figure 11.3-6, which reveals that an increased adsorption rate results in a diminished time of drainage, again owing to a cancellation of Marangoni flow.

It is thus observed that surfactant solubilized in the dispersed phase is most effective in destabilizing droplets, with the consequence that emulsions with surfactant soluble in the droplet phase are generally less stable than emulsions with surfactant soluble in the continuous phase, all other things being equal. This conclusion, which is clearly illustrated by Figure 11.3-7, is known as “Bancroft’s rule” (Davies 1957, Sherman 1968, Traykov *et al.* 1977). The interfacial mobility may be observed in Figure 11.3-7 to be higher for surfactant soluble in the droplet phase, indicating a possibly less stable emulsion film.

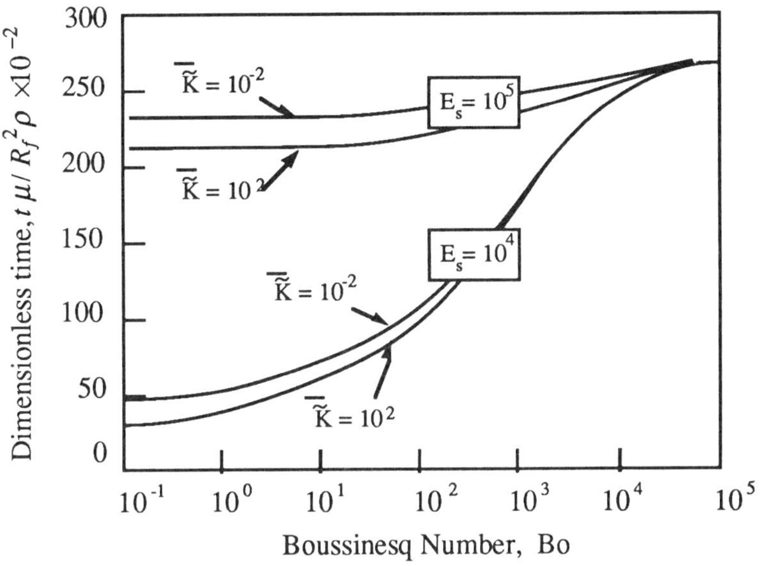

Figure 11.3-6 Dimensionless drainage time for the film to drain from a dimensionless thickness $\bar{h}_i$ to the thickness $\bar{h}_f$, versus dimensionless interfacial viscosity, at various values of the rate of surfactant adsorption. Numerical values taken from Malhotra & Wasan (1987).

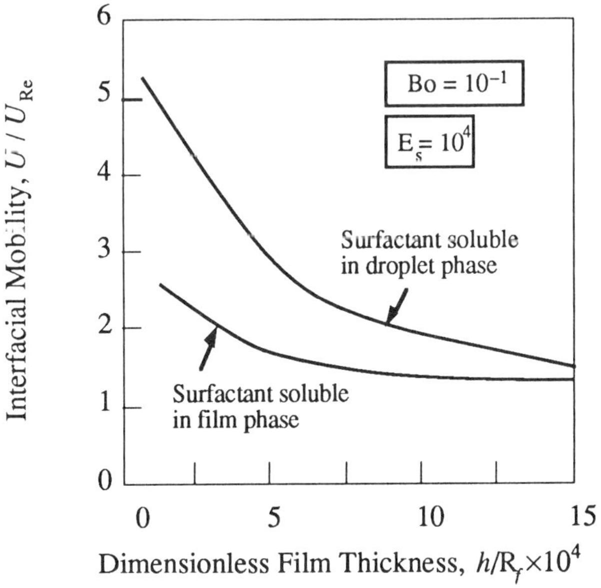

Figure 11.3-7 Interfacial mobility versus dimensionless film thickness, for surfactant soluble in film and droplet phases only. Numerical values taken from Malhotra & Wasan (1987).

11.4 Summary

Thin liquid films are formed between liquid droplets and solid or liquid surfaces, or by liquid spread upon a solid or liquid substrate. Such films are very sensitive to long-range molecular and/or structural forces, notably when the thickness of the film attains a value of 100 nanometers or less, evidenced by the presence of a disjoining pressure within the film. The magnitude and nature of the disjoining pressure will depend highly upon surfactant type, phase type, salt concentration, in addition to other physicochemical factors. It influences the geometry of the meniscus region of the thin film by altering the macroscopic contact angle which the Plateau border surfaces make with the film surfaces. The film surfaces themselves are generally planar at such thicknesses (i.e. 100 nm or less).

Interfacial rheological properties impart a major influence upon the hydrodynamics of thin liquid films on account of the large specific surface present in thin films. Interfacial-tension gradients, created by the sweeping of surfactant to the Plateau borders during film drainage, are the most significant interfacial rheological phenomenon in a draining thin liquid film. They may be responsible for a severalfold reduction in the drainage velocity of a thinning film. Their effect may be lessened by increased surface diffusion or increased adsorption from the droplet phase.

Since the surfactant mass transfer process in the thinning film is diffusion-controlled owing to the very small distance of separation between film surfaces, surfactant solubilized in the continuous (or film) phase only is advantageous for a slow draining film. Surfactant solubilized in the droplet phase, on the other hand, encourages a relatively faster draining film, as surfactant may replenish the interface more readily, diminishing the magnitude of Marangoni stresses.

Questions for Chapter 11

11.1 *i*. Why does the interfacial tension within a thin liquid film differ from the interfacial tension in the meniscus region of the film?

ii. Electrostatic and dispersion forces act not only upon the surfaces of a thin film but also upon the liquid material between these surfaces as well. One way of accounting for these forces [other than redefining the pressure as in Eq. (11.2-5)] would be to introduce a 'body force' **F** in the Navier-Stokes equations. Employing an order-of-magnitude estimate, would you expect this force to play an explicit role in the drainage of the film [not accounted for by the model that leads to (11.2-5)] ?

11.2 Consider the figure below. A fluid film is formed between two plane parallel (circular cylindrical) plates. An external force $\mathbf{F}^e$ acts inwardly upon each plate, causing the film of fluid to drain. Assuming the flow between the plates to be described by the following lubrication equations:

$$\frac{\partial p}{\partial R} = \mu \frac{\partial^2 v_R}{\partial z^2},$$

$$\frac{\partial p}{\partial z} = 0$$

and

$$\frac{1}{R}\frac{\partial v_R}{\partial R} + \frac{\partial v_z}{\partial z} = 0:$$

i. Find the velocity and pressure fields (v_R, p) in the fluid film;
ii. Derive the Reynolds relation (11.3-1).

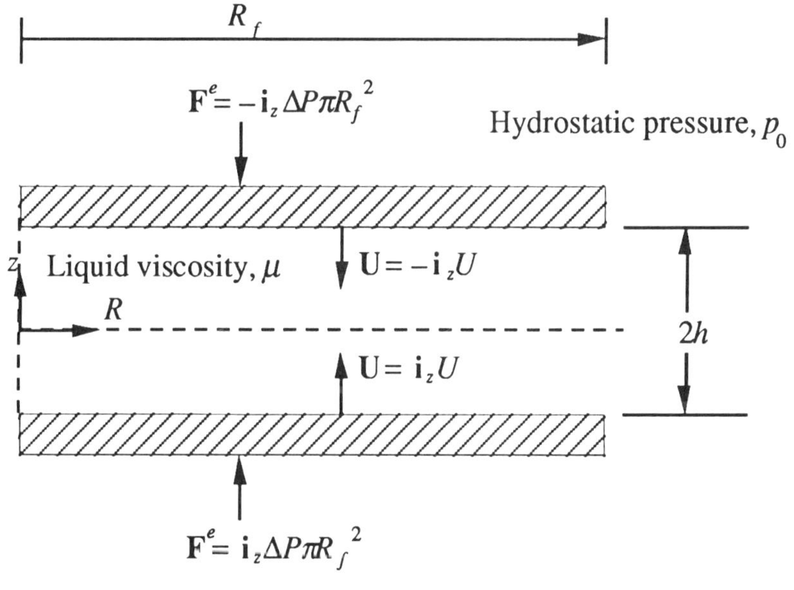

Additional Reading for Chapter 11

§11.1 Geometrical Aspects of Thin Liquid Films

Clunie, J.S., Goodman, J.F. & Ingram, B.T. 1971 Thin liquid films. *Surface & Colloid Sci.* **3**, 167–239.

Overbeek, J.Th. G. 1960 Black soap films. *J. Phys. Chem.* **64**, 1178–1183.

Mysels, K.J. 1964 Soap films and some problems in surface and colloid chemistry. *J. Phys. Chem.* **68**, 3441–3448.

Ivanov, I.B. & Toshev, B.V. 1975 Thermodynamics of thin liquid films II. Film thickness and its relation to the surface tension and the contact angle. *Colloid & Polymer Sci.* **253**, 593–599.

§11.2 Hydrostatics of Thin Liquid Films

Toshev, B.V. & Ivanov, I.B. 1975 Thermodynamics of thin liquid films I. Basic relations and conditions of equilibrium. *Colloid & Polymer Sci.* **253**, 558–565.

§11.3 Hydrodynamics of Thin Liquid Films

Zapryanov, Z., Malhotra, A.K., Aderangi, N. & Wasan, D.T. 1983 Emulsion stability: An analysis of the effects of bulk and interfacial properties on film mobility and drainage rate. *Int. J. Multiphase Flow* **9**, 105–129.

12

"If a bubble be blown with water first, made tenacious by dissolving a little soap in it, 'tis a common observation that after a while it will appear tinged with a great variety of Colours. To defend these Bubbles from being agitated by the external Air . . ., as soon as I had blown any of them, I cover'd it with a clear Glass, and by that means its Colours emerged in a very regular order, like so many concentrated Rings encompassing the top of the Bubble. And as the Bubble grew thinner by the continual subsiding of the Water, these Rings dilated slowly and overspread the whole Bubble, descending in order to the bottom of it, where they vanish'd successively. In the mean while, after all the Colours were emerged at the top, there grew in the center of the Rings a small round Black Spot . . ..

"Besides the aforesaid colour'd Rings, there would often appear small Spots of Colours, ascending and descending up and down the sides of the Bubble, by reason of some Inequalities in the subsiding Water. And sometimes small Black Spots generated at the sides would ascend up to the larger Black Spot at the top of the Bubble and unite with it."

Sir Isaac Newton (1704)

CHAPTER 12

Thin Liquid Film Stability

The stabilization of thin liquid films is a phenomenon of central importance to the film rupture between two coalescing bubbles (Vrij 1966, Ivanov & Dimitrov 1974, Sche & Fijnaut 1978) or droplets (Vrij *et al.* 1970, Gumerman & Homsy 1975, Jain *et al.* 1978, Wendel *et al.* 1981, Maldarelli *et al.* 1980), the rupture of a film formed between a bubble and solid particle (Jain & Ruckenstein 1974, 1976, Williams & Davis 1982) and the removal of a film of oil from a solid surface (Ivanov & Jain 1979, Jain & Ivanov 1980). Thin film stability has also been investigated in the context of biological cell membrane models, furnishing theoretical insight to biological processes ranging from the fusion of cells (Nicolson 1976), "flicker" phenomena of red blood cells (Burton *et al.* 1968, Brochard & Lennon 1975), the onset of microvilli in normal and neoplastic cells (Porter *et al.* 1973, Nicolson 1974, Willingham & Pastan 1975), and the engulfment of viruses within a cell (Van Oss 1978).

Most current analyses of thin film stability explicitly avoid consideration of interfacial rheological properties by invoking the assumption that film surfaces are either completely mobile or immobile, thereby focusing attention upon the limiting roles which dynamic interfacial properties play in film stabilization. As the calculated dispersion profiles for these two limiting cases differ significantly, dynamic interfacial properties have often been revealed to possess a large (if implicit) significance to the film stability process. This influence is explicitly demonstrated in the present chapter for the particular case of a thin film lying upon a solid surface. The analysis follows a brief discussion of the body force model for including long-range intermolecular forces in the hydrodynamic stability analysis.

12.1 The Body Force Model

Previous analyses of thin film stability (Felderhof 1968, Sche & Fijnaut 1978, Maldarelli & Jain 1982a, 1982b) have solved coupled continuity, Maxwell, and Euler or Navier-Stokes equations, in an attempt to describe the response of a thin liquid film (of finite thickness) within which is dissolved nonionic and/or ionic species. Dispersion and electrostatic forces are generally accounted for by means of a *body force* approach, whereby intermolecular interactions are expressed in terms of a vector body force density sensibly acting only within the thin film region. This approach, which offers an alternative to the disjoining pressure approach outlined in §11.2, provides the following two relative advantages: (i) the body force model may be applied to a nonplanar film, such as one through which waves propagate with 'short' wavelength; (ii) the body force model is easily adapted to nonsymmetrical film geometries. The body force model assigns intermolecular forces in the thin film fluid to a body force acting within the film, rather than to the film pressure [whence it appears explicitly to act at the bounding film surfaces, as in the disjoining pressure model: cf. Eqs. (11.2-7)-(11.2-9)]. The possibility of nonspherical equilibrium bubble surfaces bounding the film is allowed for by the fact that the body force (and, hence, the hydrostatic pressure) may vary in a transverse direction from film to meniscus regions [see Eq. (12.1-2) below].*

According to the body force model (Maldarelli *et al.* 1980), a body force density vector $\mathbf{F}_f$ exists in a thin liquid film owing to short- and long-range intermolecular (i.e. van der Waal, electrostatic, and structural) forces: this force vector may be defined in terms of an interaction potential Φ_f as

$$\mathbf{F}_f = -\mathbf{n}\frac{d\Phi_f}{dn}. \tag{12.1-1a}$$

Furthermore,

$$\mathbf{F}^{\mathrm{I}}_d = -\mathbf{n}\frac{d\Phi^{\mathrm{I}}_d}{dn} \equiv 0 \tag{12.1-1b}$$

and

$$\mathbf{F}^{\mathrm{II}}_d = -\mathbf{n}\frac{d\Phi^{\mathrm{II}}_d}{dn} \equiv 0 \tag{12.1-1c}$$

* The discussion following Eq. (4.2-20) at the conclusion of §4.2 identifies the relevance of the normal stress boundary condition at a fluid interface in terms of interfacial shape definition. Briefly, this condition reduces in hydrostatic equilibrium (and in the absence of an external surface-excess body force) to the Laplace condition (2.1-3). Nonspherical shapes for closed, bounded, simply-connected surfaces are permitted if the hydrostatic pressure is spatially inhomogeneous; this condition is possible in the body force model. In the disjoining pressure model, as shown in §11.2, the normal stress condition is modified by the existence of a surface-excess normal force (i.e. disjoining pressure). This latter modification is responsible in the disjoining pressure model for allowing the existence of a nonspherical droplet shape in equilibrium.

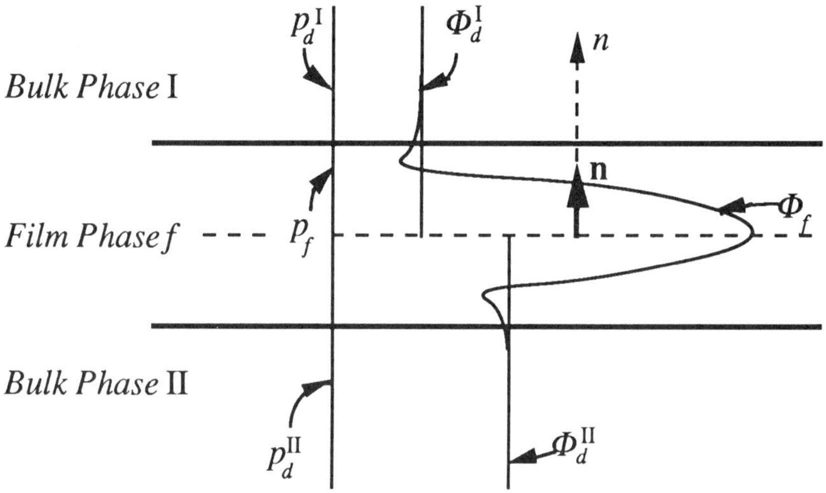

Figure 12.1-1 Pressure and body-force interaction-potential distributions in the 'body-force' model of a plane-parallel, hydrostatic liquid film.

respectively denote the absence of ('internal') body forces within the contiguous bulk phases I and II.* Here, **n** is a unit vector (tangent to the coordinate n) normal to the thin film surfaces, and Φ_d^{I} and Φ_d^{II} are the respective interaction potentials in phases I and II.

In equilibrium, the preceding body forces are counterbalanced by pressure gradients, according to the following hydrostatic equations [cf. Eqs. (4.1-4) and (4.1-13)]:

$$\nabla p_d^{\mathrm{I}} = \mathbf{F}_d^{\mathrm{I}},$$

$$\nabla p_d^{\mathrm{II}} = \mathbf{F}_d^{\mathrm{II}},$$

and

$$\nabla p_f = \mathbf{F}_f,$$

with p_d^{I}, p_d^{II}, and p_f the respective thermodynamic pressures within the bulk phases (I and II), and the film phase. It follows that the total equilibrium potentials $(\Phi + p)$ in both film and bulk phases are spatially constant:

$$p_f + \Phi_f = \text{const.}, \qquad (12.1\text{-}2)$$

$$p_d^{\mathrm{I}} + \Phi_d^{\mathrm{I}} = \text{const.}, \qquad (12.1\text{-}3)$$

---

* In general, there will be also tangential components of this body force; these will be of particular significance near the transition to the Plateau border region (i.e. at the periphery of the thin film).

and

$$p_d^{\mathrm{II}} + \Phi_d^{\mathrm{II}} = \text{const.} \tag{12.1-4}$$

Insofar as the normal stress condition (4.2-20a) requires the continuity of pressure across a planar surface in equilibrium, for the assumed plane-parallel film geometry of Figure 12.1-1, it follows that [cf. also Eq. (11.2-2)]

$$p_f = p_d^{\mathrm{I}} \quad \text{at } n = h, \tag{12.1-5 a}$$

and

$$p_f = p_d^{\mathrm{II}} \quad \text{at } n = -h. \tag{12.1-5 b}$$

Equations (12.1-5) must be reconciled with the observance of a disjoining pressure Π in a thin film, as in Eq. (11.2-6). Thus, upon letting

$$p_f + \Phi_f - p_d^{\mathrm{I}} - \Phi_d^{\mathrm{I}} = \Pi \quad \text{at } n = h$$

and

$$p_f + \Phi_f - p_d^{\mathrm{II}} - \Phi_d^{\mathrm{II}} = \Pi \quad \text{at } n = -h$$

it follows straightforwardly from Eqs. (12.1-2)–(12.1-5) that

$$\Phi_{\mathrm{f}}(h) - \Phi_d^{\mathrm{I}} = \Pi,$$

$$\Phi_{\mathrm{f}}(-h) - \Phi_d^{\mathrm{II}} = \Pi,$$

whence, from Eqs. (12.1-1), there obtains the following relation between film potential and disjoining pressure:

$$\boxed{\left.\frac{d\Phi_{\mathrm{f}}}{dn}\right|_{n=h} = \frac{d\Pi}{dh}.} \tag{12.1-6 a}$$

Moreover, upon noting the sense of the unit normal $\mathbf{n}$ as representing the *outward* normal to the lower surface and the *inward* normal to the upper surface, a corresponding relation to (12.1-6a) may be provided at the lower film surface, by

$$\left.\frac{d\Phi_{\mathrm{f}}}{dn}\right|_{n=-h} = -\frac{d\Pi}{dh}. \tag{12.1-6 b}$$

Using Eqs. (12.1-6) the thin film body force acting at the film surfaces (characteristically, only this 'surface contribution' will be retained in the results of a lubrication analysis of thin film hydrodynamics) may be expressed explicitly by employing 'DLVO' theory (Derjaguin & Landau 1941, Verwey & Overbeek 1948) to describe the disjoining pressure as a function of film thickness. Briefly, classical DLVO theory separates the disjoining pressure Π into two distinctive contributions: namely,

$$\Pi(h) \equiv \Pi^{vw} + \Pi^{e}, \tag{12.1-7}$$

where Π^{vw} is the van der Waals contribution, which contribution (in the simplest 'Hamaker formulation') may be expressed as

$$\Pi^{vw} = A / 6\pi h^3, \tag{12.1-8}$$

and Π^e is the electrostatic contribution. The latter electrostatic contribution to the disjoining pressure may be approximated (in the case of a single electrolytic component) as (Maldarelli *et al.* 1980)

$$\Pi^{e} = -\Phi_o^{DL} e^{-h\chi}, \tag{12.1-9}$$

$$\chi = \left[\frac{8\pi z^2 F^2 C_{el}}{\varepsilon_D RT}\right]^{1/2}, \tag{12.1-10}$$

$$\Phi_o^{DL} = 64 C_{el} RT \tanh\left[\frac{zF\phi_{\mathrm{I}f}}{4RT}\right]\tanh\left[\frac{zF\phi_{f\,\mathrm{II}}}{4RT}\right]. \tag{12.1-11}$$

In the above, A is the *Hamaker constant*, corresponding to the interaction of bulk-phase molecules with film molecules, χ^{-} the Debye length, C_{el} the concentration of electrolyte, ε_D the dielectric constant, R the gas constant, T the temperature, z the valence of the electrolyte, $\phi_{\mathrm{I}f}$ and $\phi_{f\mathrm{II}}$ interphase surface potentials, and F the Faraday constant.

The body force method was first employed by Felderhof (1968) in his investigation of the stability of an inviscid, plane-parallel film between two semi-infinite liquids. He found that the resultant wave motion may be described as a superposition of two fundamental modes of vibration—a symmetric, or "squeezing" mode, and an antisymmetric, or "stretching" mode. These two modes of vibration, by which the upper and lower film surfaces are respectively in-phase or out-of-phase, are depicted in Figure 12.1-2. The symmetric mode of vibration is the one normally associated with film rupture, as in this mode instabilities lead to film thinning. Later publications (e.g. Sche & Fijnaut 1978) have included the effects of viscous forces in the Felderhof (1968) analysis. Antisymmetric mode instabilities, on the other hand, which apparently occur only for sufficiently low tension film surfaces, are presently thought to be of particular importance to biological cell membranes (Maldarelli & Jain 1982a, 1982b, Jain & Maldarelli 1983) for which applications surface chemical reactions (Blumenthal *et al.* 1970, Sanfeld & Steinchen 1975, Sorensen *et al.* 1976, Velarde *et al.* 1977, Ibanes & Velarde 1977, Steinchen & Sanfeld 1977, Dagan & Pismen 1984) surface adsorption and diffusion, and surface viscosities (Sanfeld *et al.* 1979) may each acquire a large significance. (Two modes of film vibration are observed only for films formed between two liquid phases. Clearly, only a single mode of vibration can exist within a film of liquid formed between a solid and fluid phase.)

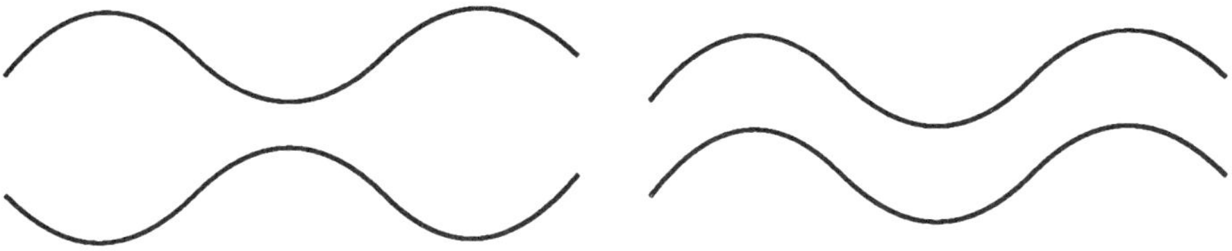

Figure 12.1-2 The vibrational modes of a thin liquid film.

As in the linear interfacial stability theory of Chapter 10, thin film stability analyses generally permit small dynamic fluctuations to occur in the equilibrium thin film geometry. These engender fluctuations in thin film fields such as velocity and pressure. The sign of the growth coefficient β (cf. §10.1) of these fluctuations may be investigated to determine the stability of the system. As noted in Chapter 10, in physical circumstances for which the growth coefficient is β is negative, the disturbances are predicted to damp: if β is positive the disturbances are predicted to grow, and the system is regarded as being unstable. Marginal stability is defined by the condition of zero growth coefficient.

The stability of a thin liquid film will generally (if not always) depend upon the wavelength of the applied disturbance. Thus, there will typically arise a particular 'critical' wavelength, k_c, at which marginal stability occurs, as well as a 'dominant' wavelength, k_d, corresponding to the largest growth coefficient, β_d. The 'time constant of rupture', τ_r, which provides an approximate measure of the time required for a small disturbance to destabilize the thin film configuration, is furnished in a linear stability analysis by the expression

$$\tau_r \equiv \frac{1}{\beta_d}. \tag{12.1-12}$$

Determination of the growth coefficient as a function of wavelength represents the principal goal of thin film stability studies.

12.2 The Stability of a Thin Film upon a Solid Surface

The stability of a nondraining, charge-neutral liquid film formed upon a planar solid surface is considered in the present section. The original analysis of Jain & Ruckenstein (1976) is followed here in most respects. The reader may also refer to the capillary wave problem of §5.6 for a close analog.

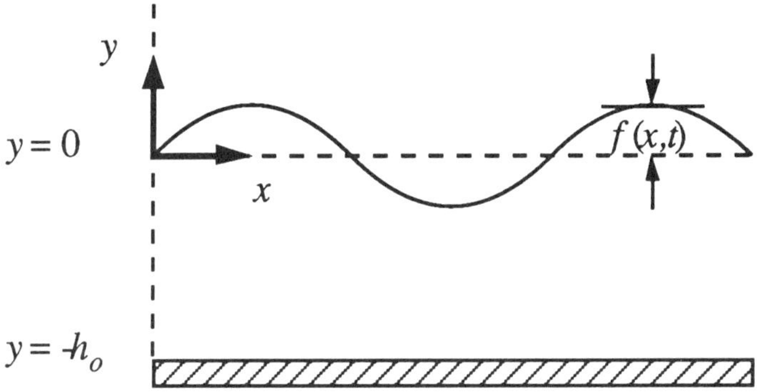

Figure 12.2-1 A capillary wave disturbance propagates along the upper surface of a thin liquid film formed at a solid boundary.

Consider a small perturbation

$$f = f_o \sin(kx + \omega t); \qquad f_o k << 1 \tag{12.2-1}$$

of a free film surface from its equilibrium, planar configuration—parallel to a solid boundary—as depicted in Figure 12.2-1.* This perturbation causes flow within both the film and semi-infinite liquid phases, which flow is assumed to obey the quasistatic linearized ('Stokes'-form of the) Navier-Stokes equations. It therefore follows from Table 4.1-1 [cf. also Eqs. (5.6-19) and (15.6-20)] that in the film phase

$$0 = -\frac{\partial p}{\partial x} - \frac{\partial \Phi_f}{\partial x} + \mu\left(\frac{\partial^2 v_x}{\partial x^2} + \frac{\partial^2 v_x}{\partial y^2}\right), \tag{12.2-2}$$

and

$$0 = -\frac{\partial p}{\partial y} - \frac{\partial \Phi_f}{\partial y} + \mu\left(\frac{\partial^2 v_y}{\partial x^2} + \frac{\partial^2 v_y}{\partial y^2}\right), \tag{12.2-3}$$

with the continuity condition provided by

$$\frac{\partial v_y}{\partial y} + \frac{\partial v_x}{\partial x} = 0. \tag{12.2-4}$$

Similarly, in the continuous phase [upon noting Eqs. (12.1-1b) and (12.1-1c)]

$$0 = -\frac{\partial \tilde{p}}{\partial x} + \tilde{\mu}\left(\frac{\partial^2 \tilde{v}_x}{\partial x^2} + \frac{\partial^2 \tilde{v}_x}{\partial y^2}\right), \tag{12.2-5}$$

---

* Gumerman & Homsy (1975) have extended this analysis by including wall and drainage effects. Williams & Davis (1982) have also shown that the inclusion of nonlinear effects (to a surfactant-free thin liquid film) may be of particular significance to the predicted rupture time, owing to the previously mentioned fact [cf. Eq. (12.1-12)] that this parameter is calculated in the linear theory by assuming wave growth well into the nonlinear regime.

$$0 = -\frac{\partial \tilde{p}}{\partial y} + \tilde{\mu}\left(\frac{\partial^2 \tilde{v}_y}{\partial x^2} + \frac{\partial^2 v_y}{\partial y^2}\right), \tag{12.2-6}$$

and

$$\frac{\partial \tilde{v}_y}{\partial y} + \frac{\partial \tilde{v}_x}{\partial x} = 0. \tag{12.2-7}$$

In the above, an overbar is used to denote continuous-phase variables.

The preceding equations are subject to no-slip conditions at the solid boundary,

$$v_x = v_y = 0 \qquad \text{at} \quad y = -h_o \tag{12.2-8}$$

and the vanishing of velocity condition far from the upper film surface,

$$\tilde{v}_x = \tilde{v}_y = 0 \qquad \text{at} \quad y \to \infty. \tag{12.2-9}$$

At the liquid-liquid interface, the continuity of velocity condition requires that

$$v_x = \tilde{v}_x \qquad \text{at} \qquad y = f, \tag{12.2-10}$$

and

$$v_y = \tilde{v}_y = \frac{\partial f}{\partial t} \qquad \text{at} \qquad y = f. \tag{12.2-11}$$

The normal stress boundary condition (4.2-20a) adopts the form [cf. also Eq. (5.6-24)]

$$p - \tilde{p} + 2\tilde{\mu}\frac{\partial \tilde{v}_y}{\partial y} - 2\mu\frac{\partial v_y}{\partial y} + \sigma\frac{\partial^2 f}{\partial x^2} = 0 \qquad \text{at} \qquad y = f, \tag{12.2-12}$$

and the tangential stress boundary condition (4.2-20b) [cf. Eq. (5.6-23)] yields

$$\mu\left(\frac{\partial v_x}{\partial y} + \frac{\partial v_y}{\partial x}\right) - \tilde{\mu}\left(\frac{\partial \tilde{v}_x}{\partial y} + \frac{\partial \tilde{v}_y}{\partial x}\right) = \frac{\partial \sigma}{\partial x}$$

$$+ (\kappa^s + \mu^s)\frac{\partial^2 v_x}{\partial x^2} \qquad \text{at} \qquad y = f. \tag{12.2-13}$$

The interfacial tension gradient in Eq. (12.2-13) couples this condition with the surfactant surface transport equation (5.2-6), as has been the case in previous chapters [cf. the examples appearing in Chapters 5, 10 and 11]. As in the other linearized formulations, attention is focused upon the particular case of small deviations from the equilibrium state, namely,

$$\rho_1^s = \rho_{1o}^s + \varepsilon\hat{\rho}_1^s \tag{12.2-14}$$

and

$$\rho_1 = \rho_{1o} + \varepsilon\hat{\rho}_1, \tag{12.2-15}$$

with ε a small dimensionless parameter (ε << 1) and "1" referring to the surfactant species. The quasistatic, low-Peclét form of the species transport equations in the contiguous bulk phases reduce from that provided in Table 5.2-1 to the following diffusion equations;

$$0 = D\left(\frac{\partial^2 \rho_1}{\partial x^2} + \frac{\partial^2 \rho_1}{\partial y^2}\right), \tag{12.2-16}$$

$$0 = \tilde{D}\left(\frac{\partial^2 \tilde{\rho}_1}{\partial x^2} + \frac{\partial^2 \tilde{\rho}_1}{\partial y^2}\right), \tag{12.2-17}$$

which are subject to the conditions

$$\frac{\partial}{\partial y}\rho_1 = 0 \qquad \text{at} \quad y = -\,h_o \tag{12.2-18}$$

$$\frac{\partial}{\partial y}\tilde{\rho}_1 = 0 \qquad \text{as} \quad y \to \infty \tag{12.2-19}$$

and [cf. Eqs. (5.2-6) and (5.6-46)]

$$\frac{\partial \rho_1^s}{\partial t} + \rho_{1o}^s\frac{\partial v_x}{\partial x} - D^s\frac{\partial \rho_1^s}{\partial x} = -\,D\frac{\partial \rho_1}{\partial y} + \tilde{D}\frac{\partial \tilde{\rho}_1}{\partial y} \qquad \text{at} \quad y = f\,. \tag{12.2-20}$$

An additional relation is required between the surface-excess species mass density and bulk-phase species densities. As was found to be the case in the thin film problem of Chapter 11, surfactant transport within the film will be diffusion-controlled (cf. §5.5), whence we may assume here equilibrium to exist at the interface. In particular, we assume a linear (i.e. Henry's law) isotherm; thus, with the aid of Eqs. (12.2-14) and (12.2-15), there obtains to leading order

$$\rho_1^s = k_a\rho_1 = \tilde{k}_a\tilde{\rho}_1, \tag{12.2-21}$$

with k_a and $\tilde{k}_a$ denoting linear adsorption coefficients.

The problem that has been defined by Eqs. (12.2-1)–(12.2-23) [with also Eq. (5.5-4)] may be solved by seeking solutions in the complex plane (whose imaginary part corresponds to the physical field quantity, with $\beta = i\omega$), of the general form

$$\begin{bmatrix} f^* \\ v_x^* \\ v_y^* \\ \tilde{v}_x^* \\ \tilde{v}_y^* \\ \mathcal{P}^* \\ \tilde{p}^* \\ \rho_1^* \\ \tilde{\rho}_1^* \\ \rho_1^{s*} \end{bmatrix} = \begin{bmatrix} f_0 \\ v_x'(y) \\ v_y'(y) \\ \tilde{v}_x'(y) \\ \tilde{v}_y'(y) \\ p'(y) + \Phi'(y) \\ \tilde{p}'(y) \\ \rho_1'(y) \\ \tilde{\rho}_1'(y) \\ \rho_1^{s\prime} \end{bmatrix} \exp[ikx + \beta t]. \qquad (12.2\text{-}23)$$

Explicitly, the following y-component complex velocity solutions obtain

$$v_y = [\cosh(ky) + C_1 \sinh(k_y) + C_2 ky \sinh(ky) + C_3 ky \cosh(ky)] e^{ikx + \beta t} \qquad (12.2\text{-}24)$$

in the film phase, and

$$\tilde{v}_y = \left[C_4 e^{-ky} + C_5 ky\, e^{-ky} + C_6 e^{ky} + C_7 ky\, e^{ky}\right] e^{ikx + \beta t}, \qquad (12.2\text{-}25)$$

in the semi-infinite phase. The remaining complex fields represented in Eq. (12.2-23) are derived in Question 12.1 and 12.2.

Substitution of Eqs. (12.2-23) into the appropriate conservation and boundary conditions ultimately furnishes a set of 8 algebraic equations in eight unknowns; explicitly, in matrix form,

$$\mathbf{A} \cdot \mathbf{C} = \mathbf{b} \qquad (12.2\text{-}26)$$

where

$$\mathbf{C} = \begin{bmatrix} C_1 \\ C_2 \\ C_3 \\ C_4 \\ C_5 \\ C_6 \\ C_7 \\ \beta^{-1} \end{bmatrix} \qquad \mathbf{b} = \begin{bmatrix} -\tilde{\mu}/\mu \\ -1 \\ 1 \\ 1 \\ \cosh(kh_o) \\ \sinh(kh_o) \\ 0 \\ 0 \end{bmatrix} \qquad (12.2\text{-}27)$$

and

$$\mathbf{A} = \begin{bmatrix} 1 & 0 & 0 & 0 & 0 & 0 & 0 & a_{18} \\ a_{21} & 1 & a_{23} & 0 & 0 & 0 & 0 & 0 \\ 0 & 0 & 0 & 1 & 0 & 0 & 0 & 0 \\ -1 & 0 & -1 & 0 & 1 & 0 & 0 & 0 \\ a_{51} & a_{52} & a_{53} & 0 & 0 & 0 & 0 & 0 \\ a_{61} & a_{62} & a_{63} & 0 & 0 & 0 & 0 & 0 \\ 0 & 0 & 0 & 0 & 0 & 1 & 0 & 0 \\ 0 & 0 & 0 & 0 & 0 & 0 & 1 & 0 \end{bmatrix}. \tag{12.2-28}$$

The matrix coefficients appearing in the above expression (12.2-28) are supplied by the expressions:

$$\begin{aligned} a_{18} &= k\sigma_{app} / 2\mu, \\ a_{21} &= a_{23} = \tilde{\mu} / \mu + (\mathrm{Bo} + \mathrm{Ma}) / 2, \\ a_{51} &= \sinh(kh_o), \\ a_{52} &= -\, kh_o \sinh(kh_o), \\ a_{53} &= (kh_o)\cosh(kh_o), \\ a_{61} &= \cosh(kh_o), \\ a_{62} &= \sinh(kh_o) + kh_o \cosh(kh_o), \\ a_{63} &= \cosh(kh_o) + kh_o \sinh(kh_o). \end{aligned} \tag{12.2-29}$$

The Boussinesq and Marangoni numbers have been defined in the above as

$$\mathrm{Bo} \equiv \frac{(\mu^s + \kappa^s)k}{\mu}, \tag{12.2-30 a}$$

$$\mathrm{Ma} \equiv \frac{kE_o}{\mu\left\{\left(D^s k^2 + \beta\right) + \dfrac{\tilde{D}^2}{\tilde{k}_a}\left(\tilde{D}k^2 + \beta\right) + \dfrac{\sqrt{D}}{k_a}\left(Dk^2 + \beta\right)\tanh\left(h_o\sqrt{\dfrac{Dk^2 + \beta}{D}}\right)\right\}}, \tag{12.2-30 b}$$

with E_o is the Gibbs elasticity [cf. Eq. (5.5-5)]. The Marangoni contribution simplifies considerably in either limit $k_a \to \infty$ (i.e. surfactant insoluble in the film phase) or $\tilde{k}_a \to \infty$ (i.e. surfactant insoluble in the continuous phase). The 'apparent interfacial tension' σ_{app} is defined so as to include the film normal body force $\mathbf{F}_f(h)$ acting at the upper film surface [cf. Eq. (12.1-1)]; explicitly,

$$\sigma_{app} \overset{\text{def}}{=} \sigma_o + \frac{1}{k^2}\left(\frac{\partial \Phi}{\partial h}\right)_{h = h_o} . \tag{12.2-31a}$$

With Eq. (12.1-6a), this relation may be expressed in terms of the disjoining pressure Π as

$$\sigma_{app} = \sigma_o + \frac{1}{k^2}\left(\frac{\partial \Pi}{\partial h}\right). \tag{12.2-31b}$$

This latter relation reveals that the apparent interfacial tension σ_{app} increases when the upper film surface is *repelled* by the film fluid (or lower solid surface). This is immediately seen by substituting Eq. (12.1-8) into Eq. (12.2-31b) (note that the assumption of a charge-neutral film eliminates the electrostatic contribution Π^{el}), to obtain

$$\boxed{\sigma_{app} = \sigma_o - \frac{1}{k^2}\left(\frac{A}{2\pi h^4}\right).} \tag{12.2-31c}$$

Thus, in the case where $A < 0$, corresponding to repulsion of the upper film surface by the film fluid or solid surface, σ_{app} is larger than the equilibrium interfacial tension σ_o.

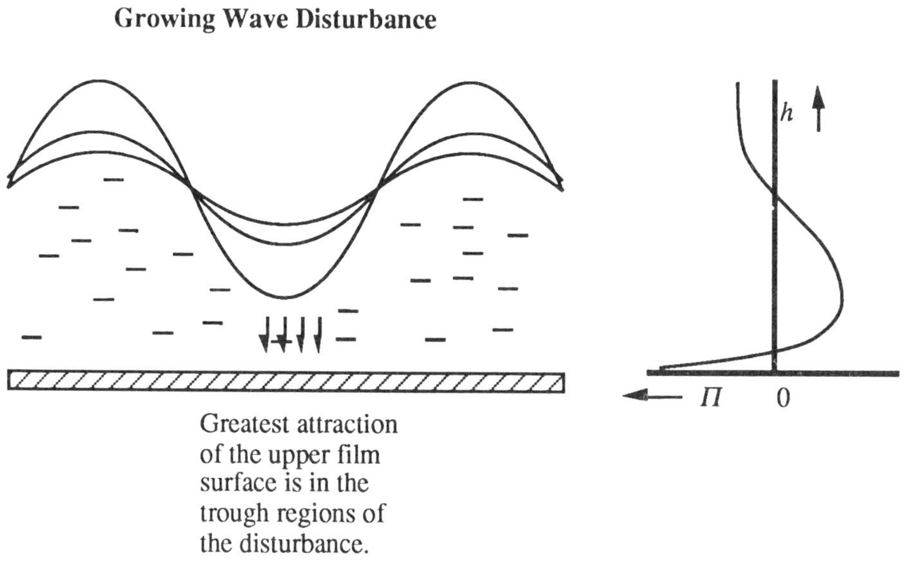
Growing Wave Disturbance
h
Π
0
Greatest attraction of the upper film surface is in the trough regions of the disturbance.

Figure 12.2-2 An unstable thin film on a solid surface. A net attractive force between the upper film surface and the film fluid causes wave disturbances to grow.

The dispersion equation governing the relation between growth coefficient β and wave number k may be obtained in the standard manner by requiring the determinant of the coefficient matrix **A**, defined by Eq. (12.2-

28), to vanish. Explicit analytical relations may be sought in the limits of a completely mobile film surface, Bo + Ma → 0—as corresponds to the absence of surfactant, and a completely immobile surface,Bo + Ma → ∞, suggesting very large interfacial-tension gradient and/or interfacial viscous stresses.

In the case of a mobile film surface, upon assuming the upper phase to be a gas, there obtains for the growth coefficient

$$\beta = -\left(\frac{\sigma_{app}}{2\mu h_o}\right) \times \left[\frac{kh_o \sinh(kh_o)\cosh(kh_o) - k^2 h_o^2}{\cosh^2(kh_o) + k^2 h_o^2}\right]. \quad (12.2\text{-}32)$$

The growth coefficient β is negative (the film being therefore stable to small disturbances) when the apparent surface tension σ_{app} is positive. As discussed previously, this corresponds to *repulsion* of the upper film surface by the film liquid or lower solid surface. Figure 12.2-2 offers an explanation of this behavior. If the net force acting upon the upper thin film surface is attractive toward the lower solid surface, the trough regions of the wave disturbance will experience a greater attraction toward the solid surface than will the crest regions of the wave. The amplitude of the wave will therefore grow and the film will ultimately be made unstable.

Marginal stability occurs when the growth coefficient is zero; as follows from Eq. (12.2-32), this is equivalent to the requirement that σ_{app} =0. Upon utilizing Eq. (12.2-31b), this condition shows that the critical wave number is given by

$$k_c = \sqrt{-\frac{1}{\sigma_o}\left(\frac{\partial \Pi}{\partial h}\right)}. \quad (12.2\text{-}33)$$

With Eq. (12.1-8), the preceding equation may be rewritten as

$$k_c = \frac{1}{h_o^2}\sqrt{A / 2\pi\sigma_o}\,. \quad (12.2\text{-}34)$$

When the film is unstable (as will occur when both A >0 and $k > k_c$ hold simultaneously) a particular wavenumber will dominate all others. The value of this wavenumber may be determined by requiring $d\beta/dk$ =0, which condition is consistent with the circumstance of maximum growth coefficient. According to Eq. (12.2-32), in the limit $kf_o << 0$ (for which the present analysis applies), the dominant wave number satisfying $d\beta/dk$ =0 is provided by

$$k_d \approx k_c / \sqrt{2} \quad (12.2\text{-}35)$$

corresponding to the following dominant growth coefficient;

$$\beta_d \approx \frac{h_o^3}{12\mu\sigma_o}\left(\frac{\partial \Pi}{\partial h}\right)^2. \qquad (12.2\text{-}36)$$

An estimate of the rupture time (12.1-12) of the film may be obtained by assuming the wavenumber k_d dominates all others; explicitly,

$$\tau_r \approx 48\pi^2\sigma_o\,\mu h_o^5 A^{-2}. \qquad (12.2\text{-}37)$$

Thus, large interfacial tension, film viscosity and/or film thickness each contribute to a large rupture time.

Interfacial-tension gradients and interfacial viscosities are also stabilizing. This may be clearly seen by considering the second analytical limit, whereby the upper film surface is completely immobile (i.e. Bo + Ma $\to\infty$). The growth coefficient in this case [corresponding to Eq. (12.2-32) for the case of a mobile film surface] is given by the expression (Jain & Ruckenstein 1976)

$$\beta = -\left(\frac{kh_o\sigma_{app}}{2\mu h_o}\right)\times\left[\frac{\sinh^2(kh_o) - k^2h_o^2}{\cosh(kh_o)\sinh(kh_o) + kh_o}\right]. \qquad (12.2\text{-}38)$$

The critical and dominant wave numbers are clearly identical to Eqs. (12.2-34) and (12.2-35). The dominant growth coefficient is found to be exactly four times smaller than for the case of a completely mobile surface [cf. Eq. (12.2-36)]; namely,

$$\beta_d \approx \frac{h_o^3}{48\mu\sigma_o}\left(\frac{\partial \Pi}{\partial h}\right)^2, \qquad (12.2\text{-}39)$$

whence also

$$\tau_r \approx 192\pi^2\sigma_o\,\mu h_o^5 A^{-2}. \qquad (12.2\text{-}40)$$

These results illustrate the potential film stabilizing effect of interfacial rheological properties.

Equations (12.2-37) and (12.2-40) represent limiting-case results for the rupture time of completely mobile and immobile film surfaces respectively. Figure 12.2-3 shows numerical results for values of the rupture time at an intermediate interfacial mobility as well. This figure may be seen to provide a close analog to Figure 11.3-4. In both cases of thin film drainage and stability, the surface dilatational elasticity will commonly contribute the largest stabilizing influence. Thus, according to typical experimental values of the dilatational elasticity (cf. Figures 6.2-3 and 6.2-4), the primary 'solidifying' of the upper thin film surface will typically result from interfacial-tension gradients, as has been discussed also in the context of Figure 11.3-4.

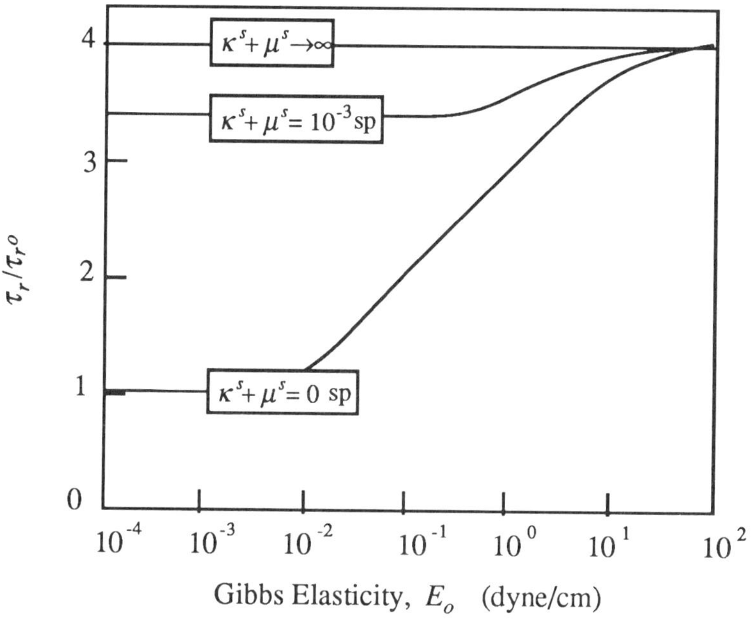

Figure 12.2-3 Dimensionless "rupture time" [expressed as the characteristic rupture time τ_r divided by the characteristic rupture time for a surfactant-free (mobile) film, the latter denoted in the figure as $\tau_r{}^o$] versus the Gibbs elasticity E_o at three interfacial viscosity values. Theoretical data are from Jain & Ruckenstein (1976).

12.3 Summary

Thin liquid films (<100 nm) formed between two fluid phases, or, equivalently, a solid and a fluid phase, may become unstable to small disturbances owing to net attractive (dispersion) forces within the film of fluid. Several factors may influence the stability of the thin film, including film viscosity, film thickness, and interfacial tension—each of whose magnitude proportionally increases the stability of the film.

Interfacial rheological properties may considerably stabilize a thin film by imparting a rigidity to liquid film surfaces. The difference between estimated rupture times for films with mobile surfaces (i.e. no surfactant, and therefore no interfacial rheological stresses) and immobile surfaces (i.e. very large interfacial rheological stresses, leading to a solid-like behavior) may be severalfold.

Questions for Chapter 12

12.1 Find explicit expressions for the complex field quantities $(\mathcal{P}^*, \tilde{p}^*, v_x^*, \tilde{v}_x^*)$ represented in Eq. (12.2-23) in terms of constants C_1–C_7 upon employing the expressions for the variables $(v_y{}^*, \tilde{v}_y{}^*)$ provided in Eqs. (12.2-24) and (12.2-25).

12.2 *i.* Substitute the solutions obtained in Question 12.1 into the conditions (12.2-8)–(12.2-13) to obtain the last six rows of the **A** matrix in Eq. (12.2-28).

ii. Obtain expressions for the complex surfactant densities $(\rho_1^{s\,*}, \rho_1{}^*, \tilde{\rho}_1{}^*)$ represented in Eq. (12.2-23) by employing Eqs. (12.2-16)–(12.2-20) together with the results of Question 12.1. Explain how these results may permit determination of the first two rows of the **A** matrix (12.2-28).

12.3 Suppose that the rate of surfactant adsorption to the liquid film surface in the film stability problem of §12.2 is the controlling step of the surfactant mass transfer process. Utilize the solutions of Question 12.1 and part *i.* of Question 12.2 with Eqs. (5.3-3), (5.3-6) and (5.3-7), with Eqs. (5.5-4) and (5.5-5), to determine the new expression for the matrix equation (12.2-26), in the following cases: *i.* surfactant is insoluble in the dispersed phase; *ii.* surfactant is insoluble in the film phase.

iii. Determine the growth coefficient β and dominant growth coefficient β_d for the case of a mobile film interface [Eqs. (12.2-32) and (12.2-36)] for both cases *i.* and *ii.*

iv. Determine the growth coefficient β and dominant growth coefficient β_d for the case of an immobile film interface [Eqs. (12.2-38) and (12.2-39)] for both cases *i.* and *ii.*

12.4 In Question 12.3, which case favors film stability: *i.* surfactant insoluble in the film phase; or *ii.* surfactant insoluble in the dispersed phase? Is this consistent with Banckroft's rule (see Figure 11.3-7 and the discussion thereof)?

Additional Reading for Chapter 12

§12.1 The Body Force Mode

Maldarelli, C., Jain, R.K., Ivanov, I.B., & Ruckenstein, E. 1980 Stability of symmetric and unsymmetric thin liquid films to short and long wavelength perturbations. *J. Colloid Interface Sci.* **78**, 118–143.

§12.2 The Stability of a Thin Film upon a Solid Surface

Krantz, W.B. & Zollars, R.L. 1976 The linear hydrodynamic stability of film flow down a vertical cylinder. *AIChE J.* **22**, 930–934.

Yih, S.M. & Seagrave, R.C. 1978 Hydrodynamic stability of thin liquid films flowing down an inclined plane with accompanying heat transfer and interfacial shear. *AIChE J.* **24**, 803–810.

Williams, M.B. & Davis, S.H. 1982 Nonlinear theory of film rupture. *J. Colloid Interface Sci.* **90**, 220–228.

13

"During the ten-year period from 1890–1990, the late Lord Rayleigh carried on many investigations of the phenomena of surface tension and the spreading of oil films on water. He found, however, that the physicists in general were not much interested in these phenomena, so that gradually he went on to other lines of investigations; but I think in these days nearly everyone is more deeply interested in the phenomenon of surface tension than they were at that time. . . .

"I can show you some of the effects of these films made with what, as a boy, we used to call camphor boats. These boats were propelled over the surface of the water by motion set up in water by pieces of camphor, which alter the surface tension.

"I will put some water in this tray and then put on the water a few little pieces of camphor. . . . These little pieces of camphor are in a constant state of motion. . . . I will now take a strip of paper and cut out of it a little boat . . .. Now you see, I have made my boat and I have a little spot in which I can put a piece of camphor. I pick up one of those pieces of camphor and put it in the little recess at the back of the boat . . . the boat immediately starts to move . . .. The reason for this movement is a change in the surface tension of water. There is a decrease in tension right back of the boat, so the water pulls the boat ahead and then it pulls it back."

Irving Langmuir (1931)

CHAPTER 13

Emulsion and Foam Stability

The long-term stability (or instability) of a foam and/or emulsion is centrally important to many industrial processes; included therein are froth flotation (Scheludko 1963, 1967), demulsification (Jones *et al.* 1957, Jones *et al.* 1978), distillation (Berg 1988), road construction (Gaestl 1967, Lane & Ottewill 1976, Scott 1976), and enhanced oil recovery (Geffen 1973, Wasan *et al.* 1979, Clint *et al.* 1981). In addition, a large number of industrial, biological, and household applications (Perri 1953, Ross 1967, Minssaiux 1974, Roberts *et al.* 1977) require an emulsion or foam which is endowed with the quality of long-term stability.

The theoretical role of dynamic interfacial properties in foam and emulsion stability has been elucidated in Chapters 11 and 12. Figures 11.3-2 through 11.3-7, as well as Figure 12.2-3, serve to illustrate most of the theoretical conclusions therein pertaining to the stabilizing influence of interfacial-tension gradients and interfacial viscosities in the evolution of thin liquid films (the latter being created by the flocculation of droplets in an emulsion, or through the intimate proximity of bubbles in a foam).

The present chapter supplements the previous theoretical considerations by providing a sampling of experimental evidence currently available for relating the stability of emulsions and foams to interfacial rheological properties. In particular, correlations are reviewed between measured values of interfacial viscosities and/or dilatational elasticities and the stability of foams and emulsions.

Insofar as the phenomenological behavior (e.g. versus surfactant composition or salinity) of interfacial elasticities and interfacial viscosities will generally not match that of static interfacial tension (cf. Figure 13.1-2), the frequent industrial need to identify surfactants capable of stabilizing or destabilizing liquid dispersions through indirect knowledge of bulk-phase and interfacial properties would appear best satisfied by including in the list of pertinent physical properties (in addition to interfacial tension) interfacial viscosity and/or interfacial elasticity values. This point will be emphasized in the first two sections of this chapter.

The most commonly appearing interfacial rheological property in the experimental data reviewed herein is the interfacial shear viscosity μ^s. This should not be construed as implying that the surface parameter μ^s—more than the Gibbs elasticity, E_o, say, or the interfacial dilatational viscosity κ^s—represents the most significant dynamic interfacial property in the stabilization process of liquid dispersions. Rather, the relative ease with which μ^s may be measured (cf. Chapter 7) together with reasonable past success in relating the magnitude of the interfacial shear viscosity with the stability of foams and/or emulsions, has resulted in what may be an overemphasis of its role in the stabilization process. Indeed, the conclusions of both Chapters 11 and 12 (cf. Figures 11.3-4 and 12.3-2 and the discussions thereof) suggest that the Gibbs elasticity, E_o [cf. Eq. (5.5-5)] may be the more important dynamic interfacial property.

One application of emulsion/foam stability that has achieved particular importance in recent years is the enhanced recovery of oil. In §13.3, special attention is devoted to a qualitative description of the role of interfacial rheological properties in enhanced oil recovery (i.e. so-called 'EOR') processes.

13.1 Experimental Studies of Emulsion Stability

An emulsion becomes unstable (regarded as the *breaking* of the emulsion) through a two-stage process known as *coagulation*. The first stage of this process consists of the *flocculation* of single (monomeric) droplets into *aggregates*, the latter comprising several droplets in intimate contact. In this stage individual droplets retain their integrity within the aggregate (whereupon they may possibly return to a nonflocculated state). Once flocculated, individual droplets may commence to coalesce through the drainage and instability of the thin liquid films separating each droplet in the flocculate. This second stage of the coagulation process represents the one most profoundly influenced by interfacial rheological properties.

The classical phenomenological theory of emulsion stability may be attributed to Smoluchowski (1916). He proposed the following relationship between the total number of droplets N (including droplets in both the monomeric and aggregate form) at an instant of time t within a sample volume of emulsion, and the initial number of droplets, N_o, in the volume:

$$N = \frac{N_o}{1 + aN_o t}. \qquad (13.1\text{-}1)$$

The above relation effectively assumes instantaneous coalescence; thus, the coefficient a may be viewed simultaneously as 'aggregation' and 'coagulation' rate constant. A generalization of the Smoluchowski relationship for finite coalescence rate is provided by the relation (Becher 1965)

$$N = \frac{N_o}{1 + aN_o t} + \frac{aN_o^2 t}{(1 + aN_o t)^2}\left\{\frac{aN_o}{K} + \left(1 - \frac{aN_o}{K}\right)e^{-Kt}\right\}, \quad (13.1-2)$$

with K the rate of coalescence. This equation properly reduces to Eq. (13.1-1) in the limit as $K \rightarrow \infty$.

The aggregation rate constant, a, is generally very large for highly concentrated emulsions, for which droplets ultimately attain a collective state of aggregation. In this limit, upon letting $a \rightarrow \infty$, Eq. (13.1-2) reduces to the following exponential decay law:

$$N = \frac{N_o}{Kt}\left(1 - e^{-Kt}\right), \quad (13.1-3)$$

as first suggested for highly concentrated emulsions by van den Tempel (1957).

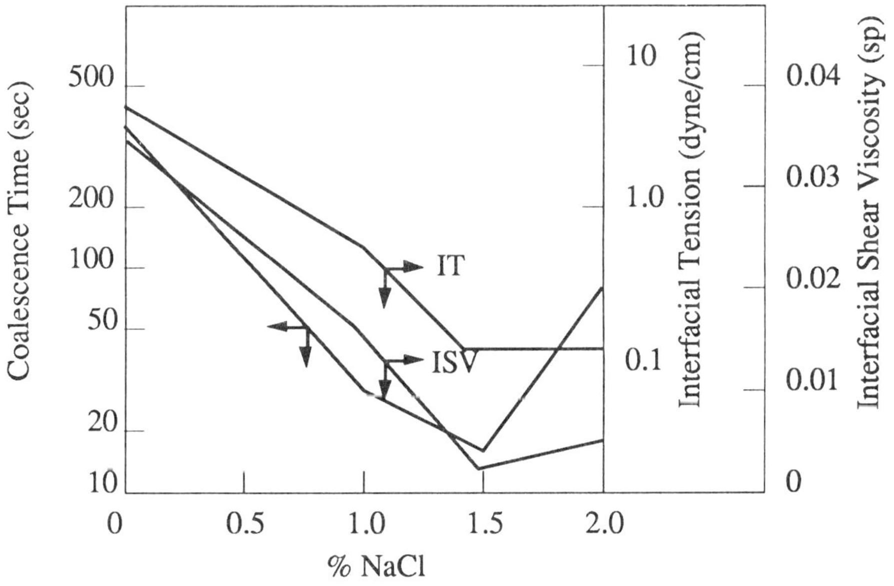

Figure 13.1-1 Coalescence (including drainage and rupture) time vs interfacial tension and interfacial shear viscosity as a function of %NaCl for the water/Salem crude oil interface with 0.5% Petrostep 420. Data are taken from Wasan *et al.* (1979).

In the current section, emulsion stability is characterized in the following three ways: (i) coalescence time (representing the rate of coalescence of two flocculated emulsion droplets); (ii) mean droplet diameter (i.e. larger droplets tend to coalesce most easily in an emulsion system, hence, the emulsion with highest stability exhibits the smallest mean droplet diameter); and (iii) droplet half-life. Other means of qualifying emulsion stability are discussed at length by Becher (1965).

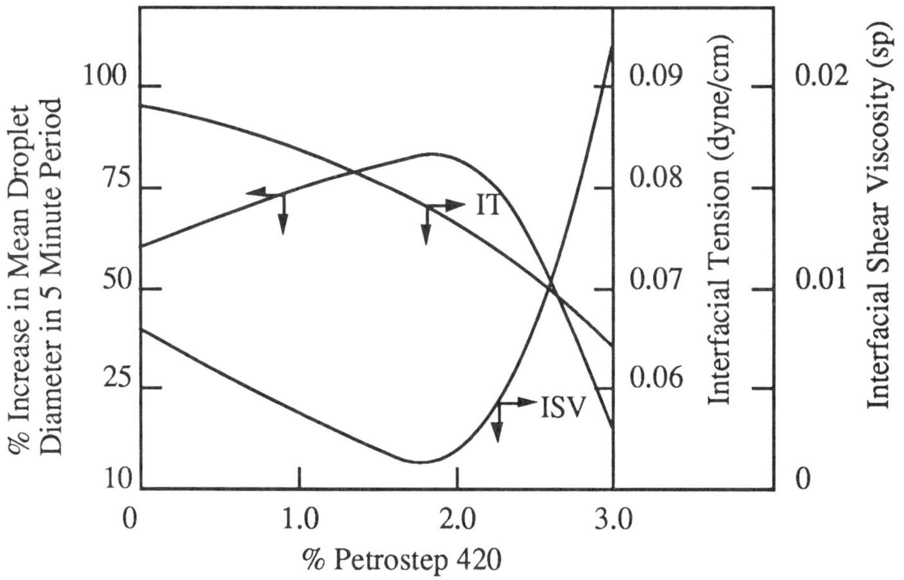

Figure 13.1-2 % Increase in the mean droplet diameter vs interfacial tension and interfacial shear viscosity as a function of %Petrostep 420 for the water/Salem crude oil interface with 0.5% Petrostep 420, at 1.5%NaCl. Data are taken from Wasan *et al.* (1979).

Cockbain and coworkers (Schulman & Cockbain 1940, Cockbain & McRoberts 1953, Cockbain 1956) have published many of the earliest data relating interfacial shear viscosity to emulsion stability. In their studies it was found that stable oil-in-water (o/w) emulsions containing lauryl alcohol, as well as various other water-soluble soaps, simultaneously exhibit low interfacial tension and high interfacial shear viscosity (Figure 13.1-1). Similar findings have been noted by other investigators (Davies & Rideal 1963). However, frequently (cf. Figures 13.1-2 and 13.1-3), maximum emulsion stability has been found to correlate with a maximum interfacial shear viscosity (as a function of surfactant concentration, for example), whereas interfacial tension for the most stable emulsions may not be minimized (cf. Figure 13.1-2). Many investigators have witnessed such strong correlation between emulsion stability and interfacial shear viscosity (Adam 1948, Bikerman 1948, Cumper & Alexander 1950, Lawerence 1952, Sherman 1953, Mysels *et al.* 1959, Morrell & Egloff 1931, Gladden & Neustadter 1972).

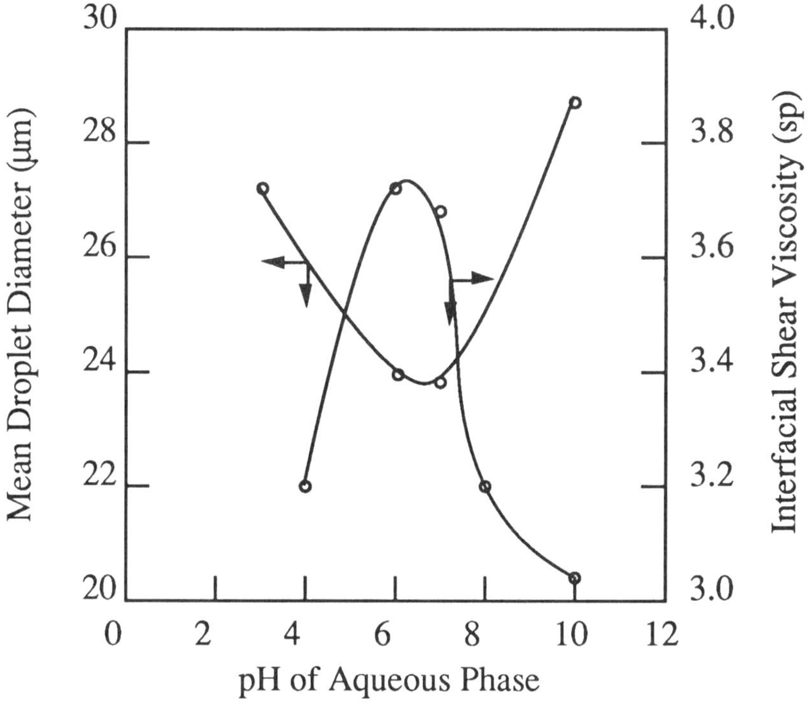

Figure 13.1-3 Mean droplet diameter vs interfacial shear viscosity as a function of aqueous phase pH for the water/mineral oil interface with 0.05 volume fraction Torulopsis petrophilum broth (biological emulsifier). Data are taken from Campanelli & Cooper (1989).

Oil slicks at sea may be dispersed by creating w/o emulsions by employing a suitable stabilizing surfactant. Morrel & Egloff (1931), Lawerence (1952), and Gladden & Neustadter (1972) have each noted that interfacial rigidity, as evidenced by a high interfacial shear viscosity (in the latter two reports particularly), is responsible for stable seawater emulsions.

The experimental studies cited above pertain to surfactant-stabilized interfaces that display Newtonian interfacial rheological behavior [cf. Eq. (4.2-17)]. Considerably less data have been collected for interfaces exhibiting non-Newtonian interfacial rheological behavior, though a few investigations have, in fact, been made. Thus, Taubman & Koretskii (1958) found high values of both the (Bingham) interfacial yield stress and interfacial shear viscosity [cf. Eq. (4.3-7)] to correlate with the stability of emulsions formed by dispersing CCl_4 droplets containing stearic acid within an aqueous solution of $AlCl_3$. Interfacial shear elasticity was also observed in the studies of Srivastava & Haydon (1964) and Boyd *et al.* (1972) to be an important stabilizing property of emulsions (see Figure 13.1-4).

As noted in the introduction to the chapter, interfacial shear viscosity correlations are most prevalent in past experimental studies which attempt to relate emulsion stability with dynamic interfacial properties. Nevertheless, as noted in Chapters 11 and 12, for sufficiently large interfacial dilatational elasticities [as arise through the Marangoni process; cf. Chapters 5, 6 and 10], interfacial viscosities play a nominal role in stabilizing a thin liquid film (cf. Figures 11.3-4 and 12.2-3). Thus, some investigations have shown interfacial viscous effects to be relatively *small* for stable emulsion systems (King 1941, Blakey & Lawerence 1954, Carless & Hallworth 1966, Neustadter *et al.* 1975). Indeed, there is evidence that the rigidity of droplet interfaces to compression, corresponding to a large interfacial elasticity, is in these cases the more important interfacial rheological parameter (King 1941, Becher 1965), though adequate experimental confirmation of interfacial dilatational elasticity effects in emulsion stability is currently lacking.

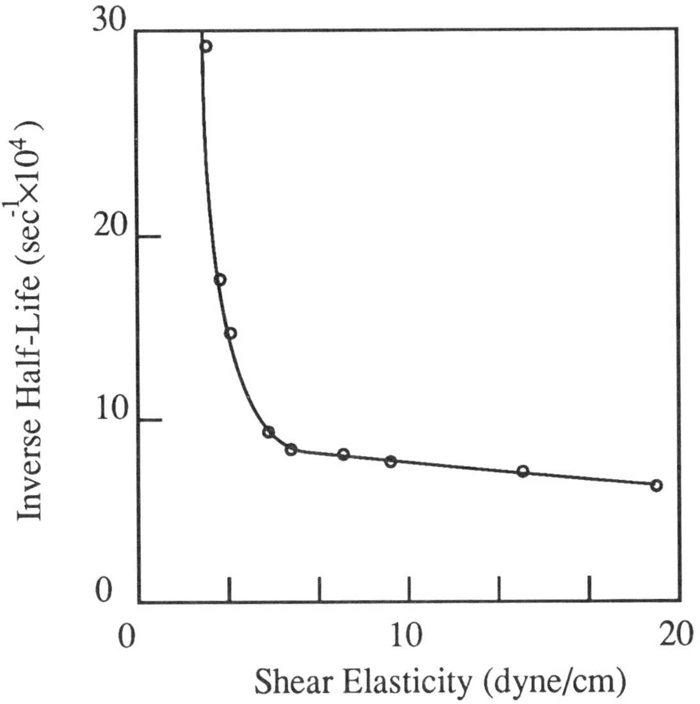

Figure 13.1-4 Inverse half-life of bovine-serum albumin-stabilized droplets at a flat interface vs interfacial shear elasticity. Data are taken from Srivastava & Haydon (1964).

13.2 Experimental Studies of Foam Stability

The 'stability' of a foam should be distinguished from the (generally unrelated) property of a foam known as its *foaminess*. The degree of foaminess of a foam may be regarded as being proportional to the initial volume of foam created upon mixing of the separate gas and liquid phases. Thus, whereas a foam may possess a high foam stability (the latter relating to the longevity of the foam, once created) it may yet exhibit a relatively small foaminess. The experimental data reviewed in this section are, therefore, distinguished as representing either foam stability or foaminess properties.

Several methods for generating and characterizing small-scale foams are discussed by Bikerman (1973).* The two used most often in the experimental studies reviewed here are: (i) the *dynamic* method, whereby single uniform-size bubbles are created at the end of a circular cylindrical capillary immersed in solution, thereafter rising to the surface of the solution to form a foam; and (ii) the *shaking* method, which involves the manual shaking of a closed cylinder of fixed liquid/gas ratio for a specified time. In each of these methods, the volume of foam initially created provides the measure of *foaminess*. The time (often represented by the half-life of bubbles formed) required for the foam to break, from the moment of its creation, serves as the measure of *foam stability*. In the dynamic method, bubbles are released at a known volumetric rate for a sufficient time to achieve a steady-state foam volume. Once steady-state has been reached, either the height of the foam, the volume of the foam, or an appropriate experimental constant (Bikerman 1973) combining foam volume with volumetric rate of gas production, in such a way as to be independent of the particular geometrical and rate conditions employed, is interpreted as a measure of foaminess. The time required for the foam height to fall a specified distance, once the production of bubbles has ceased, is viewed as a measure of the foam's stability. In the shaking method, the volume or height of foam created after a fixed duration of shaking is used as the measure of foaminess; once again, stability is measured by the rate of foam collapse.

There may be some ambiguity involved in the measurement of foam stability by the rate of fall of the foam height, as the foam often collapses in an inhomogeneous manner, leaving large holes of collapsed foam within the foam structure, while having relatively little effect upon the actual height of the foam. In this case (i.e. when foam texture becomes highly inhomogeneous), an experimentalist must make careful observations of both foam height and texture, offering a reasoned approximation of foam stability based upon each of these two criteria.

The experimental data summarized below confirm that both surface shear viscosity and surface dilatational elasticity each play an important role in foam stability (as previously suggested by the theoretical findings of Chapters 11 and 12). Unlike the comparable results for emulsion stability reviewed in §13.1, surface dilatational elasticity has frequently been either directly or indirectly implicated as the most important dynamic interfacial property for creating stable foam film surfaces. Thus, Camp & Durham (1955), Jones *et al.* (1957), Trapeznikov (1957), and Davies (1957) found

---

* As with emulsion stability, most current methods of foam stability characterization provide stability measures that depend in some way upon the instrument or technique employed. For common purposes of comparison (e.g to determine the relative stability of various foams or to elucidate the relative influence of phenomenological factors, such as interfacial tension or salinity, upon the stability of a foam) current methods of measurement are often sufficient.

that a high foam stability is not necessarily due to surface viscosity. Both Davies (1957) and Davies & Rideal (1963) have speculated that the surface dilatational elasticity may be the more important surface rheological property (though experimental evidence was not provided in these latter two studies).

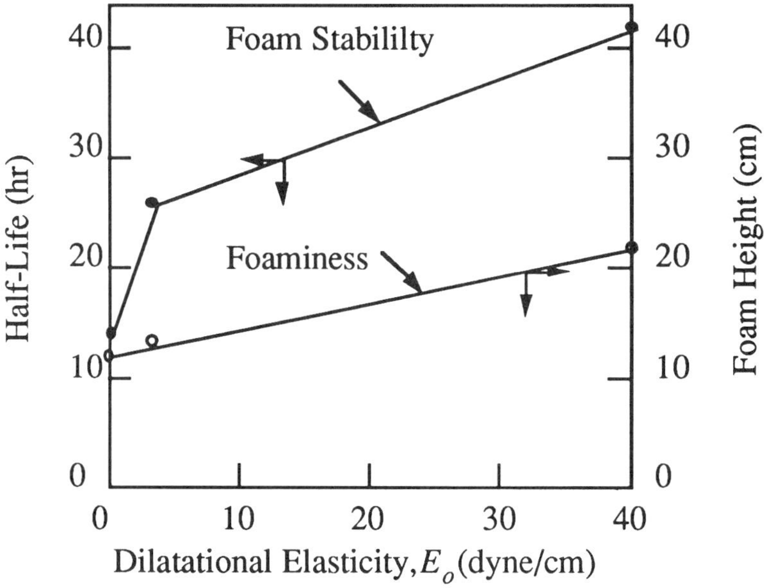

Figure 13.2-1 Foam half-life measurements of foam stability (obtained by the dynamic method) and foam height measurements of foaminess (obtained by the shaking method) vs dilatational elasticity for aqueous foams stabilized by alpha olefin sulfonates. Data are taken from Huang (1986).

Figure 13.2-1 illustrates an experimental correlation between surface dilatational elasticity and both foaminess and foam stability. The dilatational modulus [cf. Eq. (4.3-8)] is shown in the figure to increase with chain length of alpha olefin sulfonates (i.e. the stabilizing surfactant), coinciding with an increase in both foaminess and foam stability. The interfacial shear viscosity for these systems was found to be very small, on the order of 10^{-4} sp.

Correlations between the surface shear viscosity and foam stability have, nevertheless, been reported by many investigators. The early studies of Brown *et al.* (1953), Cumper (1953), Davies (1957) and Davies & Rideal (1963) each found evidence to suggest that an increase in the surface shear viscosity of foam bubble surfaces enhances the stability of the foam. Figure 13.2-2 illustrates a correlation from the latter of these studies between surface shear viscosity and foam stability. Later Graham (1976) and Graham & Phillips (1976) found similar correlations between high surface viscosity and stable foams by focusing [as did the early study of Cumper (1953)] upon protein-stabilized foams.

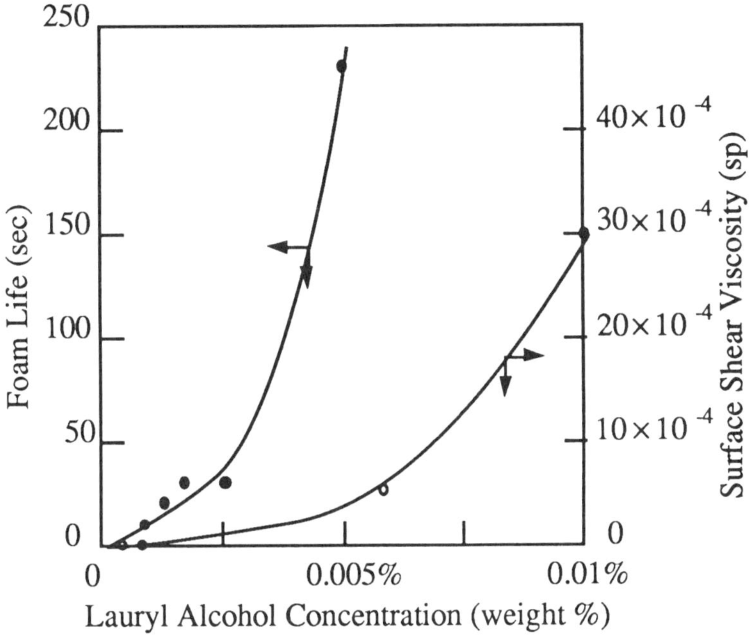

Figure 13.2-2 Correlation between surface shear viscosity and foam stability for an aqueous foam stabilized by lauryl alcohol. Data are taken from Davies & Rideal (1963).

When oil droplets are dispersed within aqueous foam lamellae (Ross 1967), the (three-phase) foam is often made unstable by the spreading of oil upon the foam film surfaces. As foam films drain, oil spread over a thin film surface (in the form of a *lens*, as depicted in Figure 13.2-3) soon spreads over the second foam film surface, creating an unstable oil film which subsequently breaks, destabilizing the foam.

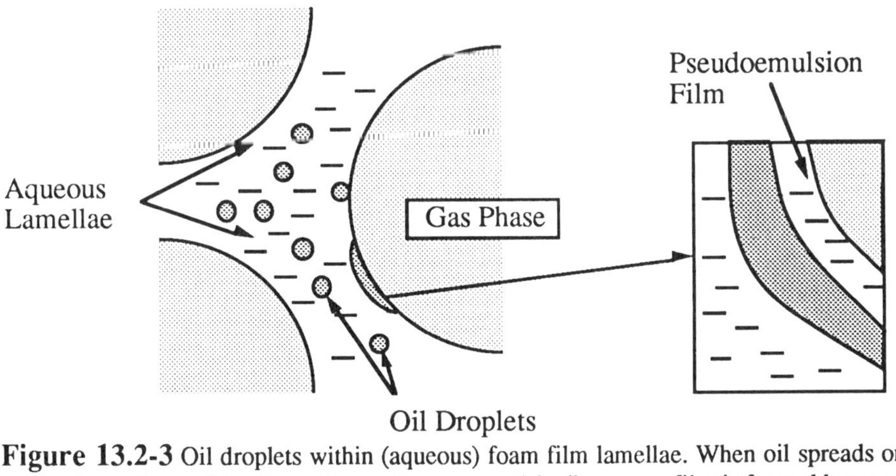

Figure 13.2-3 Oil droplets within (aqueous) foam film lamellae. When oil spreads on one of the surfaces of the foam films, a "psuedoemulsion" aqueous film is formed between the oil and gas phases.

It has been observed (Wasan *et al.* 1988) that the tendency of oil to spread along foam film surfaces depends upon the film tension of the "pseudoemulsion (oil/water/air) film" formed between dispersed oil droplets in foam lamellae and the foam film surfaces. The interfacial rheological effects (i.e. those deriving from interfacial-tension gradients and interfacial viscous stresses) include, in addition to interfacial hydrodynamic effects at the water/air film surfaces (as with aqueous foams), interfacial stresses at the oil droplet interfaces. The overall effect of interfacial rheological properties should be similar to that observed for aqueous foams; namely, increasing interfacial rigidity (both of the oil droplet interfaces and of the foam film surfaces) should increase the stability of the foam.

13.3 Interfacial Rheological Properties in Enhanced Oil Recovery

An important industrial example wherein the three-phase foam structure (water/oil/air) occurs is enhanced oil recovery. 'EOR' processes generally seek to remove residual oil either by low interfacial tension water- or foam-floods.

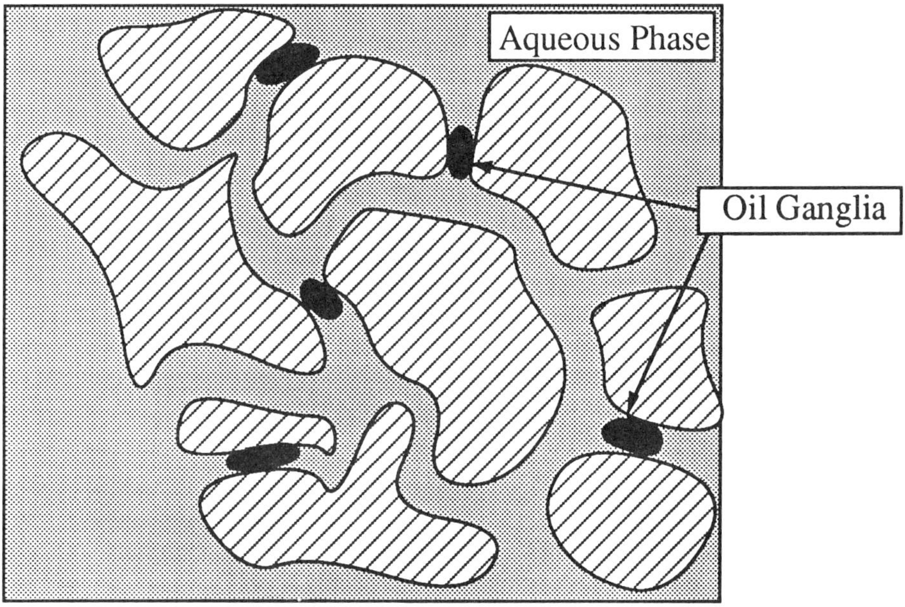

Figure 13.3-1 Oil globules entrapped within the capillaries of a porous medium, previously flooded with water (as in the secondary stage of oil recovery).

It has been estimated that 40-80% of the original oil trapped within an oil field remains following primary and secondary oil recovery processes

(Geffen 1973). The so-called "tertiary oil recovery processes" are attempts to retrieve this residual oil, most of which remains in the form of "blobs" or "ganglia" of oil trapped within narrow pores of the porous bed-rock (Figure 13.3-1). The tenacity of these ganglia to remain in place following a water flood is probably most related to the ability of water (or any other displacing fluid) to flow through the core by alternative paths, essentially bypassing the trapped ganglia.

A popular tertiary oil recovery method involves a core-flood employing an aqueous surfactant solution that is able to effect an ultra-low (oil/water) interfacial tension. In such processes, interfaces formed between the oil blobs and flooding solution easily deform on account of their low interfacial tension. This deformation leads to the displacement of oil through the capillaries of the core in an episodic "start-stop" manner, as often observed when interfaces move through pores possessing an irregular geometry (Haines 1930, Miller & Miller 1956, Heller 1968, Melrose 1970). The reason for this behavior owes to the nature of the contact angle which the oil makes with the solid bed-rock surface (Figure 13.3-2).

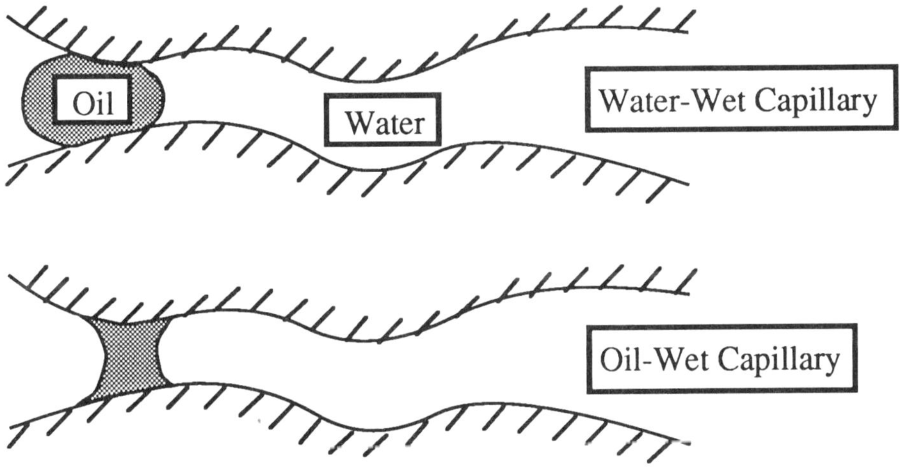

Figure 13.3-2 Equilibrium oil droplet configurations within water-wetted and oil-wetted capillaries.

In equilibrium, oil droplets within capillary pores will assume one of the geometry types depicted in Figure 13.3-2, varying in accordance with the wettability of the solid (i.e. depending upon whether the solid surface is 'oil-wetting' or 'water-wetting'). The curvature properties of the leading and trailing interfaces will be determined by: (i) the normal stress (or Laplace) condition [cf. Eq. (4.2-20a)]; and (ii) the equilibrium contact angle which the oil/water interface makes with the solid surface. As the pressure drop across an oil blob increases, its leading and trailing interfaces will deform so as to satisfy the normal stress condition at each point along the droplet interfaces. The contact angle between the droplet interface and solid surface will thus change; nevertheless, 'slip' of the oil droplet through the capillary will not occur until the contact angle of the trailing droplet interface exceeds the *advancing contact angle*, at which stage the droplet may advance through the capillary. Similarly, if a reverse pressure drop should be applied, leading to the reduction of the contact angle of the leading interface to a value less than the *receding contact angle*, the droplet will commence to recede through the capillary.

Fluid ganglia generally travel with an episodic motion owing to the nonuniform geometry of the capillary in which they move. When the advancing or receding oil droplet encounters an expansion or contraction in the capillary radius, the contact angle is suddenly altered; this can cause the droplet to stop until the fluid interfaces deform sufficiently to satisfy the conditions of oil ganglia slip noted above. A sudden change in contact angle may also result in an interfacial instability. The oil droplet thus moves through capillaries of the porous medium in a time-discontinuous manner. [An additional reason for this discontinuous motion may owe to the adsorption of surfactant upon the solid surface ahead of the moving contact line (Princen *et al.* 1988). Such adsorption alters the contact angle, resulting in an episodic motion similar to that which occurs by variations of capillary geometry.]

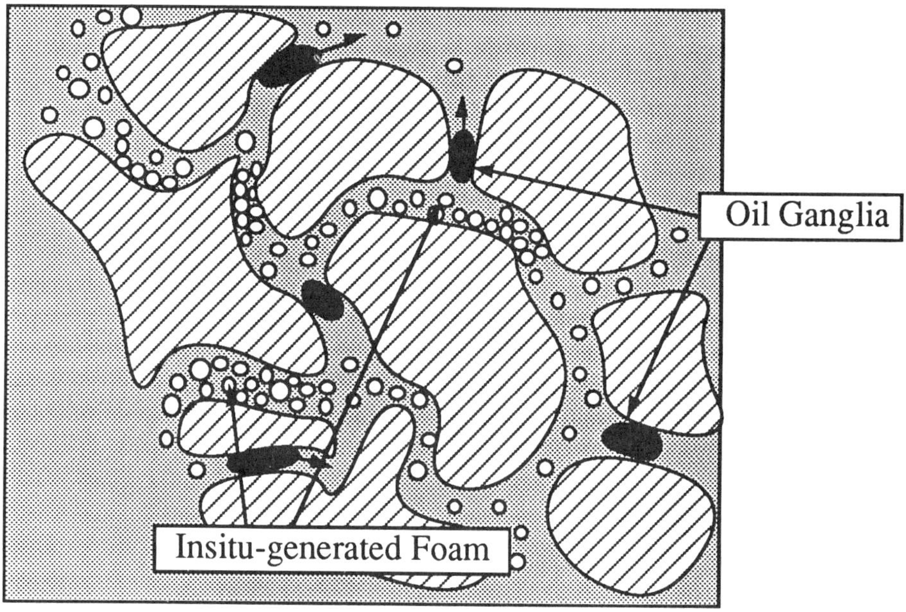

Figure 13.3-3 Foam bubbles generated insitu plug capillaries; the increased pressure drop across entrapped oil ganglia causes an enhanced displacement of oil.

Interfacial rheological stresses will clearly influence the propagation of oil through a porous core, insofar as propagation involves nonequilibrium interfacial conditions. Slattery and coworkers (Slattery 1974, 1979, Giordano & Slattery 1983, Ramamohan & Slattery 1984) have provided various qualitative and semiquantitative analyses of interfacial rheological effects in oil entrapment and displacement processes. A general conclusion of their work is that, in addition to ultra-low interfacial tension, low interfacial viscosities (and elasticities) are also favorable to oil displacement. A significant amount of work remains, however, to clarify the quantitative role of interfacial rheological properties in capillary displacement processes, not to mention the qualitative, physical role they might play in interfacial regions proximate to the moving contact line, in interfacial deformations in capillaries small enough to engender large interfacial curvature, or in the damping of interfacial instabilities during propagation of oil ganglia.

The use of an insitu-generated foam for tertiary oil recovery involves a mechanism of oil displacement which is referred to as *mobility control.* A highly viscous foam is generated within the porous medium, subsequently filling unoccupied pores: once these pores have been plugged, entrapped oil may be forced from the core. As will be demonstrated in Chapter 14, the rheological properties of a foam may be dramatically influenced by interfacial rheological properties owing to the large specific surface of foams; whence, highly viscous and/or elastic foam surfaces are desirable for improving 'mobility control' (see Figure 13.3-3).

Research in this area is also quite young. Bretherton (1961) has solved the problem of a bubble or line of bubbles propagating under the action of an imposed pressure gradient through a cylindrical capillary in which the bubbles are squeezed so as to assume a nonspherical shape. One of the few works to explicitly address the issue of interfacial rheological properties in foam flow through narrow capillaries is that of Hirasaki & Lawson (1983), who extended the work of Bretherton (1961) to include the effects of interfacial-tension gradients. They obtained an effective shear viscosity for a line of bubbles propagating within a bundle-of-tubes model of a porous medium that increases significantly owing to the existence of Marangoni stresses.

13.4 Summary

Interfacial rheology often plays an important role in the stabilization of emulsions and foams. This has been illustrated in the present chapter through a survey of current experimental correlations between interfacial rheological properties and various foam and emulsion stability measures.

The magnitude of the interfacial shear viscosity of droplet interfaces has often been observed to possess a direct correlation with the stability of emulsions. The theoretical basis for such a correlation pertains to the second stage of the coagulation process, wherein thin liquid films formed between flocculated droplets commence to drain, ultimately becoming unstable. In both thin film drainage and stability processes, interfacial shear viscosity is known to possess a drainage-hindering or stabilizing effect (Chapters 11 and 12). Likewise, interfacial dilatational elasticity may significantly hinder thin film drainage and instability, although experimental studies correlating this interfacial rheological property to emulsion stability are currently relatively few.

On the other hand, interfacial dilatational elasticity has often been observed to directly correlate with foam stability, as has also interfacial viscosity: once again, this presumably owes to the ability of dynamic interfacial properties to hinder the drainage and stability of thin foam films. The situation may be more complex when an aqueous foam is contacted with a third, nonaqueous (i.e. oil) phase. In such circumstances, interfacial rheological effects may influence both the stability of emulsion droplet interfaces within the thin film lamellae and Plateau borders and the stability of the foam film surfaces.

Questions for Chapter 13

13.1 Would you expect the rate of surfactant adsorption to foam surfaces from within the liquid films to be an important factor in the stabilization of a foam? Which dynamic interfacial property would be most influenced by surfactant adsorption rate?

13.2 Is an emulsion more or less stable when the surfactant is soluble in the dispersed phase? Why?

13.3 In which stage of the coagulation process will disjoining pressure forces play the most important role?

Additional Reading for Chapter 13

§13.1 Experimental Studies of Emulsion Stability

Malhotra, A.K. & Wasan, D.T. 1988 Interfacial rheological properties of absorbed surfactant films with applications to emulsion and foam stability. In *Thin Liquid Films*, (ed. I.B. Ivanov). New York: Marcel Dekkar.

Becher, P. 1965 *Emulsions: Theory and Practice*. New York: Reinhold Publishing Corporation.

§13.2 Experimental Studies of Foam Stability

Wasan, D.T., Nikolov, A.D., Huang, D.D., & Edwards, D.A. 1988, Foam stability: Effects of oil and film stratification. In *Surfactant-Based Mobility Control*, (ed. D.H. Smith), pp. 136–162. Washington D.C.: American Chemical Society.

§13.3 Interfacial Rheological Properties in Enhanced Oil Recovery

Giordano, R.M. & Slattery, J.C. 1983 Effect of interfacial viscosities upon displacement in capillaries with special application to tertiary oil recovery, *AIChE J.* **29**, 483–502.

R.R. Ramamohan and J.C. Slattery 1984 Effects of surface viscoelasticity in the entrapment and displacement of residual oil, *Chem. Eng. Commun.* **26**, 241–263.

14

"On the division of space with minimum partitional area. This problem is solved in foams, and the solution is interestingly seen in the multitude of film-enclosed cells obtained by blowing air through a tube in the middle of a soap solution in a large open vessel."

W. Thompson (Lord Kelvin) (1887)

CHAPTER 14

Foam Rheology

Liquid foams may be regarded as structured fluid continua that exhibit a complex rheological behavior strongly dependent upon local texture and physicochemical constitution. They are formed by dispersing a large gas volume fraction within a continuous liquid, generally in the form of polyhedral bubbles, whose dimensions typically range between 50 microns and several millimeters. The continuous liquid phase of a foam is bound within a network of thin liquid films (with thicknesses characteristically ranging from 10–1000 nanometers) and meniscus (or 'Plateau border') regions (see Chapter 18 for a fuller discussion of the 'contact lines' formed thereby). The surfaces of bubbles dispersed within a pure liquid possess a strong tendency to spontaneously minimize surface area owing to the existence of surface tension; thus, stable foams cannot be created in pure liquids. When surfactant is solubilized within the liquid phase, the adsorption of the surfactant to the bubble surfaces provides the foam a longer-term stability, by simultaneously lowering interfacial tension and giving rise to the existence of interfacial rheological properties, which latter properties render the foam surfaces 'immobile'. Applications involving foams range from shampoo and shaving cream processing, to firefighting, enhanced oil recovery, and mineral particle transport (Kitchener 1964, Kraynik 1988).

Interfacial rheological properties significantly influence the rheology of foams. This may be attributed both to the presence of surfactant adsorbed to the surfaces within foams and their large specific surface. Many currently existing theoretical studies of foam rheology avoid an explicit consideration of interfacial rheology by restricting their attention to idealized foam surfaces that exhibit a homogeneous interfacial tension and negligible interfacial viscosities and/or elasticities (Khan 1985, 1987, Khan & Armstrong 1987a, 1987b, Kraynik 1981, 1987, Kraynik & Hansen 1986, 1987, Princen 1983, 1985, Schwartz & Princen 1987, Weaire 1989).

In the present chapter, various spatially periodic foam models presently in the literature are employed for the purpose of underlining the importance of interfacial rheological properties (i.e. interfacial viscosities and interfacial-tension gradients) to composite foam rheological behavior.

In general, theoretical models such as employed herein seek to address the following principal experimental observations of foam rheological behavior: (i) foams are highly viscous; (ii) they exhibit shear thinning; (iii) they exhibit yield stress behavior; (iv) they appear to 'slip' at a solid boundary. These (and other) salient foam rheological features are theoretically probed by relating local foam structure and physicochemical characteristics to global foam rheological response. This calls for simplifying (though plausible) assumptions of local foam structure, since foams may exhibit a highly complicated local geometry.

In §14.1, the two- and three-dimensional spatially-periodic geometrical models employed in this chapter are described. This is followed in §14.2 by a discussion of the kinematics of these foam models. In §14.3, foam rheological properties are related to interfacial rheological properties for both shear and dilatational deformations.

14.1 Geometrical Models

Owing to the varied and complicated polyhedral bubble shapes encountered in real (three-dimensional) foam systems, several geometrical assumptions are necessarily made for the purpose of theoretically investigating the response of foams to deformation. In the present analysis, attention is restricted to *monodisperse*, spatially periodic foam geometries, permitting each bubble of the (unbounded) foam to be viewed as fully contained within the boundaries of a 'unit cell' whose planar or curvilinear faces connect the points of an infinite lattice.

In this chapter, foam models based upon the following three unit-cell shapes are employed: (i) a *rhomboidal dodecahedron* (corresponding to a three-dimensional 'wet' foam—i.e. possessing a relatively small dispersed-phase volume fraction); (ii) a *tetrakaidecahedron* (corresponding to a three-dimensional 'dry' foam—i.e. possessing a dispersed-phase volume fraction whose value approaches unity); and (iii) a *hexagon* (corresponding to a two-dimensional 'dry' foam).

Rhomboidal Dodecahedron Unit Cell

Imagine a monodisperse, spatially periodic 'wet' foam, formed by arranging spherical bubbles within a face-centered cubic array, as depicted in Figure 14.1-1. This array represents the closest possible packing of a monodisperse collection of spherical bubbles, hence it is the preferable packing arrangement for a monodisperse foam whose dispersed-phase volume fraction is near the value $\phi = 0.7405$ (i.e. the value for touching spheres in a face-centered cubic array). In such a packing arrangement, each bubble may be viewed as bound within a polyhedral unit cell known as a rhomboidal dodecahedron (RDH).

Figure 14.1-1 A face-centered cubic packing of spheres. Each sphere may be viewed as bound within a rhomboidal dodecahedron unit cell.

A single RDH cell is depicted in Figure 14.1-2. It is composed of: (i) 12 planar rhombus faces with each face center located at a distance $b/2$ from the cell geometric center; (ii) 24 lines at which planar cell faces meet at 120°, and (iii) 14 corners, of which 8 are formed of three lines, each pair of lines meeting at a tetrahedral angle. The total cell volume is given by $\tau_v = b^3 / \sqrt{2}$ and the total cell surface area by $A = \frac{9}{2}b^2$; additionally, $c = b\sqrt{2}/2$, $d = b/(2\sqrt{3})$, and $e = b\sqrt{6}/4$.

In equilibrium, and at dispersed-phase volume fractions $\phi > 0.7405$, bubble surfaces will flatten along the RDH cell boundaries, thereby forming planar liquid films enclosed by the flattened bubble surfaces of neighboring cells (see §11.2 for a description of the hydrostatics of such planar films). At the edges of the planar films, the bubble surfaces will meet (generally at a finite contact angle θ) spherical bubble surfaces which partially bound continuous liquid within the meniscus (or Plateau border) regions of the foam. Figure 14.1-3 provides a cross-sectional view of the RDH cell, showing thin films (of thickness h) and spherical Plateau borders. The dispersed-phase volume fraction for the RDH foam, neglecting liquid content in the thin foam films, is given by the relation (Princen *et al.* 1980),

$$\phi = 0.7405\left(-\frac{5}{\cos^3\theta} + \frac{9}{\cos^2\theta} - 3\right), \qquad (14.1\text{-}1)$$

where $0° < \theta < \theta_{max}$, say.

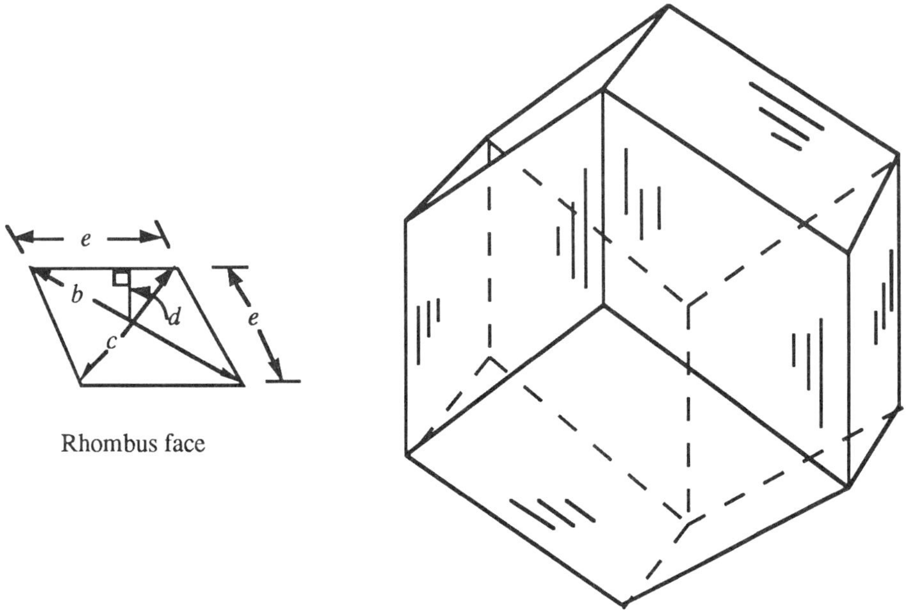

Figure 14.1-2 The rhomboidal dodecahedron.

The total (single) bubble surface area (S_b) may be decomposed into the planar portions of the bubble surface (S'_b) which form partial boundaries to the planar thin films and the spherical Plateau border surface (S''_b). Here, $S'_b = 12\pi a^2 \sin\ \theta$, with $a = b/(2\cos\theta)$ the curvature radius of the spherical surfaces.

Tetrakaidecahedron Unit Cell

At large dispersed-phase volume fractions (i.e. $\phi \rightarrow 1$), two distinct types of Plateau borders may be distinguished: (i) 'lineal Plateau borders', referring to the (contact line) region of intersection between planar foam films; and (ii) 'tetrahedral Plateau borders', referring to the (contact 'point') region of intersection of the lineal Plateau borders. Equilibrium stability criteria require that lineal Plateau borders must necessarily be formed by the intersection of (3) foam films at 120° internal angles, whereas tetrahedral Plateau borders must be formed by the intersection of (4) lineal Plateau borders at internal angles of 109.47°. A high dispersed-phase volume fraction foam created by a spatially periodic array of RDH bubbles will fully satisfy the former stability criterion, yet only 8 of 14 corners of each 'dry' RDH foam cell will satisfy the latter tetrahedral Plateau border stability criterion. Thus, the RDH cell is not a realistic foam cell for high dispersed-phase volume fractions.

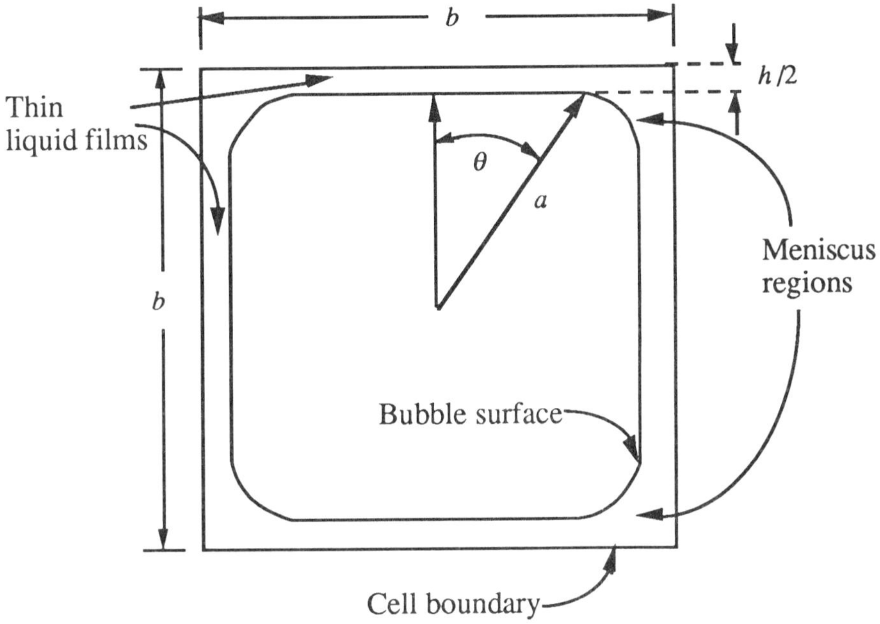

Figure 14.1-3 Cross-section of a bubble contained within the RDH cell. Planar bubble surfaces meet spherical (Plateau border) surfaces at a finite contact angle θ.

The single space-filling polyhedral structure known to minimize surface area—thereby satisfying the mechanical conditions of a stable, dry-foam unit cell—is the 'Kelvin minimal tetrakaidecahedron' [Thompson (Lord Kelvin) (1887)]. The (curvilinear) Kelvin cell may be created by slight deformations of the (planar) hexagonal faces of a tetrakaidecahedron (TKDH), which latter structure is itself a truncated octahedron, as illustrated in Figure 14.1-4. As the deviation of the hexagonal faces of the Kelvin cell from the original planar hexagonal surfaces of the TKDH cell is quite small (Princen & Levinson 1987), the polyhedron employed in this chapter to represent the unit cell of a monodisperse spatially periodic 'dry' foam is the plane-faced TKDH (of characteristic dimension f), within which is contained a single TKDH-shaped bubble. The TKDH possesses (6) square planar faces, of area f^2, each with its centerpoint a distance $(\sqrt{2})f$ from the geometrical center of the TKDH cell, and (8) hexagonal planar faces of area $(3\sqrt{3})f^2$, with their centerpoint located a distance f from the TKDH center. Hexagonal faces of the TKDH cell meet at 109.47° whereas square-hexagonal intersections meet at an internal angle of 125.26°. There are no square-square face intersections. The volume of the TKDH cell is $(8\sqrt{2})f^3$.

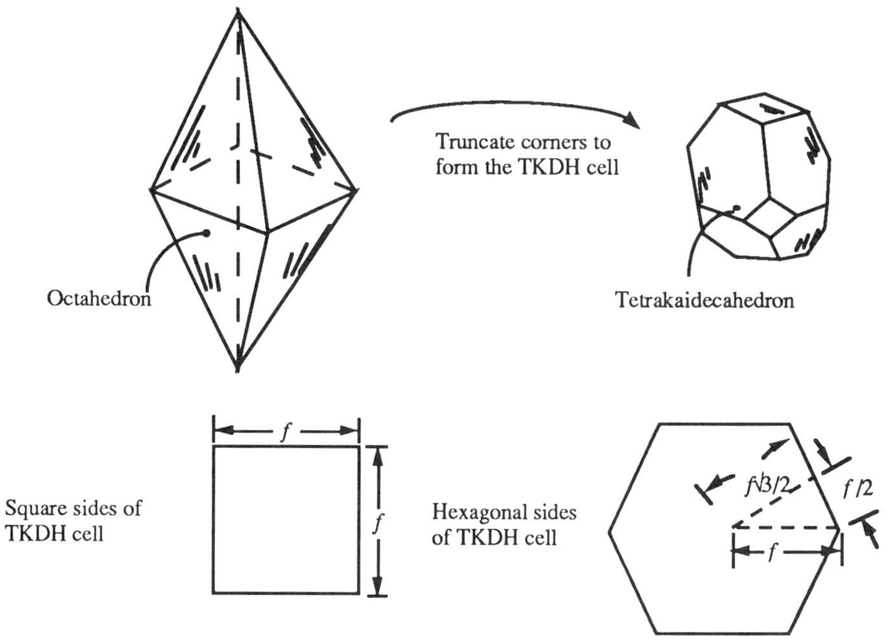

Figure 14.1-4 The tetrakaidecahedron.

Bound within each TKDH unit cell is a single TKDH bubble; thus, planar liquid films are formed between the (planar) faces of adjacent polyhedral bubbles of the foam. For each unit cell there are (8) hexagonal, planar films of thickness h and (6) square thin films of thickness $(\sqrt{2})h$. The characteristic length of each bubble (corresponding to f of the unit cell) is thus $g=f-h/2$. Whereas the geometry of meniscus regions in real foams will clearly deviate from the linear geometry indicated by the present TKDH cell model (see Figure 14.1-5), this will be of minor significance in the subsequent analysis, insofar as the TKDH bubble geometry is to be applied only for the purpose of elucidating the response of foams to a pure dilatation (for which deformation the meniscus regions will be stagnant flow regions, owing to the symmetry of the polyhedral cell).

The dispersed-phase volume fraction of the TKDH, 'dry' foam may be expressed as

$$\phi \equiv \frac{g^3}{f^3} = \frac{(f - h/2)^3}{f^3}, \tag{14.1-2}$$

or, with

$$\varepsilon \equiv \frac{h}{f} << 1, \tag{14.1-3}$$

by the approximate unit value

$$\phi = 1 + O(\varepsilon). \tag{14.1-4}$$

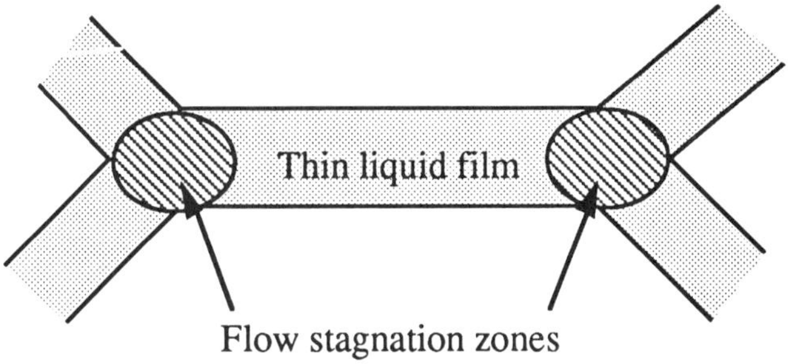

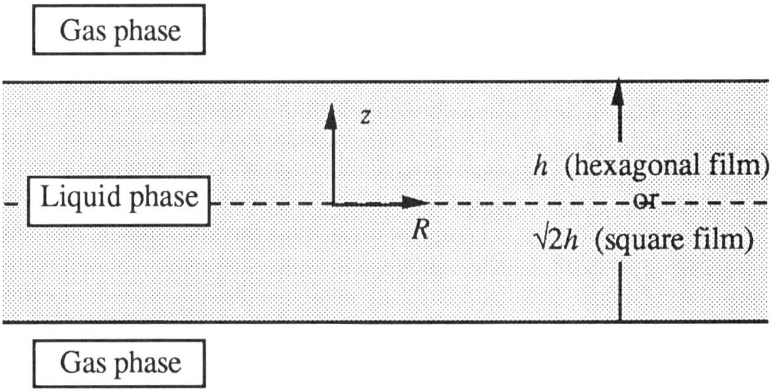

Figure 14.1-5 A cross section of a thin liquid film of the tetrakaidecahedron foam model. The indicated regions of flow stagnation correspond to the circumstances of a foam dilatation, as discussed in §14.2.

The preceding three-dimensional foam models will be employed herein for the purpose of elucidating foam rheological response to pure foam expansive or compressive motions. For the more troublesome class of shearing deformations, two-dimensional spatially periodic foam models are often employed for the conceptual simplifications they afford (Khan & Armstrong 1987a, 1987b).

Hexagonal, Two-Dimensional Unit Cell

Our two-dimensional, spatially-periodic, dry-foam model envisions the foam as composed (in equilibrium) of hexagonal cells, each of which satisfy the stability criterion of thin liquid films meeting at 120° internal angles. As illustrated in Figure 14.1-6, the position of the Plateau border at which a trio of films meets may be parameterized in the hexagonal cell model either in terms of the position vector $\mathbf{r}_o$ originating at the edge (i.e. Plateau border 'A') of one of the films and the body-fixed vectors $\mathbf{b}_1$ and $\mathbf{b}_2$, or, equivalently, in terms of the film vectors $\mathbf{g}_i$ (i = 1,2,3). The latter three vectors are provided in terms of the former by the relations,

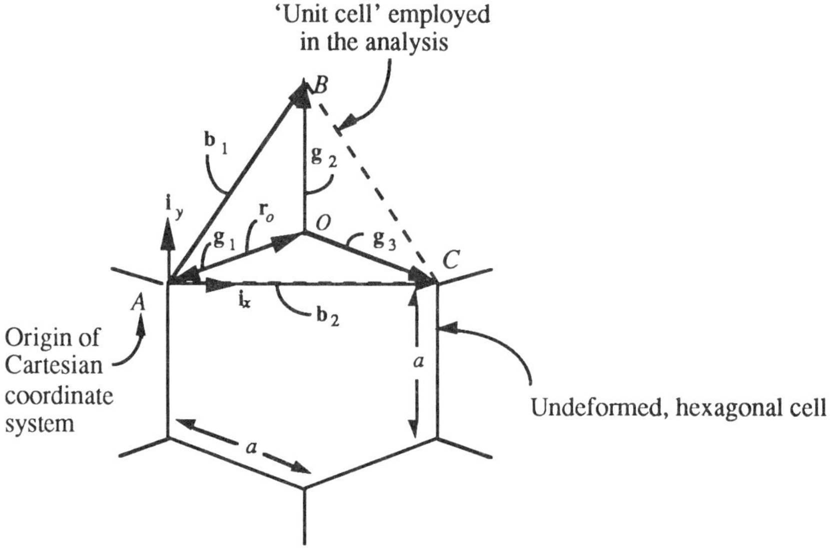

Figure 14.1-6 A hexagonal unit cell of the two-dimensional foam model.

$$\mathbf{g}_1 = -\mathbf{r}_o, \tag{14.1-5}$$

$$\mathbf{g}_2 = \mathbf{b}_1 - \mathbf{r}_o, \tag{14.1-6}$$

and

$$\mathbf{g}_3 = \mathbf{b}_2 - \mathbf{r}_o. \tag{14.1-7}$$

In the 'undeformed' state depicted in the figure, the following explicit relations hold:

$$\mathbf{b}_1^0 = \frac{\sqrt{3}}{2} a\mathbf{i}_x + \frac{3}{2} a\mathbf{i}_y, \tag{14.1-8}$$

$$\mathbf{b}_2^0 = \sqrt{3}\, a\mathbf{i}_x, \tag{14.1-9}$$

where the superscript '0' will consistently serve to denote undeformed-state variables.

It proves useful in the following also to define a unit film vector,

$$\mathbf{p}_i = \frac{1}{g_i}\mathbf{g}_i \qquad (i = 1, 2, 3;\ \text{no sum on } i), \tag{14.1-10}$$

where

$$g_i = \left(g_{ix}^2 + g_{iy}^2 \right)^{1/2} \tag{14.1-11}$$

and

$$\mathbf{g}_i = g_{ix}\mathbf{i}_x + g_{iy}\mathbf{i}_y . \tag{14.1-12}$$

In the undeformed state depicted in Figure 14.1-6, the magnitudes of the film vectors satisfy $g_i \equiv a$.

14.2 Kinematics of Spatially-Periodic Foam Media

Simple Shear: Hexagonal, Two-Dimensional Unit Cell

Our consideration of foam response to shearing deformations will be confined to the case of a two-dimensional hexagonal foam in a homogeneous simple shear field. For this purpose, the kinematical treatment of Khan & Armstrong (1987a) will be adopted. Thus, we assume a parallel displacement of hexagonal cell centers relative to the direction of the global straining motion, in addition to local equilibrium in the vicinity of the Plateau borders [meaning, explicitly, that foam films always meet at 120° angles; cf. Eq. (14.2-8)].

In these circumstances, let

$$\mathbf{E}(t, 0) = \frac{\partial \mathbf{r}_o}{\partial \mathbf{r}_o^0}, \tag{14.2-1}$$

represent a spatially homogeneous deformation dyadic field, with $\mathbf{r}_o$ denoting the position of O at any time t, and $\mathbf{r}_o^0$ the position of O at time $t = 0$. Thus,

$$\mathbf{r}_o = \mathbf{E} \cdot \mathbf{r}_o^0 . \tag{14.2-2}$$

Since films are assumed to move affinely, it follows directly that

$$\mathbf{b}_i = \mathbf{E} \cdot \mathbf{b}_i^0 . \tag{14.2-3}$$

Assuming a simple shear motion in the positive x direction, the Cartesian tensor form of the deformation dyadic $\mathbf{E}$ may be written as in the following matrix expression:

$$E_{ij} = \begin{bmatrix} 1 & \gamma \\ 0 & 1 \end{bmatrix}, \tag{14.2-4}$$

with γ the shear strain scalar. In terms of Eq. (14.2-4), together with Eq. (14.2-3), the deformed-state vectors $\mathbf{b}_i$ may be expressed in terms of the corresponding undeformed-state vectors as

$$\mathbf{b}_i = \left(b_{ix}^0 + \gamma b_{iy}^0 \right) \mathbf{i}_x + b_{iy}^0 \mathbf{i}_y . \tag{14.2-5}$$

Upon use of Eqs. (14.1-8) and (14.1-9), there obtain the explicit relations,

$$\mathbf{b}_1 = \frac{\sqrt{3}}{2} a\left[(1+\gamma\sqrt{3})\mathbf{i}_x + \sqrt{3}\,\mathbf{i}_y\right], \tag{14.2-6}$$

$$\mathbf{b}_2 = \sqrt{3}\, a\mathbf{i}_x. \tag{14.2-7}$$

The condition of local equilibrium at the lineal Plateau borders requires that film vectors $\mathbf{g}_i$ meet at 120° angles; whence, it follows that

$$\frac{\mathbf{g}_1 \times \mathbf{g}_2}{\mathbf{g}_1 \cdot \mathbf{g}_2} = \frac{\mathbf{g}_3 \times \mathbf{g}_1}{\mathbf{g}_3 \cdot \mathbf{g}_1} = \sqrt{3}\,\mathbf{i}_z. \tag{14.2-8}$$

Upon combining Eq. (14.2-8) with Eqs. (14.1-5)–(14.1-7), there obtains the following relation between the location of the Plateau border O and the deformed-state vectors $\mathbf{b}_i$:

$$\mathbf{r}_o = 2\left[\frac{1}{\sqrt{3}}(b_{1x} + b_{2x}) + (b_{1y} - b_{2y})\right]\frac{C}{D\sqrt{3}}\mathbf{i}_x + 2\left[(b_{2x} - b_{1x}) + \frac{1}{\sqrt{3}}(b_{1y} + b_{2y})\right]\frac{C}{D\sqrt{3}}\mathbf{i}_y, \tag{14.2-9}$$

where

$$C = \mathbf{b}_1 \cdot \mathbf{b}_2 + \frac{1}{\sqrt{3}}(\mathbf{b}_2 \times \mathbf{b}_1) \cdot \mathbf{i}_z \tag{14.2-10}$$

$$D = \left[\frac{1}{\sqrt{3}}(\mathbf{b}_1 + \mathbf{b}_2) \cdot \mathbf{i}_x + (\mathbf{b}_1 - \mathbf{b}_2) \cdot \mathbf{i}_y\right]^2 + \left[(\mathbf{b}_2 - \mathbf{b}_1) \cdot \mathbf{i}_x + \frac{1}{\sqrt{3}}(\mathbf{b}_1 + \mathbf{b}_2) \cdot \mathbf{i}_y\right]^2. \tag{14.2-11}$$

For the particular case of the simple shear field (14.2-4), Eqs. (14.2-6) and (14.2-7) may be substituted directly into Eq. (14.2-9) to provide the following explicit relationship between the Cartesian components (x_o, y_o) of $\mathbf{r}_o$ and the shear strain scalar γ:

$$x_o = 2a\frac{\left(\sqrt{3} + \frac{3\gamma}{2}\right)\left(3 + \frac{\gamma\sqrt{3}}{2}\right)}{\left(3 + \frac{\gamma\sqrt{3}}{2}\right)^2 + \left(\sqrt{3} - \frac{3\gamma}{2}\right)^2} \tag{14.2-12}$$

and

$$y_o = 2a\frac{\left(\sqrt{3} + \frac{3\gamma}{2}\right)\left(\sqrt{3} - \frac{3\gamma}{2}\right)}{\left(3 + \frac{\gamma\sqrt{3}}{2}\right)^2 + \left(\sqrt{3} - \frac{3\gamma}{2}\right)^2}. \tag{14.2-13}$$

Substituting Eqs. (14.1-5)–(14.1-7) into Eqs. (14.2-6), (14.2-7) and (14.2-9), furnishes the following explicit relations for the film vectors $\mathbf{g}_i$ (for a simple shear deformation):

$$\mathbf{g}_1 = -\,x_o\mathbf{i}_x - y_o\mathbf{i}_y\,, \tag{14.2-14}$$

$$\mathbf{g}_2 = \frac{\sqrt{3}}{2}\,a\left(1+\gamma\sqrt{3}\right)\mathbf{i}_x + \frac{3a}{2}\mathbf{i}_y + \mathbf{g}_1, \tag{14.2-15}$$

$$\mathbf{g}_3 = a\sqrt{3}\,\mathbf{i}_x + \mathbf{g}_1, \tag{14.2-16}$$

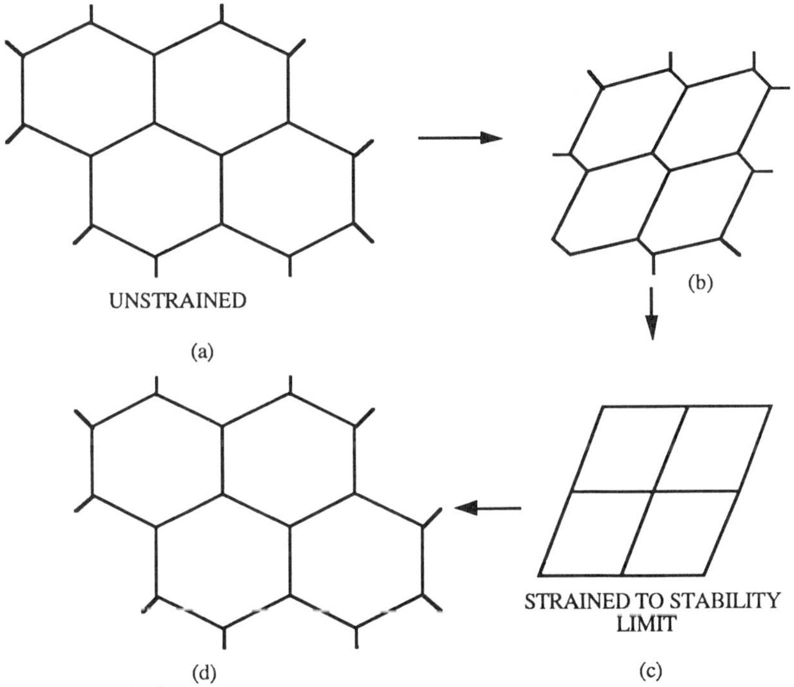

Figure 14.2-1 Cell deformation and reconstruction in a simple shear field.

$$g_1 = \frac{4\sqrt{3}\,a\left(\sqrt{3}+\frac{3}{2}\gamma\right)}{\left(3+\frac{\gamma\sqrt{3}}{2}\right)^2\left(\sqrt{3}-\frac{3}{2}\gamma\right)^2}\sqrt{\left(1+\frac{1}{4}\gamma^2\right)}, \tag{14.2-17}$$

$$g_2 = a\sqrt{\frac{x_o^2}{a^2}+\frac{y_o^2}{a^2}+\frac{3}{4}\left(1+\gamma\sqrt{3}\right)^2+\frac{9}{4}-\sqrt{3}\frac{x_o}{a}-3\frac{y_o}{a}} \tag{14.2-18}$$

$$g_3 = a\sqrt{\frac{x_o^2}{a^2}+\frac{y_o^2}{a^2}+3-2\sqrt{3}\frac{x_o}{a}}\,. \tag{14.2-19}$$

The kinematical response of the hexagonal foam model in a simple shear field may thus be viewed as embodied in the explicit relations (14.2-14)–(14.2-16). Figure 14.2-1 illustrates the two-dimensional foam evolution implied by these relations for a single deformational cycle, beginning with the undeformed state of Figure 14.1-6. Foam film instability is caused by the simultaneous elimination of two of the films contained within a single unit cell, resulting in the union of four films at each Plateau border. Khan & Armstrong (1987a) assume the foam instantaneously undergoes a 'disproportionation' process at this moment of instability, in which process the undeformed foam structure is recreated. This theoretical phenomenon of cell coalescence and disproportionation has been referred to in the literature as the process of 'hopping' (Prud'homme 1981).

Pure Dilatation: Three-Dimensional Cells

As previously noted, three-dimensional foam models are employed herein for the purpose of investigating the rheological response of foams to dilatation. The kinematical theory used for this purpose is further described by Adler & Brenner (1985) and Adler *et al.* (1985).

The global rate of deformation tensor may be defined for spatially periodic systems by integrating the local velocity vector over the bounding surface of a single unit cell: explicitly,

$$\boxed{\langle \mathbf{D} \rangle \overset{\text{def}}{=} \frac{1}{\tau_o} \oint_{\partial\tau_o} d\mathbf{S}\, \mathbf{v}\, .} \tag{14.2-20}$$

Here, $\partial\tau_o$ denotes the bounding cell surface, $d\mathbf{S}$ is an outwardly directed areal element lying on the bounding surface $\partial\tau_o$ and $\mathbf{v}$ is the local fluid velocity vector.

In the special case of a pure dilatation, the global deformation rate is characterized by an isotropic value [i.e. $\langle \mathbf{D} \rangle : \mathbf{I} = \langle \nabla \cdot \mathbf{v} \rangle$; cf. Eq. (4.1-16)]; whence,

$$\langle \mathbf{D} \rangle = \frac{1}{3} \langle \Delta \rangle \mathbf{I}, \tag{14.2-21}$$

where the global rate of foam dilatation $\langle \Delta \rangle$ satisfies the relation [cf. Eqs. (4.4-34)]

$$\langle \Delta \rangle = \frac{1}{\tau_o} \oint_{\partial\tau_o} d\mathbf{S} \cdot \mathbf{v} \equiv \frac{1}{\tau_o} \frac{d}{dt} \tau_o\, . \tag{14.2-22}$$

Equation (14.2-22) may be rewritten, upon application of the divergence theorem, as

$$\langle \Delta \rangle = \frac{1}{\tau_o} \oint_{S_b} d\mathbf{S} \cdot \mathbf{v} + \frac{1}{\tau_o} \int_{\tau_f} \nabla \cdot \mathbf{v}\, dV \ ,$$

with $d\mathbf{S}$ representing an outwardly directed areal element on the bubble surface S_b and τ_f the fluid-phase cell volume. Limiting our attention to an incompressible fluid phase, the second term in the above vanishes; thus,

$$\langle\Delta\rangle = \frac{1}{\tau_o}\oint_{S_b} d\mathbf{S}\cdot\mathbf{v}\,. \qquad (14.2\text{-}23)$$

Rhomboidal Dodecahedron Unit Cell

As will be confirmed *a posteriori* by satisfaction of the normal stress interfacial boundary condition, isotropic expansion/compression of each symmetrical RDH cell results in a uniform, radial expansion of the bubble contained within the cell. Thus, from Eq. (14.2-23) it follows directly that (cf. Figures 14.1-2 and 14.1-3)

$$\boxed{\mathbf{v}^o\cdot\mathbf{n} = \begin{cases} v_s & \text{on } S_b'', \\ v_s\cos\theta & \text{on } S_b'. \end{cases}} \qquad (14.2\text{-}24\text{a})$$

Here, $\mathbf{v}^o\!\cdot\mathbf{n}$ represents the normal bubble surface velocity, with

$$v_s \equiv \frac{\langle\Delta\rangle\tau_o}{S_b'\cos\theta + S_b''}\,. \qquad (14.2\text{-}24\text{b})$$

Tetrakaidecahedron Unit Cell

Whereas similar arguments as employed above for the RDH cell expansion apply equally well to the expansion of a TKDH cell, for the condition $\phi \rightarrow 1$ (in which condition the TKDH model is assumed to apply), virtually the entire closed surface of each TKDH bubble may be seen to bound planar liquid films owing to the fact that the meniscus regions reduce (in the limit $\phi \rightarrow 1$) to 'contact lines' of infinitesimal volume. [In contrast, the greatest percentage of bubble surface area in the RDH cell (for which $\phi \approx 0.74$) lies upon the spherical Plateau border surfaces.] Thus, flow within the thin liquid films bounded below and above by the planar surfaces of each TKDH bubble [which flow results from the foam expansion (14.2-22)], will be shown to be of greater significance to the rheological behavior of the foam than the outward movement of the bubble surfaces.

According to Eq. (14.2-22), the rate of foam expansion may be related to the size of individual foam lamellae by the expression (cf. Figure 14.1-4)

$$\langle\Delta\rangle = 3\dot{f}\,/\,f\,, \qquad (14.2\text{-}25)$$

where,

$$\tau_o = 8\sqrt{2} f^3, \tag{14.2-26}$$

is the instantaneous cell volume,

$$\tau_b = 8\sqrt{2}\, g^3, \tag{14.2-27}$$

the instantaneous dispersed-phase (i.e. 'bubble') volume, and

$$\tau_f = \tau_o - \tau_b, \tag{14.2-28}$$

the (constant) fluid-phase volume. Limiting subsequent attention to the case

$$\langle \Delta \rangle = \text{const.}, \tag{14.2-29}$$

it follows from Eq. (14.2-25) that the characteristic dimension f of bubble lamellae will grow exponentially with time and rate of expansion, according to the relation

$$f = f_o \exp\left[\frac{\langle \Delta \rangle}{3} t\right]. \tag{14.2-30}$$

The rate of thinning (or thickening) of foam films with finite $\langle \Delta \rangle$ may be determined from the above by noting that, on account of the incompressibility of continuous fluid,

$$\tau_f = \text{const.} = 8\sqrt{2} f^3 - 8\sqrt{2}(f - h/2)^3. \tag{14.2-31}$$

This provides the following leading order relation for the film thickness [cf. Eq. (14.1-3)];

$$h(t) = \frac{\tau_f}{12\sqrt{2} f^2} + O(\varepsilon). \tag{14.2-32}$$

Upon employing Eqs. (4.2-30) and (4.2-32) there obtains

$$h(t) = \frac{\tau_f}{12\sqrt{2} f_o^2} \exp\left[-\tfrac{2}{3}\langle \Delta \rangle t\right] + O(\varepsilon)$$

$$\equiv h_o \exp\left[-\tfrac{2}{3}\langle \Delta \rangle t\right] + O(\varepsilon). \tag{14.2-33}$$

Here, h_o denotes the film thickness at $t = 0$. Differentiating the preceding relation shows that the velocity of thinning depends linearly upon $\langle \Delta \rangle$ and film thickness h; namely,

$$\frac{dh}{dt} = -\frac{2}{3} \Delta h + O(\varepsilon). \tag{14.2-34}$$

14.3 Foam Dynamics and Rheological Properties

The instantaneous, global pressure tensor $\langle \mathbf{P} \rangle$ of a spatially periodic medium will be defined in the following by the cellular relation (Adler & Brenner 1985):

$$\langle \mathbf{P} \rangle \overset{\text{def}}{=} \frac{1}{\tau_o} \oint_{\partial \tau_o} \mathbf{x} \, d\mathbf{S} \cdot \mathbf{P} \,. \tag{14.3-1}$$

Here, **x** represents a local, cellular position vector, which for convenience is chosen to originate at the cell's geometric center, and **P** is the local pressure tensor.

Simple Shear: Two-Dimensional Foam Model

The hexagonal cell model is applied herein to the case of a dry foam $\phi \rightarrow 1$. Unlike the three-dimensional TKDH model (which also applies to the case of a dry foam $\phi \rightarrow 1$), the two-dimensional hexagonal cell model fails to realistically account for liquid flow in the thin foam films owing to the one-dimensionality of these films. Thus, the following analysis of foam response to shear deformation assumes all (equilibrium and nonequilibrium) forces to derive from within the pair of surfaces comprising the one-dimensional foam films.

In these circumstances, Eq. (14.3-1) is to be replaced by the discrete relation,

$$\langle \mathbf{P} \rangle = \frac{1}{\tau_o} \sum_{i=1}^{6} \mathbf{x}_i \mathbf{F}_i \,, \tag{14.3-2}$$

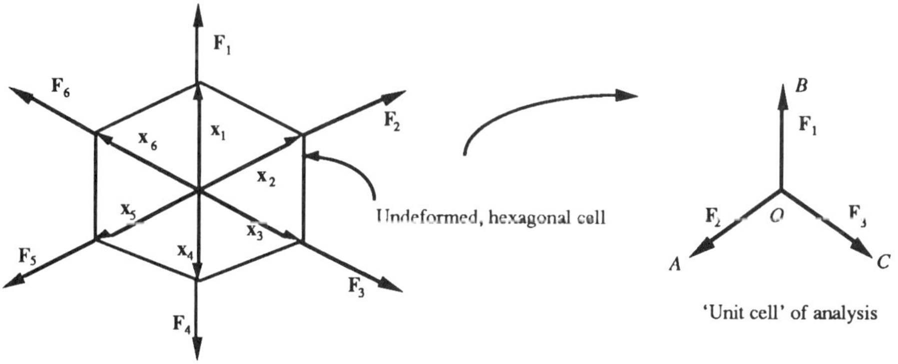

Figure 14.3-1 Replacement of original unit cell by an equivalent cell for the evaluation of Eq. (14.3-2).

where the vectors $\mathbf{F}_i$ represent outwardly directed forces from the bounding cell surface, originating at the six discrete Plateau borders of a single hexagonal foam cell (Figure 14.3-1). The position vectors $\mathbf{x}_i$ originate at the center of the hexagonal cell, and terminate at the Plateau borders. The hexagonal cell volume is given by

$$\tau_o = \frac{6\sqrt{3}}{4} a^2. \tag{14.3-3}$$

As indicated by a comparison of Figures 14.1-6 and 14.3-1, Eq. (14.3-2) may be represented equivalently by the sum,

$$\langle \mathbf{P} \rangle = \frac{2}{\tau_o} \sum_{i=1}^{3} \mathbf{g}_i \mathbf{F}_i , \tag{14.3-4}$$

where the vectors $\mathbf{F}_i$ are envisioned as acting outward along each film i from a common Plateau border, in the same direction as the film vectors $\mathbf{g}_i$, the latter set of vectors ($i = 1,2,3$) being depicted in Figure 14.1-6.

The film forces $\mathbf{F}_i$ may be expressed in terms of the surface-excess pressure tensor [cf. Eq. (4.2-15)] acting along the two surfaces comprising the *ith* foam film; thus,

$$\mathbf{F}_i = 2\mathbf{p}_i \cdot \mathbf{P}^s_i , \tag{14.3-5}$$

where the unit film vectors $\mathbf{p}_i$ are defined as in Eq. (14.1-10). Upon assuming the surface-excess stress tensor to satisfy the Newtonian form (4.2-17), it follows directly that

$$\mathbf{P}^s_i = \mathbf{p}_i \mathbf{p}_i (\sigma + \kappa^s \nabla_s \cdot \mathbf{v}^o_i),$$

since the one-dimensional film surfaces are incapable of undergoing shear deformation. Recalling the interpretation of $\nabla_s \cdot \mathbf{v}^o_i$ as the rate of surface expansion [cf. Eqs. (3.4-7) and (3.4-20)], there obtains the expression

$$\nabla_s \cdot \mathbf{v}^o_i = \frac{1}{g_i} \frac{dg_i}{dt} , \tag{14.3-6}$$

in terms of the film lengths g_i defined in Eq. (14.1-11) [cf. Eqs. (14.2-17)–(14.2-19)]. Combining the above relations yields

$$\mathbf{F}_i = 2\left[\sigma + \kappa^s \frac{1}{g_i} \frac{dg_i}{dt}\right] \mathbf{p}_i . \tag{14.3-7}$$

Upon substitution of Eqs. (14.3-7) and (14.1-10) into Eq. (14.3-4), there follows the equation

$$\boxed{\langle \mathbf{P} \rangle = \frac{4}{\tau_o} \sum_{i=1}^{3} \left[\sigma + \kappa^s \dot{\gamma} \frac{1}{g_i} \frac{dg_i}{d\gamma}\right] \mathbf{g}_i \mathbf{g}_i / g_i ,} \tag{14.3-8}$$

with $\dot{\gamma}$ representing the shear rate.

It should be recognized that Eq. (14.3-8) defines an *instantaneous* global pressure tensor. A corresponding time-independent foam pressure tensor may be determined by integrating Eq. (14.3-8) over a periodic time interval of the deformation process. An instantaneous, apparent, foam shear viscosity may also be determined from Eq. (14.3-8) in a straightforward manner (Question 14.2).

Pure Dilatation: Three-Dimensional Foam Models

The global foam pressure tensor (14.3-1) is evaluated below to leading order in the liquid-phase volume fraction $(1-\phi)$ by examining the dynamic response of the three-dimensional RDH and TKDH foam models to a pure foam dilatation (14.2-21). In the relatively dilute RDH case, the leading-order, local foam rheological contribution to the global pressure tensor is that contributed by the interfacial rheological properties of bubble surfaces. In the TKDH case, shear stresses originating in the thin foam films which bound the TKDH bubbles contribute the primary contribution to the global foam stress tensor. Interfacial rheological properties influence the magnitude of these film stresses by altering the mobility of the foam surfaces, as in the thin film analyses of Chapters 11 and 12.

Rhomboidal Dodecahedron Unit Cell

Upon employing the divergence theorem for a closed volume with a surface of discontinuity (Question 3.2), together with the identity (4.4-35d), Eq. (14.3-1) may ultimately be rewritten as

$$\langle \mathbf{P} \rangle = \frac{1}{\tau_o} \int_{S_b} \mathbf{x} \, d\mathbf{S} \cdot \mathbf{P} + \frac{1}{\tau_o} \int_{\tau_f} \mathbf{P} \, dV \ , \qquad (14.3\text{-}9)$$

where we have used the fact that

$$\nabla \cdot \mathbf{P} = \mathbf{0}$$

for quasistatic creeping flows.

The second integral appearing in Eq. (14.3-9) diminishes at a rate $O(1-\phi)$ with increasing dispersed-phase volume fraction ϕ; in contrast, the first integral, representing an integration of the *liquid*-phase pressure tensor over the surface of a single bubble, is independent of ϕ. Thus, to leading order in the liquid-phase volume fraction,

$$\langle \mathbf{P} \rangle = \frac{1}{\tau_o} \int_{S_b} \mathbf{x} \, d\mathbf{S} \cdot \mathbf{P} + O(1-\phi). \qquad (14.3\text{-}10)$$

The interfacial-stress boundary condition (4.2-9) [particularly note in this relation the definition of the 'jump' condition (3.6-12b) appearing on the left-hand side] allows the above equation to be rewritten as:

$$\langle \mathbf{P} \rangle = \frac{1}{\tau_o} \int_{S_b} \mathbf{x} dS \left[\mathbf{n} \cdot \mathbf{P}_b - \nabla_s \cdot \mathbf{P}^s - \mathbf{F}^s \right] + O(1-\phi),$$

$$= \frac{1}{\tau_o} \int_{S_b} \mathbf{x} dS \left[-\mathbf{n}\, p_b - \nabla_s \cdot \mathbf{P}^s - \mathbf{F}^s \right] + O(1-\phi). \quad (14.3\text{-}11)$$

Here, $\mathbf{n}$ is the unit outward normal of the bubble surfaces, p_b the bubble pressure, $\mathbf{P}^s$ the surface-excess pressure tensor (4.2-15) and $\mathbf{F}^s$ the surface-excess force. The latter areal vector field density is sensible particularly along the (relatively small) proportion S_b' of total bubble surface S_b which forms a partial boundary to planar foam films; this force arises on account of the existence of a disjoining pressure [cf. Eq. (11.2-7)].

Assuming the validity of the Newtonian interfacial rheological expression (4.2-17), it follows, upon substitution of Eq. (14.2-24) into this relation and subsequent substitution of the result [with Eqs. (4.2-15) and (3.3-8)] into Eq. (14.3-11), that

$$\nabla_s \cdot \mathbf{P}^s = \begin{cases} -\frac{2}{a}\left(\sigma + \frac{2}{a}\kappa^s v^s\right)\mathbf{n} & \text{on } S_b'', \\ \mathbf{0} & \text{on } S_b'. \end{cases} \quad (14.3\text{-}12)$$

Furthermore, upon utilizing Eq. (11.2-7), the surface-excess force vector may be expressed as

$$\mathbf{F}^s = \begin{cases} \mathbf{0} & \text{on } S_b'', \\ -\Pi\,\mathbf{n} & \text{on } S_b'. \end{cases} \quad (14.3\text{-}13)$$

Substitution of the preceding expressions into Eq. (14.3-11) yields

$$\langle \mathbf{P} \rangle = \frac{1}{\tau_o} \int_{S_b''} a\mathbf{nn} dS \left[-p_b + \frac{2}{a}(\sigma + \kappa^s v^s) \right]$$

$$+ \frac{1}{\tau_o} \int_{S_b'} (b/2)\mathbf{nn} dS \left[-p_b + \Pi \right], \quad (14.3\text{-}14)$$

which relation employs the bubble geometry of the RDH model discussed in §14.2 (see Figure 14.1-2 and the discussion thereof). Performing the trace of this expression reveals that

$$\mathbf{I}:\langle \mathbf{P} \rangle = \frac{aS_b''}{\tau_o}\left[-p_b + \frac{2}{a}\left(\sigma + \frac{2}{a}\kappa^s v^s\right)\right] + \frac{bS_b'}{2\tau_o}\left[-p_b + \Pi \right]$$

$$\equiv 3\left[-p_b + \frac{2\sigma}{a} \right] + \frac{4S_b''}{a\tau_o}\kappa^s v^s, \quad (14.3\text{-}15)$$

in which the latter step has been accomplished upon assuming the validity of Eq. (11.2-4).

Integration of Eq. (14.3-14) reveals (as is clear also upon intuitive grounds) that $\langle \mathbf{P} \rangle$ is an isotropic tensor for the pure dilatation (14.2-21).

Thus, as in Eq. (4.4-37), $\langle \mathbf{P} \rangle$ may be decomposed in terms of a global foam pressure $\langle p \rangle$ and global foam dilatational viscosity $\langle \kappa \rangle$, as

$$\langle \mathbf{P} \rangle = \mathbf{I}[-\langle p \rangle + \langle \kappa \rangle \langle \Delta \rangle], \tag{14.3-16}$$

whence it follows that

$$\mathbf{I}{:}\langle \mathbf{P} \rangle = 3[-\langle p \rangle + \langle \kappa \rangle \langle \Delta \rangle]. \tag{14.3-17}$$

Upon comparison of Eqs. (14.3-15) and (14.3-17), equating like powers in $\langle \Delta \rangle$ and utilizing Eq. (14.2-24b), it follows that, in terms of the characteristic dimension b of the RDH polyhedron (see Figure 14.1-2), and to leading order in the liquid-phase volume fraction,

$$\langle \kappa \rangle = \frac{4\kappa^s}{3(b/2)} \left[\frac{\cos\theta}{1 + (S'_b / S''_b)\cos\theta} \right] + O(1-\phi), \tag{14.3-18}$$

as well as

$$\langle p \rangle = p_b - \frac{2\sigma}{(b/2)} \cos\theta + O(1-\phi). \tag{14.3-19}$$

Additionally, Eqs. (11.2-4), (11.2-10) and (11.2-11) may be employed to express the Plateau border curvature radius a [$= (b/2)\cos\theta$] as a function of foam film contact angle θ and film thickness h, yielding the relation

$$a = \frac{h}{\dfrac{\sigma_f}{2\sigma} - \cos\theta}. \tag{14.3-20}$$

By approximating the Plateau border surface area by the area of the bounding *cell surface* not bordering a thin liquid film (i.e. $S''_b \approx A - S'_b$, where $A = (9/2)b^2$ is the total area of the surface bounding the RDH cell), Eq. (14.3-18) may be expressed by the formula

$$\boxed{\langle \kappa \rangle = \frac{4\kappa^s}{3h} \left[\frac{\dfrac{\sigma_f}{2\sigma} - \cos\theta}{1 + \dfrac{\cos\theta \tan^2\theta}{(3/2\pi) - \tan^2\theta}} \right] + \ldots} \tag{14.3-21}$$

Equation (14.3-21) furnishes a semi-quantitative estimate of the dilatational viscosity of a 'wet' (i.e. low dispersed-phase volume fraction) foam. It shows a direct proportionality between interfacial dilatational viscosity and foam viscosity $\langle \kappa \rangle$ and an inverse dependence upon thin film thickness. The proportionality between foam viscosity and foam film contact angle may be an artifice of the RDH cell geometry (though it does ultimately indicate an increase in the dilatational viscosity $\langle \kappa \rangle$ with increasing dispersed-phase volume fraction ϕ, as expected). The absence of an interfacial-tension gradient contribution in Eq. (14.3-21) owes to the fact that the thin films (in which regions interfacial-tension gradients play their most important role) of a wet foam occupy relatively miniscule liquid regions between adjacent bubbles, in comparison to the meniscus regions. Interfacial-tension gradients will, however, play a significant role in the rheology of 'dry' foams [cf. Eq. (14.3-63)], for which the greatest areal fraction of each closed bubble surface furnishes upper and lower boundaries to thin liquid films. This latter case is considered below.

Tetrakaidecahedron Unit Cell

As demonstrated in §14.2 [cf. Eqs. (14.2-25)–(14.2-34)], the contraction or expansion of cellular units within a spatially periodic foam leads to a thinning or thickening of thin liquid films formed between the planar faces of neighboring polyhedral bubbles. In the following, a lubrication hydrodynamic analysis of the thin film motion caused by the normal film surface velocity (14.2-34) is employed, together with the restriction that the total liquid volume within the films is for all times conserved (as is strictly the case for an isotropic unit cell expansion), to determine the leading-order contribution to the global dilatational viscosity of a TKDH foam undergoing a pure dilatational motion (14.2-21).

The quasistatic Stokes and continuity equations in the liquid phase within a thin liquid film of thickness h (see Figure 14.1-5) are given by*

$$\nabla \cdot \mathbf{P} = \mathbf{0} \tag{14.3-22}$$

and

$$\nabla \cdot \mathbf{v} = 0, \tag{14.3-23}$$

with **P** the liquid-phase Newtonian pressure tensor [cf. Eqs. (4.1-13) and (4.1-21)]. These are subject to the following boundary conditions:

$$\mathbf{v} \cdot \mathbf{n} = 0 \qquad \text{at} \quad z = 0, \tag{14.3-24}$$

$$\mathbf{n} \cdot \nabla \mathbf{v} = \mathbf{0} \qquad \text{at} \quad z = 0, \tag{14.3-25}$$

$$\mathbf{v} \cdot \mathbf{n} = -V \qquad \text{at} \quad z = h/2, \tag{14.3-26}$$

and [cf. Eq. (4.2-9)]

$$-\mathbf{n} \cdot \|\mathbf{P}\| = \mathbf{F}^s + \nabla_s \cdot \mathbf{P}^s \qquad \text{at} \quad z = h/2. \tag{14.3-27}$$

Here, $V=(1/2)dh/dt$ is given by Eq. (14.2-34), **n** is the unit surface normal of the bubble surface (pointing inward to the liquid phase, as in Figure 11.2-1) and the surface-excess pressure tensor $\mathbf{P}^s$ is assumed to obey

* Whereas our explicit attention is presently to a (hexagonal) film of thickness h, the results of this section are generalized to (square) films, whose thickness is $\sqrt{2}h$, later in the section [cf. Eq. (14.3-55)].

Newtonian constitutive behavior [cf. Eqs. (4.2-15), (4.2-17) and (4.2-18)]; explicitly,

$$\mathbf{P}^s = \mathbf{I}_s[\sigma + (\kappa^s - \mu^s)\nabla_s \cdot \mathbf{v}] + \mu^s\left(\mathbf{I}_s \cdot \nabla\mathbf{v} + \mathbf{I}_s \cdot (\nabla\mathbf{v})^\dagger\right), \qquad (14.3\text{-}28)$$

with μ^s the surface shear viscosity, κ^s the surface dilatational viscosity and σ the surface tension.

As in the thin film drainage problem of §11.3, simultaneous consideration of the surfactant transport problem is required in order to specify an appropriate constitutive relation for the surface tension gradient arising in the tangential component of the condition (14.3-27).

The quasistatic convective-diffusion equation (5.1-5) may be expressed in an invariant form as

$$\mathbf{v} \cdot \nabla\rho_1 = D\nabla^2\rho_1, \qquad (14.3\text{-}29)$$

for surfactant species '1'. This equation is subject to the following boundary conditions:

$$\mathbf{n} \cdot \nabla\rho_1 = \mathbf{0} \quad \text{at} \quad z = 0, \qquad (14.3\text{-}30)$$

and [cf. Eq. (5.2-6)]

$$\nabla_s \cdot (\mathbf{v}\,\rho_1^s) = D^s\nabla_s^2\rho_1^s + \mathbf{n} \cdot \|\mathbf{j}_1\| \quad \text{at} \quad z = h/2, \qquad (14.3\text{-}31)$$

which latter surface transport process is assumed to be 'diffusion-controlled' (cf. §5.4), as was shown to be the case in the film drainage problem of Chapter 11 (see the discussion following Figure 11.3-5).

A general solution to Eqs. (14.3-22) and (14.3-23) in circular cylindrical coordinates (R, θ, z) is provided by the relations

$$v_R = \frac{1}{\mu}\sum_{n=1}^{\infty}\{(A_n + C_n z)e^{-\lambda_n z} + (B_n - D_n z)e^{\lambda_n z}\}J_1(\lambda_n R), \qquad (14.3\text{-}32)$$

$$v_z = \frac{1}{\mu}\sum_{n=1}^{\infty}\left\{\left[A_n + C_n\left(z + \frac{1}{\lambda_n}\right)\right]e^{-\lambda_n z} + \left[B_n - D_n\left(z - \frac{1}{\lambda_n}\right)\right]e^{\lambda_n z}\right\}J_0(\lambda_n R), \qquad (14.3\text{-}33)$$

and

$$p = \sum_{n=1}^{\infty}2\{C_n e^{-\lambda_n z} + D_n e^{\lambda_n z}\}J_0(\lambda_n R), \qquad (14.3\text{-}34)$$

where $J_1(\lambda_n R)$ and $J_0(\lambda_n R)$ are, respectively, first- and zeroth-order Bessel functions of the first kind respectively and A_n, B_n, C_n, D_n and λ_n are coefficients to be determined. Applying boundary conditions (14.3-24), (14.3-25) and (14.3-26) to Eqs. (14.3-32)–(14.3-34) yields

$$v_R = -12V\frac{Rz^2}{h^3} + \frac{1}{\mu}\sum_{n=1}^{\infty} A_n\Big\{\left(e^{-\lambda_n z} + e^{\lambda_n z}\right) - z\frac{\left(e^{-\lambda_n z} - e^{\lambda_n z}\right)\left(e^{-\lambda_n z} - e^{\lambda_n z}\right)}{\left(h/2 + \frac{1}{\lambda_n}\right)e^{-\lambda_n z} + \left(h/2 - \frac{1}{\lambda_n}\right)e^{\lambda_n z}}\Big\}J_1(\lambda_n R), \quad (14.3\text{-}35)$$

$$v_z = 8V\frac{z^3}{h^3} + \frac{1}{\mu}\sum_{n=1}^{\infty} A_n\Big\{\left(e^{-\lambda_n z} - e^{\lambda_n z}\right) - \left(e^{-\lambda_n z} - e^{\lambda_n z}\right) \times \frac{\left(z + \frac{1}{\lambda_n}\right)e^{-\lambda_n z} + \left(z - \frac{1}{\lambda_n}\right)e^{\lambda_n z}}{\left(h/2 + \frac{1}{\lambda_n}\right)e^{-\lambda_n z} + \left(h/2 - \frac{1}{\lambda_n}\right)e^{\lambda_n z}}\Big\}J_0(\lambda_n R), \quad (14.3\text{-}36)$$

and

$$p = -12V\frac{R^2}{h^3} + 24V\frac{z^2}{h^3} - \frac{1}{\mu}\sum_{n=1}^{\infty} 2A_n\Big(e^{-\lambda_n z} - e^{\lambda_n z}\Big) \times \frac{\left(e^{-\lambda_n z} + e^{\lambda_n z}\right)}{\left(h/2 + \frac{1}{\lambda_n}\right)e^{-\lambda_n z} + \left(h/2 - \frac{1}{\lambda_n}\right)e^{\lambda_n z}}\Big\}J_0(\lambda_n R). \quad (14.3\text{-}37)$$

Introducing the following dimensionless quantities

$$\begin{aligned} \tilde{z} &\equiv 2z/h & \tilde{A}_n &\equiv 2\varepsilon A_n / V\mu, \\ \tilde{R} &\equiv R/f & \tilde{\lambda}_n &\equiv f\lambda_n, \\ \tilde{v}_R &\equiv \varepsilon v_R / V & \tilde{p} &\equiv p\frac{\varepsilon^3 f}{\mu V}, \\ \tilde{v}_z &\equiv v_z / V & \tilde{\sigma} &\equiv \frac{\varepsilon^2 \sigma}{\mu V}, \end{aligned} \quad (14.3\text{-}38)$$

permits Eqs. (14.3-35)–(14.3-37) to be expressed to leading order in ε as

$$\tilde{v}_R = -\frac{3}{2}\tilde{R}\tilde{z}^2 + \sum_{n=1}^{\infty}\tilde{A}_n\left(1 - 3\tilde{z}^2\right)J_1(\tilde{\lambda}_n\tilde{R}) + O(\varepsilon), \quad (14.3\text{-}39)$$

$$\tilde{v}_z = \tilde{z}^3 + \sum_{n=1}^{\infty}\tilde{\lambda}_n\tilde{A}_n\left(\tilde{z}^3 - \tilde{z}\right)J_0(\tilde{\lambda}_n\tilde{R}) + O(\varepsilon), \quad (14.3\text{-}40)$$

$$\tilde{p} = -\frac{3}{2}\tilde{R}^2 + \sum_{n=1}^{\infty} 6\tilde{A}_n\frac{1}{\tilde{\lambda}_n}J_0(\tilde{\lambda}_n\tilde{R}) + O(\varepsilon), \quad (14.3\text{-}41)$$

or, equivalently,

$$\tilde{v}_R = -\frac{3}{2}\tilde{R}\tilde{z}^2 + \left(1 - 3\tilde{z}^2\right) B(\tilde{R}) + O(\varepsilon), \qquad (14.3\text{-}42)$$

$$\tilde{v}_z = \tilde{z}^3 + \left(\tilde{z}^3 - \tilde{z}\right)\frac{1}{\tilde{R}}\frac{d}{d\tilde{R}}\tilde{R}B(\tilde{R}) + O(\varepsilon), \qquad (14.3\text{-}43)$$

$$\tilde{p} = -\frac{3}{2}\tilde{R}^2 + 6\int_{m=0}^{\tilde{R}} B(m)\,dm + O(\varepsilon). \qquad (14.3\text{-}44)$$

The remaining undetermined variable B may be determined by the tangential stress condition (14.3-27) [cf. Eq. (11.3-12) for an explicit representation of this condition in circular cylindrical coordinates], which condition, to leading order in ε, may be expressed in non-dimensional form as,

$$\frac{\partial \tilde{v}_R}{\partial \tilde{z}} = \frac{\partial \tilde{\sigma}}{\partial \tilde{R}} + O(\varepsilon) \quad \text{at } \tilde{z} = 1. \qquad (14.3\text{-}45)$$

As in previous examples of this type (cf. §§5.6, 11.3 and 12.2), we assume small perturbations of the surfactant mass densities ρ_1 and ρ_1^s from their respective equilibrium values ρ_{1o} and ρ_{1o}^s; namely,

$$\tilde{\rho}_1 = \rho_1 / \rho_{1o} = 1 + O(\varepsilon), \qquad (14.3\text{-}46)$$

and

$$\tilde{\rho}_1^s = \rho_1^s / \rho_{1o}^s = 1 + O(\varepsilon), \qquad (14.3\text{-}47)$$

whence the approximation (5.5-4) holds, yielding the following leading-order constitutive expression for the dimensionless interfacial tension gradient:

$$\frac{\partial \tilde{\sigma}}{\partial \tilde{R}} = \left(\frac{\partial \tilde{\sigma}}{\partial \tilde{\rho}_1^s}\right)_o \frac{\partial \tilde{\rho}_1^s}{\partial \tilde{R}} + O(\varepsilon). \qquad (14.3\text{-}48)$$

It furthermore follows from Eqs. (14.3-29) and (14.3-30) that, to leading order in ε,

$$\frac{\partial \tilde{\rho}_1}{\partial \tilde{z}} = O(\varepsilon), \qquad (14.3\text{-}49)$$

which, with Eq. (14.3-31), suggests the leading-order relation,

$$\frac{1}{\tilde{R}}\frac{\partial}{\partial \tilde{R}}\tilde{R}\tilde{v}_R = \frac{1}{\mathrm{Pe}_s}\frac{1}{\tilde{R}}\frac{\partial}{\partial \tilde{R}}\tilde{R}\frac{\partial \tilde{\rho}_1^s}{\partial \tilde{R}} + O(\varepsilon), \qquad (14.3\text{-}50)$$

where

$$\mathrm{Pe}_s \equiv \frac{Vf}{\varepsilon D^s} \tag{14.3-51}$$

is the surface Peclét number.

Finally, upon employing Eqs. (14.3-42), (14.3-45), (14.3-48) and (14.3-50), we obtain

$$B = -\frac{1}{2}\mathrm{N}_\sigma \tilde{R}, \tag{14.3-52}$$

wherein

$$\mathrm{N}_\sigma \equiv \frac{1 + (1/2)\tilde{E}_o \,\mathrm{Pe}_s}{1 + (1/3)\tilde{E}_o \,\mathrm{Pe}_s}, \tag{14.3-53}$$

with

$$\tilde{E}_o = -\left(\frac{\partial \tilde{\sigma}}{\partial \tilde{\rho}_1^s}\right) \tag{14.3-54}$$

defining a dimensionless Gibbs elasticity number [cf. Eq. (5.5-5)]. The relation (14.3-52) may be substituted into Eqs. (14.3-42)–(14.3-44) to furnish explicit leading-order hydrodynamic expressions for the velocity and pressure fields within the thin foam film. Insofar as these results are expressed in a dimensionless form, they are equally applicable to hexagonal and square films, provided the dimensionless variable $\tilde{z}$ is defined appropriately. Thus, the scaling used in the representation (14.3-38) applies to the case of a hexagonal film (see Figure 14.1-5), whereas

$$\tilde{z} \equiv 2z \;/\sqrt{2}\,h \tag{14.3-55}$$

furnishes the equivalent dimensionless variable for square films (whose thickness is $\sqrt{2}\,h$ rather than h).

To determine the global foam dilatational viscosity $\langle\kappa\rangle$ we evaluate below the global viscous energy dissipation function

$$\begin{aligned}\langle\Phi\rangle &= \langle\kappa\rangle\langle\Delta\rangle^2 \\ &\equiv \frac{1}{\tau_o}\int_{\tau_o} 2\mu \mathbf{D}:\mathbf{D}\,dV + \frac{1}{\tau_o}\int_{S_b}(\mathbf{P}^s - \sigma \mathbf{I}_s):\nabla_s \mathbf{v}\,dA\,.\end{aligned} \tag{14.3-56}$$

The first term on the right-hand side may be seen to express the total viscous energy dissipation arising due to flow in the liquid (film) phase (Bird *et al.* 1960, Adler & Brenner 1985) and the second term corresponds to the total (cellular) energy dissipation arising on account of flow in the bubble surface (see Question 16.3).

Upon noting the definition (4.1-16) of the deformation rate tensor and utilizing the leading-order expressions (14.3-42) and (14.3-33) for the velocity field within a characteristic thin liquid film, it follows that,

$$\tilde{\mathbf{D}} = \left(\frac{\partial \tilde{V}_r}{\partial \tilde{z}}\right)(\mathbf{i}_z \mathbf{i}_R + \mathbf{i}_R \mathbf{i}_z) + O(\varepsilon), \qquad (14.3\text{-}57)$$

furnishes a leading-order approximation to the dimensionless deformation rate tensor. Introducing the dimensionless quantity

$$\langle \tilde{\kappa} \rangle \equiv \langle \kappa \rangle \varepsilon \,/\, \mu, \qquad (14.3\text{-}58)$$

and observing that from Eq. (14.2-34),

$$\langle \Delta \rangle = \begin{cases} -\dfrac{3V}{h} & \text{(hexagonal films)}, \\ -\dfrac{3V}{\sqrt{2}\,h} & \text{(square films)}, \end{cases} \qquad (14.3\text{-}59)$$

it ultimately follows that

$$\langle \tilde{\kappa} \rangle = \frac{1 - N_\sigma + N_\sigma^2}{\sqrt{2}} \iiint_{\text{hexagonal}} d\tilde{x}\, d\tilde{y}\, d\tilde{z} \left(\tilde{x}^2 + \tilde{y}^2\right) \tilde{z}^2$$
$$+ \left(1 - N_\sigma + N_\sigma^2\right) \iiint_{\text{square}} d\tilde{x}\, d\tilde{y}\, d\tilde{z} \left(\tilde{x}^2 + \tilde{y}^2\right) \tilde{z}^2 + O(\varepsilon). \qquad (14.3\text{-}60)$$

In this expression (dimensionless) Cartesian coordinates $(\tilde{x}, \tilde{y}, \tilde{z})$ have been employed (with $\tilde{x} = x/f$, $\tilde{y} = y/f$) originating at the centerpoint of each film. In addition, the interfacial viscous dissipation terms have been observed to represent lower-order effects relative to the dissipation in the thin films. As may be surmized from Figure 14.1-4 [noting also that there exist eight hexagonal and six square (half) films per TKDH cell], the explicit integrations required for the TKDH unit cell are:

$$\iiint_{\text{hexagonal}} d\tilde{x}\, d\tilde{y}\, d\tilde{z} = 8 \times 12 \times \int_{\tilde{x}=0}^{\sqrt{3}/2} d\tilde{x} \int_{\tilde{y}=0}^{\tilde{x}/\sqrt{3}} d\tilde{y} \int_{\tilde{z}=0}^{1} d\tilde{z}, \qquad (14.3\text{-}61)$$

and

$$\iiint_{\text{square}} d\tilde{x}\, d\tilde{y}\, d\tilde{z} = 6 \times \int_{\tilde{x}=0}^{1} d\tilde{x} \int_{\tilde{y}=0}^{1} d\tilde{y} \int_{\tilde{z}=0}^{1} d\tilde{z}. \qquad (14.3\text{-}62)$$

Substitution of Eqs. (14.3-61) and (14.3-62) into Eq. (14.3-60), upon integration of the former, yields

$$\boxed{\langle \kappa \rangle = \mu \frac{4\left(4 + \sqrt{2}\right)}{3\varepsilon \sqrt{2}} \left(1 - N_\sigma + N_\sigma^2\right) + O(1),} \qquad (14.3\text{-}63)$$

for the leading order (dimensional) global dilatational viscosity of a 'dry', monodisperse, spatially periodic, TKDH foam. The dilatational viscosity $\langle \kappa \rangle$ arises due to two primary mechanisms: (i) viscous flow within the thin films, and (ii) interfacial-tension gradients acting along the foam bubble surfaces. The effect of interfacial-tension gradients is to increase the foam viscosity as they impede flow near the surfaces of the thin foam films by contributing to a larger film stress. As in the case of a wet foam [cf. Eq. (14.3-21)], the foam dilatational viscosity is seen to be inversely proportional to thin film thickness.

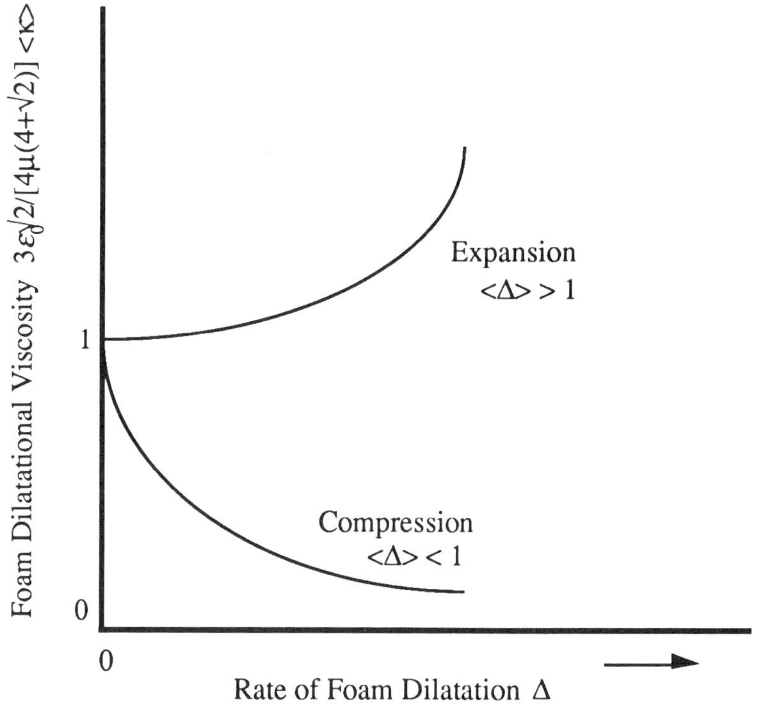

Figure 14.3-1 Dimensionless foam dilatational viscosity vs foam dilatation rate for a dry foam TKDH model. Here, $\varepsilon_0 \equiv h_0/f_0$ represents a reference state film size.

Figure 14.3-1 displays the dependence of the foam dilatational viscosity upon the rate of foam expansion, for $N_\sigma = 1$. The foam viscosity $\langle\kappa\rangle$ increases for a foam expansion, as the foam films thin, and velocity gradients occuring over the thickness of the film increase in magnitude, whereas for opposite reasons the foam viscosity $\langle\kappa\rangle$ decreases as the foam is compressed.

14.4 Summary

Foam rheological behavior may be greatly influenced by the rheological nature of foam bubble surfaces. In the present chapter, this fact has been demonstrated for shear and dilatational foam deformations by employing various two- and three-dimensional spatially periodic foam models.

Interfacial viscosities have been observed to be most important for 'wet' foams, for which liquid is contained primarily within the meniscus (i.e. Plateau border) regions of the foam. The deformation of bubble surfaces thus produces an interfacial viscous stress that becomes increasingly important as the surface-to-liquid volume ratio increases [cf. Eqs. (14.3-9) and (14.3-10)]. A result for the leading-order dilatational viscosity of a 'wet' (i.e. relatively low dispersed-phase volume fraction) foam (composed of periodic rhomboidal dodecahedron unit cells) has been calculated [cf. Eq. (14.3-21)].

Interfacial tension gradients were demonstrated to be important in the dry foam limit ($\phi \rightarrow 1$) owing to the increased importance of liquid flow in the thin liquid films of the foam, in which most of the liquid of a dry foam is bound. Interfacial-tension gradients inhibit liquid flow near the film surfaces thereby increasing the viscous nature of the foam. The leading-order dilatational viscosity of a dry foam composed of a spatially periodic array of tetrakaidecahedron bubbles has been determined [cf. Eq. (14.3-68)].

A two-dimensional model was also employed to demonstrate the importance of interfacial viscous effects in the shear deformation of a (dry) foam. This led to the result (14.3-8) for the foam pressure tensor in terms of the dilatational viscosity of bubble surfaces.

Questions for Chapter 14

14.1 Condition (14.2-8) establishes a stability condition for the evolution of the two-dimensional hexagonal foam structure of Figure 14.1-6 in the case in which local equilibrium is assumed at the Plateau borders. In general, the angles at which three films meet at a representative Plateau border must be determined by requiring the summation of forces [cf. Eq. (14.3-7)] acting along each film at the Plateau border to vanish; explicitly,

$$\sum_{i=1}^{3} \mathbf{F}_i = \mathbf{0}.$$

Use this condition with Eqs. (14.3-7), (14.1-5)–(14.1-7), (14.1-10), (14.1-11), (14.2-6) and (14.2-7) to obtain two first-order, nonlinear, coupled ordinary differential equations for the x and y components of $\mathbf{r}$ ($=x\mathbf{i}_x+y\mathbf{i}_y$) as a function of time and shear rate $\dot{\gamma}$ [i.e. as provided by Eqs. (14.2-12) and (14.2-13)].

14.2 As in Eq. (14.3-46), a global (instantaneous) foam *shear* viscosity $\langle \mu \rangle$ may be defined for the two-dimensional foam model considered in this chapter upon equating global and local viscous energy dissipation. This gives,

$$\langle \Phi \rangle = 2\langle \mu \rangle \dot{\gamma}^2$$
$$\equiv 2\frac{1}{\tau_o} \sum_{i=1}^{3} \boldsymbol{\tau}_i^s : \nabla_s \mathbf{v}_i^o \, .$$

where

$$\boldsymbol{\tau}_i^s = \mathbf{P}_i^s - \sigma \mathbf{p}_i \mathbf{p}_i$$

and

$$\tau_o = \frac{6a^2\sqrt{3}}{4} .$$

i. Using Eq. (14.3-6) together with Eqs. (14.2-17)–(14.2-19), derive an explicit expression for $\langle \mu \rangle$ for the case of negligibly small foam shear strain (i.e., $\gamma = 0$).

ii. Yield stress phenomena may be incorporated into the definition of an apparent foam shear viscosity $\langle \mu \rangle_{app}$ by equating the 'apparent' global viscous energy dissipation with the total work required to deform the local film structure; explicitly (see Question 16.1),

$$2\langle \mu \rangle_{app} \dot{\gamma}^2 = 2\frac{1}{\tau_o} \sum_{i=1}^{3} \mathbf{P}_i^s : \nabla_s \mathbf{v}_i^o$$
$$\equiv 2\frac{1}{\tau_o}\sigma \sum_{i=1}^{3} \nabla_s \mathbf{v}_i^o + 2\frac{1}{\tau_o} \sum_{i=1}^{3} \boldsymbol{\tau}_i^s : \nabla_s \mathbf{v}_i^o .$$

Derive a definition for $\langle \mu \rangle_{app}$ using this relation (observing that the surface tension term appearing in the resulting relation may be identified with the yield stress of the two-dimensional foam).

14.3 *i.* Substitute the value (14.3-53) into the relations (14.3-42)–(14.3-44) to obtain explicit solutions for the velocity and pressure fields within a characteristic thinning/thickening film of the expanding/contracting TKDH foam.

ii. Find the limiting velocity and pressure fields in the two limiting circumstances wherein the interfacial tension gradients are negligibly small and very large.

14.4 *i.* Derive the expressions (14.2-12) and (14.2-13) by substituting Eqs. (14.2-14)–(14.2-16) into Eq. (14.2-8).

ii. Find the results (14.2-17)–(14.2-19) by substituting Eqs. (14.2-12)–(14.2-16) into Eq. (14.1-11).

Additional Reading for Chapter 14

§14.1 Geometrical Models

Lissant, K.J. 1966 The geometry of high-internal-phase-ratio emulsions. *J. Colloid Interface Sci.* **22**, 463–468.

Princen, H.M. & Levinson, P. 1987 The surface area of Kelvin's minimal tetrakaidecahedron: The ideal foam cell? *J. Colloid Interface. Sci.* **120**, 172–175.

Princen, H.M., Aronson, M.P. & Moser, J.C. 1980 Highly concentrated emulsions II. Real systems. The effect of film thickness and contact angle on the volume fraction in creamed emulsions. *J. Colloid Interface. Sci.* **75**, 247–270.

§14.2 Kinematics of Spatially-Periodic Foam Media

Khan, S.A. & Armstrong, R.C. 1987 Rheology of foams I. Theory for dry foams. *J. Non-Newtonian Fluid Mech.* **22**, 1–22.

§14.3 Foam Dynamics and Rheological Properties

Edwards, D.A. & Wasan, D. T. 1990 Foam dilatational rheology I. Dilatational viscosity. *J. Colloid Interface Sci.* **139**, 479–487.

PART II

MICROMECHANICAL THEORY OF INTERFACIAL TRANSPORT PROCESSES

15

"But we know from observation that it is only within very small distances of such a surface that any mass is sensibly affected by its vicinity,—a natural consequence of the exceedingly small sphere of molecular action,—and this fact renders possible a simple method of taking account of the variations in the densities of the component substances and of energy and entropy, which occur in the vicinity of surfaces of discontinuity. We may use this term, for the sake of brevity, without implying that the discontinuity is absolute, or that the term distinguishes any surface with mathematical precision. It may be taken to denote the nonhomogeneous film which separates the homogeneous or nearly homogeneous masses."

J. Willard Gibbs (1906)

CHAPTER 15

A Surface-Excess Theory of Interfacial Transport Processes

In Part I of this text we addressed the basic theory, methods of measurement, and practical applications of interfacial rheology while adhering throughout to a 'macroscale' point of view, wherein the interface between two separately homogeneous 'bulk' fluids is regarded as a two-dimensional deformable surface. This macroscale entity was assumed endowed with 'surface' properties, such as interfacial tension, interfacial viscosity, and interfacial diffusivity, which properties appear in boundary conditions [cf. Eqs. (4.2-20) and (5.2-6)] imposed upon the normal component of bulk-phase flux fields at the macroscale interface. Bulk transport processes occuring near or at the fluid interface were demonstrated (cf. §§4.4, 5.6, 10.3, 10.4, 11.3, 12.2, 14.3) to be sensibly influenced by the existence of such surface properties, notably for systems possessing a large specific surface, and for interfaces upon which are adsorbed surface-active agents.

In Part II, which follows, a rigorous, *microscale* theory of interfacial transport processes will be developed. The presentation will be seen to parallel the *macroscale* interfacial treatment of Chapters 3 to 5. Thus, the current chapter develops a generic surface-excess theory of interfacial transport processes. This development provides an analog of Chapter 3, wherein a purely macroscale theory of interfacial transport processes was addressed. Subsequently, in Chapter 16. we focus upon the particular circumstance of *linear and angular momentum transport*, as in the macroscale theory of Chapter 4. Chapter 17 applies the generic theory developed in the present chapter to *species transport,* thereby providing a complement to Chapter 5.

The primary utility of the surface-excess theory which follows is to elucidate the three-dimensional, microscale (ultimately, molecular) origins of the two-dimensional, macroscale phenomenological interfacial properties introduced in Part I. Thus, beginning with a three-dimensional, microcontinuum, *physical* view of the highly inhomogeneous interfacial transition zone, a rational, *mathematical* reconciliation of the macro- and microscale views is provided in this chapter, utilizing singular perturbation methods. Thereby, the generic surface-excess balance law (3.6-16) is now rederived [cf. Eq. (15.7-20)], but with an explicit microscale interpretation of the surface-excess fields appearing therein [cf. Eqs. (15.6-18) and (15.6-36)]. This allows the derivation in Chapters 16 and 17 of explicit surface-excess balance equations for the transport of linear momentum [cf. Eq. (16.1-13)], angular momentum [cf. Eq. (16.2-8)], species [cf. Eq. (17.2-4)], energy (Question 16.1) and entropy (Question 16.2). By adopting specific (illustrative) constitutive models of the microscale fields appearing in the corresponding definitions of surface-excess fields, appropriate constitutive equations for these (macroscale) surface-excess fields may be developed, as is demonstrated in §§16.3 and 17.2.

An extension of the surface-excess formalism to three-phase contact "lines" is outlined in Chapter 19, at the conclusion of which further generalizations are discussed.

The generic surface-excess balance equation (15.7-18) developed in the present chapter extends the comparable macroscale result (3.6-16) by including additional interfacial transport mechanisms deriving from the *nonmaterial* nature of fluid interfaces. The conceptual framework necessary to analyze nonmaterial interfaces will be seen to require only relatively little extension of those theoretical concepts already introduced in Part I.

We begin by reviewing conventional microscale geometric views of interfacial fluid systems , following which we review existing theoretical approaches towards modeling the macroscale kinematical and dynamical behavior of such systems. This leads subsequently, in §15.2, to a general overview of the asymptotic scheme employed in this chapter towards rationalizing such behavior. In §15.3 the fictitious (induced) nonlocal coordinate parameterization of space employed in the analysis is discussed, drawing upon geometrical topics previously introduced in Chapter 3, as well as in Appendices A and B. Section 15.4 addresses kinematical preliminaries, whereas §15.5 provides a detailed formulation of the matched asymptotic expansion scheme which underlies Part II of this book. This is followed in §15.6 by the derivation of surface-excess field properties, leading in §15.7 to the development of a generic surface-excess conservation equation for a *nonmaterial* interface.

15.1 The Microscale View of a Fluid Interface

The two-dimensional, macroscale view of interfaces pursued in Part I of this text (cf. §3.1), is universally recognized as representing only an approximate representation of the true physical state of the system, corresponding to the existence of a large disparity between the macrolength scale L and microlength scale l (over which length the microscale interfacial inhomogeneity exists); explicitly, $l/L \ll 1$. In actuality, the interfacial region between two immiscible fluids is a highly inhomogeneous, three-dimensional transition region across which rapid changes in material properties (and, concomitantly, fields, such as mass density, species concentration, etc.) may occur. In this diffuse, three-dimensional, microscale view, the relevant continuum fields are assumed to vary continuously throughout the entire fluid-filled space, albeit some fields extremely steeply within the interfacial region in the direction of the macroscale surface normal.

Although this microscale continuum description embodies the underlying physics of the problem—and thus presents the 'true' or 'exact' physical description* of the interfacial region—the macroscale view, with its simpler and more tractable equations, is generally the more useful of the two in applications, as has been convincingly documented in Part I. This state of affairs is a consequence of the fact that many experiments are, by design or default, unable to resolve the fine-scale detail existing within the interfacial transition region; instead, such coarse observations provide only gross information regarding the 'average' or macroscale effects of phenomena existing within the interfacial region upon those 'bulk fluid' phenomena existing in regions proximate to the interface.

In order to proceed from the 'exact' microscale description of the pertinent phenomena to a formulation of the macroscale interfacial constitutive equations, the constitutive form adopted by the microscale continuum fields within the interfacial transition region must be hypothesized. One important difference between the interfacial region and the contiguous bulk-fluid regions, which must be incorporated into any rational interfacial model, is the local inhomogeneity within the transition region created by the existence of short-range intermolecular and physicochemical forces. Two types of approaches—statistical-mechanical and continuum-mechanical—have been employed to model the interfacial transition region. In both types of approaches the inhomogeneities are generally incorporated by assuming that within the interfacial transition region all microscale continuum fields are locally transversely isotropic with respect to the direction normal to the dividing surface. Far away from the interface the microscale fields are assumed to become locally isotropic.

Statistical-Mechanical Approaches

Equilibrium statistical-mechanical methods (Kirkwood & Buff 1949, Irving & Kirkwood 1950, Buff 1955, Hill 1959, Defay *et al.* 1966, Rowlinson & Widom 1982) have frequently been used to investigate the diffuse interfacial transition region in equilibrium systems. In many of these schemes an anisotropic microscale pressure tensor is postulated in the interfacial region; cf. Question 16.5. (Far from the interfacial region the hydrostatic microscale pressure tensor is assumed to possess its standard isotropic form, characterized by a scalar pressure field.) This anisotropic pressure tensor is then determined from the singlet and pair molecular

---

* 'True' in a continuum sense. Of course, the discrete molecular view of matter is even 'truer'. However, in our hierarchy of views, the inhomogeneous continuum rather than molecular view will be adopted as embodying the 'first principles' physics underlying the traditional macroscale view of interfacial phenomena.

distribution functions. Using these microscale models, the macroscale constitutive equation for the surface-excess pressure tensor has been shown (Buff 1955) to be uniquely characterized and quantified by an interfacial tension, itself a surface-excess macroscale quantity.

Gradient theory (Van der Waals & Kohnstamm 1908, Lovett *et al.* 1972, Evans 1979, Davis & Scriven 1982) has also been used to model the interfacial transition region. In this approach, the *equilibrium* properties of the interface are assumed to be expressible solely in terms of the microscale fluid density and its normal gradients proximate to the interfacial region.

In recent years, nonequilibrium statistical mechanical arguments have been employed to derive the equations of motion for *dynamical* interfacial systems (Ronis *et al.* 1978, Ronis & Oppenheim 1978, Davis 1987).

Continuum-Mechanical Approaches

Macroscale constitutive equations for the momentum transport processes accompanying moving and deforming interfaces have been developed from a fluid-mechanical viewpoint (Eliassen 1963, Slattery 1967, Goodrich 1981a, Goodrich 1981b) by supposing the pertinent microscale fields within the interfacial transition region to be transversely isotropic with respect to the direction normal to the dividing surface, while possessing linear constitutive forms similar to those of three-dimensional bulk fluids. In particular, the stress tensor is assumed to possess a transversely isotropic Newtonian form.

Our continuum-mechanical scheme will not initially confine itself to a particular constitutive model of the three-dimensional fluid proximate to the interface, although specific constitutive choices will eventually be addressed in §§16.3 and 17.2. In the former section, transversely isotropic Newtonian behavior will be assumed for the interfacial deviatoric stress tensor [cf. Eq. (16.3-6)]; nevertheless, the pressure itself will be assumed *isotropic*, as in conventional hydrodynamic descriptions. Within the interfacial transition region, pressure gradients in equilibrium systems will be supposed balanced by forces of an intermolecular origin (jointly with any other external forces that may be acting on the system). Interfacial tension will thus appear as the macroscopic manifestation of short-range intermolecular forces arising from physicochemical inhomogeneities existing within the interfacial transition region. This *isotropic* view of the equilibrium pressure tensor is in marked contrast to the more conventional statistical-mechanical view, wherein the intermolecular forces are implicitly incorporated into an *anisotropic* equilibrium pressure tensor. Fundamentally, however, both views arrive at physically indistinguishable macroscale. [Brenner (1979) compares the two theories in depth: cf. Question 16.5.]

15.2 Philosophy of the Microscale Interfacial Theory

Perturbation Scheme

A matched asymptotic expansion scheme is employed herein to investigate the macroscale transport equations characterizing moving and deforming nonmaterial interfaces. In order to better understand the asymptotic approach utilized, consider a system consisting of two fluids whose common 'boundary' is a thin ($l \ll L$) three-dimensional transition region. Let the characteristic length l of the microscale be chosen as the effective 'thickness' of this interfacial transition region. Imagine an experimental probe able to discern field changes occurring over the small length scale l, while still sampling a sufficiently large region of space (and time) to permit the random motions of individual molecules to average out and, hence, enabling the system to be modeled as a continuum. Use of this *microscale probe* permits experimental mapping of each continuum field characterizing the system throughout the entire fluid space. Each such field, when observed from the microscale, is assumed to be strictly continuous throughout this space, including the interfacial transition region (cf. Figures 3.1-1 and 3.6-3).

Suppose that the size of the probe is incrementally increased, permitting the probe to sample even larger representative fluid regions. At some stage, the probe will clearly become too coarse to detect any steep field changes occurring over the length scale l of the interfacial transition region, although its aperture will still be sufficiently small to detect and record those more gradual field changes occurring within the separate bulk-fluid regions. Since this now *macroscale probe* is, by definition, able *only* to discern continuum field changes occurring over the length scale L characteristic of changes within the bulk-fluid regions, the interfacial region will *appear* to be a singular two-dimensional surface to the 'macroscale observer' employing the probe. Use of this macroscale probe permits a systematic mapping of any physical continuum field characterizing the system to be performed throughout the entire space. Although all relevant physical fields will, by definition, be separately continuous within each of the two bulk-fluid regions, such fields may appear (to a macroscale observer) to change discontinuously upon crossing the interface (see Figure 3.6-3).

As l is generally of the order of nanometers and L millimeters or centimeters, these disparate lengths will henceforth be assumed to satisfy the strong inequality $\varepsilon \equiv l/L \ll 1$. The essence of the asymptotic approach in the micromechanical theory consists of recognizing that the diffuse, three-dimensional, interfacial continuum transition region seen by the microscale observer will *only truly become a two-dimensional macroscale singular surface in the formal mathematical limit* $\varepsilon \to 0$.

These and other basic elements of our interfacial theory are illustrated through the following intuitive 'thought' experiment, to which reference will be made in later sections.

Imagine, momentarily, a particularly ingenious engineer, who (being thoroughly acquainted with the interfacial theory presented thus far in the text—though as yet unaware of the present distinction between macro- and micro-interfacial points of view) appoints for himself the task of experimentally verifying a previously known equilibrium adsorption isotherm of a certain oil/water/surfactant system. Having selected for his experiment a circular cylindrical vessel (whose proportions are those indicated in Figure 15.2-1), he proceeds to pour into the container a volume $\bar{V}_{\mathrm{II}}$ of water, in which he has previously dissolved a total mass M_{II}^{A} of surfactant A. Next, he very carefully pours over the water a total volume $\bar{V}_{\mathrm{I}}$ of some immiscible oil, in which he has previously dissolved a total mass M_{I}^{A} of the same surfactant. After a considerable lapse of time, reasonably certain that a state of physical equilibrium has been reached, he decides to measure the spatial distribution of surfactant within the two-phase system.

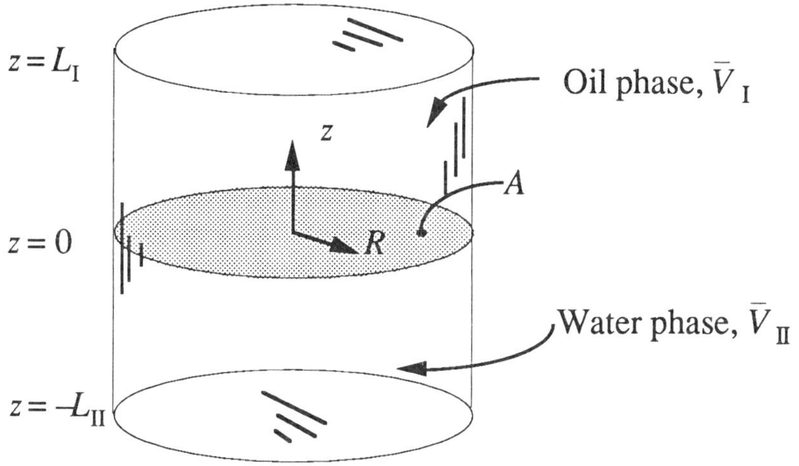

Figure 15.2-1 Circular cylindrical vessel containing oil and water phases.

For this purpose, the investigator—being intent upon obtaining the most precise measurements possible—elects to use a very fine scale (i.e. l-scale) probe. The probe is inserted into the water phase very near to the container bottom, $z = -L_{\mathrm{II}}$, where a mass density $\bar{\rho}_{\mathrm{II}}^{A}$ is observed. Gradually the probe is moved toward the interface, during which time this same mass density $\bar{\rho}_{\mathrm{II}}^{A}$ is continuously recorded. Suddenly, very near to the interface, the probe begins to detect a mass density $\rho^{A}(z)$, which varies extremely sharply in the axial direction z. Nevertheless, the investigator fails to observe any discontinuity in $\rho^{A}(z)$. Eventually, having traversed a lengthscale l, the probe has been carried across (what visually appears to be) the 'interface' and is soon well into the oil phase, where it now records an essentially constant mass density, $\bar{\rho}_{\mathrm{I}}^{A}$. This latter value continues to be observed until the probe encounters the upper boundary, $z = L_{\mathrm{I}}$.

The engineer begins to ponder the meaning of his measurements as he contemplates the exasperating reality that there *is* no (two-dimensional, discontinuous) interface; his data have been faithfully reproduced in Figure 15.2-2.

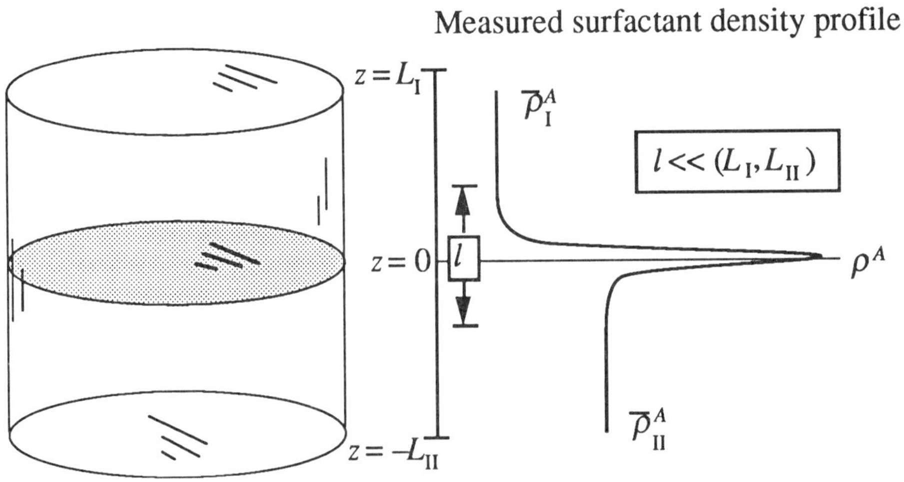

Figure 15.2-2 Continuous surfactant mass density profile across the interface.

Convinced that he has been severely misled by the (macroscale) discontinuous interfacial perspective of Part I, the engineer succeeds in convincing himself that 'true', physical, interfacial boundaries are, in fact, three-dimensional, continuous transition zones existing between relatively homogeneous 'bulk' phases. Bearing this novel discovery in mind, and having repeatedly performed the preceding measurements with similar results, he concludes that despite the presence of strong inhomogeneities in $\rho^{A}(z)$, the continuous (micro-) fluid system is, nevertheless, in a state of equilibrium; thus, he rationalizes the existence of the conservative micropotential $E(z)$, whose gradients give rise to an adsorptive force acting upon surfactant species in the vicinity of the interfacial transition zone (i.e. over the observed scale l). In particular, he demonstrates that the measured surfactant density values may be interpreted in terms of the following Boltzmann distribution of surfactant:

$$\rho^{A} = C \exp[-E(z)], \tag{15.2-1}$$

with C a constant, and where

$$\bar{\rho}_{\mathrm{II}}^{A} \equiv \lim_{z \to -L_{\mathrm{II}}} \rho^{A} \tag{15.2-2}$$

furnishes the measured value $\bar{\rho}_{\mathrm{II}}^{A}$ in the oil phase, and

$$\bar{\rho}_{\mathrm{I}}^{A} \equiv \lim_{z \to L_{\mathrm{I}}} \rho^{A} \tag{15.2-3}$$

yields the measured value $\bar{\rho}_{\mathrm{I}}^{A}$ in the water phase.

Reflecting upon these facts, and, particularly, having recognized that the ratio

$$\bar{\rho}_{\mathrm{I}}^{A} / \bar{\rho}_{\mathrm{II}}^{A} = K, \tag{15.2-4}$$

defines a partition coefficient

$$K = \exp(\bar{E}_{\mathrm{II}} - \bar{E}_{\mathrm{I}}), \tag{15.2-5}$$

with

$$\bar{E}_{\mathrm{I}} \equiv \lim_{z \to L_{\mathrm{I}}} E(z); \quad \bar{E}_{\mathrm{II}} \equiv \lim_{z \to -L_{\mathrm{II}}} E(z), \tag{15.2-6}$$

[which coefficient exhibits a value extremely near to that expected for surfactant A according to the 'macroscale' interface partition relation (5.3-1)], our engineer learns to associate the mathematical limits $\bar{\rho}_{\mathrm{I}}^{A}$ and $\bar{\rho}_{\mathrm{II}}^{A}$ with their corresponding 'bulk'-phase mass density values. The 'interface', he concludes, exists only according to a visual observation which is capable of resolving the macroscale vessel lengths L_{I} and L_{II}, though not the microscale length l; it does not possess a *true* physical existence, rather only an asymptotically limiting one.

Recalling the surface-excess formalism of Chapter 3 [cf. Eqs. (3.6-7) and (3.6-8)], he proceeds to make the following calculation:

$$M^{A} = M_{\mathrm{I}}^{A} + M_{\mathrm{II}}^{A} \tag{15.2-7}$$

represents the total amount of surfactant contained in the vessel: in contrast to this 'true' value,

$$\bar{M}^{A} = \bar{M}_{\mathrm{I}}^{A} + \bar{M}_{\mathrm{II}}^{A}, \tag{15.2-8}$$

with

$$\bar{M}_{\mathrm{I}}^{A} = \int_{V_{\mathrm{I}}} \bar{\rho}_{\mathrm{I}}^{A}\, dV \;; \quad \bar{M}_{\mathrm{II}}^{A} = \int_{V_{\mathrm{II}}} \bar{\rho}_{\mathrm{II}}^{A}\, dV , \tag{15.2-9}$$

denotes the amount of surfactant A in the 'bulk' phases I and II (assuming the values $\bar{\rho}_{\mathrm{I}}^{A}$ and $\bar{\rho}_{\mathrm{II}}^{A}$ to be valid up to the coordinate surface $z = 0$, which surface has been 'visually' located 'at' the interface). Therefore, the difference

$$(M^{A})^{s} = M^{A} - \bar{M}^{A}, \tag{15.2-10}$$

is identified as the total 'surface-excess' amount of surfactant 'at' the interface. Explicitly,

$$(M^A)^s = \int_A dA \int_{z=0}^{L_{\rm I}} (\rho^A - \bar{\rho}_{\rm I}^A)dz + \int_A dA \int_{z=-L_{\rm II}}^{0} (\rho^A - \bar{\rho}_{\rm II}^A)dz \ , \qquad (15.2\text{-}11)$$

with A denoting the circular cross-sectional area of the container. Upon identifying the 'interface' with the coordinate surface $z = 0$ (on which surface the symbol A will be understood henceforth to apply—the area A being clearly the same for all z-coordinate surfaces) the surface-excess areal mass density may be defined [as in Eq. (3.6-7b)] by

$$(M^A)^s = \int_A dA \rho_A^s \ , \qquad (15.2\text{-}12)$$

with

$$\rho_A^s = \int_{z=0}^{L_{\rm I}} (\rho^A - \bar{\rho}_{\rm I}^A)dz + \int_{z=-L_{\rm II}}^{0} (\rho^A - \bar{\rho}_{\rm II}^A)dz \ . \qquad (15.2\text{-}13)$$

Alas, he is disturbed by the result. For if, indeed, Eq. (15.2-13) represents the proper definition of the surface-excess surfactant density ρ_A^s in terms of the true, three-dimensional, microscale density ρ^A, it cannot possibly be an intensive, physical property of the system, insofar as it depends upon the dimensions of the container. This unsettling fact leads our engineer to recognize that the integrands in the preceding relation are only truly sensible (in the vicinity of $z = 0$) over the scale l, which is the scale of sharpest variation of E. Thus, upon scaling the z-coordinate as

$$\begin{aligned} \tilde{z} &= z \,/\, l, \\ &\equiv z \,/\, L\varepsilon, \end{aligned} \qquad (15.2\text{-}14)$$

with the parameter ε and macroscale length L satisfying

$$\varepsilon \equiv l \,/\, L; \quad L \,/\, L_{\rm I} = O(1); \quad L \,/\, L_{\rm II} = O(1), \qquad (15.2\text{-}15)$$

he obtains

$$\rho_A^s = L\varepsilon \int_0^{\infty} (\rho^A - \bar{\rho}_{\rm I}^A)d\tilde{z} + L\varepsilon \int_{-\infty}^{0} (\rho^A - \bar{\rho}_{\rm II}^A)d\tilde{z}$$

$$- L\varepsilon \int_{L_{\rm I}/\varepsilon L}^{\infty} (\rho^A - \bar{\rho}_{\rm I}^A)d\tilde{z} - L\varepsilon \int_{-\infty}^{-L_{\rm II}/\varepsilon L} (\rho^A - \bar{\rho}_{\rm II}^A)d\tilde{z} \ . \qquad (15.2\text{-}16)$$

Upon taking the limit $\varepsilon \to 0$ (corresponding to physical observation, as in Figure 15.2-2), the latter two integrals in the above expression (which depend explicitly upon the vessel dimensions $L_{\rm I}$ and $L_{\rm II}$) vanish asymptotically; thus, he deduces

$$\boxed{\rho_A^s = \lim_{\varepsilon \to 0} \left[L\varepsilon \int_0^{\infty} (\rho^A - \bar{\rho}_{\mathrm{I}}^A) d\tilde{z} + L\varepsilon \int_{-\infty}^{0} (\rho^A - \bar{\rho}_{\mathrm{II}}^A) d\tilde{z} \right],} \tag{15.2-17}$$

as an appropriate definition of the intensive surface-excess mass density [cf. Eq. (15.6-18) for a more formal, generic definition]. Thereby he concludes that *only* in the limit $\varepsilon \to 0$ may the property ρ_A^s (at least asymptotically) be regarded as a 'physical' property of the interface, independent of the vessel distances L_{I} and L_{II}.

Finally, upon comparison of Eqs. (15.2-1)–(15.2-3) and (15.2-7), he discovers the existence of a (macroscale) linear adsorption isotherm:

$$\bar{\rho}_\alpha^A = K_\alpha^A \rho_A^s \qquad (\alpha = \mathrm{I}, \mathrm{II}), \tag{15.2-18}$$

where the 'adsorption' coefficient

$$K_\alpha^A = \frac{\exp(-\bar{E}_\alpha)}{\lim\limits_{\varepsilon \to 0} \left[L\varepsilon \int\limits_{\tilde{z}=0}^{\infty} [\exp(-E) - \exp(-\bar{E}_{\mathrm{I}})] d\tilde{z} + L\varepsilon \int\limits_{\tilde{z}=-\infty}^{0} [\exp(-E) - \exp(-\bar{E}_{\mathrm{II}})] d\tilde{z} \right]}, \tag{15.2-19}$$

is found to exhibit values extremely near to the (macroscale) linear adsorption coefficients expected for surfactant A relative to the oil and water phases.

The preceding informal example* serves to underline several of the physical bases of our 'zeroth-order' asymptotic theory of interfacial transport processes. In particular, owing to the existence of the two greatly disparate length scales l and L over which significant continuum field changes occur, it has been demonstrated that interfacial systems may ultimately be viewed as physically continuous regions which exhibit *mathematical* discontinuities in the asymptotic limit $l/L \to 0$; such physical systems are amenable to mathematical analysis via the method of matched asymptotic expansions (Nayfeh 1973). In such an asymptotic scheme (which formalizes the preceding, intuitive scheme), ε plays the role of a small parameter in the

---

* An analogous 'thought' experiment is developed in Question 15.6 for the case of a spherical droplet of radius a. It is demonstrated therein (as a natural consequence of requiring that surface-excess densities be independent of extensive parameters—such as the characteristic linear dimension L of the vessel in which the interface is contained) that interfacial theory imposes a lower bound upon the permissible radius of curvature a of the droplet. In particular, by requiring $\varepsilon \to 0$ as in Eq. (15.2-17), it proves necessary to impose the additional requirement that $a/L = O(1)$, or, equivalently, $l/a = O(\varepsilon)$. Unless this condition is met, 'surface-excess' quantities will depend *inter alia* upon the conditions of the particular experiment employed for the purpose of their measurement.

perturbation analysis of the exact, microscale transport equations. Considering the premise upon which our theoretical analysis is based, only the first-order terms in this perturbation expansion, found by taking the limit as $\varepsilon \rightarrow 0$, will prove essential. This limits our results to situations for which the curvature H of the interface is a strictly macroscale quantity, i.e. $H \leq O(L^{-1})$.

Such a perturbation analysis will be seen to provide a systematic and rational method for deriving—from knowledge of the comparable microscale, three-dimensional conservation and constitutive equations (together with the accompanying volumetric phenomenological material functions)—the pertinent macroscale, two-dimensional, interfacial conservation and constitutive transport equations originally introduced on an *ad hoc* basis in Chapters 4 and 5 (together with expressions permitting *a priori* calculations of the macroscale interfacial phenomenological coefficients appearing in the latter). Trivially, one concurrently obtains the bulk conservation and constitutive equations and accompanying phenomenological coefficients characterizing the immiscible fluids lying on either side of the singular surface. In general, these macroscale phenomenological coefficients will be discontinuous across the interface, as has been discussed repeatedly in Part I. Our methods will prove to be valid not only for material interfaces, but for nonmaterial phase interfaces too.

15.3 Fictitious, Induced, Nonlocal Coordinate System

The subsequent analysis employs a parent surface-fixed coordinate parameterization of the microscale interfacial region (Figure 15.3-1) and the surrounding space in terms of 'fictitious', *semi-orthogonal* curvilinear coordinates (q^1, q^2, q^3) (cf. Appendix B). As noted in Appendix B, the term "semi-orthogonal" refers to a three-dimensional curvilinear coordinate system for which two of the three curvilinear coordinates [namely the *transverse* coordinate pair (q^1, q^2)] need not be orthogonal to one another, whereas the third coordinate (namely the *normal* coordinate q^3) is orthogonal to the other two. We choose to describe this parameterization as 'fictitious' in order to emphasize that this particular parameterization [being a special case of generally nonorthogonal curvilinear coordinates (q^1, q^2, q^3), which special case affords a considerably simplified treatment of the nonlocal projection operations described in the subsection at the conclusion of the present section] represents neither a completely general mapping of space nor, to the order of error allowed in the asymptotic theory, a physically restrictive one. This latter fact will be evidenced by the observed independence of all the macroscale results of our subsequent surface-excess theory upon the particular *nonlocal* (i.e. extrinsic) parameterization employed.

Eventually, we will associate a particular coordinate parent surface, namely $q^3 =$ constant, with the (macroscale) interface; however, for the time being, such an identification is irrelevant. What follows in the remainder of this section are *exact* geometrical relations pertaining to the parent surface parameterization via this nonlocal, semi-orthogonal scheme. As no (physical) interface exists at the microscale level of description (all microscale fields necessarily being *continuous*), it is important to distinguish conceptually between the parent surface, which is a purely *mathematical* construction that exists at all scales—micro or macro—and the *physical* 'interface', which will be seen to 'exist' only at the macroscale level.

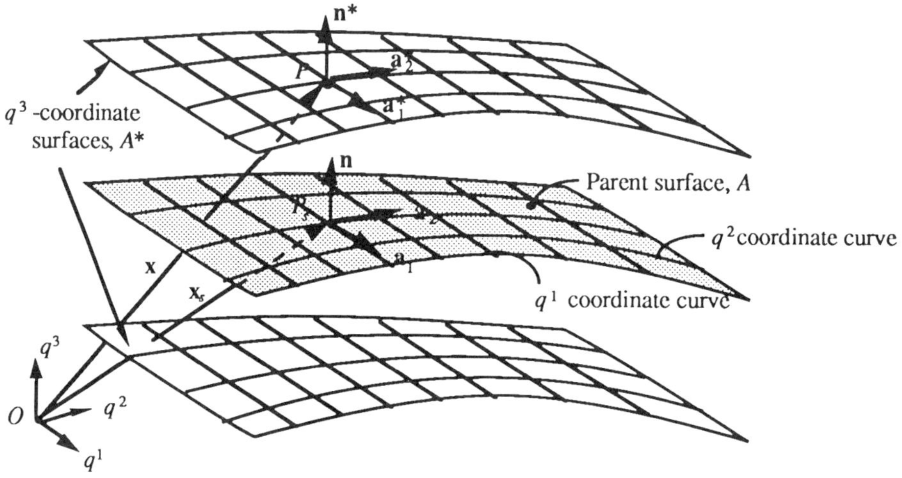

Figure 15.3-1 Coordinate parameterization of space in terms of parent surface-fixed, semi-orthogonal curvilinear coordinates (q^1, q^2, q^3). The surfaces q^3 = const. represent a family of coordinate surfaces, of which the surface A (for which $q^3 = q^3_0$ = const.), is termed the *parent surface*. This latter surface will ultimately be identified with the interface. As described in the text, the position vector $\mathbf{x}$ locates a point P in space; whereas the vector $\mathbf{x}_s$ locates a point P_s on the parent surface A.

As depicted in Figure 15.3-1, the parent surface and surrounding space is herein regarded as parameterized by a series of (generally nonparallel) q^3-coordinate surfaces. Of the infinite family of q^3-coordinate surfaces, the particular one of these that will ultimately be identified with the (macroscale) interface, will henceforth be termed the *parent surface*. It will be designated by q^3 = const. $\equiv q^3_0$, say, or, generically, by the symbol A.* All other familial surfaces q^3 = const., will be designated by the symbol A^*.†

* The parent surface A is subsequently chosen [cf. Eqs. (15.6-8) and (15.6-9), as well as Eqs. (15.6-26) and (15.6-27)] to coincide with the macroscale physical interface.

† An asterisk (*) will be used throughout this chapter to distinguish geometrical and/or kinematical variables defined at an arbitrary point (q^1, q^2, q^3) in physical space from the corresponding variables defined at a point (q^1, q^2, q^3_0), say, on the parent surface.

(Ultimately, in §15.5, these familial surfaces will be regarded as comprising the so-called interfacial transition region.)

As in Part I, a position vector $\mathbf{x}$ will be used to locate points in space relative to a space-fixed observer, situated at point O. A space-fixed point will be characterized throughout by the requirement that $\mathbf{x}$ = const. Points lying on the instantaneous parent surface A will be identified by assigning to each a position vector $\mathbf{x}_s$ measured relative to O. In general, these vectors may be parameterized respectively as

$$\mathbf{x} = \mathbf{x}(q^1, q^2, q^3, t) \tag{15.3-1}$$

[with the explicit dependence upon time t arising by consequence of the fact that the position $\mathbf{x}$ of the space point (q^1, q^2, q^3) may vary in time relative to the space-fixed origin O owing to the general time-dependence of the surface-fixed coordinates (q^1, q^2, q^3)] and

$$\mathbf{x}_s = \mathbf{x}_s(q^1, q^2, t). \tag{15.3-2}$$

From this latter relation it follows upon use of the chain rule that

$$d\mathbf{x}_s = \left(\frac{\partial \mathbf{x}_s}{\partial q^\alpha}\right)_t dq^\alpha + \left(\frac{\partial \mathbf{x}_s}{\partial t}\right)_{q^1, q^2} dt, \tag{15.3-3}$$

$(\alpha = 1,2)$. Equivalently,

$$d\mathbf{x}_s = \mathbf{a}_\alpha dq^\alpha + \mathbf{u}_n dt, \tag{15.3-4}$$

where we have employed Eq. (3.2-2) and defined

$$\boxed{\mathbf{u}_n \overset{\text{def}}{=} \left(\frac{\partial \mathbf{x}_s}{\partial t}\right)_{q^1, q^2}} \tag{15.3-5}$$

to be the so called *parent surface coordinate velocity*.* The phrase "surface-fixed point" will consistently refer to a point whose $(q^1, q^?)$ values remain unchanged in time (see Figure 15.4-2). Such a surface-fixed point will, in general, possess an extrinsic velocity $\mathbf{u}_n(\mathbf{x}_s)$ relative to a space-fixed observer [cf. example 2 of §3.7 and footnote below Eq. (15.4-6)].

For a review of the geometrical properties of the q^3-coordinate surfaces A and A^* induced by the surface parameterization (15.3-2), the interested reader may refer to §3.2 and Appendix B. Thus, $\mathbf{n}$ denotes a unit normal vector defined at point $\mathbf{x}_s(q^1, q^2)$ lying on A [cf. Eq. (3.2-5)], whereas $\mathbf{n}^*$ denotes a unit normal vector at (q^1, q^2) on A^*. Similarly, $\mathbf{b}$ denotes the curvature dyadic defined at $\mathbf{x}_s$ on A [cf. Eq. (3.2-10)], whereas $\mathbf{b}^*$ denotes the comparable curvature dyadic on A^*; etc.

---

* Determination of this parent surface coordinate velocity [for cases other than a *material* macroscale interface, for which special case $\mathbf{u}_n$ is later demonstrated to be equivalent to the material velocity of the macroscale interface, to within an error of $O(\varepsilon)$] is illustrated in example 3 of §3.7, as well as Questions 15.3, 15.5 and 17.3.

A differential element of arc length dn along the q^3-coordinate curve is related to the differential displacement dq^3 by the expression [cf. Eq. (B.1-10e)]

$$dn = \frac{1}{h_3^*} dq^3 \qquad (h_3^* > 0), \tag{15.3-6}$$

where $h_3^*(q^1, q^2, q^3)$ denotes the scale factor of the q^3-coordinate curve [cf. Eqs. (A.1-27) and (A.1-30)]. Accordingly, the displacement $n \equiv n(q^1, q^2, q^3)$ from the parent surface along the q^3-coordinate curve may be determined from the integral [cf. Eq. (3.6-4a)]†

$$n = \int_{q_0^3}^{q^3} \frac{1}{h_3^*} dq^3 . \tag{15.3-7}$$

Because of the role played in the subsequent theory by the intrinsic variable n, the coordinate variable set $(\mathbf{x}_s, q^3)$ appearing in the arguments of the various field variables will often be replaced by the alternative variable set $(\mathbf{x}_s, n)$. However, when replacing q^3 by n, it must be borne in mind that the surfaces of constant n are not necessarily equivalent to surfaces of constant q^3.*

For the sake of definiteness, the normal coordinate n may be constructed by following a normal trajectory from the surface point $\mathbf{x}_s$. However, since we will subsequently be interested only in the locations of these points in space that are proximate to the parent surface and that are separated by distances of $O(l)$, to the order of error allowed in the asymptotic theory the normal trajectories n may be represented as *perpendicular straight lines* emanating from the surface point $\mathbf{x}_s$: this fact obtains, despite the obvious possibility that curves of field gradients [which curves are ultimately associated with the coordinate n in later sections: cf. Eqs. (15.5-1)–(15.5-5)] may, in fact, follow curvilinear rather than rectilinear paths through the interfacial region. At least to within $O(\varepsilon)$, all (nonpathological) trajectories from the parent surface A will be well approximated by rectilinear, orthogonal trajectories. This fact will eventually be made explicit by the requirement (15.5-28).

---

† To avoid circumlocution, we have deleted the prime that would normally have appeared in the dummy variable $q^{3'}$ occuring in the integrand of Eq. (15.3-7). This scheme will be consistently followed throughout the subsequent text.

* Surfaces of constant n are only equivalent to q^3-coordinate surfaces in circumstances for which h_3^* is independent of q^1, q^2 and t. Such conditions hold for the special case of *parallel surfaces* (Kosinski 1986, Eliassen 1963), for which $h_3^* = 1$ (see Figure 15.3-3). This fact partially accounts for the special significance of the parallel surface parameterization.

Owing to the above considerations, we will henceforth refer to the parameterization ($\mathbf{x}_s$, n) as a *local parameterization* of space, seeking thereby to denote a parameterization of those points lying in immediate proximity to the interface. [In contrast, the *nonlocal* parameterization (q^1, q^2, q^3) represents a complete mapping of *all* points in space.] These local and nonlocal parameterizations will, in the macroscale, asymptotic limit $\varepsilon \to 0$, be found to correspond respectively to *intrinsic* and *extrinsic* parameterizations of the interface.

Since the q^3-coordinate curves are defined to follow normal trajectories to the q^3-coordinate surfaces, a unit tangent vector to a q^3-coordinate curve is equivalent to $\mathbf{n}^*$, the unit normal vector to the q^3-coordinate surface. Thus, the curvature κ^* as well as the principal unit normal vector $\mathbf{p}^*$ of a q^3-coordinate curve may both be derived from the properties of $\mathbf{n}^*$ via the relation (Aris 1962)

$$\mathbf{n}^* \cdot \nabla \mathbf{n}^* = \frac{\partial \mathbf{n}^*}{\partial n} = \kappa^* \mathbf{p}^*. \tag{15.3-8}$$

The coordinate parameterization of physical space by the variables (q^1, q^2, q^3) will be unique at each point so long as the Jacobian

$$J = \left(\frac{\partial \mathbf{x}}{\partial q^1}\right) \cdot \left[\left(\frac{\partial \mathbf{x}}{\partial q^2}\right) \times \left(\frac{\partial \mathbf{x}}{\partial q^3}\right)\right] \tag{15.3-9}$$

is non-vanishing at that point. The existence of the Jacobian (except possibly at isolated points) is assured by virtue of the fact that each and every point in space is assigned some value (q^1, q^2, q^3). (A unique parameterization of physical space will, as previously intimated, be strictly necessary only within the interfacial transition region, $n = O(l)$. As such, J can, in fact, vanish at points far from the interfacial transition region without negating the subsequent theory.)

Nonlocal Projection Operations

The exact projection (Figure 15.3-2) of a differential areal element dA^* lying upon an arbitrary q^3-coordinate surface onto the parent surface is given by the expression [cf. Eqs. (15.A-2) and (15.A-22) appearing in the appendix at the end of this chapter]

$$dA^* = M^*(\mathbf{x}_s, q^3)dA. \tag{15.3-10}$$

Here, dA is the differential element of the parent surface found by projecting the element dA^* onto the parent surface along an envelope determined by the q^3-coordinate curves; moreover, $M^*(q^1, q^2, q^3)$ is the *areal magnification factor*

$$M^* = \exp\left(- \int_{q_0^3}^{q^3} 2H^* \frac{1}{h_3^*} dq^3 \right), \tag{15.3-11}$$

with the mean curvature H^* of the coordinate surface A^* defined as in Eq. (3.2-13).

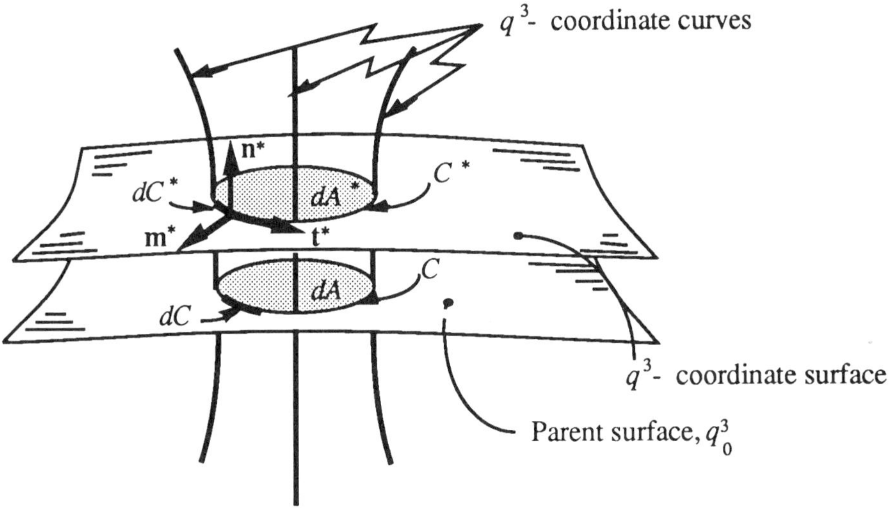

Figure 15.3-2 Projection of the differential areal and length elements dA^* and dC^* onto the parent surface along an envelope determined by the q^3-coordinate curves. Also shown are the unit normal vector $\mathbf{n}^*$ to the q^3-coordinate surface, the unit tangent vector $\mathbf{t}^*$ to the curve C^*, and the unit surface vector $\mathbf{m}^*=\mathbf{t}^*\times\mathbf{n}^*$ lying normal to the curve C^* in the tangent plane of the q^3-coordinate surface.

The relationship governing the projection onto the parent surface of a differential element of arc length dC^* lying along a curve C^* on an arbitrary q^3-coordinate surface is given by the expression [cf. Eqs. (15.A-24), (15.A-33), and (15.A-34)]

$$dC^* = N^*(\mathbf{x}_s, q^3)\, dC. \tag{15.3-12}$$

Here, dC represents a differential element of length on the parent surface found by projecting the element dC^* onto the parent surface along lines corresponding to the q^3-coordinate curves. In the above, $N^*(q^1, q^2, q^3)$ is the *lineal magnification factor*

$$N^* = \exp\left(- \int_{q_0^3}^{q^3} \kappa_n(\mathbf{t}^*) \frac{1}{h_3^*} dq^3 \right), \qquad (15.3\text{-}13)$$

with the scalar

$$\kappa_n^*(\mathbf{t}^*) = \mathbf{t}^*\mathbf{t}^* : \mathbf{b}^* \qquad (15.3\text{-}14)$$

the normal curvature of the q^3-coordinate surface in the direction of $\mathbf{t}^*$, the latter being the unit tangent vector to the curve C^*. Also, related to this curve is the unit vector $\mathbf{m}^* = \mathbf{t}^* \times \mathbf{n}^*$, normal to the curve and lying in the tangent plane to the q^3-coordinate surface. This vector may be written as [cf. (15.A-43)]

$$\mathbf{m}^* = \mathbf{m} - \int_{q_0^3}^{q^3} (\mathbf{t}^*\mathbf{t}^* + \mathbf{n}^*\mathbf{n}^*) \cdot (\nabla \mathbf{n}^*) \cdot \mathbf{m}^* \frac{1}{h_3^*} dq^3 . \quad (15.3\text{-}15)$$

Two examples, or special subsets of semi-orthogonal surface-fixed coordinate systems that often prove useful in applications may be identified, namely: (i) a parallel-surface parameterization, for which $q^3 \equiv n$; and (ii) an orthogonal curvilinear coordinate system, for which q^1 and q^2 are now orthogonal to one another. These are described below.

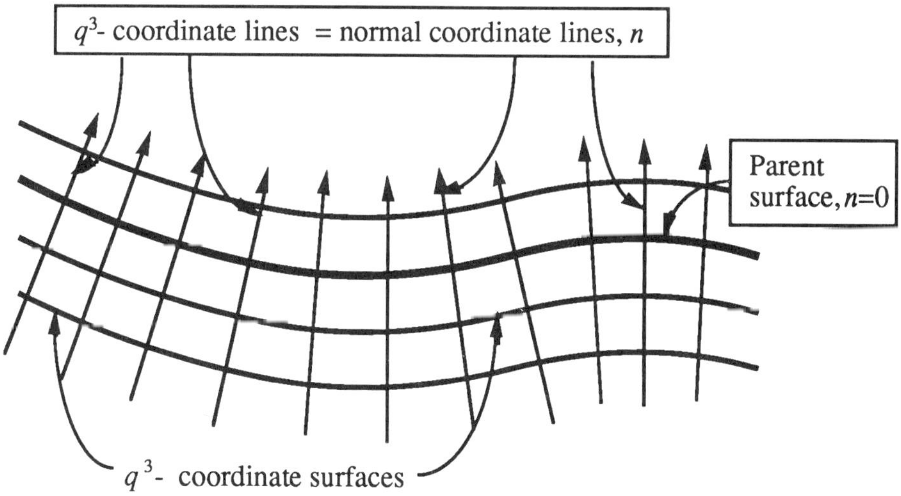

Figure 15.3-3 The parallel surface parameterization of the interfacial region represents a special case of semi-orthogonal coordinates for which q^3-coordinate curves are now rectilinear and identical with n coordinate lines. Each q^3-coordinate surface is parallel to every other q^3-coordinate surface.

Parallel surface parameterization

Once an arbitrary parameterization (q^1, q^2) of the parent surface has been performed, a family of parallel surfaces (Kosinski 1986, Eliassen 1963) q^3 =const. may by constructed to parameterize the remainder of physical space (Figure 15.3-3). (This, for example, would be a suitable parameterization for spherical interfaces.) In particular, given the parameterization $\mathbf{x}_s(q^1, q^2, t)$ for the parent surface, the q^3-coordinate curves can be obtained by choosing h_3=1. Thus, application of Eq. (15.3-7) yields the relation q^3=n. In addition, using Eq. (15.3-8), and Eqs. (15.A-13), and (15.A-16), the curvature κ^* of the q^3-coordinate curve can be shown to be zero. This parameterization of physical space will yield a family of parallel q^3-coordinate surfaces, in which each member of the family is separated from another by a fixed distance Δn = const.

The geometric construction of the parallel surface parameterization is performed by drawing a straight line normal to the parent surface; this straight line is the q^3-coordinate curve. A q^3-coordinate surface is defined to lie at a constant distance n from the coordinate parent surface, such distances being measured along the normal lines.

Orthogonal coordinate parameterization

Here, (q^1, q^2, q^3) constitute a mutually orthogonal system of curvilinear coordinates. The parameterization of physical space using such an orthogonal system is less restrictive in terms of the variety of possible applications thereof than is the preceding parallel surface parameterization. Appendices A and B may be consulted for further details, as well as the textbook by Happel & Brenner (1983), where a compilation of many useful orthogonal curvilinear coordinate systems and formulas may be found. The mutual orthogonality of the q^1- and q^2-coordinate curves required of orthogonal curvilinear coordinate systems can be achieved by choosing the q^1- and q^2-coordinate curves to be the lines of curvature, as is made clear by Eqs. (B.1-20)–(B.1-22).

15.4 Kinematics

Several different velocity fields will be defined in the following (see Figure 15.4-1). Thus, the mass-average velocity of a material particle P, initially located at the space-fixed point $\mathbf{x}(0)$, is defined as in Eq. (3.4-2):

$$\mathbf{v} \overset{\text{def}}{=} \left(\frac{\partial \mathbf{x}}{\partial t}\right)_{\mathbf{x}(0)} . \tag{15.4-1}$$

This extrinsic microscale velocity field is assumed everywhere continuous (in particular, throughout the interfacial transition region).

In addition, the extrinsic velocity $\mathbf{u}$ of a point P_s permanently affixed to the parent surface A (and thus generally nonmaterial), initially located on the parent surface at the surface-fixed point $\mathbf{x}_s(0)$, is defined by the expression [cf. Eq. (3.4-5)]

$$\mathbf{u} \overset{\text{def}}{=} \left(\frac{\partial \mathbf{x}_s}{\partial t}\right)_{\mathbf{x}_s(0)} . \tag{15.4-2}$$

Use of the chain rule allows the convected time derivative [following the point $\mathbf{x}_s(0)$] of a generic function $f \equiv f(q^1, q^2, t)$ to be expressed as in Eq. (3.4-9); explicitly,

$$\left(\frac{\partial f}{\partial t}\right)_{\mathbf{x}_s(0)} = \left(\frac{\partial f}{\partial t}\right)_{q^1, q^2} + \mathbf{u} \cdot \nabla_s f . \tag{15.4-3}$$

With the choice $f \equiv \mathbf{x}_s$, we thereby obtain

$$\mathbf{u} = \left(\frac{\partial \mathbf{x}_s}{\partial t}\right)_{q^1, q^2} + \mathbf{u} \cdot \mathbf{I}_s \tag{15.4-4}$$

upon noting the representation of the surface idemfactor $\mathbf{I}_s$ provided in Question 3.14. Furthermore, upon employing the decomposition $\mathbf{I}=\mathbf{I}_s+\mathbf{nn}$, it follows directly, upon comparison with Eq. (15.3-8), that

$$\boxed{\mathbf{u} \cdot \mathbf{nn} = \mathbf{u}_n = u_n \mathbf{n} = \left(\frac{\partial \mathbf{x}_s}{\partial t}\right)_{q^1, q^2, q^3_0}} \tag{15.4-5}$$

is the (extrinsic) velocity of the intrinsic coordinates (q^1, q^2) lying on the parent surface $q^3 = q^3_0$. Clearly, with q^3_0 replaced by q^3 this relation applies equally well to any coordinate surface A^* as it does to A; thus,

$$\boxed{\mathbf{u}_n^* \equiv u_n^* \mathbf{n}^* \overset{\text{def}}{=} \left(\frac{\partial \mathbf{x}}{\partial t}\right)_{q^1, q^2, q^3}} \tag{15.4-6}$$

provides a definition of the extrinsic 'coordinate velocity' of any coordinate trio (q^1, q^2, q^3) in space. These relations reveal that any 'surface-fixed' coordinate pair (q^1, q^2) will move solely along a q^3-coordinate trajectory, normal to the q^3-coordinate surface to which they are affixed. All other motions of the surface point P_s relative to a space-fixed observer (specifically, those motions within the two-dimensional surface space) are hence

intrinsically* observable to a surface-fixed observer, who measures motion relative to himself, i.e. relative to the 'surface-fixed' coordinate pair (q^1, q^2) (see Figure. 15.4-2).

---

* According to an intrinsic observer, the rate of change $\dot{q}^\alpha$ $(\alpha = 1,2)$ of the coordinates of a point P_s whose initial position is denoted by the coordinate pair $q^\alpha(0)$ $(\alpha = 1,2)$ is given by the expression

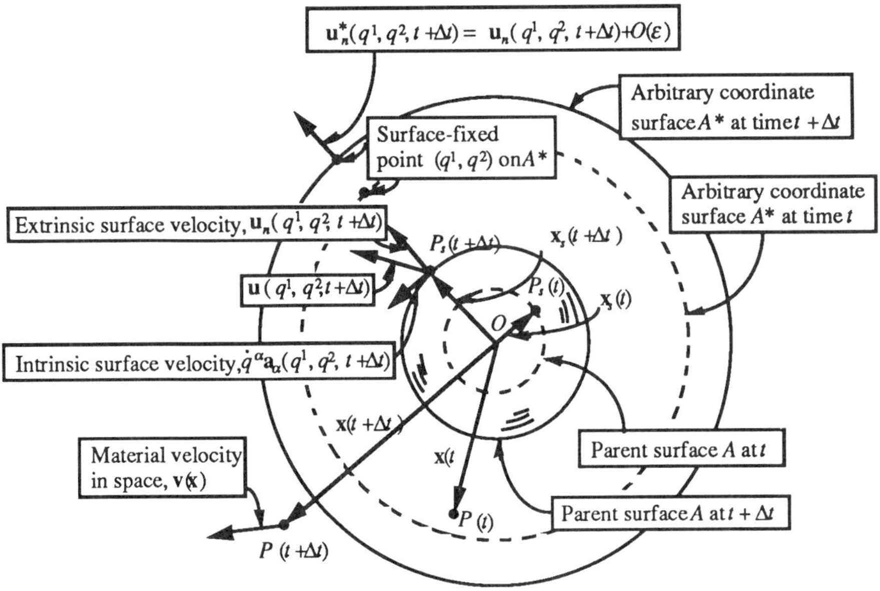

Figure 15.4-1 Summary of the velocities employed in Chapter 15 for the special case of an expanding and rotating spherical parent surface. Explicitly, $q^1 = \theta$, $q^2 = \phi$, $q^3 = r(t)$ (see Figure B.1-6). Three distinct velocities are referred to in the figure: (i) the mass-average (material) velocity $\mathbf{v}(\mathbf{x})$ of the particle P instantaneously located at the point $\mathbf{x}$ [cf. Eq. (15.4-1)]. In the most general circumstance that the q^3-coordinate surfaces are nonmaterial, the spatial trajectory of the particle P will generally cross q^3-coordinate surfaces. (ii) The surface velocity $\mathbf{u}(\mathbf{x}_s)$ of a point P_s within the parent surface A [$q^3 = r_0(t)$] [cf. Eq. (15.4-2)]. The velocity vector $\mathbf{u}$ of this generally nonmaterial point may be decomposed into intrinsic ($\mathbf{a}_\alpha \dot{q}^\alpha$) and extrinsic ($\mathbf{u}_n$) vectors [see the footnote following Eq. (15.4-6)]. The extrinsic velocity vector $\mathbf{u}_n$ ['parent surface coordinate velocity'; cf. Eq. (15.3-5)] points in a direction normal to the parent surface [cf. Eq. (15.4-5)]. (iii) The coordinate velocity $\mathbf{u}_n^*$ of the surface-fixed pair (q^1, q^2) on an arbitrary q^3-coordinate surface A^* [cf. Eq. (15.4-6)]. This coordinate velocity is induced by the extrinsic ($\mathbf{u}_n$) motion of the parent surface, by which motion the coordinate velocity $\mathbf{u}_n^*$ may be calculated to leading order in ε [see the footnote following Eq. (15.7-5)].

$$\dot{q}^\alpha \overset{\text{def}}{=} \left(\frac{\partial q^\alpha}{\partial t}\right)_{q^1(0),\ q^2(0)}$$

From Eq. (15.3-4), together with Eq. (15.4-2), it follows directly from the above that

$$\mathbf{u} = \mathbf{a}_\alpha \dot{q}^\alpha + \mathbf{u}_n.$$

Thus, $\mathbf{a}_\alpha \dot{q}^\alpha$ represents the purely intrinsic velocity of the point P_s within the surface, whereas $\mathbf{u}_n$ represents the purely extrinsic velocity of P_s in space (see Figure 15.4-1).

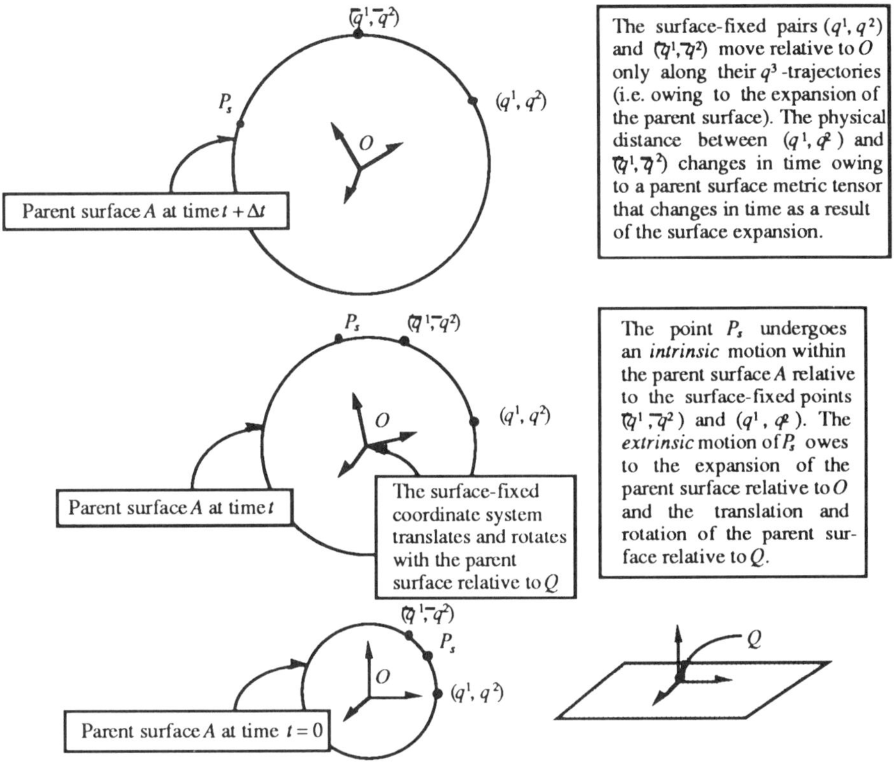

$(\bar{q}^1, \bar{q}^2)$
P_s
(q^1, q^2)
O
Parent surface A at time t + Δt
The surface-fixed pairs (q^1, q^2) and $(\bar{q}^1, \bar{q}^2)$ move relative to O only along their q^3-trajectories (i.e. owing to the expansion of the parent surface). The physical distance between (q^1, q^2) and $(\bar{q}^1, \bar{q}^2)$ changes in time owing to a parent surface metric tensor that changes in time as a result of the surface expansion.
Parent surface A at time t
The surface-fixed coordinate system translates and rotates with the parent surface relative to Q
The point P_s undergoes an intrinsic motion within the parent surface A relative to the surface-fixed points $(\bar{q}^1, \bar{q}^2)$ and (q^1, q^2). The extrinsic motion of P_s owes to the expansion of the parent surface relative to O and the translation and rotation of the parent surface relative to Q.
Parent surface A at time t = 0
Q

Figure 15.4-2 Intrinsic and extrinsic motions of a surface point P_s within a translating, expanding and rotating spherical parent surface. Explicitly, $q^1 = \theta$, $q^2 = \phi$, $q^3 = r(t)$ (see Figure B.1-6). The coordinate pairs (q^1, q^2) and $(\bar{q}^1, \bar{q}^2)$ represent 'surface-fixed' points on the parent surface. Intrinsic motion of P_s within the parent surface is measured relative to such surface-fixed points. The rotation and translation of the surface-fixed coordinate system (together with the parent surface) hence does not engender any intrinsic surface motion. The expansion of the spherical parent surface results both in an extrinsic motion of *all* points on the parent surface relative to O, and a physical separation of surface-fixed points relative to each other on account of the time-dependence of the surface metric tensor, the latter time-dependence resulting as a consequence of the surface expansion [cf. Eqs. (3.4-13) and (3.4-14), as well as Question 3.2].

It will prove useful in the following to define a coordinate-fixed time derivative as

$$\frac{\delta}{\delta t} \stackrel{\text{def}}{=} \left.\frac{\partial}{\partial t}\right)_{q^1,\, q^2,\, q^3} \equiv \left.\frac{\partial}{\partial t}\right)_{\mathbf{x}} + \mathbf{u}_n{}^* \cdot \nabla, \qquad (15.4\text{-}7)$$

(i.e. relative to an observer moving along q^3-coordinate curves with velocity $\mathbf{u}_n$*). In addition, define a reference surface-fixed time derivative as

$$\frac{\delta_s}{\delta t} \stackrel{\text{def}}{=} \left.\frac{\partial}{\partial t}\right)_{q^1,\, q^2,\, q_0^3} . \tag{15.4-8}$$

Employing these definitions, the Reynolds transport theorem (cf. Question 3.5) for the time derivative of the volume integral of a generic function $\mathbf{f} \equiv \mathbf{f}(q^1, q^2, q^3, t)$ over the coordinate-fixed control volume $V(t)$ is given by the expression

$$\frac{d}{dt}\int_V \mathbf{f}\, dV = \int_V \left[\frac{\delta\, \mathbf{f}}{\delta t} + (\nabla \cdot \mathbf{u}_n^*)\mathbf{f}\right] dV . \tag{15.4-9}$$

The corresponding surface Reynolds transport theorem [cf. Eq. (3.4-17)] for the time derivative of the area integral of a function $\mathbf{f} \equiv \mathbf{f}(q^1, q^2, t)$ over an area $A(t)$ on the coordinate reference surface is given by the expression

$$\frac{d}{dt}\int_A \mathbf{f}\, dA = \int_A \left[\frac{\delta_s\, \mathbf{f}}{\delta t} + (\nabla_s \cdot \mathbf{u}_n)\mathbf{f}\right] dA . \tag{15.4-10}$$

15.5 Singular Perturbation Analysis

Consider a generic, continuum, tensorial, microscale field $\lambda = \lambda(\mathbf{x})$ (e.g. a volumetric density of some extensive physical property $\mathcal{P}$ as in §3.5), defined at every point $\mathbf{x}$ of the three-dimensional system. (For simplicity and focus, the explicit time dependence of λ, if any, has been suppressed in its argument.) This field, which may equivalently be represented as

$$\lambda = \lambda(\mathbf{x}_s, n), \tag{15.5-1}$$

[since, as in Eq. (3.6-4b), $\mathbf{x} = \mathbf{x}_s \oplus n$; see also Eq. (15.3-7)], is assumed to be a continuous function of position — in particular of the distance n — throughout all of the domain occupied by microfluid. Within the two bulk-fluid regions, and for a fixed value of $\mathbf{x}_s$, the field λ will generally change only gradually with the distance n; however, within the interfacial transition region, λ may undergo rapid changes with n.*

The respective characteristic lengths l and L of the interfacial and bulk-fluid regions permit us to define the nondimensional normal distance

* For example, if λ represents the mass density ρ in an immiscible oil-water system of significantly different bulk densities, then ρ will be sensibly constant in both the bulk phases. However, in proximity to the interface, its magnitude will change quite rapidly with n, varying from the bulk density, $\bar{\rho}_{\text{water}}$, of water to that of oil, $\bar{\rho}_{\text{oil}}$, within a distance of order of $l \approx 100$ Å (cf. Figure 15.2-2).

$$\tilde{n} = n/l, \tag{15.5-2a}$$

which is characteristic of *microscale* distances, and

$$\bar{n} = n/L \equiv \varepsilon\tilde{n}, \tag{15.5-2b}$$

which is characteristic of *macroscale* distances. In the latter equation the characteristic length ratio

$$\varepsilon \equiv l/L << 1 \tag{15.5-3}$$

is assumed to satisfy the indicated inequality. (As gradients of the field λ *within* the interface will always be supposed to scale with L, there exists no explicit need to define a dimensionless distance, $\Delta\bar{\mathbf{x}}_s$, say, within the interface.)

The density λ may change significantly within the diffuse transition region of thickness l, i.e. with $O(1)$ changes in $\tilde{n}$; explicitly, $\partial\lambda/\partial\tilde{n}$ may be of $O(1)$ for $|\tilde{n}|$ of $O(1)$. The functional form

$$\lambda = \lambda(\mathbf{x}_s, \tilde{n}; \varepsilon) \tag{15.5-4}$$

is thus a useful representation of Eq. (15.5-1) for those positions $\mathbf{x}$ lying within the interfacial transition zone. In this form, changes in the value of λ occurring over the microscale distance $|\tilde{n}| \equiv O(1)$ can be distinguished. Those gradual changes in λ, if any, occurring within the bulk-fluid regions $|\tilde{n}| >> 1$ will occur only over distances of the order of the depth L of the bulk phases, i.e. over distances $|\tilde{n}| \equiv O(\varepsilon^{-1})$.*

Sensible changes in the value of λ within the bulk-fluid regions, $|\bar{n}| \equiv O(1)$, occur only over the much larger length scale $|\Delta\bar{n}| \equiv O(1)$. The functional form

$$\lambda = \lambda(\mathbf{x}_s, \bar{n}; \varepsilon) \tag{15.5-5}$$

is thus a more useful alternative to Eq. (15.5-4) for those positions $\mathbf{x}$ lying within these bulk-fluid regions. In this form, changes in λ over macroscale distances — corresponding to $|\Delta\bar{n}| \equiv O(1)$ — can be distinguished. Any rapid changes in λ within the interfacial transition region occur over

---

* Here and hereafter, the gage symbol "$\equiv O$" is used to indicate "exactly equal to the specific order". In other words, the statement $f(\varepsilon) \equiv O[g(\varepsilon)]$ as $\varepsilon \to 0$ is true if there exist two positive numbers A and B independent of ε, and an $\varepsilon_o > 0$ such that $A|g(\varepsilon)| \le |f(\varepsilon)| \le B|g(\varepsilon)|$ for all $|\varepsilon| \le \varepsilon_o$. For the statement $f(\varepsilon) = O[g(\varepsilon)]$ (with the equality symbol replacing the equivalence symbol) as $\varepsilon \to 0$ to be true, however, the only constraint on $f(\varepsilon)$ is that $|f(\varepsilon)| \le B|g(\varepsilon)|$ for all $|\varepsilon| \le \varepsilon_o$.

distances characterized by $|\Delta \bar{n}| \equiv O(\varepsilon)$. As $\varepsilon \to 0$, such changes will appear to occur *discontinuously* when fields described by the functional form (15.5-5) are expanded via a regular perturbation expansion scheme that utilizes ε as the small parameter.

A small parameter expansion with respect to ε can be applied to the field λ. Separate expansions must be performed within the interfacial transition region and within the two bulk-fluid domains lying on either side of this region. In the terminology of singular perturbation theory (Nayfeh 1973) the interfacial region, defined by $|\tilde{n}| \equiv O(1)$, constitutes the *inner* region, whereas the two bulk-fluid regions, defined by $|\bar{n}| \equiv O(1)$, constitute the *outer* regions.

The inner perturbation expansion of λ in the small parameter ε, valid within the interfacial transition region, is performed on the basis of the functional form (15.5-4), keeping $\tilde{n}$ (and $\mathbf{x}_s$) fixed, namely*

$$\lambda = \lambda(\mathbf{x}_s, \tilde{n}; \varepsilon) = \tilde{\lambda}(\mathbf{x}_s, \tilde{n}) + \tilde{\lambda}_R(\mathbf{x}_s, \tilde{n}; \varepsilon), \qquad (15.5\text{-}6)$$

wherein

$$\tilde{\lambda}(\mathbf{x}_s, \tilde{n}) \equiv \lim_{\substack{\varepsilon \to 0 \\ \tilde{n} \text{ fixed}}} \lambda(\mathbf{x}_s, \tilde{n}; \varepsilon) \qquad (15.5\text{-}7)$$

represents the leading-order term in this expansion. All remaining higher-order terms are contained in the *remainder function* $\tilde{\lambda}_R$, which necessarily satisfies the constraint

---

* For simplicity, we depart here from standard notation (Kevorkian & Cole 1981), which would require that what is here called $\tilde{\lambda}$ be called $\tilde{\lambda}_0$; that is, ordinarily one deals with the *complete* inner expansion to *all* orders in ε, namely

$$\tilde{\lambda}(\mathbf{x}_s, \tilde{n}; \varepsilon) \overset{\text{def}}{=} \tilde{\lambda}(\mathbf{x}_s, \bar{n} = \tilde{n}\varepsilon; \varepsilon),$$

whence

$$\tilde{\lambda}(\mathbf{x}_s, \tilde{n}; \varepsilon) = \tilde{\lambda}_0(\mathbf{x}_s, \tilde{n}) + \tilde{f}_1(\varepsilon)\tilde{\lambda}_1(\mathbf{x}_s, \tilde{n}) + \tilde{f}_2(\varepsilon)\tilde{\lambda}_2(\mathbf{x}_s, \tilde{n}) + \ldots,$$

where $\tilde{f}_n(\varepsilon) \to 0$ and $\tilde{f}_{n+1}/\tilde{f}_n \to 0$ as $\varepsilon \to 0$. Since, however, we will not address these higher-order terms here, we can confound $\tilde{\lambda}(\mathbf{x}_s, \tilde{n}; \varepsilon)$ with its leading-order term, $\tilde{\lambda}_0(\mathbf{x}_s, \tilde{n})$ [and simply refer to the latter as $\tilde{\lambda}(\mathbf{x}_s, \tilde{n})$]. The remaining, higher-order terms are assigned to the function $\tilde{\lambda}_R(\mathbf{x}_s, \tilde{n}; \varepsilon)$.

$$\lim_{\substack{\varepsilon \to 0 \\ \tilde{n} \text{ fixed}}} \lambda_R(\mathbf{x}_s, \tilde{n}; \varepsilon) = 0. \tag{15.5-8}$$

The term $\tilde{\lambda}$ may be used to asymptotically approximate λ within the inner region; explicitly,*

$$\lambda = \tilde{\lambda}[1 + o(1)] \qquad \text{for } |\tilde{n}| = O(1). \tag{15.5-9}$$

Perturbation expansions of λ in terms of the small parameter ε, valid within the two outer regions, are performed on the basis of the functional form (15.5-5), while keeping $\bar{n}$ (and $\mathbf{x}_s$) fixed in the expansion:†

$$\lambda = \lambda(\mathbf{x}_s, \bar{n}; \varepsilon) = \bar{\lambda}_\mathrm{I}(\mathbf{x}_s, \bar{n}) + \bar{\lambda}_{\mathrm{I}R}(\mathbf{x}_s, \bar{n}; \varepsilon) \quad (\bar{n} > 0), \tag{15.5-10}$$

$$\lambda = \lambda(\mathbf{x}_s, \bar{n}; \varepsilon) = \bar{\lambda}_\mathrm{II}(\mathbf{x}_s, \bar{n}) + \bar{\lambda}_{\mathrm{II}R}(\mathbf{x}_s, \bar{n}; \varepsilon) \quad (\bar{n} < 0), \tag{15.5-11}$$

wherein

$$\bar{\lambda}_\mathrm{I}(\mathbf{x}_s, \bar{n}) \equiv \lim_{\substack{\varepsilon \to 0 \\ \bar{n} \text{ fixed}}} \lambda(\mathbf{x}_s, \bar{n}; \varepsilon) \qquad (\bar{n} > 0), \tag{15.5-12}$$

$$\bar{\lambda}_\mathrm{II}(\mathbf{x}_s, \bar{n}) \equiv \lim_{\substack{\varepsilon \to 0 \\ \bar{n} \text{ fixed}}} \lambda(\mathbf{x}_s, \bar{n}; \varepsilon) \qquad (\bar{n} < 0), \tag{15.5-13}$$

constitute the lowest-order terms in each of the respective expansions. All remaining higher-order terms are embodied within the remainder functions $\bar{\lambda}_{\mathrm{I}R}$ and $\bar{\lambda}_{\mathrm{II}R}$, each of which is assumed to possess the property that

$$\lim_{\substack{\varepsilon \to 0 \\ \bar{n} \text{ fixed}}} \bar{\lambda}_{\alpha R}(\mathbf{x}_s, \bar{n}; \varepsilon) = 0 \qquad (\alpha = \mathrm{I}, \mathrm{II}). \tag{15.5-14}$$

The terms $\bar{\lambda}_\mathrm{I}$ and $\bar{\lambda}_\mathrm{II}$ can be used to approximate λ within the appropriate outer regions; explicitly,

---

* The gage symbol "o" may be defined by the statement $f(\varepsilon) = o[g(\varepsilon)]$ as $\varepsilon \to \varepsilon_o$ (with $\varepsilon_o > 0$) if, given any function $\delta(n) > 0$, keeping n fixed, there exists a neighborhood N_δ of ε_o such that $|f(\varepsilon)| \le \delta(n)|g(\varepsilon)|$ for all ε in N_δ. More informally, the equivalent notation $f \ll g$ may be used.

† *Vis-a-vis* notation, similar remarks to those made in the previous footnote apply here too.

$$\lambda = \bar{\lambda}_{\text{I}}[1 + o(1)] \qquad \text{for } |\bar{n}| >> O(\varepsilon) \qquad (\bar{n} > 0), \quad (15.5\text{-}15)$$

$$\lambda = \bar{\lambda}_{\text{II}}[1 + o(1)] \qquad \text{for } |\bar{n}| >> O(\varepsilon) \qquad (\bar{n} < 0). \quad (15.5\text{-}16)$$

In general, the outer field $\bar{\lambda}$ may be discontinuous across the interface:

$$\|\bar{\lambda}\| \equiv \lim_{\bar{n} \to 0+} \bar{\lambda}_{\text{I}}(\mathbf{x}_s, \bar{n}) - \lim_{\bar{n} \to 0-} \bar{\lambda}_{\text{II}}(\mathbf{x}_s, \bar{n}) \neq 0, \qquad (15.5\text{-}17)$$

following the jump notation introduced in Eq. (3.6-12b).

Throughout this chapter, a tilde will consistently be used to denote inner variables (both dependent and independent), and an overbar to denote outer variables. In addition, only the lowest-order terms in ε will be retained in the respective perturbation expansions of λ in terms of these two classes of variables.

Products of the dependent variables appear in the subsequent microscale theory. Since the limit of the product of two functions is equal to the product of the limits so long as all limits exist, we may write, for the outer limit,

$$\overline{\mathbf{ab}} = \bar{\mathbf{a}}\bar{\mathbf{b}}, \qquad (15.5\text{-}18)$$

where $\mathbf{a}(\mathbf{x}_s, \bar{n}; \varepsilon)$ and $\mathbf{b}(\mathbf{x}_s, \bar{n}; \varepsilon)$ are arbitrary functions (of any tensorial order) possessing definable outer limits $\bar{\mathbf{a}}$ and $\bar{\mathbf{b}}$.* Similarly, for the inner limit,

$$\widetilde{\mathbf{ab}} = \tilde{\mathbf{a}}\tilde{\mathbf{b}}, \qquad (15.5\text{-}19)$$

where $\mathbf{a}(\mathbf{x}_s, \tilde{n}; \varepsilon)$ and $\mathbf{b}(\mathbf{x}_s, \tilde{n}; \varepsilon)$ are arbitrary functions possessing definable inner limits $\tilde{\mathbf{a}}$ and $\tilde{\mathbf{b}}$.

Lying between the inner, interfacial transition region and the two outer, bulk-fluid regions are two *intermediate* regions, within which both the outer and inner expansions are assumed to be simultaneously (asymptotically) valid (see Figure 15.5-1). With r, say, a bounded constant of $O(1)$ (i.e. independent of ε), the dimensionless normal distance

---

* Explicitly, Eq. (15.5-18) states that

$$\lim_{\substack{\varepsilon \to 0 \\ \bar{n} \text{ fixed}}} [\mathbf{a}(\mathbf{x}_s, \bar{n}; \varepsilon)\mathbf{b}(\mathbf{x}_s, \bar{n}; \varepsilon)] = \lim_{\substack{\varepsilon \to 0 \\ \bar{n} \text{ fixed}}} [\mathbf{a}(\mathbf{x}_s, \bar{n}; \varepsilon)] \lim_{\substack{\varepsilon \to 0 \\ \bar{n} \text{ fixed}}} [\mathbf{b}(\mathbf{x}_s, \bar{n}; \varepsilon)]$$

A similar interpretation applies for the identity (15.5-19), in which $\bar{n}$ is replaced by $\tilde{n}$.

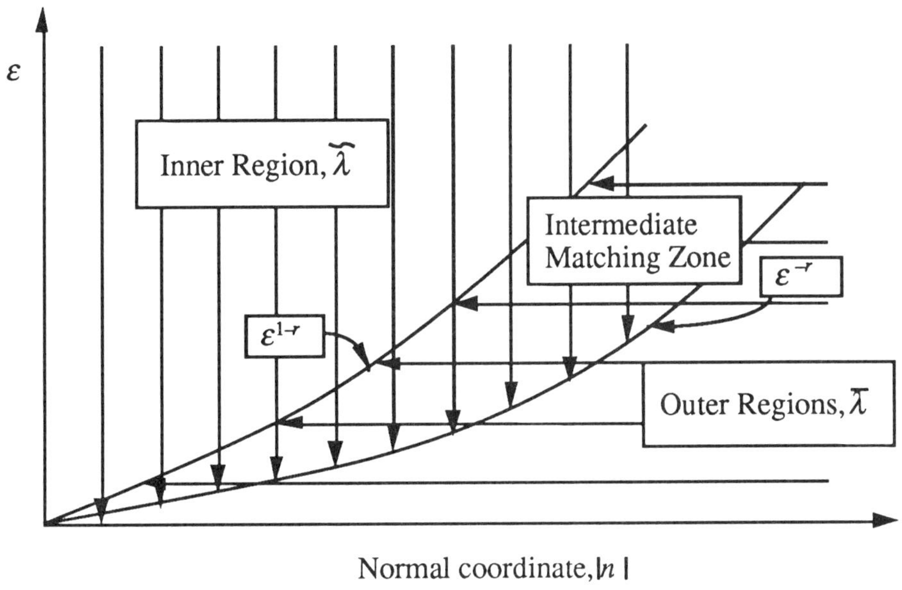

Figure 15.5-1 Regions of validity of the inner ($\tilde{\lambda}$) and outer ($\bar{\lambda}$) limits of the generic field λ. The inner region corresponds to the 'interfacial region' and the outer regions to the contiguous 'bulk-phase' regions. The region of overlap—where both $\tilde{\lambda}$ and $\bar{\lambda}$ are simultaneously valid and, hence, in which region the matching conditions (15.5-21) and (15.5-22) apply—is called the 'intermediate region'.

$$\hat{n} \overset{\text{def}}{=} \varepsilon^{-r}\bar{n} \equiv \varepsilon^{1-r}\tilde{n} \qquad (0 < r < 1) \qquad (15.5\text{-}20)$$

will be taken to be of $O(1)$ within the intermediate regions. The two limiting values $r = 0$ and 1 in the latter respectively correspond to $\hat{n}$ being equivalent to $\bar{n}$ for $(r = 0)$ and $\tilde{n}$ for $(r = 1)$. These r limits may become tighter in specific applications, depending on the exact functional form of the field λ. Since the outer and inner expansions are both assumed valid within the two intermediate regions corresponding to $|\hat{n}| = O(1)$, the pair of *matching conditions*

$$\boxed{\lim_{\substack{\varepsilon \to 0 \\ \hat{n}\ \text{fixed}}} \bar{\lambda}_{\mathrm{I}}(\mathbf{x}_s, \bar{n} = \varepsilon^{r}\hat{n}) = \lim_{\substack{\varepsilon \to 0 \\ \hat{n}\ \text{fixed}}} \tilde{\lambda}(\mathbf{x}_s, \tilde{n} = \varepsilon^{r-1}\hat{n}) \qquad (\hat{n} > 0)}$$

(15.5-21)

and

$$\lim_{\substack{\varepsilon \to 0 \\ \hat{n}\text{ fixed}}} \bar{\lambda}_{II} \mathbf{x}_s,\ \bar{n} = \varepsilon^r \hat{n}) = \lim_{\substack{\varepsilon \to 0 \\ \hat{n}\text{ fixed}}} \tilde{\lambda}(\mathbf{x}_s,\ \tilde{n} = \varepsilon^{r-1} \hat{n}) \qquad (\hat{n} < 0) \tag{15.5-22}$$

for the inner and outer expansions must hold for *all* $\mathbf{x}_s$. Of course, as $\varepsilon \to 0$ with $\hat{n}$ fixed, $\bar{n} \equiv O(\varepsilon^r) \to \pm 0$ whereas $\tilde{n} \equiv O(\varepsilon^{r-1}) \to \pm \infty$.

Table 15.5-1 tabulates order-of-magnitude, scaling values for the dimensionless normal distances within each of the three microscale domains, as derived from the respective definitions (15.5-2a), (15.5-2b) and (15.5-20) of $\tilde{n}$, $\bar{n}$ and $\hat{n}$.

Table 15.5-1
ORDERS-OF-MAGNITUDE OF INTERFACIAL SCALES
($0<r<1$)

Scaled Normal Distance	Outer Regions	Intermediate Regions	Inner Region
$\bar{n}$	$O(1)$	$O(\varepsilon^r)$	$O(\varepsilon)$
$\hat{n}$	$O(\varepsilon^{-r})$	$O(1)$	$O(\varepsilon^{1-r})$
$\tilde{n}$	$O(\varepsilon^{-1})$	$O(\varepsilon^{r-1})$	$O(1)$

One further expression relating the inner and outer limits of a generic macroscale quantity may be established as follows: Let $\mathbf{f}^*(\mathbf{x}_s, n)$ (with the asterisk possessing its usual significance in terms of the q^3-coordinate family surfaces relative to the parent surface q_0^3) represent a dimensionless quantity that is defined to be macroscale in the sense that

$$\frac{\partial \mathbf{f}^*}{\partial \bar{n}} = O(1) \quad \text{everywhere.} \tag{15.5-23}$$

This function may be written in the integral form

$$\mathbf{f}^*(\mathbf{x}_s,\ n) = \mathbf{f}(\mathbf{x}_s) + \int_0^{\bar{n}} \frac{\partial \mathbf{f}^*}{\partial \bar{n}'}(\mathbf{x}_s,\ \bar{n}';\varepsilon)\, d\bar{n}', \tag{15.5-24}$$

wherein

$$\mathbf{f}(\mathbf{x}_s) \stackrel{\text{def}}{=} \mathbf{f}^*(\mathbf{x}_s,\ 0) \tag{15.5-25}$$

is the value of $\mathbf{f}^*$ on the parent surface. Integration of Eq. (15.5-24) using the order-of-magnitude relation (15.5-23) yields the expression

$$\mathbf{f}^*(\mathbf{x}_s,\ n) = \mathbf{f}(\mathbf{x}_s) + \bar{n}\, O(1) \equiv \mathbf{f}(\mathbf{x}_s) + \tilde{n}\, O(\varepsilon). \tag{15.5-26}$$

Form the respective inner [(15.5-7)] and outer [(15.5-10) and (15.5-11)] limits of the above equation to obtain the appropriate limiting expression:

$$\tilde{\mathbf{f}}^*(\mathbf{x}_s, \bar{n}) = \mathbf{f}(\mathbf{x}_s, 0) = \lim_{\bar{n} \to 0+} \bar{\mathbf{f}}_{\mathrm{I}}^*(\mathbf{x}_s, \bar{n}) = \lim_{\bar{n} \to 0-} \bar{\mathbf{f}}_{\mathrm{II}}^*(\mathbf{x}_s, \bar{n}). \tag{15.5-27}$$

Table 15.5-2
IDENTIFICATION OF f* AND f

$\mathbf{f}^*(\mathbf{x}_S, n)$	$\mathbf{f}(\mathbf{x}_S)$	$\partial\mathbf{f}^*/\partial \bar{n}$	$\partial\mathbf{f}^*/\partial \bar{n}$ is $O(1)$ by:
$\mathbf{n}^*$	$\mathbf{n}$	$(\kappa^* L)\mathbf{p}^*$	Eq. (15.5-28)
$\mathbf{x}L$	$\mathbf{x}_S L$	$\mathbf{n}^*$	unit vector
$\mathbf{v}/U$	$\mathbf{v}^o/U$ †	$(\partial\mathbf{v}/\partial \bar{n})U^{-1}$	material interface

This relation will prove important in the subsequent analysis of surface-excess transport processes. Table 15.5-2 provides a correlation between the three different choices of pertinent geometric and kinematic parameters $\mathbf{f}^*$ appearing in the left-hand column, and the comparable quantities appearing in the remaining three columns.

Curvature Restrictions and Asymptotic Approximations to the Exact Projection Operations

As previously observed [cf. the paragraphs immediately following Eq. (15.3-7)], the normal coordinate n may be (locally) constructed by drawing linear trajectories from the parent surface $n = 0$. Thus, in the neighborhood of the interface, the q^3-coordinate curves are assumed to be completely defined at the macroscale, in the sense that the curvature κ_n^* of all q^3-coordinate curves [cf. Eqs. (B.1-21) and (B.1-22)] must satisfy the relation

$$|\kappa_n^*| L = O(1) \tag{15.5-28}$$

within the vicinity of the transition zone. This scaling assures that the magnitude $|\kappa_n^*|^{-1}$ of the radius of curvature of the interface is of $O(L)$, and hence is a macroscale quantity.

In addition, as has been illustrated in Question 15.6 (and again in Question 16.5), the asymptotic restriction $\varepsilon \to 0$ necessary for the existence of a truly macroscale, physical, interfacial theory, imposes a further restriction upon the curvature radii of the (macroscale) interface. Owing to

† This corresponds to the identification (3.6-17), revealing that $(\partial\mathbf{v}/\partial \bar{n})U^{-1} = O(1)$ for material interfaces, with U a characteristic interfacial velocity.

this fact, the two principal curvatures κ_1^* and κ_2^* of the surfaces A^* are assumed to satisfy the scaling relation [cf. footnote following Eq. (15.2-19)]

$$|\kappa_\alpha^*| L = O(1) \qquad (\alpha = 1, 2) \tag{15.5-29}$$

within the interfacial transition zone. This latter condition will limit the subsequent theory to interfaces possessing relatively small curvatures. [Comments and/or exercises pertaining to the inapplicability of the present asymptotic surface-excess theory to an interface whose curvature is of the same order, namely $O(l)$, as the thickness of the interfacial transition region, are provided at various points in Part II: cf. the footnote following Eq. (15.2-19) and Questions 15.6 and 16.5.]

Order-of-magnitude values within the inner and intermediate regions may now be established for each of the geometric parameters characterizing the family of q^3-coordinate surfaces. Thus, upon utilizing Eqs. (15.5-28), (3.2-12a), (3.2-13), (3.2-14) and (15.3-8), the following scalings are found to apply within both the inner and intermediate regions:

$$\mathbf{b}^* = O(1), \tag{15.5-30}$$

$$H^* L = O(1), \tag{15.5-31}$$

$$K^* L^2 = O(1) \tag{15.5-32}$$

and

$$\mathbf{n}^* \cdot (\nabla \mathbf{n}^*) L = O(1). \tag{15.5-33}$$

Substitution of Eqs. (15.4-2) and (15.3-6) into (15.3-11) yields

$$M^* = \exp\left[-\varepsilon \int_0^{\tilde{n}} 2H^* L d\tilde{n}\right]. \tag{15.5-34}$$

Similar substitution into Eq. (15.3-13) yields

$$N^* = \exp\left[-\varepsilon \int_0^{\tilde{n}} L\kappa_n^*(\mathbf{t}^*) d\,\tilde{n}\right], \tag{15.5-35}$$

where, from Eqs. (15.3-14) and (15.4-23),

$$L\kappa_n^*(\mathbf{t}^*) = \mathbf{t}^*\mathbf{t}^* : (\mathbf{b}^* L) = O(1) \tag{15.5-36}$$

within both the inner and intermediate regions. Perform Taylor series expansions of Eqs. (15.5-34) and (15.5-35), and subsequently employ the order-of-magnitude expressions (15.5-31) and (15.5-36) together with the scalings tabulated in Table 15.5-1 to obtain

$$M^* = 1 + O(\varepsilon), \qquad N^* = 1 + O(\varepsilon), \tag{15.5-37}$$

valid within the inner region, $\tilde{n} = O(1)$, and

$$M^* = 1 + O(\varepsilon^r), \qquad N^* = 1 + O(\varepsilon^r) \qquad (15.5\text{-}38)$$

($0<r<1$), valid within the two intermediate regions, $\hat{n} = O(1)$.

Similar estimates may be derived for $\mathbf{m}^*$. In this context, substitute Eq. (3.2-10) into Eq. (15.3-15) to obtain

$$\mathbf{m}^* = \mathbf{m} - \varepsilon \int_0^{\tilde{n}} (\mathbf{t}^*\mathbf{t}^* + \mathbf{n}^*\mathbf{n}^*) \cdot (-\mathbf{b} + \mathbf{n}^*\mathbf{n}^* \cdot \nabla\mathbf{n}^*) \cdot \mathbf{m}^* L d\tilde{n}. \quad (15.5\text{-}39)$$

Expand Eq. (15.5-31) via a Taylor series and employ the order-of-magnitude expressions (15.5-30) and (15.5-32), together with the appropriate values cited in Table 15.5-1. This scheme furnishes the estimate

$$\mathbf{m}^* = \mathbf{m} + O(\varepsilon), \qquad (15.5\text{-}40)$$

valid within the inner region, and

$$\mathbf{m}^* = \mathbf{m} + O(\varepsilon^r) \qquad (15.5\text{-}41)$$

($0<r<1$), valid within the two intermediate regions.

15.6 Surface-Excess Properties

Imagine a macroscale experimentalist who sets about the task of investigating the behavior of a generic field variable λ across a fluid interface contained within the coordinate-fixed volume $V(t)$ of fluid depicted in Figure 15.6-1. Owing to the relatively coarse scale of resolution of the macroscale experimental probe (see Figure 15.6-2), our experimentalist is able to directly discern the experimental values of only the outer, bulk-fluid fields. This perspective (in contrast to that of the more refined microscale experimentalist of §15.2) leads him to conclude *physically* that λ is given at each point $\mathbf{x}$ ($\bar{n} \neq 0$) by its bulk-fluid value(s) $\bar{\lambda}$; namely (see Question 15.7)

$$\lambda \Rightarrow \bar{\lambda} \equiv \begin{cases} \bar{\lambda}_{\mathrm{I}} & \text{for } \bar{n} > 0, \\ \bar{\lambda}_{\mathrm{II}} & \text{for } \bar{n} < 0. \end{cases} \qquad (15.6\text{-}1)$$

Whereas such a macroscale observer will inevitably suppose that each of these separate bulk values $\bar{\lambda}$ is valid right up to the two-dimensional macroscale interface $\bar{n} = 0$, in actuality we recognize these values to be strictly valid only for $|\bar{n}| >> O(\varepsilon)$. Changes occurring *within* the diffuse interfacial transition region cannot then be explicitly observed by such a macroscale experimentalist; they will, however, be implicitly relevant in

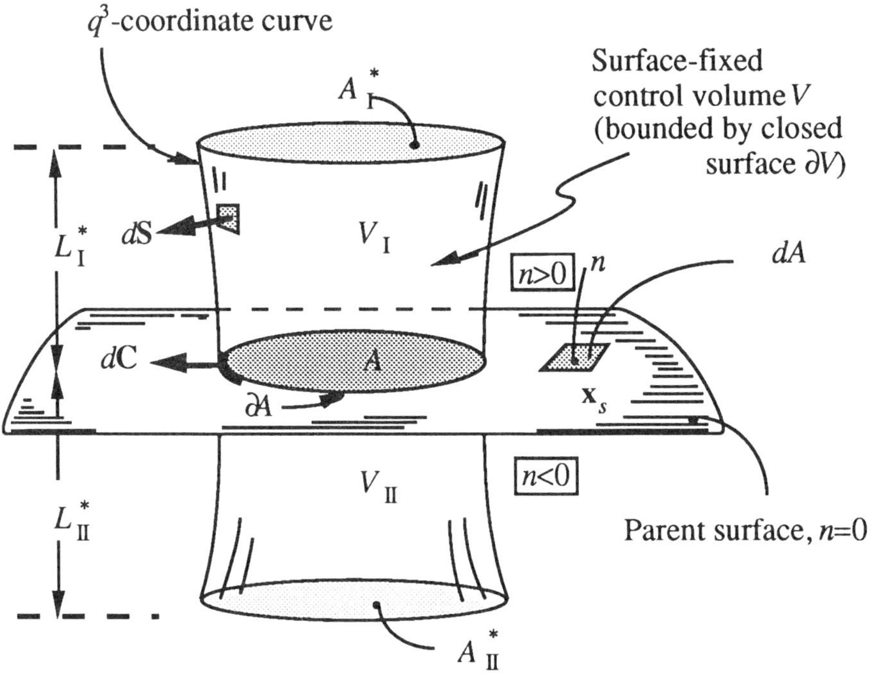

Figure 15.6-1 Parent surface-fixed control volume $V(t)$ in the shape of a pillbox, used to obtain surface-excess properties. Such a control volume denotes a *nonmaterial* fluid domain whose boundary points move strictly along q^3-coordinate curves with velocity $\mathbf{u}_n^*$, induced by the extrinsic motion $\mathbf{u}_n$ of the interface [cf. Eqs. (15.3-5) and (15.4-6) and the discussion thereof]. The parent surface (n=0) lies within the interfacial transition region. The control volume $V=V_I \oplus V_{II} \oplus A$ straddles this region and extends well into the two fully-developed bulk-fluid regions. The construction of the parent surface-fixed control volume $V(t)$ in terms of the familial surface-fixed curvilinear coordinates (q^1, q^2, q^3) is as follows: The latter coordinates are chosen such that q^3-coordinate curves constitute the sides of the control volume, with the top and bottom 'caps' (denoted respectively by the symbols A_I^*, A_{II}^*) instantaneously coinciding with q^3-coordinate surfaces. The control volume formed in this way is designated by the symbol $V(t)$. The areal element A lying on the parent surface n=0 and contained within V is bounded by the closed curve ∂A. Thus, the closed surface ∂V which bounds V externally consists of A_I^*, A_{II}^* and the q^3-coordinate curve envelope constituting the sides of the volume. The volumes V_I and V_{II} represent the two regions of the control volume V lying respectively above (n>0) and below (n<0) the parent coordinate surface n=0. Associated with these two volumes are the closed surfaces ∂V_I and ∂V_{II} which respectively bound V_I and V_{II} externally. The closed surface ∂V_I consists of the two ends A_I^* and A, including sides determined by the q^3-coordinate curve envelope lying above the parent surface n=0. Similarly, the closed surface ∂V_{II} consists of the two ends A_{II}^* and A, together with sides determined by the q^3-coordinate curve envelope lying below n=0.

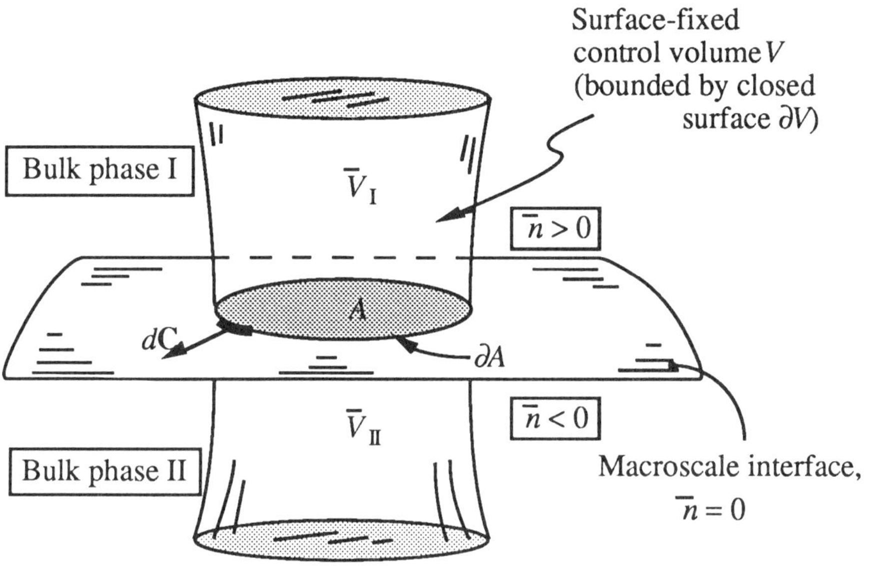

Figure 15.6-2 Macroscale view of the surface-fixed control volume $V(t)$ of Figure 15.6-1 (see also Figure 3.6-1). The interface divides the volume $V(t)$ into the separately continuous volumes $\bar{V}_I$ and $\bar{V}_{II}$, respectively containing bulk-fluid phases I and II. The symbols used in the figure correspond to those of Figure 3.6-1. Thus, the domain $V(t)$ may be decomposed as $V(t) = \bar{V} \oplus A$, with $\bar{V} = \bar{V}_I \oplus \bar{V}_{II}$. Bounding the volume $V(t)$ is the closed surface $\partial V(t)$, which may similarly be decomposed as $\partial V(t) = \partial \bar{V} \oplus \partial A$ [cf. Eqs. (3.6-1)–(3.6-3)].

determining the (macroscale) interfacial jump boundary condition at $\bar{n} = 0$ relating fields $\bar{\lambda}_I(\mathbf{x}_s, 0) \equiv \bar{\lambda}(\mathbf{x}_s, 0+)$ and $\bar{\lambda}_{II}(\mathbf{x}_s, 0) \equiv \bar{\lambda}(\mathbf{x}_s, 0-)$.

Differences existing between the macroscale and microscale views of transport phenomena can be formally reconciled by assigning surface-excess areal densities λ^s of extensive physical properties to each point $\mathbf{x}_s$ of the two-dimensional interface (as well as at each instant of time in unsteady-state circumstances).*

Appropriate expressions for such surface-excess densities are developed in this section for both three-dimensional volumetric and areal flux density microscale fields. The symbol $\mathbf{j}$ is used in the following† to designate

---

* The basic surface-excess procedure has been discussed previously in §3.6; cf. Eqs. (3.6-6)–(3.6-11) and the discussion pertaining thereto. See also the example outlined in §15.2.

† Owing to the present focus upon a *nonmaterial*, parent surface-fixed control volume V, the flux field $\mathbf{j}$ includes both a diffusive flux ϕ and a convective flux relative to the

flux fields in order to avoid any confusion with volumetric fields (designated by the symbol λ). Nonetheless, all of the relations discussed thus far for the field λ apply equally well to $\mathbf{j}$.

Volumetric Fields

Let the continuous variable $\lambda \equiv \lambda(\mathbf{x}, t)$ denote the microscale volumetric density of some extensive property $\mathcal{P}$. The total amount $\Lambda \equiv \Lambda(t)$ of $\mathcal{P}$ instantaneously contained within the surface-fixed control volume $V(t)$ is then given by the expression

$$\Lambda = \int_{V(t)} \lambda dV \,. \tag{15.6-2}$$

A macroscale experimentalist whose instruments traverse the volume $V(t)$ will, however, resolve and hence observe only the macroscale field $\bar{\lambda} \equiv \bar{\lambda}(\mathbf{x}, t)$ [cf. Eq. (15.6-1)], thereby concluding that the total amount $\bar{\Lambda} \equiv \bar{\Lambda}(t)$ of $\mathcal{P}$ instantaneously contained within the control volume is given rather by [cf. Eq. (3.6-6)]

$$\bar{\Lambda} = \int_{\bar{V}_{\mathrm{I}}(t)} \bar{\lambda}_{\mathrm{I}}\, dV + \int_{\bar{V}_{\mathrm{II}}(t)} \bar{\lambda}_{\mathrm{II}}\, dV \equiv \int_{\bar{V}(t)} \bar{\lambda}\, dV \,. \tag{15.6-3}$$

Such a macroscale observer will assign to the interface any disparity that exists between the two amounts Λ and $\bar{\Lambda}$, thus obtaining [cf. Eq. (15.2-10)]

$$\Lambda^s \stackrel{\text{def}}{=} \lim_{\varepsilon \to 0} (\Lambda - \bar{\Lambda}) \equiv \lim_{\varepsilon \to 0} \int_{V(t)} [\lambda(\mathbf{x}_s, n) - \bar{\lambda}(\mathbf{x}_s, \bar{n})] dV \tag{15.6-4}$$

for the *surface-excess* amount Λ^s of $\mathcal{P}$ on A (upon again explicitly suppressing the time in the arguments of the pertinent functions). A continuous, *surface-excess areal density field*, $\lambda^s \equiv \lambda^s(\mathbf{x}_s)$, say, (representing the surface-excess amount of the property $\mathcal{P}$ per unit area) may then be defined at each point $\mathbf{x}_s$ of the interfacial domain A by the relation [cf. Eqs. (15.2-12) and (15.2-17)]

$$\Lambda^s \equiv \int_A \lambda^s(\mathbf{x}_s) dA. \tag{15.6-5}$$

---

familial surface-fixed coordinate system (q^1, q^2, q^3) previously described in §15.3. This contrasts with the approach of Chapter 3, wherein generic conservation laws were derived by considering transport relative to a *material*-fixed control volume V; hence, only the diffusive flux ϕ appeared there [cf. Eq. (3.5-6)].

The field λ^s is a *macroscale* field, whose functional dependence upon the volumetric field λ may be established from Eq. (15.6-4) as follows.

Identification of the Parent Surface with the Macroscale Interface

A differential volume element dV may be represented by

$$dV = dA^* dn, \tag{15.6-6}$$

with dA^* a differential areal element on a q^3-coordinate familial surface. Upon employing the projection relation (15.3-10), this becomes

$$dV = M^* dAdn \tag{15.6-7}$$

in terms of the projected differential areal element dA lying on the interfacial parent coordinate surface n=0. Use of the above relation permits a separation of the volume in Eq. (15.6-4) into a two-fold integration: (i) along a familial q^3-coordinate curve, and (ii) over the parental reference surface; explicitly,

$$\int_V (\lambda - \bar{\lambda}) dV = \int_A dA \int_{n=-L_{\text{II}}^*}^{n=L_{\text{I}}^*} M^*(\mathbf{x}_s, n)\left[\lambda(\mathbf{x}_s, n) - \bar{\lambda}(\mathbf{x}_s, \bar{n})\right] dn. \tag{15.6-8}$$

Here, the lengths $L_{\text{I}}^*(\mathbf{x}_s) = \int_{n=0}^{n_{\text{I}}^*} dn$ and $L_{\text{II}}^*(\mathbf{x}_s) = \int_{n=n_{\text{II}}^*}^{0} dn$ represent distances between the parent surface and the familial surface end caps A_1^* and A_2^*, respectively, such distances being measured along q^3-coordinate curves (see Figure 15.6-1), the trajectories of such curves corresponding to $(q^1, q^2) \equiv \mathbf{x}_s$ remaining constant during the integration over n. In sequence, formally identify the *mathematical* parent surface A with the *physical*, macroscale interface in the $\varepsilon \to 0$ limit (while recognizing that the areal element A may be chosen arbitrarily), substitute Eq. (15.6-8) into Eq. (15.6-4), and employ the definition (15.6-5) for λ^s so as to obtain

$$\boxed{\lambda^s(\mathbf{x}_s) = \lim_{\varepsilon \to 0} \int_{n=-L_{\text{II}}^*}^{n=L_{\text{I}}^*} M^*(\mathbf{x}_s, n)\left[\lambda(\mathbf{x}_s, n) - \bar{\lambda}(\mathbf{x}_s, \bar{n})\right] dn\,.} \tag{15.6-9}$$

This important relation sheds significant light upon a question frequently raised in the literature (Gibbs 1957, Buff 1956, Murphy 1966, Melrose 1968) regarding the 'curvature-dependence' of interfacial properties. The appearance of the areal magnification factor M^* [cf. Eq. (15.3-11)] in the above quadrature formula would seem to indicate the possibility of the surface-excess density λ^s depending upon the curvature of the parent surface [i.e. the macroscale interface]. However, in the matched asymptotic surface-excess theory employed herein this curvature dependence is completely illusory, since the areal magnification factor M^* has been previously demonstrated to be independent of the curvature of the interface in the $\varepsilon \to 0$ limit [cf. Eqs. (15.5-37) and (15.5-38)]. Hence, upon performing the limit in Eq. (15.6-9), the surface-excess areal density λ^s is shown below to be asymptotically independent of interface curvature [cf. Eq. (15.6-18)]. [See the discussion immediately following Eq. (15.6-27) and Question 16.5 for related comments.]

This expression may be simplified by decomposing the above integral into the sum of three separate integrals, respectively corresponding to contributions from the two outer regions and the inner interfacial transition region:

$$\lambda^s(\mathbf{x}_s) = \lim_{\varepsilon \to 0} \left[\int_{n=l_{\mathrm{I}}^*}^{n=L_{\mathrm{I}}^*} M^*(\lambda - \bar{\lambda}_{\mathrm{I}})dn + \int_{n=-L_{\mathrm{II}}^*}^{n=-l_{\mathrm{II}}^*} M^*(\lambda - \bar{\lambda}_{\mathrm{II}})dn + \int_{n=-l_{\mathrm{II}}^*}^{n=l_{\mathrm{I}}^*} M^*(\lambda - \bar{\lambda})dn \right]. \quad (15.6\text{-}10)$$

Here, the lengths l_{I}^* and l_{II}^* represent distances between the parent surface and the upper and lower intermediate regions, respectively, such distances being measured along q^3-coordinate curves. (By definition, an intermediate region lies simultaneously, albeit asymptotically, in the domain of validity of both the inner and outer expansions, as illustrated in Figure 15.5-1.)

Within the upper outer region, $l_{\mathrm{I}}^* \le n \le L_{\mathrm{I}}^*$, the representation (15.5-10) is applicable. This, together with the requirement (15.5-14) imposed upon $\bar{\lambda}_{\mathrm{I}R}$, reduces the first integral in Eq. (15.6-10) to the form

$$\lim_{\varepsilon \to 0} \int_{n=l_{\mathrm{I}}^*}^{n=L_{\mathrm{I}}^*} M^*(\lambda - \bar{\lambda}_{\mathrm{I}})dn = \lim_{\varepsilon \to 0} \int_{n=l_{\mathrm{I}}^*}^{n=L_{\mathrm{I}}^*} M\bar{\lambda}_{\mathrm{I}R}\, dn = 0. \quad (15.6\text{-}11)$$

The final equality in the above expression is obtained by recognizing that M^* remains finite as $\varepsilon \to 0$ since M^* is independent of ε within each of the outer regions. Similarly, in the lower outer region, $l_{\mathrm{II}}^* \le -n \le L_{\mathrm{II}}^*$ we obtain

$$\lim_{\varepsilon \to 0} \int_{n=-L_{\text{II}}^{*}}^{n=-l_{\text{II}}^{*}} M^{*}(\lambda - \tilde{\lambda}_{\text{II}})dn = \lim_{\varepsilon \to 0} \int_{n=-L_{\text{II}}^{*}}^{n=-l_{\text{II}}^{*}} M^{*}\tilde{\lambda}_{\text{II}\,R}\,dn = 0. \quad (15.6\text{-}12)$$

Within the inner region, $-l_{\text{II}}^{*} \le n \le l_{\text{I}}^{*}$, Eq. (15.5-6) may be used to represent λ, thereby obtaining the expression

$$\lim_{\varepsilon \to 0} \int_{n=-l_{\text{II}}^{*}}^{n=l_{\text{I}}^{*}} M^{*}(\lambda - \bar{\lambda})dn =$$

$$\lim_{\varepsilon \to 0} \int_{n=-l_{\text{II}}^{*}}^{n=l_{\text{I}}^{*}} M^{*}(\tilde{\lambda} - \bar{\lambda})dn + \int_{n=-l_{\text{II}}^{*}}^{n=l_{\text{I}}^{*}} M^{*}\tilde{\lambda}_{R}\,dn \qquad (15.6\text{-}13)$$

for the last integral in Eq. (15.6-10). The condition (15.5-8) imposed upon $\tilde{\lambda}_R$ assures that

$$\lim_{\varepsilon \to 0} \int_{n=-l_{\text{II}}^{*}}^{n=l_{\text{I}}^{*}} M^{*}\tilde{\lambda}_{R}\,dn = 0. \qquad (15.6\text{-}14)$$

Use of Eqs. (15.5-1)–(15.5-3), (15.5-20), (15.5-29) and (15.5-30) yields

$$\lim_{\varepsilon \to 0} \int_{n=-l_{\text{II}}^{*}}^{n=l_{\text{I}}^{*}} M^{*}(\mathbf{x}_s, n)\left[\tilde{\lambda}(\mathbf{x}_s, \tilde{n}) - \bar{\lambda}(\mathbf{x}_s, \tilde{n})\right]dn =$$

$$\lim_{\varepsilon \to 0} \varepsilon L \int_{\tilde{n}=-c_{\text{II}}\varepsilon^{r-1}}^{\tilde{n}=c_{\text{I}}\varepsilon^{r-1}} [1 + O(\varepsilon^{r})]\left[\tilde{\lambda}(\mathbf{x}_s, \tilde{n}) - \bar{\lambda}(\mathbf{x}_s, n = \varepsilon\tilde{n})\right]d\tilde{n}\,,$$

(15.6-15)

with c_{I} and c_{II} positive numbers whose values may depend upon $\mathbf{x}_s$, but are independent of ε. Passage to the limit in the latter half of Eq. (15.6-15) furnishes the asymptotic relation

$$\lim_{\varepsilon \to 0} \int_{n=-l_{\text{II}}^{*}}^{n=l_{\text{I}}^{*}} M^{*}[\tilde{\lambda} - \bar{\lambda}]dn$$

$$\approx \varepsilon L \int_{\tilde{n}=-\infty}^{\tilde{n}=\infty} \left[\tilde{\lambda}(\mathbf{x}_s, \tilde{n}) - \bar{\lambda}(\mathbf{x}_s, 0)\right]d\tilde{n}\,, \qquad (15.6\text{-}16)$$

wherein

$$\bar{\lambda}(\mathbf{x}_s, 0) \equiv \begin{cases} \lim\limits_{n \to 0+} \bar{\lambda}_{\text{I}}(\mathbf{x}_s, n) & \text{for } \tilde{n} > 0, \\ \lim\limits_{n \to 0-} \bar{\lambda}_{\text{II}}(\mathbf{x}_s, n) & \text{for } \tilde{n} < 0. \end{cases} \qquad (15.6\text{-}17)$$

Substitution of Eqs. (15.6-11)-(15.6-14) and (15.6-16) into Eq. (15.6-10) thereby yields the operational definition

$$\boxed{\lambda^{s}(\mathbf{x}_s) = \varepsilon L \int_{\tilde{n}=-\infty}^{\tilde{n}=\infty} \left[\tilde{\lambda}(\mathbf{x}_s, \tilde{n}) - \bar{\lambda}(\mathbf{x}_s, 0)\right] d\tilde{n}} \qquad (15.6\text{-}18)$$

for the surface-excess areal density field λ^{s} in terms of the inner and outer limits $\tilde{\lambda}$ and $\bar{\lambda}$, respectively, of the *exact* volumetric density field $\lambda(\mathbf{x})$. The aim of achieving a wholly macroscale theory is realizable only if the integral in Eq. (15.6-18) is convergent. In this context we note that the matching conditions (15.5-21) and (15.5-22) provide necessary, but insufficient, conditions for convergence.

Flux Fields

A similar expression for the surface-excess lineal flux density field $\mathbf{j}^s$ of the extensive property $\mathcal{P}$ may be derived from the continuous, microscale field $\mathbf{j}$ of that property as follows: The total efflux $\mathbf{J}$ of $\mathcal{P}$ through the bounding surface $\partial V(t)$ of the parent surface-fixed control volume $V(t)$ is given by the expression

$$\mathbf{J} = \int_{\partial V(t)} d\mathbf{S} \cdot \mathbf{j}, \qquad (15.6\text{-}19)$$

where $d\mathbf{S}$ is a directed surface element on ∂V possessing the direction of the outward-drawn normal to ∂V (see Figure 15.6-1). A macroscale experimentalist whose instruments traverse the volume V will, however, resolve and hence observe only the macroscale field $\bar{\mathbf{j}}$ [cf. Eq. (15.6-1)], thus concluding that the total efflux $\mathbf{J}$ of $\mathcal{P}$ through the macroscale surface $\partial V(t)$ bounding the volume $V(t)$ is given rather by the expression

$$\bar{\mathbf{J}} = \int_{\partial \bar{V}(t)} d\mathbf{S} \cdot \bar{\mathbf{j}}. \qquad (15.6\text{-}20)$$

Such a macroscale observer will suppose that any discrepancy existing between the exact and macroscale fluxes $\mathbf{J}$ and $\bar{\mathbf{J}}$ is caused by a flux field along the macroscale interface, thereby obtaining

$$\mathbf{J}^{s} \overset{\text{def}}{=} \lim_{\varepsilon \to 0} (\mathbf{J} - \bar{\mathbf{J}}) = \lim_{\varepsilon \to 0} \int_{\partial V(t)} d\mathbf{S} \cdot [\mathbf{j}(\mathbf{x}_s, n) - \bar{\mathbf{j}}(\mathbf{x}_s, \bar{n})] \qquad (15.6\text{-}21)$$

for the *surface-excess flux* $\mathbf{J}^s$ of $\mathcal{P}$ assigned to the closed curve ∂A. A continuous, surface-excess lineal flux-density field $\mathbf{j}^s$ may be defined at every point $\mathbf{x}_s$ such that [cf. Eq. (3.6-10b)]

$$\mathbf{J}^{s} = \int_{\partial A} d\mathbf{C} \cdot \mathbf{j}^{s}(\mathbf{x}_s), \qquad (15.6\text{-}22)$$

where $d\mathbf{C}=\mathbf{m}dC$ is a directed line element on ∂A (see Figure 15.6-2). The field $\mathbf{j}^s$ is a *macroscale* field, one whose functional dependence upon the areal efflux $\mathbf{j}$ can be established [cf. Eq. (15.6-36)] from Eq. (15.6-21) as in the following paragraphs.

Identification of the Parent Surface with the Macroscale, Physical Interface

A differential element of surface $d\mathbf{S}$ lying upon the sides of the control volume may be represented as

$$d\mathbf{S} = \mathbf{m}^* dC^* dn, \tag{15.6-23}$$

with dC^* a differential lineal element lying on the q^3-coordinate familial surface, and $\mathbf{m}^*$ the unit outer normal vector to the curve C^* and lying in the tangent plane to the q^3-coordinate surface at $\mathbf{x}_s$ (see Figure 15.3-4). Use of the projection relation (15.3-12) thereby gives

$$d\mathbf{S} = \mathbf{m}^* N^* dC dn, \tag{15.6-24}$$

in terms of the projected differential lineal element dC lying on the coordinate reference surface n=0. The integral in Eq. (15.6-21) may be separated into two areal integrals on the end caps A_{I}^* and A_{II}^*, together with an integral along the lateral q^3-coordinate curve envelope. Use of Eq. (15.6-24) allows this latter integral to be partitioned into an integration along the q^3-coordinate curve, followed by an integration along the curve ∂A. Performing these operations gives

$$\int_{\partial V} d\mathbf{S}\cdot(\mathbf{j}-\bar{\mathbf{j}}) = \int_{\partial A} dC \int_{n=-L_{\mathrm{II}}^*}^{n=L_{\mathrm{I}}^*} N^*(\mathbf{x}_s, n)\mathbf{m}^*\cdot[\mathbf{j}(\mathbf{x}_s, n)-\bar{\mathbf{j}}(\mathbf{x}_s, \bar{n})]\,dn$$
$$+\int_{\partial A_{\mathrm{I}}^*} dA^*\mathbf{n}^*\cdot\left(\mathbf{j}-\bar{\mathbf{j}}_{\mathrm{I}}\right) - \int_{\partial A_{\mathrm{II}}^*} dA^*\mathbf{n}^*\cdot\left(\mathbf{j}-\bar{\mathbf{j}}_{\mathrm{II}}\right). \tag{15.6-25}$$

Since A_{I}^* and A_{II}^* respectively lie within the upper and lower outer regions, Eqs. (15.5-10) and (15.5-11) may be used to represent $\mathbf{j}$ in the latter two integrals appearing above. This, together with the requirement imposed by Eq. (15.5-14) upon $\bar{\mathbf{j}}_{\alpha R}$, makes

$$\lim_{\varepsilon\to 0}\int_{\partial A_{\alpha}^*} dA^*\mathbf{n}^*\cdot\left(\mathbf{j}-\bar{\mathbf{j}}_{\alpha}\right) = \lim_{\varepsilon\to 0}\int_{\partial A_{\alpha}^*} dA^*\mathbf{n}^*\cdot\bar{\mathbf{j}}_{\alpha R} = 0 \quad (\alpha = \mathrm{I, II}). \tag{15.6-26}$$

Formally identify the parent coordinate surface A with the interface in the limit $\varepsilon\to 0$ (while recognizing that the areal element A may be chosen arbitrarily), substitute Eqs. (15-6-25) and (15.6-26) into (15.6-21), and subsequently utilize the definition (15.6-22) of $\mathbf{j}^s$ to obtain

$$\boxed{\mathbf{m}\cdot\mathbf{j}^s(\mathbf{x}_s) = \lim_{\varepsilon\to 0}\int_{n=-L^*_{\mathrm{II}}}^{n=L^*_{\mathrm{I}}} N^*(\mathbf{x}_s, n)\mathbf{m}^*\cdot[\mathbf{j}(\mathbf{x}_s, n) - \bar{\mathbf{j}}(\mathbf{x}_s, \bar{n})]\,dn\,.}$$

(15.6- 27)

Similar to the formula (15.6-27) (below which related comments have been made as are made here) the above relation appears to indicate the possible dependence of the surface-excess flux $\mathbf{j}^s$ upon the curvature of the interface, owing to the existence of the lineal magnification factor N^* [cf. Eq. (15.3-13)] in the integrand. Again, this dependence is illusory in the $\varepsilon \to 0$ limit, owing to the fact that Eqs. (15.5-37)–(15.5-41) apply. Thus, as demonstrated below, the surface-excess flux $\mathbf{j}^s$ is asymptotically independent of the curvature of the interface.

The preceding expression may be decomposed into three separate contributions, corresponding respectively to integration over the two outer regions and the inner interfacial transition region; explicitly,

$$\mathbf{m}\cdot\mathbf{j}^s(\mathbf{x}_s) = \lim_{\varepsilon\to 0}\left[\int_{n=l^*_{\mathrm{I}}}^{n=L^*_{\mathrm{I}}} N^*\mathbf{m}^*\cdot(\mathbf{j}-\bar{\mathbf{j}}_{\mathrm{I}})\,dn + \int_{n=-L^*_{\mathrm{II}}}^{n=-l^*_{\mathrm{II}}} N^*\mathbf{m}^*\cdot(\mathbf{j}-\bar{\mathbf{j}}_{\mathrm{II}})\,dn + \int_{n=-l^*_{\mathrm{II}}}^{n=l^*_{\mathrm{I}}} N^*\mathbf{m}^*\cdot(\mathbf{j}-\bar{\mathbf{j}})\,dn\right]. \qquad (15.6\text{-}28)$$

Equation (15.5-10) may be used to represent $\mathbf{j}$ within the upper outer region. This, together with the requirement (15.5-14) demanded of $\bar{\mathbf{j}}_{\mathrm{I}R}$, enables the first integral appearing in Eq. (15.6-28) to be written as

$$\lim_{\varepsilon\to 0}\int_{n=l^*_{\mathrm{I}}}^{n=L^*_{\mathrm{I}}} N^*\mathbf{m}^*\cdot(\mathbf{j}-\bar{\mathbf{j}}_{\mathrm{I}})\,dn = \lim_{\varepsilon\to 0}\int_{n=l^*_{\mathrm{I}}}^{n=L^*_{\mathrm{I}}} N^*\mathbf{m}^*\cdot\bar{\mathbf{j}}_{\mathrm{I}R}\,dn = 0. \qquad (15.6\text{-}29)$$

In obtaining the final equality in the preceding expression, it has been recognized that N^* remains finite as $\varepsilon\to 0$, since N^* is independent of ε within the two outer regions. Similarly, upon employing Eqs. (15.5-11) and (15.5-14), the second integral in Eq. (15.6-28) adopts the form

$$\lim_{\varepsilon\to 0}\int_{n=-L^*_{\mathrm{II}}}^{n=-l^*_{\mathrm{II}}} N^*\mathbf{m}^*\cdot(\mathbf{j}-\bar{\mathbf{j}}_{\mathrm{II}})\,dn = \lim_{\varepsilon\to 0}\int_{n=-L^*_{\mathrm{II}}}^{n=-l^*_{\mathrm{II}}} N^*\mathbf{m}^*\cdot\bar{\mathbf{j}}_{\mathrm{II}R}\,dn = 0. \qquad (15.6\text{-}30)$$

Equation (15.5-6) may be used to represent **j** within the inner region, thereby obtaining

$$\lim_{\varepsilon\to 0}\int_{n=-L^{\bullet}_{\rm II}}^{n=-l^{\bullet}_{\rm II}} N^*\mathbf{m}^*\cdot(\mathbf{j}-\bar{\mathbf{j}})dn$$

$$=\lim_{\varepsilon\to 0}\left[\int_{n=-L^{\bullet}_{\rm II}}^{n=-l^{\bullet}_{\rm II}} N^*\mathbf{m}^*\cdot(\tilde{\mathbf{j}}-\bar{\mathbf{j}})dn+\int_{n=-L^{\bullet}_{\rm II}}^{n=-l^{\bullet}_{\rm II}} N^*\mathbf{m}^*\cdot\tilde{\mathbf{j}}_R\,dn\right] \quad (15.6\text{-}31)$$

for the last integral in Eq. (15.6-28). The condition (15.5-8) required of $\tilde{\mathbf{j}}_R$ assures that

$$\lim_{\varepsilon\to 0}\int_{n=-L^{\bullet}_{\rm II}}^{n=-l^{\bullet}_{\rm II}} N^*\mathbf{m}^*\cdot\bar{\mathbf{j}}_R\,dn=0. \quad (15.6\text{-}32)$$

Jointly, Eqs. (15.5-1)–(15.5-3), (15.5-20), (15.5-29), (15.5-30), (15.5-32), and (15.5-33) combine to furnish the expression

$$\lim_{\varepsilon\to 0}\int_{n=-L^{\bullet}_{\rm II}}^{n=-l^{\bullet}_{\rm II}} N^*(\mathbf{x}_s,n)\mathbf{m}^*(\mathbf{x}_s,n)\cdot\left[\tilde{\mathbf{j}}(\mathbf{x}_s,\tilde{n})-\bar{j}(\mathbf{x}_s,\bar{n})\right]dn$$

$$=\lim_{\varepsilon\to 0}\left\{\varepsilon L\int_{n=-c_{\rm II}\varepsilon^{r-1}}^{n=c_{\rm I}\varepsilon^{r-1}}[1+O(\varepsilon^r)]\,[\mathbf{m}^*+O(\varepsilon^r)]\right.$$

$$\left.\cdot\left[\tilde{\mathbf{j}}(\mathbf{x}_s,\tilde{n})-\bar{\mathbf{j}}(\mathbf{x}_s\bar{n}=\varepsilon\tilde{n})\right]d\tilde{n}\right\}. \quad (15.6\text{-}33)$$

Passage to the limit on the right-hand side of the above equation thereby produces the relation

$$\lim_{\varepsilon\to 0}\int_{n=-L^{\bullet}_{\rm II}}^{n=-l^{\bullet}_{\rm II}} N^*\mathbf{m}^*\cdot(\tilde{\mathbf{j}}-\bar{\mathbf{j}})dn$$

$$\approx\varepsilon L\int_{n=-\infty}^{n-\infty}\mathbf{m}^*\cdot\left[\tilde{\mathbf{j}}(\mathbf{x}_s,\tilde{n})-\bar{\mathbf{j}}(\mathbf{x}_s,0)\right]d\tilde{n}, \quad (15.6\text{-}34)$$

where the notation of Eq. (15.6-17) (with λ=**j**) has been employed.

Introduce Eqs. (15.6-29)-(15.6-32) and (15.6-34) into (15.6-28) to obtain

$$\mathbf{m}\cdot\mathbf{j}^s(\mathbf{x}_s)=\varepsilon L\int_{n=-\infty}^{n=\infty}\mathbf{m}^*\cdot\left[\tilde{\mathbf{j}}(\mathbf{x}_s,\tilde{n})-\bar{\mathbf{j}}(\mathbf{x}_s,0)\right]d\tilde{n}. \quad (15.6\text{-}35)$$

Use of the identity $\mathbf{m}^*=\mathbf{m}\cdot\mathbf{I}_s$ yields the operational formula*

---

* Since the arbitrary vector **m** lies only in the tangent plane, the actual solution of Eq. (15.6-35) is

$$\mathbf{j}^S(\mathbf{x}_s)=\varepsilon L\int_{-\infty}^{\infty}\left[\tilde{\mathbf{j}}(\mathbf{x}_s,\tilde{n})-\bar{\mathbf{j}}(\mathbf{x}_s,0)\right]d\tilde{n}+\mathbf{nF},$$

$$\mathbf{j}^s(\mathbf{x}_s) = \varepsilon L \int_{n=-\infty}^{n=\infty} \mathbf{I}_s \cdot [\tilde{\mathbf{j}}(\mathbf{x}_s, \tilde{n}) - \bar{\mathbf{j}}(\mathbf{x}_s, 0)] d\tilde{n} \quad (15.6\text{-}36)$$

for computing the two-dimensional surface-excess flux density field $\mathbf{j}^s$ from knowledge of the inner and outer limits of the exact, three-dimensional areal flux density field $\mathbf{j}(\mathbf{x})$. In order to achieve a macroscale theory this integral must be convergent, as in the case of the comparable integral (15.6-18). In this context we note that the matching conditions (15.5-21) and (15.5-22) constitute necessary, but insufficient, conditions for convergence. In any event, all integrals appearing in the subsequent theory will be supposed convergent, as was the case with the integral (15.6-18).

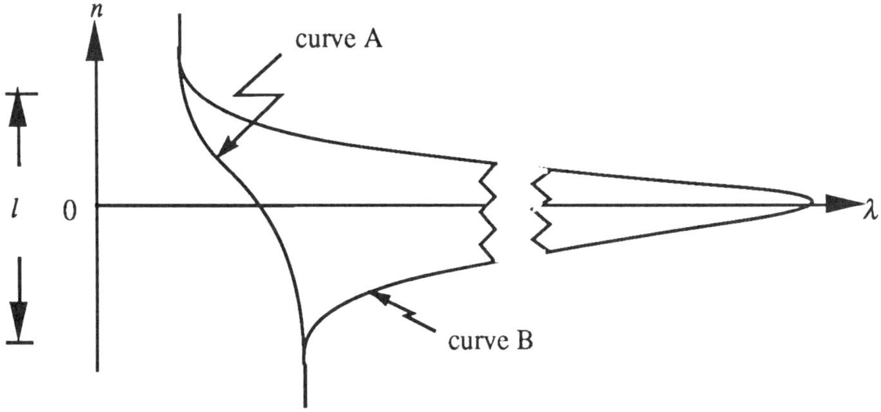

Figure 15.6-3 Schematic showing the two main types of behavior for a microscale field λ within the interfacial region. Curve B depicts a field which attains a relatively large maximum within the transition region, whereas curve A depicts a field possessing no large maximum value.

Discussion

Since the operational expressions (15.6-18) and (15.6-36) for computing the surface-excess fields λ^s and $\mathbf{j}^s$ from the given microscale data are multiplied by the small quantity ε, nonnegligible values for λ^s and $\mathbf{j}^s$

---

in which $\mathbf{F}$ is an arbitrary three-dimensional tensor of rank one less than that of $\mathbf{j}^s$. In subsequent applications it is only the directional quantity $\mathbf{I}_s \cdot \mathbf{j}^s$ rather than $\mathbf{j}^s$ itself that appears. We have anticipated this by suppressing the $\mathbf{nF}$ term that would otherwise arise via use of the projection operator $\mathbf{I}_s$ on the RHS of the integrand in Eq. (15.6-36). Note that whereas $\mathbf{n} \cdot \mathbf{j}^s = 0$ identically with the choice (15.6-36), this does not imply that the expression $\mathbf{j}^s \cdot \mathbf{n} = 0$ is generally true for situations in which $\mathbf{j}$ is of rank greater than unity.

will only arise when $|\tilde{\lambda}_{max}|/|\tilde{\lambda}|>>1$ and $|\tilde{\mathbf{j}}_{max}|/|\tilde{\mathbf{j}}|>>1$, respectively. For example, consider the pair of postulated field behaviors depicted in Fig. 15.6-3. If the form of the microscale density field λ is such that the latter does not attain a large maximum value within the interfacial region (as illustrated by curve A), the derived surface-excess field will be $O(\varepsilon)$, whence it may be concluded that since $\varepsilon<<1$, $\lambda^s_A=0$ for all practical purposes. Physical examples of fields typified by curve A, and hence possessing vanishing surface-excess densities, are the total mass density ρ and the mass density ρ_i of conventional surface *inactive* solutes; thus, $\rho^s=0$ and $\rho^s_i=0$. On the other hand, if λ does attain a large maximum value within the interfacial region, as illustrated by curve B, the surface-excess field will be sensible; i.e. $\lambda^s_B\neq 0$. One example of such a field is the mass density ρ_i of a surface-*active* species *i*; hence, $\rho^s_i\neq 0$ for surface-active solutes.

Large values of $|\lambda|$ (or $|\mathbf{j}|$) within the interfacial transition region may be expected to exist whenever surface-active substances are adsorbed at the interface and conversely to be absent for conventional, surface-inactive solutes. Thus, the surface-excess densities λ^s and $\mathbf{j}^s$ may be expected to be sensible only when the interface contains adsorbed surfactants, as has been repeatedly demonstrated throughout Part I.

Placement of the dividing surface

According to traditional interfacial transport theories (Eliassen 1963, Murphy 1966, Slattery 1967, Deemer & Slattery 1978), which aim at an "exact" description of interfacial (and, concomitantly, bulk) transport phenomena, the location of the dividing surface must be precisely defined in order to provide "exact" definitions of the requisite surface-excess properties. However, in the asymptotic theory presented here, a *precise* location for the coordinate surface proves both unnecessary and irrelevant. Rather, the only constraint placed upon the location of the coordinate parent reference surface—the latter representing the position assigned to the (macroscale) interface in the limit $\varepsilon\to 0$—is that it must be located somewhere within the inner interfacial region; additionally, the principal curvatures κ_1 and κ_2 of the parent surface must be macroscale in size [cf. Eq. (15.5-29)].

As proof that the precise location of the coordinate surface is inconsequential in the unambiguous definition of surface-excess fields, and hence properties, consider the consequences of a shift by the amount of $|\Delta\tilde{n}|=O(1)$ in the location of the choice of parent surface. Changing the position of this coordinate surface from $n_0=0$ to $n_0=a$, where both values of n_0 lie within the inner region [i.e. $a = O(\varepsilon)$], will result in the concomitant change

$$\Delta\lambda^s = \varepsilon L\left\{\int_0^a [\bar{\lambda}_{\mathrm{I}}(0) - \bar{\lambda}_{\mathrm{II}}(0)]d\tilde{n} + \int_a^{+\infty} [\bar{\lambda}_{\mathrm{I}}(0) - \bar{\lambda}_{\mathrm{I}}(a)]d\tilde{n} + \int_{-\infty}^a [\bar{\lambda}_{\mathrm{II}}(0) - \bar{\lambda}_{\mathrm{II}}(a)]d\tilde{n}\right\} \quad (15.6\text{-}37)$$

in the value of the surface-excess quantity λ^s. By definition, the respective orders-of-magnitude of the various quantities appearing in these integrals are

$$\bar{\lambda}_{\mathrm{I}}(0) - \bar{\lambda}_{\mathrm{II}}(0) = O(1), \quad (15.6\text{-}38)$$

and

$$\bar{\lambda}_{\alpha}(0) - \bar{\lambda}_{\alpha}(a) = O(\varepsilon) \qquad (\alpha = \mathrm{I, II}). \quad (15.6\text{-}39)$$

Thus, upon using Table 15.5-1, the difference $\Delta\lambda^s$ derived from Eq. (15.6-37), namely

$$\Delta\lambda^s = O(1)O(\varepsilon) + O(\varepsilon)O(\varepsilon^r) + O(\varepsilon)O(\varepsilon^r) \equiv O(\varepsilon) \quad (15.6\text{-}40)$$

becomes negligible as $\varepsilon \to 0$. Accordingly, varying the position of the coordinate reference surface within the interfacial transition region will be without sensible effect upon the numerical value of the $O(1)$ surface-excess quantity λ^s, at least to the order of approximation of our theory. Similar remarks apply to the insensibility of $\mathbf{j}^s$ to the exact placement of the dividing surface within the interfacial transition region.

Although most prior theories attempt to precisely define the requisite position of the dividing surface (Eliassen 1963, Murphy 1966, Kirkwood & Buff 1949), such as to render these theories "exact", such a definition is not necessarily possible. In particular, Schofield & Henderson (1982) conclude that, from a statistical mechanical point of view, the position of the "surface of tension" (i.e. the particular dividing surface for which interfacial 'bending stress' vanishes; cf. Question 16.5) is ill-defined, and hence " . . . could be placed anywhere in the interfacial region by choosing different contours in the definition of the pressure tensor."

15.7 Generic Conservation Equations

The following development of macroscale, generic transport equations parallels the earlier development of macroscale transport equations in §§3.5 and 3.6. There are, however, crucial differences between the two approaches; primary among these is the present perspective of a microscale observer, who mathematically 'discovers' a macroscale surface-excess transport law in the mathematical limit $\varepsilon \to 0$ (and, it may be added, *only* in this limit, corresponding to the circumstances under which the surface $\bar{n} = 0$ becomes a mathematically singular surface). In addition, with the purpose of broadening our perspective to include the possibility of nonmaterial phase interfaces, our attention will be focused on a parent surface-fixed (rather than material-fixed) control volume V. An immediate consequence of this fact is the present use [cf. Eqs. (15.6-19)–(15.6-22)] of the notation $\mathbf{J}$ [rather than $\mathbf{\Phi}$, cf. Eqs. (3.6-9)–(3.6-10)] for the efflux through the bounding surface ∂V. Thus, whereas $\mathbf{\Phi}$ corresponds to the purely diffusive flux of material through a boundary, $\mathbf{J}$ denotes a combined convective and diffusive flux [cf. Eq. (15.7-2)].

By pursuing the current rigorous, asymptotic approach, operational formulas of the surface-excess fields λ^s and $\mathbf{j}^s$ in terms of quadratures of their volumetric microscale counterparts, namely $\lambda(\mathbf{x})$ and $\mathbf{j}(\mathbf{x})$, in proximity to the interface are furnished by Eqs. (15.6-18) and (15.6-36). This contrasts with the entirely macroscale definitions of these same fields provided by Eqs. (3.6-7b) and (3.6-10b).

Microscale Equations

We begin with the generic volumetric balance equation (3.5-9) developed in Chapter 3 for the microscale transport relative to a space-fixed coordinate system of a continuous, three-dimensional fluid property $\mathcal{P}$ whose volumetric density is $\psi(\mathbf{x}, t)$; explicitly,

$$\frac{\partial \psi}{\partial t} + \nabla \cdot (\mathbf{v}\,\psi + \phi) - \pi - \zeta = \mathbf{0}\,, \tag{15.7-1}$$

where the terms appearing in this equation possess the physical interpretations assigned them in §3.5.

For our purposes, it proves convenient to rewrite Eq. (15.7-1) utilizing the familial coordinate-fixed time derivative (15.4-7), while employing the following definition of the convective-diffusive flux vector $\mathbf{j}$ relative to the familial coordinate system (q^1, q^2, q^2) described in §§15.3 and 15.4:

$$\mathbf{j} = (\mathbf{v} - \mathbf{u}_n{}^*)\psi + \phi. \tag{15.7-2}$$

This gives

$$\frac{\delta\psi}{\delta t} + (\nabla\cdot\ \mathbf{u}_n{}^*\,)\psi + \nabla \cdot \mathbf{j} - \pi - \zeta = \mathbf{0}\,. \tag{15.7-3}$$

Equation (15.7-3) will henceforward be viewed as the 'true' generic balance equation, valid at each point $\mathbf{x}$, and applicable not only within the highly inhomogeneous interfacial fluid region $\tilde{n} = O(1)$, but equally at all points in the relatively homogeneous fluid regions $\bar{n} = O(1)$.

Macroscale Bulk-Fluid Equations

The generic conservation equation for the intensive, macroscale volumetric density $\overline{\psi}$ is found by taking the outer limit of Eq. (15.7-3):

$$\lim_{\substack{\varepsilon \to 0 \\ \bar{n} \text{ fixed}}} \left[\frac{\delta \psi}{\delta t} + (\nabla \cdot \mathbf{u}_n^*)\psi + \nabla \cdot \mathbf{j} - \pi - \zeta\right] = \mathbf{0} . \tag{15.7-4}$$

Substitution of Eqs. (15.5-10), (15.5-11) and (15.5-18) into Eq. (15.7-4) yields

$$\left[\frac{\delta \overline{\psi}}{\delta t} + (\nabla \cdot \mathbf{u}_n^*)\overline{\psi} + \nabla \cdot \overline{\mathbf{j}} - \overline{\pi} - \overline{\zeta}\right]$$

$$+ \lim_{\substack{\varepsilon \to 0 \\ \bar{n} \text{ fixed}}} \left[\frac{\delta \overline{\psi}_{\alpha R}}{\delta t} + (\nabla \cdot \mathbf{u}_n^*)\overline{\psi}_{\alpha R} + \nabla \cdot \overline{\mathbf{j}}_{\alpha R} - \overline{\pi}_{\alpha R} - \overline{\zeta}_{\alpha R}\right] = \mathbf{0}$$

$$(\alpha = \text{I, II}) , \tag{15.7-5}$$

which holds strictly only for points within the outer (and intermediate) regions, namely $|\bar{n}| \geq O(\varepsilon^r)$. However, for $\varepsilon \to 0$, the preceding constraint imposed upon $\bar{n}$ reduces to the domain $|\bar{n}| > 0$. In this limit, $\overline{\mathbf{u}}_n^* = \mathbf{u}_n^*$ as follows from the scaling relation provided in the footnote below, upon use of Eqs. (15.5-23)–(15.5-27).†

Further progress requires additional assumptions regarding the *mathematical* nature of the generic densities ψ and $\mathbf{j}$ for the particular class of *physical* problems of interest to us here. In particular, it will be supposed that the temporal and spatial differential operations can be interchanged with the outer limit operation such that, with use of Eq. (15.5-14), we obtain

$$\lim_{\substack{\varepsilon \to 0 \\ \bar{n} \text{ fixed}}} \frac{\delta \overline{\psi}_{\alpha R}}{\delta t} = \mathbf{0} \quad (\alpha = \text{I, II}) \tag{15.7-6}$$

and

$$\lim_{\substack{\varepsilon \to 0 \\ \bar{n} \text{ fixed}}} \nabla \cdot \overline{\mathbf{j}}_{\alpha R} = \mathbf{0} \quad (\alpha = \text{I, II}) . \tag{15.7-7}$$

---

† Note that $\mathbf{u}_n^*$ satisfies the scaling relation

$$\frac{L}{U} \nabla \cdot \mathbf{u}_n^* = O(1),$$

where U is a characteristic interfacial velocity. This corresponds to the restriction [cf. Eqs. (15.6-8) and (15.6-9) and the pertinent discussion thereof, noting also the definition (15.3-10)] that the familial coordinate surfaces A^* move together with the interface—and thus, to within $O(\varepsilon)$, with the parental surface velocity $\mathbf{u}_n$.

These assumptions impose no significant constraints, and serve to eliminate pathological density functions. In particular, if $\bar{\lambda}_{\alpha R}$ (with $\lambda \equiv \psi, \mathbf{j}$) can be written in the form

$$\bar{\lambda}_{\alpha R}(\mathbf{x}_s, \bar{n}, t; \varepsilon) = \bar{f}_{\alpha 1}(\varepsilon)\bar{\lambda}_{\alpha R_1}(\mathbf{x}_s, \bar{n}, t) + \bar{f}_{\alpha 2}(\varepsilon)\bar{\lambda}_{\alpha R_2}(\mathbf{x}_s, \bar{n}, t) + \ldots,$$

Equations (15.7-6) and (15.7-7) will then hold automatically. Since rapid changes in λ occur only with respect to the variable $\bar{n}$, the main restriction imposed upon the continuum fields by Eq. (15.7-7) arises due to the constraint imposed upon the normal gradient.

Substitution of Eqs. (15.5-14), (15.7-6) and (15.7-7) into (15.7-5) gives

$$\frac{\delta\bar{\psi}}{\delta t} + (\nabla \cdot \mathbf{u}_n^*)\bar{\psi} + \nabla \cdot \bar{\mathbf{j}} - \bar{\pi} - \bar{\zeta} = \mathbf{0} \quad (\bar{n} \neq 0). \qquad (15.7\text{-}8)$$

This equation is separately valid within each of the two macroscale bulk-fluid phases $\bar{V}_{\text{I}}$ and $\bar{V}_{\text{II}}$ (Figure 15.6-2), respectively corresponding to $\bar{n} > 0$ and $\bar{n} < 0$, although not at the interface $\bar{n} = 0$, across which any one or more of the macroscale fields (denoted by the presence of an overbar) may be discontinuous.

Equation (15.7-8) is, of course, identical to Eq. (3.6-15). This is immediately evident upon substitution of the relations (15.4-7) and (15.7-2) into (15.7-8). [Note that the absence of an overbar on the macroscale velocity field $\mathbf{v}$ in Eq. (3.6-15) corresponds to the fact that for *material* fluid interfaces the macroscale velocity vector is continuous across the interface (3.6-17). For *nonmaterial* interfaces, $\bar{\mathbf{v}}$ will generally be discontinuous at $\bar{n} = 0$.]

(Macroscale) Interfacial Equation

The generic conservation equation governing the total amount of $\mathcal{P}$ contained within the surface-fixed control volume $V(t)$, as observed by a microscale observer, is provided by the expression

$$\frac{d}{dt}\int_{V(t)} \psi \, dV + \int_{\partial V(t)} d\mathbf{S} \cdot \mathbf{j} = \int_{V(t)} (\pi + \zeta) dV. \qquad (15.7\text{-}9)$$

A macroscale observer (Figure 15.6-2) will, however, conclude from Eq. (15.7-7) that the relationships

$$\frac{d}{dt}\int_{\bar{V}_\alpha(t)} \bar{\psi}_\alpha \, dV + \int_{\partial\bar{V}_\alpha(t)} d\mathbf{S} \cdot \bar{\mathbf{j}}_\alpha$$

$$= \int_{\bar{V}_\alpha(t)} (\bar{\pi}_\alpha + \bar{\zeta}_\alpha) dV \qquad (\alpha = \text{I, II}) \qquad (15.7\text{-}10)$$

hold within the two separate regions $\bar{V}_{\mathrm{I}}$ and $\bar{V}_{\mathrm{II}}$ (bounded externally by the closed surfaces $\partial\bar{V}_{\mathrm{I}}$ and $\partial\bar{V}_{\mathrm{II}}$, respectively, as in Figure 15.7-1). In order to establish the macroscale jump boundary condition that such an observer would impose at the interface to effect closure, the macroscale balance equations (15.7-10) are subtracted from the exact, microscale equation (15.7-9) and the macroscale limit subsequently taken [note the similarities between the present procedure and the intuitive approach of §15.2; cf. Eqs. (15.2-10), (15.2-11) and (15.2-17)]; explicitly,

$$\lim_{\varepsilon \to 0}\left\{\left[\frac{d}{dt}\int_{V(t)}\psi\, dV + \int_{\partial V(t)} d\mathbf{S}\cdot\mathbf{j} - \int_{V(t)}(\pi+\zeta)dV\right] - \left[\frac{d}{dt}\int_{\bar{V}_{\mathrm{I}}\oplus\bar{V}_{\mathrm{II}}}\bar{\psi}\, dV + \int_{\partial\bar{V}_{\mathrm{I}}\oplus\partial\bar{V}_{\mathrm{II}}} d\mathbf{S}\cdot\bar{\mathbf{j}} - \int_{\bar{V}_{\mathrm{I}}\oplus\bar{V}_{\mathrm{II}}}(\bar{\pi}+\bar{\zeta})dV\right]\right\} = 0. \qquad (15.7\text{-}11)$$

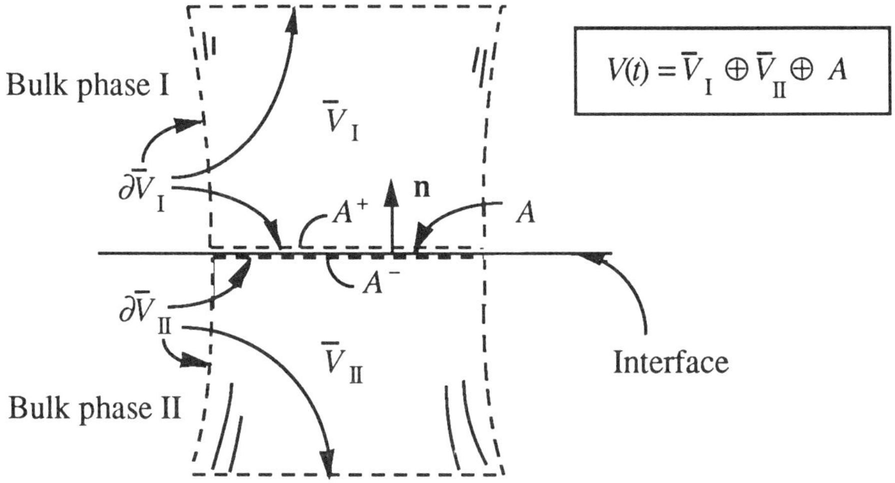

Figure 15.7-1 The asymptotic, macroscale view of the volume $V(t)$, showing the exclusion of the surface A from each of the separately continuous volumes $\bar{V}_{\mathrm{I}}$ and $\bar{V}_{\mathrm{II}}$ containing the bulk-phase fluids I and II. The surfaces A^+ and A^- (respectively representing the + and – sides of A, and represented by the dashed lines on either side of the solid line A) are bounding surfaces of the macroscale fluid volumes $\bar{V}_{\mathrm{I}}$ and $\bar{V}_{\mathrm{II}}$.

Upon utilizing Eqs. (15.6-3), (3.6-1), (3.6-2) and (3.6-3), the preceding may be rewritten as

$$\lim_{\varepsilon \to 0} \left\{ \left[\frac{d}{dt} \int_{V(t)} \psi \, dV + \int_{\partial V(t)} d\mathbf{S} \cdot \mathbf{j} - \int_{V(t)} (\pi + \zeta) dV \right] - \left[\frac{d}{dt} \int_{V(t)} \bar{\psi} \, dV + \int_{\partial V(t)} d\mathbf{S} \cdot \bar{\mathbf{j}} - \int_{V(t)} (\bar{\pi} + \bar{\zeta}) dV \right. \right.$$
$$\left. \left. - \int_{A(t)} dA \mathbf{n} \cdot \left[\bar{\mathbf{j}}_{\mathrm{I}}(\mathbf{x}_s, 0) - \bar{\mathbf{j}}_{\mathrm{II}}(\mathbf{x}_s, 0) \right] \right] \right\} = 0. \quad (15.7\text{-}12)$$

The final term in Eq. (15.7-12) arises owing to the fact that the surface A bounds ∂V_{I} and ∂V_{II} but not ∂V. Its derivation is considered in Question 15.2.

The preceding expression can be simplified by separately examining the comparable micro- and macro-terms appearing in the above integrands. Towards this goal, employ Eqs. (15.6-4) and (15.6-5) to transform the volumetric terms containing ψ, π and ζ into relations involving their surface-excess areal density counterparts, ψ^s, π^s and ζ^s, via the generic expression

$$\lim_{\varepsilon \to 0} \int_{V(t)} [\lambda(\mathbf{x}_s, n) - \bar{\lambda}(\mathbf{x}_s, \bar{n})] dV = \int_{A(t)} \lambda^s(\mathbf{x}_s) dA . \quad (15.7\text{-}13)$$

The applicability of the generic equation (15.7-12) to the function ψ depends upon the validity of the assumption

$$\lim_{\substack{\varepsilon \to 0 \\ \tilde{n} \text{ fixed}}} \frac{\delta \tilde{\psi}_{\alpha R}}{\delta t} = 0 \qquad (\alpha = \mathrm{I}, \mathrm{II}), \quad (15.7\text{-}14)$$

namely, that the time-derivative and limit operations can be interchanged within the inner region. With use of Eqs. (15.6-21) and (15.6-22), those integrals in Eq. (15.7-12) containing the diffusive flux may be written in the form

$$\lim_{\varepsilon \to 0} \left[\int_{\partial V(t)} d\mathbf{S} \cdot \mathbf{j}(\mathbf{x}_s, n) - \int_{\partial V(t)} d\mathbf{S} \cdot \bar{\mathbf{j}}(\mathbf{x}_s, \bar{n}) \right] = \int_{\partial A(t)} d\mathbf{C} \cdot \mathbf{j}^s(\mathbf{x}_s), \quad (15.7\text{-}15)$$

involving the surface-excess lineal flux density $\mathbf{j}^s$. Use of the above results, together with the definition (15.5-17) of the jump discontinuity, permits Eq. (15.7-12) to be written as

$$\frac{d}{dt} \int_{A(t)} \psi^s(\mathbf{x}_s) dA + \int_{\partial A(t)} d\mathbf{C} \cdot \mathbf{j}^s(\mathbf{x}_s) + \int_{A(t)} \mathbf{n} \cdot \|\bar{\mathbf{j}}\| \, dA$$
$$- \int_{A(t)} [\pi^s(\mathbf{x}_s) + \zeta^s(\mathbf{x}_s)] \, dA = 0. \quad (15.7\text{-}16)$$

The divergence theorem for a surface [cf. Eq. (3.3-7)] provides

$$\int_{\partial A} d\mathbf{C} \cdot \mathbf{j}^s = \int_A \nabla_s \cdot (\mathbf{I}_s \cdot \mathbf{j}^s) dA . \quad (15.7\text{-}17)$$

Upon utilizing the surface analog of (15.4-10) Reynolds transport theorem together with the preceding relation, and noting that A may be chosen arbitrarily, Eq. (15.7-16) furnishes the pointwise interfacial conservation equation

$$\frac{\delta_s \psi^s}{\delta t} - 2Hu_n \psi^s + \nabla_s \cdot (\mathbf{I}_s \cdot \mathbf{j}^s) - \pi^s - \zeta^s = -\mathbf{n} \cdot \|\bar{\mathbf{j}}\|, \quad (15.7\text{-}18)$$

valid at every point $\mathbf{x}_s$ of the macroscale interface $\bar{n} = 0$. This equation, which relates the surface-excess density fields on the left-hand side to the 'jump' $\mathbf{n} \cdot \|\bar{\mathbf{j}}\|$ [cf. Eq. (15.5-17)] in the generally discontinuous macroscale bulk-fluid areal flux density across the interface, constitutes the interfacial boundary condition imposed upon $\mathbf{n} \cdot \|\bar{\mathbf{j}}\|$ across the (nonmaterial) interface, $\bar{n} = 0$. All fields appearing in Eq. (15.7-18) are *macroscale* fields; thus, this interfacial equation is to be understood as describing strictly macroscale phenomena, in the same sense (and at the same macroscale level of description) as Eq. (15.7-8).

Equation (15.7-18) may be expressed in an alternate form in light of the expression (15.7-2) for $\mathbf{j}$. Application of the product relations (15.5-18) and (15.5-19), together with the scaling relation provided in the footnote pertaining to Eq. (15.7-5) [with Eq. (15.5-27)], permits Eq. (15.7-2) to be rewritten in either of the two alternative forms,

$$\tilde{\mathbf{j}} = (\tilde{\mathbf{v}} - \mathbf{u}_n)\tilde{\psi} + \tilde{\phi},$$

and

$$\bar{\mathbf{j}} = (\bar{\mathbf{v}} - \mathbf{u}_n)\bar{\psi} + \bar{\phi}.$$

Substitute these results into Eq. (15.6-36) to obtain

$$\mathbf{I}_s \cdot \mathbf{j}^s = (\mathbf{I}_s \cdot \mathbf{v}\,\psi)^s + \mathbf{I}_s \cdot \phi^s. \quad (15.7\text{-}19)$$

Use of this relation in Eq. (15.7-18) yields the canonical form,

$$\boxed{\begin{aligned}&\frac{\delta_s \psi^s}{\delta t} - 2Hu_n \psi^s + \nabla_s \cdot (\mathbf{I}_s \cdot \mathbf{v}\,\psi)^s \\ &\quad + \nabla_s \cdot (\mathbf{I}_s \cdot \phi^s) - \pi^s - \zeta^s = -\mathbf{n} \cdot \|(\bar{\mathbf{v}} - \mathbf{u})\bar{\psi} + \bar{\phi}\|,\end{aligned}} \quad (15.7\text{-}20)$$

of the generic conservation equation governing interfacial transport phenomena for nonmaterial interfaces.

The explicit representation of the latter equation in terms of a *mass-average* or apparently 'material' surface velocity, $\mathbf{V}$, say [corresponding to $\mathbf{v}^o$ in Part I; cf. Eq. (3.4-6)] will be addressed in the following chapter (cf. Appendix 16.A). Presently, it is apparent that Eq. (15.7-20) is quite similar to Eq. (3.6-16), particularly for the case of a material interface, for which

$$\bar{\mathbf{v}}(0+) = \bar{\mathbf{v}}(0-) \equiv \mathbf{v}^o = \mathbf{u},$$

whereupon Eq. (15.7-20) may be demonstrated to reduce to Eq. (3.6-16) [cf. Eqs. (16.1-13) and (16.1-14)]. In general, however, for nonmaterial interfaces the surface velocity $\mathbf{u}_n$ must be determined from the physical nature of the particular interfacial flow. This was illustrated in example 3 of §3.7, and is considered again in Question 15.3.

Identification of Generic Quantities

Table 3.5-1 provides an identification of the generic mathematical field quantities appearing in Eq. (15.7-1) for several important *physical* cases of transport phenomena. The analogous generic identifications of the surface-excess properties appearing in the balance law (15.7-20) must be derived from the appropriate microscale definitions (15.6-18) and (15.6-36) of the surface-excess variables, having specified the nature of the microscale, interfacial fluid. [The results, to be derived in the following two chapters, are summarized (for material interfaces) in Table 3.6-1.]

Having identified the physical microscale density fields, constitutive equations for ϕ, π and ζ are needed. Together, these ultimately furnish corresponding macroscale conservation and constitutive equations governing the bulk and interfacial transport processes. The macroscale conservation equations for the two outer bulk-fluid regions are determined by Eq. (15.7-7), with macroscale densities defined in Eqs. (15.5-12) and (15.5-13). Due to the separability condition (15.5-18), the bulk-fluid equations reduce to the familiar conservation and constitutive equations for three-dimensional fluids, albeit characterized by different phenomenological coefficients on the two sides of the interface. These latter bulk equations have previously been considered in §§4.1 and 5.1 for the cases of mass, linear and angular momentum, and species transport.

15.8 Constitutive Equations

To finally adapt our asymptotic view to the formulation of *macroscale* interfacial constitutive equations, comparable constitutive forms governing the appropriate *microscale* continuum fields within the respective interfacial transition region must be hypothesized. An important distinction between the interfacial and bulk fluid regions, which needs to be implicitly incorporated into any rational interfacial model, is the inherent inhomogeneity within the transition region created by the existence of short-range intermolecular and/or physicochemical forces. Such inhomogeneities are generally accounted for (Eliassen 1963, Kirkwood & Buff 1949, Goodrich 1981a, Goodrich 1981b) by assuming that, within the interfacial transition region, all microscale continuum fields are locally transversely isotropic with respect to the direction normal to the (macroscale) interface, and that these fields possess linear constitutive forms analogous to those of their relatively homogeneous, bulk-phase counterparts. These microscale fields are assumed to become locally isotropic at distances far from the interface.

This microscale model will be adopted in the following chapters, principally to demonstrate explicitly the procedure for determining (*macroscale*) interfacial constitutive equations from the corresponding volumetric *microscale* constitutive forms via the asymptotic scheme developed in the present and preceding chapters. Examples of this are provided in §§16.3 and 17.2, respectively pertaining to the cases of linear momentum and species transport. (Of course, the asymptotic approach formulated here is not restricted to any *specific* microscale constitutive model.)

15.9 Summary

A generic surface-excess theory of interfacial transport processes for nonmaterial interfaces has been developed in the present chapter. The theory begins with a microscale description of an interface as a three-dimensional, continuous (though highly inhomogeneous) transition region between two otherwise relatively homogeneous 'bulk-fluid' phases. Employing the method of matched asymptotic expansions (§15.5), relations between this continuous, microscale *physical* view and the discontinuous, macroscale *mathematical* view of interfaces have been developed, particularly in the limit wherein the interfacial length scale l is much smaller than the experimental length scale L characterizing the apparatus dimensions or probe size. Thus, definitions of generic surface-excess fields [cf. Eqs. (15.6-18) and (15.6-36)], assigned to a (parent) reference surface that moves and deforms with the macroscale interface, have been derived in terms of the corresponding three-dimensional, microscale generic fields. These have been related through a generic surface-excess transport equation (15.7-20). This equation provides a generalization of the generic interfacial transport law developed in Part I [cf. Eq. (3.6-16)] to *nonmaterial* interfaces.

In the following two chapters this generic surface-excess balance law will be applied to the important special cases of linear momentum and species transport respectively.

Appendix 15.A

Projections onto the Parent Surface, $q^3 \equiv q_0^3$

The projection of a differential areal element (refer to the caption of Figure 3.4-3), namely

$$dA^* = \sqrt{a^*}\ \mathrm{e}_{\alpha\beta}\ dq^\alpha dq^\beta, \qquad (15.A-1)$$

lying on a familial q^3-coordinate surface onto the parental coordinate surface $q^3 \equiv q_0^3$ along an envelope determined by the q^3-coordinate curves (see Figure 15.3-3) is

$$dA^* = M^*\ dA, \qquad (15.A-2)$$

in which

$$dA = \sqrt{a}\ \mathrm{e}_{\alpha\beta}\ dq^\alpha dq^\beta, \qquad (15.A-3)$$

and

$$M^* = \sqrt{a^*/a}\ . \qquad (15.A-4)$$

Here, the magnification factor M^* may be seen to correspond to the surface Jacobian (3.4-13) (see also Question 3.4). Equation (15.A-4) permits the following explicit calculation of M^* [cf. Eq. (15.A-22)]:

According to Eq. (3.2-5),

$$a^* = \left|\mathbf{a}_1^* \times \mathbf{a}_2^*\right|^2 = a_{11}^*\ a_{22}^* - a_{12}^*\ a_{21}^*. \qquad (15.A-5)$$

Differentiation of $\sqrt{a^*}$ with respect to an arbitrary variable ξ gives

$$\frac{\partial\sqrt{a^*}}{\partial\xi} = \frac{1}{2\sqrt{a^*}}\left(\frac{\partial a^*}{\partial\xi}\right). \qquad (15.A-6)$$

From Eq. (15.A-5),

$$\begin{aligned}\frac{\partial a^*}{\partial\xi} &= a_{22}^*\left(\mathbf{a}_1^*\cdot\frac{\partial\mathbf{a}_1^*}{\partial\xi}\right) + a_{11}^*\left(\mathbf{a}_2^*\cdot\frac{\partial\mathbf{a}_2^*}{\partial\xi}\right)\\ &\quad - a_{12}^*\left(\mathbf{a}_1^*\cdot\frac{\partial\mathbf{a}_2^*}{\partial\xi} + \mathbf{a}_2^*\cdot\frac{\partial\mathbf{a}_1^*}{\partial\xi}\right)\\ &= (a_{22}^*\,\mathbf{a}_1^* - a_{12}^*\,\mathbf{a}_2^*)\cdot\frac{\partial\mathbf{a}_1^*}{\partial\xi} + (a_{11}^*\,\mathbf{a}_2^* - a_{21}^*\,\mathbf{a}_1^*)\cdot\frac{\partial\mathbf{a}_2^*}{\partial\xi}.\end{aligned} \qquad (15.A-7)$$

As follows from Eq. (3.2-7), the above relation may be expressed as

$$\frac{1}{2}\frac{\partial a^*}{\partial\xi} = a^*\,\mathbf{a}^{\alpha*}\cdot\frac{\partial\mathbf{a}_\alpha^*}{\partial\xi}, \qquad (15.A-8)$$

whence Eq. (15.A-6) becomes

$$\frac{\partial\sqrt{a^*}}{\partial\xi} = \sqrt{a^*}\,\mathbf{a}^{\alpha *}\cdot\frac{\partial\mathbf{a}_\alpha{}^*}{\partial\xi}. \qquad (15.A-9)$$

In the following, ξ will be taken to be q^3.

Use of the definitions (3.2-2) and (3.2-5) (in addition to acknowledging that $\mathbf{g}_1\cdot\mathbf{g}_3=0$ and $\mathbf{g}_2\cdot\mathbf{g}_3=0$ for semi-orthogonal coordinates, as discussed in the introductory remarks of Appendix B), produces the expression

$$\frac{\partial\mathbf{a}_\alpha{}^*}{\partial q^3} = \frac{\partial}{\partial q^3}\left(\frac{\partial\mathbf{x}}{\partial q^\alpha}\right) = \frac{\partial\mathbf{g}_3{}^*}{\partial q^\alpha}. \qquad (15.A-10)$$

The identity

$$\mathbf{g}_3{}^* = \frac{1}{h_3{}^*}\mathbf{n}^*,$$

yields the expression

$$\frac{\partial\mathbf{g}_3{}^*}{\partial q^\alpha} = \mathbf{n}^*\frac{\partial}{\partial q^\alpha}\left(\frac{1}{h_3{}^*}\right) + \frac{1}{h_3{}^*}\frac{\partial\mathbf{n}^*}{\partial q^\alpha}. \qquad (15.A-11)$$

Upon employing Eq. (3.2-11) together with the known symmetry of the curvature dyadic, we obtain

$$\frac{\partial\mathbf{n}^*}{\partial q^\alpha} = \mathbf{a}_\alpha{}^*\cdot(\nabla_s\mathbf{n}^*) = (\nabla_s\mathbf{n}^*)\cdot\mathbf{a}_\alpha{}^* \qquad (15.A-12)$$

Differentiation with respect to q^i of the normalization relation $\mathbf{n}^*\cdot\mathbf{n}^*=1$ gives

$$\mathbf{n}^*\cdot\frac{\partial\mathbf{n}^*}{\partial q^i} = 0 \qquad (i = 1, 2, 3). \qquad (15.A-13)$$

Substitute Eq. (15.A-11) into (15.A-10), dot the result with $\mathbf{n}^*$, and employ Eq. (15.A-13) (with $i=\alpha$) to simplify the result. This yields

$$\frac{\partial}{\partial q^\alpha}\left(\frac{1}{h_3{}^*}\right) = \mathbf{n}^*\cdot\frac{\partial\mathbf{a}_\alpha{}^*}{\partial q^\alpha}. \qquad (15.A-14)$$

The orthogonalization condition $\mathbf{a}_\alpha{}^*\cdot\mathbf{n}^*=0$ may be differentiated with respect to q^3 to give

$$\mathbf{n}^*\cdot\frac{\partial\mathbf{a}_\alpha{}^*}{\partial q^\alpha} = -\mathbf{a}_\alpha{}^*\cdot\frac{\partial\mathbf{n}^*}{\partial q^\alpha}. \qquad (15.A-15)$$

From the above relations, we obtain

$$\frac{\partial}{\partial q^\alpha}\left(\frac{1}{h_3{}^*}\right) = -\mathbf{a}_\alpha{}^*\cdot\frac{\partial\mathbf{n}^*}{\partial q^\alpha}, \qquad (15.A-16)$$

whence

$$\mathbf{n}^*\frac{\partial}{\partial q^\alpha}(\ln h_3{}^{*-1}) = -(\nabla_n{}^*\mathbf{n}^*)\cdot\mathbf{a}_\alpha{}^*, \qquad (15.A-17)$$

wherein

$$\nabla_n^* = \mathbf{n}^* \mathbf{n}^* \cdot \nabla . \qquad (15.A-18)$$

Substitute Eqs. (15.A-12) and (15.A-17) into Eq. (15.A-11). Use of the resulting expression in Eq. (15.A-10) eventually yields

$$\frac{\partial \mathbf{a}_\alpha^*}{\partial q^3} = \frac{1}{h_\alpha^*}(\nabla_s^* \mathbf{n}^* - \nabla_n^* \mathbf{n}^*) \cdot \mathbf{a}_\alpha^* . \qquad (15.A-19)$$

Dot multiply the above equation with $\mathbf{a}_\alpha^*$, and recall that $\mathbf{a}_\alpha^* \cdot \mathbf{n}^* = 0$, to obtain

$$\mathbf{a}_\alpha^* \cdot \frac{\partial \mathbf{a}_\alpha^*}{\partial q^3} = \frac{1}{h_\alpha^*}\nabla_s^* \mathbf{n}^* : \mathbf{a}_\alpha^* \mathbf{a}^{\alpha *} . \qquad (15.A-20)$$

Employ the definition for the mean curvature given by Eq. (3.2-13), and subsequently substitute the resulting expression into Eq. (15.A-9) to obtain the differential equation

$$\frac{\partial \sqrt{a^*}}{\partial q^3} = -2H^* \sqrt{a^*} \frac{1}{h_3^*} , \qquad (15.A-21)$$

which can be integrated from the parental coordinate surface ($q^3 = q_0^3$) to an arbitrary familial q^3-coordinate surface to eventually produce the relation [cf. Eq. (15.3-11)]

$$\boxed{M^* = \sqrt{a^* / a} = \exp\left(-\int_{q_0^3}^{q^3} 2H^* \frac{1}{h_3^*} dq^3 \right).} \qquad (15.A-22)$$

Similar to the above, consider the projection of a differential lineal element

$$dC^* = \frac{1}{h_c^*} ds , \qquad (15.A-23)$$

lying on a familial q^3-coordinate surface (with ds a differential element of arc length along the curve C^*, and h_c^* the local scale factor of this curve) onto the parent surface along lines determined by the q^3-coordinate curves, such that

$$dC^* = N^* \, dC , \qquad (15.A-24)$$

where

$$dC = \frac{1}{h_c} ds \qquad (15.A-25)$$

and

$$N^* = h_c / h_c^* . \qquad (15.A-26)$$

An explicit expression for N^* may be determined by considering the derivative

$$\frac{\partial}{\partial q^3}\left(\frac{1}{h_c{}^*}\right) = \frac{h_c{}^*}{2}\left[\frac{\partial}{\partial q^3}\left(\frac{1}{h_c{}^*}\right)^2\right]. \qquad (15.A-27)$$

The tangent vector to the curve C^* may be defined as

$$\mathbf{t}_c{}^* \stackrel{\text{def}}{=} \frac{\partial \mathbf{x}}{\partial s} \equiv \frac{1}{h_c{}^*}\mathbf{t}^*, \qquad (15.A-28)$$

with $\mathbf{t}^*$ a *unit* tangent vector along the curve C^*. Thus,

$$\frac{1}{2}\frac{\partial}{\partial q^3}\left(\frac{1}{h_c{}^*}\right)^2 = \mathbf{t}_c{}^* \cdot \frac{\partial \mathbf{t}_c{}^*}{\partial q^3}. \qquad (15.A-29)$$

Use of Eqs. (3.2-2), (3.2-5) and (15.A-28) yields

$$\frac{\partial \mathbf{t}_c{}^*}{\partial q^3} = \frac{\partial}{\partial q^3}\left(\frac{\partial \mathbf{x}}{\partial s}\right) = \frac{\partial}{\partial s}\left(\frac{\partial \mathbf{x}}{\partial q^3}\right) = \frac{\partial}{\partial s}\left(\frac{\mathbf{n}^*}{h_3{}^*}\right). \qquad (15.A-30)$$

Upon employing the expression

$$\frac{\partial}{\partial s}\left(\frac{\mathbf{n}^*}{h_3{}^*}\right) = \mathbf{t}_c{}^* \cdot \nabla\left(\frac{\mathbf{n}^*}{h_3{}^*}\right) = \frac{1}{h_3{}^*}\mathbf{t}_c{}^* \cdot \nabla(\mathbf{n}^*) + \mathbf{t}_c{}^* \cdot \nabla\left(\frac{1}{h_3{}^*}\right)\mathbf{n}^*, \qquad (15.A-31)$$

together with Eqs. (15.A-28), (15.A-29) and (15.A-30), and subsequently noting that $\mathbf{t}^* \cdot \mathbf{n}^* = 0$, the derivative (15.A-27) becomes

$$\frac{\partial}{\partial s}\left(\frac{1}{h_c{}^*}\right) = \mathbf{t}_c{}^*\mathbf{t}_c{}^*\mathbin{:}\nabla\mathbf{n}^*\frac{1}{h_c{}^*}\frac{1}{h_3{}^*} = \kappa_n{}^*\frac{1}{h_c{}^*}\frac{1}{h_3{}^*}, \qquad (15.A-32)$$

wherein

$$\kappa_n{}^*(\mathbf{t}^*) = -\mathbf{t}^*\mathbf{t}^*\mathbin{:}\nabla_s\mathbf{n}^* \equiv \mathbf{t}^*\mathbf{t}^*\mathbin{:}\mathbf{b} \qquad (15.A-33)$$

is the curvature of the q^3-coordinate curve. Integration of Eq. (15.A-32) from the parental surface q_0^3 to a familial q^3-coordinate surface yields [cf. Eq. (15.3-13)]

$$\boxed{N^* = h_c / h_c{}^* = \exp\left(-\int_{q_0^3}^{q^3}\kappa_n(\mathbf{t}^*)\frac{1}{h_3{}^*}dq^3\right).} \qquad (15.A-34)$$

The unit surface vector $\mathbf{m}^*$ that forms the outward normal to the curve C^* will vary with q^3 according to the equation

$$\frac{\partial \mathbf{m}^*}{\partial q^3} = \frac{\partial \mathbf{t}^*}{\partial q^3} \times \mathbf{n}^* + \mathbf{t}^* \times \frac{\partial \mathbf{n}^*}{\partial q^3}. \qquad (15.A-35)$$

Upon employing Eqs. (15.A-28) and (15.A-30)–(15.A-32), one obtains

$$\frac{\partial \mathbf{t}^*}{\partial q^3} = -\mathbf{t}_c^*(h_c^*)^2\frac{\partial}{\partial q^3}\left(\frac{1}{h_c^*}\right) + h_c^*\frac{\partial \mathbf{t}_c^*}{\partial q^3}$$
$$= -\mathbf{t}^*\,\mathbf{t}^*:\nabla\mathbf{n}^*\frac{1}{h_3^*}\mathbf{t}^* + \mathbf{t}^*\cdot(\nabla\mathbf{n}^*)\frac{1}{h_3^*} + \mathbf{t}^*\cdot\left(\nabla\frac{1}{h_3^*}\right)\mathbf{n}^*$$

(15.A - 36)

and

$$\frac{\partial \mathbf{n}^*}{\partial q^3} = \mathbf{n}^*\cdot\nabla\mathbf{n}^*. \qquad (15.A - 37)$$

Use of the relation $\mathbf{m}^* = \mathbf{t}^*\times\mathbf{n}^*$, in conjunction with the vector identity

$$\mathbf{a}\times(\mathbf{b}\times\mathbf{c}) = \mathbf{b}(\mathbf{a}\cdot\mathbf{c}) - \mathbf{c}(\mathbf{a}\cdot\mathbf{b}), \qquad (15.A - 38)$$

(in which **a**, **b** and **c** are arbitrary vectors), yields the relation

$$(\mathbf{t}^*\cdot\nabla\mathbf{n}^*)\times\mathbf{n}^* = \mathbf{m}^*(\mathbf{t}^*\,\mathbf{t}^*:\nabla\mathbf{n}^*) - \mathbf{t}^*(\mathbf{m}^*\,\mathbf{t}^*:\nabla\mathbf{n}^*) \qquad (15.A - 39)$$

and

$$\mathbf{t}^*\times(\mathbf{n}^*\cdot\nabla\mathbf{n}^*) = -\mathbf{n}^*(\mathbf{m}^*\,\mathbf{n}^*:\nabla\mathbf{n}^*). \qquad (15.A - 40)$$

In the above equation, it was noted that

$$\mathbf{n}^*\,\mathbf{n}^*:\nabla\mathbf{n}^* = 0. \qquad (15.A - 41)$$

Substitute Eqs. (15.A-36), (15.A-37) and (15.A-39)–(15.A-41) into (15.A-35) to obtain

$$\frac{\partial \mathbf{m}^*}{\partial q^3} = -(\mathbf{t}^*\,\mathbf{t}^* + \mathbf{n}^*\,\mathbf{n}^*)\cdot(\nabla\mathbf{n}^*)\cdot\mathbf{m}^*\frac{1}{h_3^*}. \qquad (15.A - 42)$$

Integration of the above expression yields

$$\boxed{\mathbf{m}^* = \mathbf{m} - \int_{q_0^3}^{q^3}(\mathbf{t}^*\,\mathbf{t}^* + \mathbf{n}^*\,\mathbf{n}^*)\cdot(\nabla\mathbf{n}^*)\cdot\mathbf{m}^*\frac{1}{h_3^*}dq^3.} \qquad (15.A - 43)$$

Questions for Chapter 15

15.1 What physical interpretation would you associate with the familial q^3-coordinate surfaces A^* in the geometrical parameterization of §15.3? Is this association necessary (i.e. must the coordinate surfaces A^* possess a physical interpretation)?

15.2 Making reference to Fig. 15.7-1, demonstrate that

$$\int_{\partial V_{\mathrm{I}}(t)\oplus\partial V_{\mathrm{II}}(t)} d\mathbf{S}\cdot\mathbf{j} = \int_{\partial V(t)} d\mathbf{S}\cdot\mathbf{j} + \int_{A} dA\mathbf{n}\cdot\|\mathbf{j}\|,$$

as in Eq. (15.7-12).

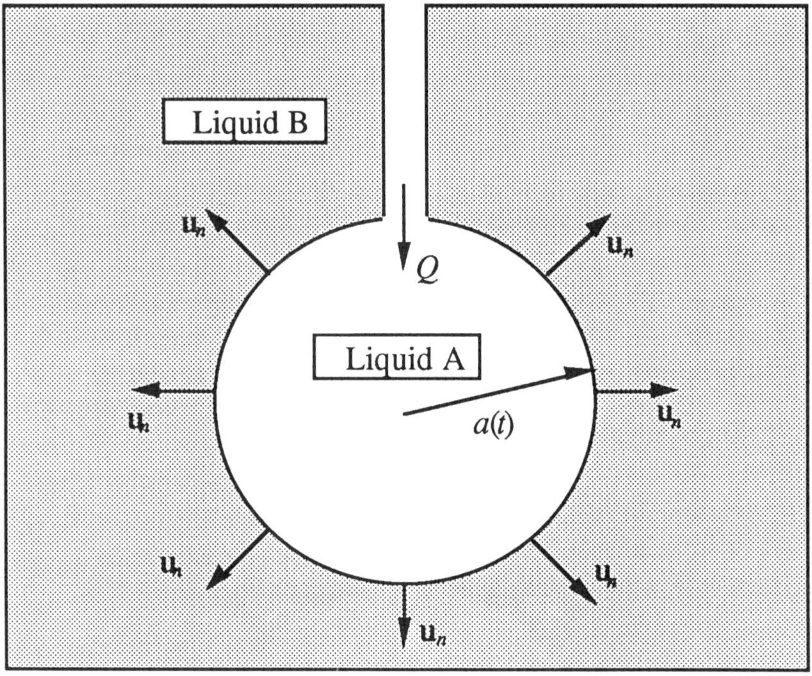

15.3 A liquid A flows out of a narrow capillary tube and into liquid B at a volumetric rate Q, thereby creating a spherical droplet of instantaneous radius a, as shown in the figure above. The mass densities (ρ) of liquids A and B are identical, and the droplet sufficiently small to enable it to be regarded as spherical throughout its expansion. In the process of expanding, however, the droplet liquid A is observed at a given moment in time to dissolve within the continuous liquid B at an instantaneous volumetric rate f. Determine the mass-average velocity $\mathbf{v}$ in both liquid phases as well as the (nonmaterial) surface velocity $\mathbf{u}_n$ of the spherical droplet.

15.4 Suppose that an *insoluble* surfactant is adsorbed on the macroscale droplet interface in the preceding question. For the pure radial expansion discussed, the generic surface-excess balance equation (15.7-20) can be shown to furnish the expression:

$$\frac{\partial \rho_1^s}{\partial t} + \frac{2}{a} u_r \rho_1^s = 0,$$

where ρ_1^s represents the surface-excess mass density of surfactant (i.e. species '1').

i. Determine the instantaneous value of ρ_1^s for the conditions described in Question 15.3, given that no surfactant is dissolved in either bulk phase, and that the total mass of surfactant in the interfacial region is *m*. Assume also that the net flow of mass across the interface fA = const. and Q = const.

ii. How would this result change if the macroscale interface were material?

15.5 Upon neglecting the surface-excess mass density, employ Eq. (15.7-20) to show that the equality

$$\mathbf{n}\cdot\bar{\mathbf{v}}(\mathbf{x}_s, 0+) = u_n(\mathbf{x}_s) + \frac{\bar{\rho}(\mathbf{x}_s, 0-)}{\bar{\rho}(\mathbf{x}_s, 0+)}[\mathbf{n}\cdot\bar{\mathbf{v}}(\mathbf{x}_s, 0-) - u_n(\mathbf{x}_s)]$$

holds for a nonmaterial interface. How would you explain the existence of a velocity discontinuity across a macroscale interface in *microscale* terms. [Hint: Consider mass continuity within the inhomogeneous interfacial region.]

15.6 An experiment is to be performed analogous to the the experiment outlined in §15.2. In this case the water phase (II) is dispersed within the oil phase (I), thereby existing in an equilibrium state as a spherical droplet of radius a, as depicted below. The respective volumes of water ($\bar{V}_{II}$) and oil ($\bar{V}_I$) are contained within a spherical vessel of radius R. As in §15.2, total masses M_I^A and M_{II}^A of surfactant A are dissolved within the phases I and II prior to contact of the two phases.

i. In terms of the spherical coordinates depicted in the figure below, provide an expression for the equilibrium, micro-(continuous) distribution of surfactant $\rho^A(r)$ that will be measured by a microscale observer.

ii. Show by steps analogous to Eqs. (15.2-7)–(15.2-11) that the following expression obtains for the total surface-excess amount of surfactant mass 'at' the interface:

$$(M^A)^s = \int_A dA \int_{r=a}^{R} \left(\frac{r}{a}\right)^2 (\rho^A - \bar{\rho}_I^A)dr + \int_A dA \int_{z=0}^{a} \left(\frac{r}{a}\right)^2 (\rho^A - \bar{\rho}_{II}^A)dr\ .$$

[In this expression, the surface area $A = 4\pi a^2$ corresponds to the area of the $r \equiv a$ coordinate surface, which coordinate surface has been visually located (by a macroscale observer) 'at' the interface.]

iii. Demonstrate that the term $(r/a)^2$ in the preceding expression is equivalent in the present case of spherical coordinates to the magnification factor M^* defined in Eq. (15.3-11).

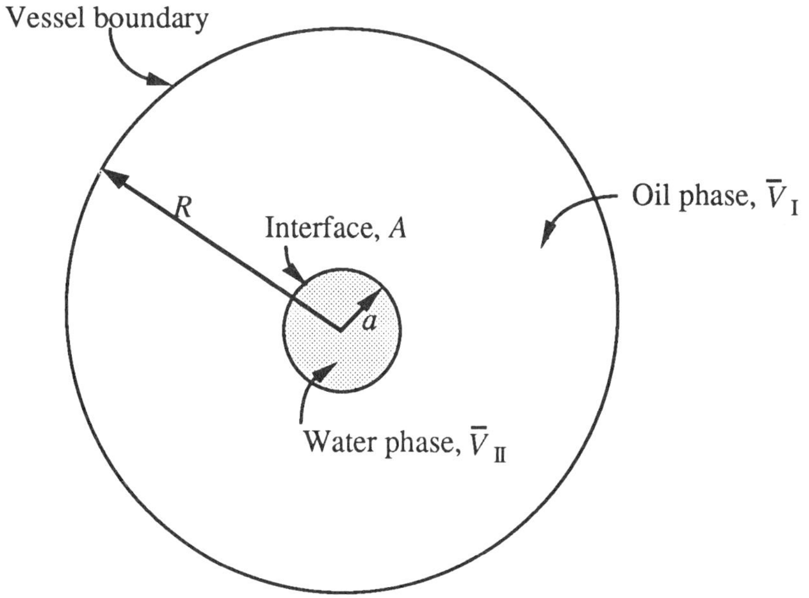

iv. Follow steps analogous to those in Eqs. (15.2-14)–(15.2-17) to obtain an expression for the surface-excess mass density of surfactant analogous to Eq. (15.2-17). (For this purpose, define a normal coordinate as $n = r - a$.)

v. What is the scaling restriction placed upon the radius a of the spherical droplet by the preceding definition of the surface-excess mass density?

15.7 Suppose the *exact* mass density profile across a fluid interface separating two (macroscopically immiscible) fluid phases I and II satisfies the following microscale, continuous relation:

$$\rho = \frac{1}{2}(\bar{\rho}_{\mathrm{I}} + \bar{\rho}_{\mathrm{II}}) + \frac{1}{2}(\bar{\rho}_{\mathrm{I}} - \bar{\rho}_{\mathrm{II}})\tanh\left(\frac{n}{l}\right).$$

A macroscale experimentalist wishes to confirm his visual perception that the interface is proximate to the coordinate surface $n = 0$ by measuring the 'bulk-phase' mass densities $\bar{\rho}_{\mathrm{I}}$ and $\bar{\rho}_{\mathrm{II}}$ on either side of the surface $n = 0$. For this purpose, he employs a probe to extract a cubic volume $V_p = L^3$ ($L >> l$) of fluid. By measuring the total mass M_p contained within the measurement probe volume V_p, the experimentalist is able to deduce the mass density of the fluid at the 'point' where the probe was inserted.

i. Prove that if the probe is inserted into the fluid-fluid system on the phase I side of the interface, so as to occupy the distance $n = 0$ to $n = +L$, the measured mass density of the fluid at this location will be observed to be the 'bulk-phase' value $\bar{\rho}_{\mathrm{I}}$ (utilizing the fact that $L >> l$, assuming no variation of the *true*, continuous mass density in directions tangential to the interface and choosing the cross-sectional area of the probe tangent to the interface to be constant for all positions $0 \le n \le L$).

ii. Similarly, show that if the probe is inserted in the fluid-fluid system on the phase II side of the interface, so as to occupy the distance $n = 0$ to $n = -L$, the measured mass density of the fluid at this location will be observed to be the 'bulk-phase' value $\bar{\rho}_{\text{II}}$ (again, utilizing the fact that $L >> l$, assuming no variation of the *true*, continuous mass density in directions tangential to the interface and choosing the cross-sectional area of the probe tangent to the interface to be constant for all positions $-L \le n \le 0$).

iii. According to the preceding macroscale density measurements, where is the interface located? If the macroscale experimentalist were to insert his probe *near* to the locations cited in *i*. and *ii*. above, only now displaced a distance $\pm l$ in a normal direction from the coordinate surface $n=0$, would the results of *i*. and *ii*. be altered? What conclusion can you reach regarding the ability of a macroscale experimentalist to precisely locate the macroscale, two-dimensional interface?

Additional Reading for Chapter 15[*]

§15.1 The Microscale View of a Fluid Interface

Eliassen, J.D. 1963 *Interfacial Mechanics*. Ph.D. dissertation. University of Minnesota, Minneapolis, Minnesota.

Ono, S. & Kondo, S. 1960 Molecular theory of surface tension in liquids. *Encyclopedia of Physics* (ed. S. Flugge), Vol. 10, pp. 134–280.

§15.3 Coordinate Parameterization of the Microscale Interfacial Region

Mavrovouniotis G. & Brenner, H. 1991 A micromechanical investigation of interfacial transport processes: I. Interfacial conservation equations. (Submitted for publication.)

§15.4 Kinematics

Yarin, A.L. 1979 Stability of a jet of viscoelastic liquid in the presence of a mass flux at its surface. *J. Eng. Phys.*. **37**, 904–910.

§15.5 Singular Perturbation Analysis

Nayfeh, A.H. 1973 *Perturbation Methods*. New York: Wiley.

[*] Chapters 15–17 represent an amplification of material previously published in the PhD thesis of G. Mavrovouniotis (1989). [See the *Preface* for further clarifying comments pertaining to the evolution of the micromechanical theory outlined herein.]

De Bruijn 1958 *Asymptotic Methods in Analysis*. Amsterdam: North-Holland Publishing Co.

Van Dyke, M. 1964 *Perturbation Methods in Fluid Mechanics*. London: Academic Press.

Kaplun, S. 1967 *Fluid Mechanics and Singular Perturbations* (eds. P.A. Lagerstrom, L.N. Howard, C.S. Liu). New York: Academic Press.

Cole, J.D. 1968 *Perturbation Methods in Applied Mathematics*. Waltham, Massachusetts: Blaisdell Publishing Co.

§15.6 Surface-Excess Properties

Gibbs, J.W. 1957 *The Collected Works of J. Willard Gibbs*, Vol. I. New Haven, Connecticut: Yale University Press, pp. 219–331.

§15.7 Generic Conservation Equations

Goodrich, F.C. 1981 The theory of capillary excess viscosities. *Proc. Roy. Soc. Lond. A* **374**, 341–370.

Deemer, A.R. & Slattery, J.C. 1978 Balance equations and structural models for phase interfaces. *Int. J. Multiphase Flow* **4**, 171–192.

16

"It has already been asserted, by Mr. Monge and others, that the phenomena of capillary tubes are referable to the cohesive attraction of the superficial particles only of the fluids employed, and that the surface must consequently be formed into curves of the nature of linteariæ, which are supposed to be the results of a uniform tension of a surface, resisting the pressure of a fluid, either uniform, or varying according to a given law."

Thomas Young (1805)

CHAPTER 16

Surface-Excess Transport of Momentum

The generic surface-excess balance equation (15.7-20) is employed in the current chapter to derive the surface-excess linear [cf. Eq. (16.1-16)] and angular [cf. Eq. (16.2-8)] momentum equations for nonmaterial fluid interfaces. These equations may be seen as generalizations of the macroscale interfacial transport relations (4.2-3) and (4.2-7), previously derived in Chapter 4. The surface-excess theory outlined in Chapter 15 is also employed for the purpose of *deriving* a (macroscale) constitutive relation [cf. Eq. (16.3-24)] for the surface-excess pressure tensor, beginning with an explicit micromechanical, constitutive model of the interfacial transition zone—namely, one in which the microscale fluid is regarded as being incompressible and its microscale pressure tensor [cf. Eqs. (16.3-1) and (16.3-6)] regarded as being both transversely isotropic and linearly related to the rate of interfacial fluid deformation. This procedure stands in marked contrast to the purely phenomenological, entirely macroscale approach of Chapter 4. There, the two-dimensional surface-excess stress tensor was assumed *a priori* to display a known constitutive behavior [e.g. Eq. (4.2-17) for Newtonian interfaces], similar to the constitutive approach for the stress tensor of three-dimensional continuum fluids [e.g. Eq. (4.1-15)], as if there existed some compelling *a priori physical* reason for an interface to display Newtonian behavior. The asymptotic microscale surface-excess theory of Part II is thus demonstrated to provide a rigorous rational scheme for deriving expressions describing the constitutive behavior of surface-excess fields, based upon plausible micromcechanical models.

The Microscale→Macroscale linear constitutive example briefly outlined above and brought to fruition in §16.3 is to be regarded as purely illustrative of the general scheme. Other microscale constitutive models will, of course, lead to other types of macroscale constitutive behavior.

The example of a spherical gas bubble expanding radially within a quiescent fluid, previously considered in Chapter 4 from a purely phenomenological macroscale interfacial point of view of the interface (cf.

example 2 of §4.4), is reconsidered herein from the microscale perspective. The micro- and macro-approaches to the problem yield identical macroscale results in the asymptotic limit as the interfacial length scale l becomes negligibly small relative to the macroscale L.

Surface-excess energy and entropy balances are derived in Questions 16.1 and 16.2, together yielding a surface-excess entropy inequality that is found (Question 16.3) to place a constitutive restriction upon the surface-excess stress tensor, in accord with more general continuum-mechanical principles (Moeckel 1975).

We begin in §§16.1 and 16.2 with a derivation of the explicit forms of the surface-excess mass, linear and angular momentum equations. This is followed in §16.3 by a derivation of the Newtonian surface constitutive model for the surface-excess stress tensor. Section 16.4 addresses the example of a surfactant-adsorbed expanding gas bubble from the microscale interfacial perspective.

16.1 Conservation of Mass and Linear Momentum

Surface-Excess Mass Conservation

Upon referring to the explicit identifications made in Table 3.5-1 for the continuous three-dimensional fields (in the present microscale context), together with the surface-excess definitions (15.6-18), (15.6-36) and (15.7-19), the generic interfacial conservation equation (15.7-20) becomes for the mass density $(\psi \equiv \rho)$]*

$$\frac{\partial \rho^s}{\partial t} - 2Hu_n\rho^s + \nabla_s \cdot (\rho \mathbf{v})^s = -\mathbf{n} \cdot \|(\bar{\mathbf{v}} - \mathbf{u})\bar{\rho}\|, \qquad (16.1\text{-}1)$$

in which

$$\boxed{\rho^s(\mathbf{x}_s) \stackrel{\text{def}}{=} \varepsilon L \int_{-\infty}^{\infty} [\tilde{\rho}(\mathbf{x}_s, \tilde{n}) - \bar{\rho}(\mathbf{x}_s, 0)]\, d\tilde{n}} \qquad (16.1\text{-}2)$$

is the surface-excess mass density and

$$(\rho \mathbf{v})^s (\mathbf{x}_s) \stackrel{\text{def}}{=} \varepsilon L \int_{-\infty}^{\infty} [\tilde{\rho}\tilde{\mathbf{v}}(\mathbf{x}_s, \tilde{n}) - \bar{\rho}\bar{\mathbf{v}}(\mathbf{x}_s, 0)]\, d\tilde{n} \qquad (16.1\text{-}3)$$

---

* The standard partial time derivative symbol $\partial/\partial t$ appears in Eq. (16.1-1) rather than the previous symbol $\delta_s/\delta t$ appearing in the generic equation (15.7-20). With the exception of Chapter 15, where the symbol $\delta_s/\delta t$ is employed for the purpose of emphasizing the use of parental surface-fixed coordinates [cf. Eqs. (15.4-7) and (15.4-8)], the present use of $\partial/\partial t$ is consistent with notation employed throughout the text for time differentiation relative to a 'fixed point'. Thus, when the partial time derivative $\partial/\partial t$ appears in an interfacial transport equation, it is henceforth understood to represent time differentiation relative to an interfacial 'surface-fixed point', i.e. at fixed (q^1, q^2) surface coordinate values. See also similar comments following Eq. (3.6-16).

the surface-excess linear momentum vector density. Equation (16.1-1) represents the surface-excess mass balance equation for a nonmaterial interface.

An explicit form of Eq. (16.1-1) requires decomposition of the surface-excess linear momentum density $(\rho \mathbf{v})^s$ into the product of surface-excess mass density ρ^s and apparent 'mass-average' surface velocity vector $\mathbf{V}$, say [cf. Eq. (16.A-1)]; thus,

$$(\rho \mathbf{v})^s \equiv \rho^s \mathbf{V}, \tag{16.1-4}$$

As shown in Appendix 16.A following this chapter, the desired decomposition (16.1-4) is provided for nonmaterial interfaces [satisfying the physical assumptions embodied in the kinematical conditions (16.A-5) and (16.A-14)] by the assignation

$$\boxed{\mathbf{V} = \mathbf{v}_s + \mathbf{u} \cdot \mathbf{n}\mathbf{n},} \tag{16.1-5}$$

with

$$\mathbf{v}_s(\mathbf{x}_s) \equiv \mathbf{I}_s \cdot \overline{\mathbf{v}}(\mathbf{x}_s, 0+) = \mathbf{I}_s \cdot \overline{\mathbf{v}}(\mathbf{x}_s, 0-) = \mathbf{I}_s \cdot \mathbf{v}^o(\mathbf{x}_s) \tag{16.1-6}$$

the tangential mass-average surface velocity vector [cf. Eq. (3.4-12)]. Note that for the class of nonmaterial fluid interfaces considered in this text [satisfying Eqs. (16.A-5) and (16.A-14)], the tangential bulk-phase material velocity is continuous across the macroscale interface. Thus, in the remainder of Part II, only the normal component of the mass-average velocity can suffer possible discontinuities; explicitly, in general,

$$\mathbf{n}\mathbf{n} \cdot \overline{\mathbf{v}}(\mathbf{x}_s, 0+) \neq \mathbf{n}\mathbf{n} \cdot \overline{\mathbf{v}}(\mathbf{x}_s, 0-). \tag{16.1-7}$$

Upon employing Eqs. (16.1-4) and (16.1-5), Eq. (16.1-1) may thus be written in the form

$$\boxed{\frac{\partial \rho^s}{\partial t} - 2Hu_n \rho^s + \nabla_s \cdot (\rho^s \mathbf{v}_s) = -\mathbf{n} \cdot \|(\overline{\mathbf{v}} - \mathbf{u})\overline{\rho}\|.} \tag{16.1-8}$$

In the limit of a material interface, namely,

$$\overline{\mathbf{v}}(\mathbf{x}_s, 0+) = \overline{\mathbf{v}}(\mathbf{x}_s, 0-) = \mathbf{u}(\mathbf{x}_s) = \mathbf{V}(\mathbf{x}_s) \equiv \mathbf{v}^\circ(\mathbf{x}_s), \tag{16.1-9}$$

Eq. (16.1-8) reduces to its material counterpart (4.2-1).

The surface-excess mass density terms appearing on the left-hand side of Eq. (16.1-8) may normally be neglected as being of $O(\varepsilon)$ relative to the nonmaterial 'source' terms appearing on the right-hand side, as follows upon observing that the integrand in Eq. (16.1-2) is, in most physical circumstances, of $O(1)$. Thus, with [cf. Eq. (4.2-8a)]*

$$\rho^s = 0, \qquad (16.1\text{-}10)$$

the surface-excess mass conservation equation (16.1-8) reduces to

$$\mathbf{n} \cdot \|(\bar{\mathbf{v}} - \mathbf{u})\bar{\rho}\| = 0, \qquad (16.1\text{-}11)$$

thereby providing a central condition to be satisfied at a nonmaterial interface.

Surface-Excess Linear Momentum Conservation

Upon noting the explicit identifications made in Table 3.5-1, together with the surface-excess definitions (15.6-18), (15.6-36) and (15.7-19), the generic interfacial conservation equation (15.7-20) adopts the following form for the interfacial transport of the linear momentum density ($\psi \equiv \rho\mathbf{v}$):

$$\frac{\partial}{\partial t}(\rho\,\mathbf{v})^s - 2Hu_n(\rho\,\mathbf{v})^s + \nabla_s \cdot [(\rho\,\mathbf{v}\,\mathbf{v})^s - \mathbf{I}_s \cdot \mathbf{P}^s]$$
$$- \mathbf{F}^s = -\mathbf{n} \cdot \|\bar{\rho}(\bar{\mathbf{v}} - \mathbf{u})\bar{\mathbf{v}} - \bar{\mathbf{P}}\|, \qquad (16.1\text{-}12)$$

wherein

$$\boxed{\mathbf{P}^s(\mathbf{x}_s) \overset{\text{def}}{=} \varepsilon L \int_{-\infty}^{\infty} \mathbf{I}_s \cdot [\tilde{\mathbf{P}}(\mathbf{x}_s, \tilde{n}) - \bar{\mathbf{P}}(\mathbf{x}_s, 0)]d\tilde{n}\,,} \qquad (16.1\text{-}13)$$

furnishes an explicit microscale formulation of the surface-excess pressure tensor† [cf. Eq. (15.6-36)], and

$$\boxed{\mathbf{F}^s(\mathbf{x}_s) \overset{\text{def}}{=} \varepsilon L \int_{-\infty}^{\infty} [\tilde{\mathbf{F}}(\mathbf{x}_s, \tilde{n}) - \bar{\mathbf{F}}(\mathbf{x}_s, 0)]d\tilde{n}\,,} \qquad (16.1\text{-}14)$$

provides a corresponding microscale formulation of the surface-excess body-force vector density [cf. Eq. (15.6-18)].

---

* Questions pertaining to the algebraic sign of ρ^s, and particularly, raising the possibility of *negative* surface-excess mass densities (Slattery 1990), are irrelevant in the (macroscale) asymptotic surface-excess theory of this text, insofar as such questions regard macroscopically-insensible [i.e. $O(\varepsilon)$] surface-excess magnitudes. Whereas in all practical applications of the text the assignation $\rho^s = 0$ has been made [cf. Eq. (4.2-8a)], the symbol ρ^s nevertheless appears in general surface-excess balance equations [cf. Eq. (16.1-16)], thereby preserving the complete two-dimensional analogy existing between these equations and their three-dimensional counterparts.

† See the footnote below Table 2.3-1 for a clarification of the terminology 'surface-excess' as applied to intensive (two-dimensional) field quantities, such as (16.1-13), which are not defined strictly as quadratures of differences between corresponding micro- and macroscale (three-dimensional) fields.

Employ Eq. (16.1-4) and note from Eq. (16.1-13) that

$$\mathbf{I}_s \cdot \mathbf{P}^s = \mathbf{P}^s,$$

[cf. the footnote following Eq. (4.2-3)], whereupon the surface-excess linear momentum equation (16.1-12) adopts the alternative form

$$\frac{\partial}{\partial t}(\rho^s \mathbf{V}) - 2Hu_n \rho^s \mathbf{V} + \nabla_s \cdot (\rho^s \mathbf{v}_s \mathbf{V} - \mathbf{P}^s) - \mathbf{F}^s = -\mathbf{n} \cdot \|\bar{\rho}(\bar{\mathbf{v}} - \mathbf{u})\bar{\mathbf{v}} - \bar{\mathbf{P}}\|. \quad (16.1\text{-}15)$$

Use of Eq. (16.1-8), together with the definition (3.4-11)* of the surface-convected time derivative, permits Eq. (16.1-15) to be written explicitly as

$$\boxed{\rho^s \frac{D_s \mathbf{V}}{Dt} - \nabla_s \cdot \mathbf{P}^s - \mathbf{F}^s = -\mathbf{n} \cdot \|\bar{\rho}(\bar{\mathbf{v}} - \mathbf{u})(\bar{\mathbf{v}} - \mathbf{V}) - \bar{\mathbf{P}}\|.} \quad (16.1\text{-}16)$$

Equations (16.1-8) and (16.1-16) represent the respective generalizations of Eqs. (4.2-1) and. (4.2-4) to nonmaterial interfaces. Thus, in the special circumstance of a material interface, for which Eq. (16.1-9) obtains, Eqs. (16.1-8) and (16.1-16) respectively reduce identically to their material interface counterparts of Chapter 4, thereby justifying the surface-excess identifications of Table 3.6-1.

16.2 Conservation of Angular Momentum

With the choice $\psi \equiv \mathbf{x} \times \rho \mathbf{v} + \rho \mathbf{a}$ for the angular momentum density (Table 3.5-1), the generic balance equation (15.7-26) becomes

$$\begin{aligned}
&\frac{\partial}{\partial t}(\rho^s \mathbf{x}_s \times \mathbf{V}) + \frac{\partial}{\partial t}(\rho^s \mathbf{a}^s) + \nabla_s \cdot (\rho^s \mathbf{v}_s \mathbf{x}_s \times \mathbf{V}) \\
&+ \nabla_s \cdot (\rho^s \mathbf{v}_s \mathbf{a}^s) - 2Hu_n \rho^s \mathbf{x}_s \times \mathbf{V} - 2Hu_n \rho^s \mathbf{a}^s \\
&- \nabla_s \cdot (\mathbf{P}^s \times \mathbf{x}_s) - \nabla_s \cdot \mathbf{C}^s - \mathbf{x}_s \times \mathbf{F}^s - \mathbf{G}^s \\
&+ \mathbf{x}_s \times \mathbf{n} \cdot \|\bar{\rho}(\bar{\mathbf{v}} - \mathbf{u})\bar{\mathbf{v}} - \bar{\mathbf{P}}\| + \mathbf{n} \cdot \|\bar{\rho}(\bar{\mathbf{v}} - \mathbf{u})\bar{\mathbf{a}} - \bar{\mathbf{C}}\| = \mathbf{0}. \quad (16.2\text{-}1)
\end{aligned}$$

In obtaining the above expression, the following definitions have been employed (in addition to those offered in §16.1):

* When the kinematical condition embodied in Eq. (16.1-6) obtains, the definition (3.4-11) holds even though the interface is nonmaterial (16.1-7). This owes to the fact that $\mathbf{v}^o \cdot \nabla_s \equiv \mathbf{v}_s \cdot \nabla_s$.

$$\boxed{\mathbf{a}^s(\mathbf{x}_s) \stackrel{\text{def}}{=} \varepsilon L \frac{1}{\rho^s} \int_{-\infty}^{\infty} [\tilde{\rho}\tilde{\mathbf{a}}(\mathbf{x}_s, \tilde{n}) - \bar{\rho}\bar{\mathbf{a}}(\mathbf{x}_s, 0)] d\tilde{n}} \quad (16.2\text{-}2)$$

is the surface-excess internal angular momentum density pseudovector;

$$\boxed{\mathbf{G}^s(\mathbf{x}_s) \stackrel{\text{def}}{=} \varepsilon L \int_{-\infty}^{\infty} [\tilde{\mathbf{G}}(\mathbf{x}_s, \tilde{n}) - \bar{\mathbf{G}}(\mathbf{x}_s, 0)] d\tilde{n}} \quad (16.2\text{-}3)$$

is the surface-excess body-couple pseudovector density; and

$$\boxed{\mathbf{C}^s(\mathbf{x}_s) \stackrel{\text{def}}{=} \varepsilon L \int_{-\infty}^{\infty} \mathbf{I}_s \cdot [\tilde{\mathbf{C}}(\mathbf{x}_s, \tilde{n}) - \bar{\mathbf{C}}(\mathbf{x}_s, 0)] d\tilde{n}} \quad (16.2\text{-}4)$$

is the surface-excess couple-stress tensor, the latter clearly satisfying

$$\mathbf{I}_s \cdot \mathbf{C}^s = \mathbf{C}^s. \quad (16.2\text{-}5)$$

Upon use of the general dyadic identity (valid for any dyadic $\mathbf{P}^s$)

$$-\nabla_s \cdot (\mathbf{P}^s \times \mathbf{x}_s) = \mathbf{x}_s \times (\nabla_s \cdot \mathbf{P}^s) + \mathbf{P}^s_{\times}, \quad (16.2\text{-}6)$$

wherein the surface-excess antisymmetric stress pseudovector $\mathbf{P}^s_{\times}$ is defined in Eq. (4.2-6), Eq. (16.2-1) may be rearranged to the form

$$\mathbf{x}_s \times \Big[\frac{\partial}{\partial t}(\rho^s \mathbf{V}) - 2Hu_n \rho^s \mathbf{V} + \nabla_s \cdot (\rho^s \mathbf{v}_s \mathbf{V})$$
$$- \nabla_s \cdot \mathbf{P}^s - \mathbf{F}^s + \mathbf{n} \cdot \|\bar{\rho}(\bar{\mathbf{v}} - \mathbf{u})\bar{\mathbf{v}} - \bar{\mathbf{P}}\| \Big]$$
$$+ \frac{\partial}{\partial t}(\rho^s \mathbf{a}^s) - 2Hu_n \rho^s \mathbf{a}^s + \nabla_s \cdot (\rho^s \mathbf{v}_s \mathbf{a}^s)$$
$$- \nabla_s \cdot \mathbf{C}^s - \mathbf{P}^s_{\times} - \mathbf{G}^s + \mathbf{n} \cdot \|\bar{\rho}(\bar{\mathbf{v}} - \mathbf{u})\bar{\mathbf{a}} - \bar{\mathbf{C}}\| = \mathbf{0}.$$

As the term in square brackets vanishes in consequence of Eq. (16.1-15), one thereby obtains

$$\frac{\partial}{\partial t}(\rho^s \mathbf{a}^s) - 2Hu_n \rho^s \mathbf{a}^s + \nabla_s \cdot (\rho^s \mathbf{v}_s \mathbf{a}^s)$$
$$- \nabla_s \cdot \mathbf{C}^s - \mathbf{P}^s_{\times} - \mathbf{G}^s + \mathbf{n} \cdot \|\bar{\rho}(\bar{\mathbf{v}} - \mathbf{u})\bar{\mathbf{a}} - \bar{\mathbf{C}}\| = \mathbf{0}. \quad (16.2\text{-}7)$$

Equivalently, from Eq. (16.1-8), employing Eq. (3.4-11),

$$\boxed{\rho^s \frac{D_s \mathbf{a}^s}{Dt} - \nabla_s \cdot \mathbf{C}^s - \mathbf{P}^s_{\times} - \mathbf{G}^s + \mathbf{n} \cdot \|\bar{\rho}(\bar{\mathbf{v}} - \mathbf{u})\bar{\mathbf{a}} - \bar{\mathbf{C}}\| = \mathbf{0}.} \quad (16.2\text{-}8)$$

Equations (16.2-7) and (16.2-8) represent the respective generalizations of Eqs. (4.2-7) and (4.2-8) to nonmaterial interfaces. When the material interface condition (16.1-14) holds, the preceding surface-excess angular momentum relations may be observed to justify the identifications of the surface-excess fields appearing in Table 3.6-1.

16.3 A Constitutive Law for the Surface-Excess Pressure Tensor

A macroscale constitutive equation is derived in the present section via use of Eq. (16.1-13) for the surface-excess pressure tensor $\mathbf{P}^s$. For this purpose, attention is restricted to an incompressible, transversely-isotropic microscale fluid satisfying the (microscale) kinematical constraint (16.3-8) appropriate to a (macroscale) material interface. In these model circumstances it will be shown that the Boussinesq-Scriven interfacial stress tensor (4.2-17) obtains. Thus, not only does our example illustrate the general microscale asymptotic method for deriving macroscale interfacial constitutive relations, but equally it provides a microscale rationale for the most common mode of macroscale interfacial rheological behavior, as embodied in the classical Boussinesq-Scriven constitutive model.

The three-dimensional microscale stress tensor is henceforth assumed to satisfy the decomposition [cf. Eq. (4.1-13)]

$$\mathbf{P} = -\,p\,\mathbf{I} + \boldsymbol{\tau}, \tag{16.3-1}$$

with p the equilibrium pressure and $\boldsymbol{\tau}$ the viscous stress tensor, assumed herein to be symmetric (namely, $\boldsymbol{\tau} = \boldsymbol{\tau}^{\dagger}$). The microscale equilibrium pressure term p is assumed to be everywhere isotropic, such that the microscale anisotropy assumed to exist within the interfacial region is fully attributable to short-range intermolecular forces identified with the microscale external force density vector $\mathbf{F}$ rather than with the microscale pressure p, as is often done (refer to the caption of Figure 16.3-1).

Upon employing Eq. (16.1-13), the surface-excess pressure tensor $\mathbf{P}^s$ may be written as

$$\mathbf{P}^s = \sigma\,\mathbf{I}_s + \boldsymbol{\tau}^s, \tag{16.3-2}$$

where

$$\boxed{\sigma(\mathbf{x}_s) \stackrel{\text{def}}{=} \varepsilon L \int_{-\infty}^{\infty} [\bar{p}(\mathbf{x}_s, 0) - \tilde{p}(\mathbf{x}_s, \tilde{n})]\,d\tilde{n}} \tag{16.3-3}$$

provides a surface-excess formulation of the interfacial tension in terms of the microscale pressure field $p(\mathbf{x})$ (see Figure 16.3-1), and

$$\boxed{\boldsymbol{\tau}^s(\mathbf{x}_s) \stackrel{\text{def}}{=} \varepsilon L \int_{-\infty}^{\infty} \mathbf{I}_s \cdot [\tilde{\boldsymbol{\tau}}(\mathbf{x}_s, \tilde{n}) - \bar{\boldsymbol{\tau}}(\mathbf{x}_s, 0)]\,d\tilde{n}} \tag{16.3-4}$$

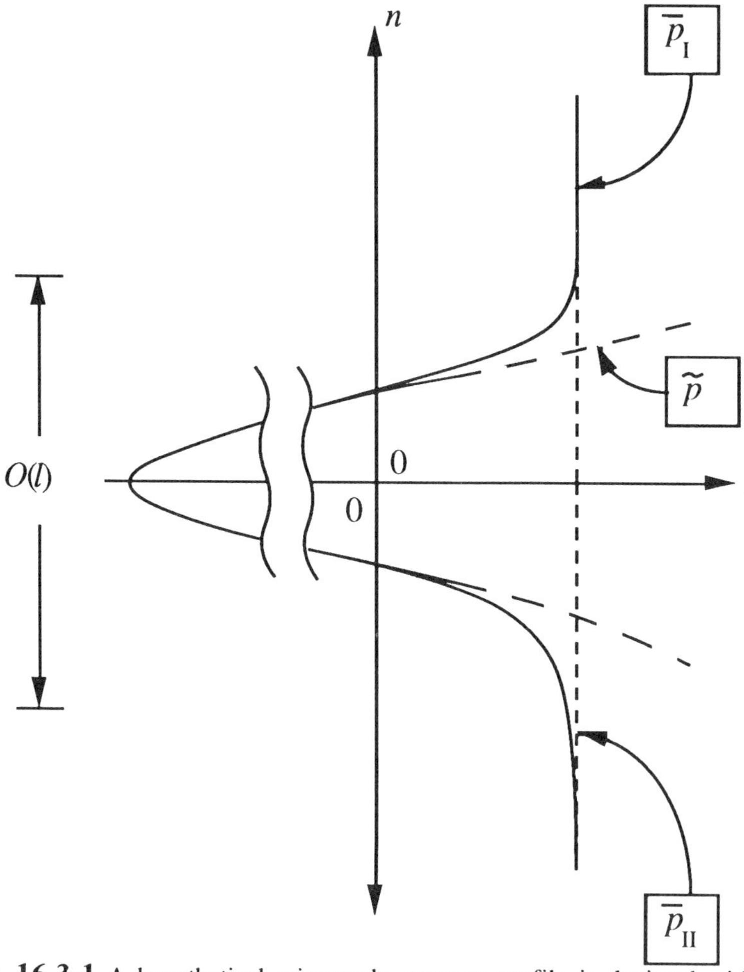

Figure 16.3-1 A hypothetical microscale pressure profile is depicted within the interfacial transition region separating bulk-phase fluids I and II. Negative values of the microscale pressure scalar p within the interfacial region are required in order to satisfy the assumption of a positive interfacial tension, σ, according to the surface-excess formulation (16.3-3). Classical statistical-thermodynamic theories of interfaces [see Rowlinson & Widom (1982)] predict interfacial tension from microscale data by employing formulas [analogous to (16.3-3)] which assume the existence of a *tangential* microscale pressure p_T: statistical-thermodynamic models [see, e.g., the review of Ono & Kondo (1960)] have demonstrated that the scalar p_T achieves large negative values (as depicted above for p) within a fluid interfacial transition region owing to the highly inhomogeneous nature of this region. A relation between the two scalars p_T and p, for the particular case of a spherical interface, is provided in Question 16.5 [note also the alternative definition of interfacial tension (i.e. γ) provided in this question in terms of the tangential microscale pressure, p_T].

similarly formulates the surface-excess viscous stress tensor in terms of the microscale deviatoric stress field $\tau(\mathbf{x})$.

Within the diffuse interfacial region, the existence of short-range, anisotropic, intermolecular forces (attributable to the presence of steep mass density and/or species concentration gradients in a direction normal to the interface) is assumed to endow the microscale viscous stress tensor τ with local axial symmetry about the local direction of the normal vector $\mathbf{n}^*$ at the point $\mathbf{x}$.* In addition, τ is supposed linearly proportional to the rate-of-deformation tensor $\mathbf{D}$ [cf. Eq. (4.1-16)] by analogy with comparable three-dimensional bulk fluids. Upon supposing the microscale fluid to be incompressible, namely

$$\nabla \cdot \mathbf{v} = \mathbf{D}:\mathbf{I} = \left(\mathbf{D}:\mathbf{I}_s^*\right) + \left(\mathbf{D}:\mathbf{n}^*\mathbf{n}^*\right) = 0, \qquad (16.3\text{-}5)$$

the viscous stress tensor can thus be written in the transversely isotropic form (Eliassen 1963, Goodrich 1981b)†

$$\begin{aligned}\tau = {} & \eta\left(\mathbf{D}:\mathbf{I}_s^*\right)\mathbf{I}_s^* + 2\mu\left(\mathbf{I}_s^* \cdot \mathbf{D} \cdot \mathbf{I}_s^*\right) \\ & + 2\mu'\left(\mathbf{I}_s^* \cdot \mathbf{D} \cdot \mathbf{n}^*\mathbf{n}^* + \mathbf{n}^*\mathbf{n}^* \cdot \mathbf{D} \cdot \mathbf{I}_s^*\right) \\ & - 2\mu''\left(\mathbf{D}:\mathbf{I}_s^*\right)\mathbf{n}^*\mathbf{n}^*, \end{aligned} \qquad (16.3\text{-}6)$$

with the microscale phenomenological viscosity functions η, μ, μ' and μ'' each functions of $\mathbf{x}$ in general. Steep surfactant concentration gradients existing in the interfacial region are postulated to impart a strong functional dependence upon normal distance n to each of the four phenomenological viscosity coefficients appearing in Eq. (16.3-6) [see the paragraphs following Eq. (16.3-25) for a fuller discussion]. Within the two bulk-fluid regions, the effect of the interfacial inhomogeneity upon the constitutive form of τ is assumed negligible. In particular, the microscale fluid is supposed locally isotropic within these regions; explicitly,

$$\bar{\mu}' = \bar{\mu}'' \equiv \bar{\mu}, \text{ say,} \qquad \text{and} \qquad \bar{\eta} = 0. \qquad (16.3\text{-}7)$$

Thus, upon formally passing to the outer limit [cf. Eqs. (15.5-12) and (15.5-13)] in Eq. (16.3-6), we obtain

$$\bar{\tau} = 2\bar{\mu}\,\bar{\mathbf{D}},$$

which is the usual constitutive equation for an incompressible, isotropic Newtonian fluid, as in Eq. (4.1-21).

Consistent with our assumption of a *material* fluid interface (cf. the third entry in Table 15.5-2),the microscale velocity vector $\mathbf{v}$ in the following analysis will be taken to be a macroscale quantity, in the sense that

---

* The parent surface-fixed coordinate system (q^1, q^2, q^3) discussed in §15.3 is employed here.

† To obtain such a relation from the theory of anisotropic fluids (Stokes 1985), the so-called director inertia must be negligible.

$$\left(\frac{1}{U}\right)\frac{\partial \mathbf{v}}{\partial \bar{n}} = O(1) \tag{16.3-8}$$

everywhere; i.e. $O(1)$ changes in $\mathbf{v}$ are assumed to occur only over macroscale distances. Such a supposition implies that $\mathbf{v}$ is continuous across the interface when viewed from the macroscale, as shown by the following argument: Upon employing Eqs. (15.5-23)–(15.5-27) [with $\mathbf{f}^* \equiv \mathbf{v}(\mathbf{x}_s, n)$ and $\mathbf{f} \equiv \mathbf{v}(\mathbf{x}_s, 0)$], the respective inner and outer limits of $\mathbf{v}$ are identical, and are given by the single entity $\mathbf{v}^o$ in the following relation:

$$\begin{aligned}\tilde{\mathbf{v}}(\mathbf{x}_s, \tilde{n}) &= \mathbf{v}(\mathbf{x}_s, 0) \\ &= \lim_{\bar{n} \to 0+} \bar{\mathbf{v}}_{\mathrm{I}}(\mathbf{x}_s, \bar{n}) \\ &= \lim_{\bar{n} \to 0-} \bar{\mathbf{v}}_{\mathrm{II}}(\mathbf{x}_s, \bar{n}) \\ &\equiv \mathbf{v}^o(\mathbf{x}_s).\end{aligned} \tag{16.3-9}$$

Formally, our assumption of a *material* macroscale interface requires that the normal components of the bulk-phase velocities $\bar{\mathbf{v}}_{\mathrm{I}}$ and $\bar{\mathbf{v}}_{\mathrm{II}}$ be both continuous across the interface and equivalent to the normal surface velocity u_n; explicitly,

$$\mathbf{n} \cdot \mathbf{u} \equiv \mathbf{n} \cdot \mathbf{v}^o. \tag{16.3-10}$$

Owing to the assumed, relatively homogeneous, transverse physicochemical properties of the interfacial region, macroscale discontinuities can arise only in the normal component of macroscale flux fields across the interface [cf. Eq. (15.7-20)]. Thus, transverse or lateral gradients of the microscale velocity vector $\mathbf{v}$ within the interfacial region are explicitly assumed in the subsequent theory to be macroscale in nature. This assures that no discontinuity at $\bar{n} = 0$ can exist in the tangential components of the macroscale stress tensor; thus,

$$\left(\frac{L}{U}\right)\frac{\partial}{\partial \bar{n}}(\nabla_s^* \mathbf{v}) = O(1). \tag{16.3-11}$$

Upon employing Eqs. (15.4-23)-(15.4-27) [with $\mathbf{f}^* \equiv \nabla_s \mathbf{v}(\mathbf{x}_s, n)$ and $\mathbf{f} \equiv \nabla_s \mathbf{v}^o(\mathbf{x}_s)$], we may then write

$$\widetilde{(\nabla_s^* \mathbf{v})} = \lim_{\bar{n} \to 0+} \overline{(\nabla_s^* \mathbf{v})}_{\mathrm{I}} = \lim_{\bar{n} \to 0-} \overline{(\nabla_s^* \mathbf{v})}_{\mathrm{II}} \equiv \nabla_s \mathbf{v}^o \tag{16.3-12}$$

for the respective inner and outer limits of the tangential contribution to the microscale velocity gradient. In contrast with the above condition of continuity, the normal gradient of $\mathbf{v}$ may appear to be discontinuous across the interface in the macroscale asymptotic limit.

In order to establish the constitutive form adopted by the macroscale interfacial viscous stress tensor $\boldsymbol{\tau}^s$ from the definition (16.3-4) together with the microscale constitutive equation (16.3-6) for $\boldsymbol{\tau}$, inner and outer expansions of the various deformational terms appearing in Eq. (16.3-6) must be calculated. Use of Eq. (16.3-12), jointly with the definition (4.1-16) of **D**, shows that the inner limit of $\mathbf{D}:\mathbf{I}_s^*$ may be written in the form

$$\widetilde{\mathbf{D}:\mathbf{I}_s^*} = \widetilde{(\nabla\mathbf{v}):\mathbf{I}_s^*} = \widetilde{\nabla_s^*\cdot\mathbf{v}} \equiv \nabla_s\cdot\mathbf{v}^o. \tag{16.3-13}$$

Similarly, the outer limit of $\mathbf{D}:\mathbf{I}_s^*$ is given by the relation

$$\overline{(\mathbf{D}:\mathbf{I}_s^*)}_\alpha = \overline{(\nabla_s^*\cdot\mathbf{v})}_\alpha \qquad (\alpha = \mathrm{I, II}). \tag{16.3-14}$$

Form the respective interfacial limits $\bar{n}\to 0\pm$ of the above equation and employ Eq. (16.3-12) to obtain

$$\lim_{\bar{n}\to 0+}\overline{(\mathbf{D}:\mathbf{I}_s^*)}_{\mathrm{I}} = \lim_{\bar{n}\to 0-}\overline{(\mathbf{D}:\mathbf{I}_s^*)}_{\mathrm{II}} \equiv \nabla_s\cdot\mathbf{v}^o. \tag{16.3-15}$$

The inner and outer limits of the term $\mathbf{I}_s^*\cdot\mathbf{D}\cdot\mathbf{I}_s^*$ can similarly be reduced through use of the product relations (15.5-18) and (15.5-19), together with the definition of the deformation dyadic **D** (4.1-16) and the limiting relation (16.3-12); this yields

$$\begin{aligned}\widetilde{(\mathbf{I}_s^*\cdot\mathbf{D}\cdot\mathbf{I}_s^*)} &= \tfrac{1}{2}\left[(\nabla_s\mathbf{v}^o)\cdot\mathbf{I}_s + \mathbf{I}_s\cdot(\nabla_s\mathbf{v}^o)^{\dagger}\right]\\ &= \lim_{\bar{n}\to 0+}\overline{(\mathbf{I}_s^*\cdot\mathbf{D}\cdot\mathbf{I}_s^*)}_{\mathrm{I}}\\ &= \lim_{\bar{n}\to 0-}\overline{(\mathbf{I}_s^*\cdot\mathbf{D}\cdot\mathbf{I}_s^*)}_{\mathrm{II}}.\end{aligned} \tag{16.3-16}$$

As shown later in this section, the term appearing in Eq. (16.3-6) involving the quantity $\mathbf{I}_s^*\cdot\mathbf{D}\cdot\mathbf{n}^*\mathbf{n}^*$ does not contribute to the macroscale constitutive equation for τ^s. Rather than be sidetracked into a lengthy proof of this assertion at this point, we may anticipate the result, and assume that

$$\tau^s\cdot\mathbf{n} = \mathbf{0}, \tag{16.3-17}$$

postponing a proof of this statement until later in this section [cf. the subsection following Eq. (16.3-26)].

Given the limiting forms (16.3-12), (16.3-15) and (16.3-16), together with the assertion (16.3-17), we are now in a position to calculate the constitutive form adopted by the interfacial viscous stress tensor τ^s. This may be done by substituting the microscale constitutive equation (16.3-6) for τ into the integral formulation (16.3-4) of τ^s. Thus, consider the integral arising when the first term appearing in Eq. (16.3-6) is substituted into (16.3-4). Application of Eqs. (16.3-13) and (16.3-15) to this integral yields

$$\varepsilon L \int_{-\infty}^{+\infty} \mathbf{I}_s \cdot \{\widetilde{[\eta\,(\mathbf{D}:\mathbf{I}_s^*)\,\mathbf{I}_s^*]\,(\mathbf{x}_s, \tilde{n})} - \overline{[\eta\,(\mathbf{D}:\mathbf{I}_s^*)\,\mathbf{I}_s^*]}\,(\mathbf{x}_s, 0)\} d\tilde{n} \equiv \eta^s (\nabla_s \cdot \mathbf{v}^o)\,\mathbf{I}_s \quad (16.3\text{-}18)$$

as the contribution of this term to the constitutive equation for τ^s. Here, the interfacial viscosity coefficient η^s is given by the surface-excess relation

$$\boxed{\eta^s(\mathbf{x}_s) \overset{\text{def}}{=} \varepsilon L \int_{-\infty}^{+\infty} [\tilde{\eta}(\mathbf{x}_s, \tilde{n}) - \bar{\eta}(\mathbf{x}_s, 0)]\,d\tilde{n}\,.} \quad (16.3\text{-}19)$$

Similarly, the integral arising when the second term in Eq. (16.3-6) is substituted into the definition (16.3-4) can be written in the form

$$\varepsilon L \int_{-\infty}^{+\infty} \mathbf{I}_s \cdot \{\widetilde{[\,2\mu\,(\mathbf{I}_s^* \cdot \mathbf{D} \cdot \mathbf{I}_s^*)\,\mathbf{I}_s^*]\,(\mathbf{x}_s, \tilde{n})} - \overline{[\,2\mu\,(\mathbf{I}_s^* \cdot \mathbf{D} \cdot \mathbf{I}_s^*)\,\mathbf{I}_s^*]}\,(\mathbf{x}_s, 0)\} d\tilde{n} \equiv 2\mu^s\,\mathbf{D}_s\,, \quad (16.3\text{-}20)$$

where

$$\boxed{\mu^s(\mathbf{x}_s) \overset{\text{def}}{=} \varepsilon L \int_{\tilde{n}=-\infty}^{+\infty} [\,\tilde{\mu}(\mathbf{x}_s, \tilde{n}) - \bar{\mu}(\mathbf{x}_s, 0)]\,d\tilde{n}\,,} \quad (16.3\text{-}21)$$

is an interfacial viscosity coefficient [i.e. the *interfacial shear viscosity* of Newtonian interfaces; cf. Eq. (16.3-24)] and

$$\mathbf{D}_s = \tfrac{1}{2}\left[(\nabla_s \mathbf{v}^o) \cdot \mathbf{I}_s + \mathbf{I}_s \cdot (\nabla_s \mathbf{v}^o)^{\dagger}\right] \quad (16.3\text{-}22)$$

is the interfacial rate-of-deformation tensor [cf. Eq. (4.2-18) and Appendix 4.C]. Due to the constraint (16.3-17), none of the other terms in (16.3-6) contribute to τ^s.

The above results combine to yield the purely macroscale expression

$$\tau^s = \eta^s (\nabla_s \cdot \mathbf{v}^o)\,\mathbf{I}_s + 2\mu^s \mathbf{D}_s \quad (16.3\text{-}23)$$

as the operational form of the interfacial constitutive equation for τ^s. Substitution of the above relation into Eq. (16.3-2) furnishes the interfacial constitutive equation for the total stress tensor, namely

$$\mathbf{P}^s = \sigma \mathbf{I}_s + 2\mu^s \left(\mathbf{D}_s - \tfrac{1}{2}\mathbf{I}_s \mathbf{I}_s : \mathbf{D}_s\right) + \kappa^s \mathbf{I}_s \mathbf{I}_s : \mathbf{D}_s\,, \quad (16.3\text{-}24)$$

where μ^s is the interfacial shear viscosity, and

$$\boxed{\kappa^s \overset{\text{def}}{=} \eta^s + \mu^s} \quad (16.3\text{-}25)$$

the interfacial dilatational viscosity.

The macroscale expression (16.3-24) for the interfacial pressure tensor $\mathbf{P}^s$ reproduces identically the Newtonian interfacial rheological constitutive relation embodied in Eqs. (4.2-15) and (4.2-17), previously proposed on a purely phenomenological, continuum-mechanical basis by analogy with conventional three-dimensional Newtonian fluids (albeit modified to reflect the generally non-Euclidean nature of the curved interfacial 'space'). Equation (16.3-24) has been obtained strictly for material interfaces [cf. Eqs. (16.3-9) and (16.3-10)], and for which the three-dimensional microscale fluid in proximity to the interfacial region possesses the linear, transversely-isotropic form (16.3-6). An important condition upon the validity of (16.3-24) is contained in Eq. (16.3-17); as demonstrated in subsequent paragraphs, this condition [indeed, as is clearly true for the kinematical conditions (16.3-12)–(16.3-16)] is appropriate only in the asymptotic limit as $\varepsilon \rightarrow 0$. We make this $\varepsilon \rightarrow 0$ remark explicit because other microscale→macroscale interfacial theories [e.g Eliassen (1963), Goodrich (1981b)] claim to be *exact,* rather than asymptotic, whence the validity of the relation $\boldsymbol{\tau}^s \cdot \mathbf{n} = \mathbf{0}$ remains an open question in their theories. In turn, the answer to this question impacts upon the question of whether or not the interfacial stress tensor $\boldsymbol{\tau}^s$ (or $\mathbf{P}^s$) is *symmetric*. And this is an important issue in connection with the moment-of-momentum equation for interfaces, as discussed in the footnote following Eq. (4.2-3).

In general, the microscale shear viscosity fields $\mu(\mathbf{x}) \equiv \mu(\mathbf{x}_s, n)$ and $\eta(\mathbf{x}) \equiv \eta(\mathbf{x}_s, n)$ required in the respective evaluation of μ^s and η^s via Eqs. (16.3-19) and (16.3-21), may possess a steep gradient in the neighborhood of the interface when adsorbed surfactants are present; that is, since the microscale, three-dimensional concentration $\rho_A(\mathbf{x})$ of the surfactant species ($\equiv A$), say, varies steeply within the interfacial region [with $\partial\rho_A/\partial n=0$ in the neighborhood of $n=0$; cf. Eq. (17.2-6)], the same is supposed true of the viscosities. This fact implicitly assumes here a functional dependence of μ and η upon ρ_A (with $\partial\mu/\partial\rho_A > 0$ and $\partial\eta/\partial\rho_A > 0$) in the manner of, say, the analog of the Einstein equation (Einstein 1905; cf. Einstein 1956; see also Adler *et al.* 1990) for the viscosity of suspensions. Thus, the monolayer coverage achieved by adsorbed surfactants at the interface can be expected to yield very large three-dimensional viscosities within the diffuse interfacial region proximate to $n=0$ (such as depicted in curve B of Figure 15.6-3). The existence of such large microscale viscosities thus accounts for the concomitant existence of sensible interfacial viscosities.*

* The interfacial viscosity formulas (16.3-19) and (16.3-21) may be rewritten in order to explicitly reveal the microscale physicochemical conditions necessary for the existence of sensible values of μ^s and η^s, as discussed above. Thus, upon employing Laplace's method for the asymptotic evaluation of the generic quadrature formula,

If the microscale pressure tensor **P** is everywhere locally isotropic, then $\mu=\mu'=\mu''$ and $\eta=0$ in Eq. (16.3-6). In such circumstances the interfacial shear and dilatational viscosities can be shown to be identical, i.e. $\mu^s = \eta^s$, and to be given explicitly in terms of the prescribed microscale viscosity data $\mu(\mathbf{x}_s, n)$ as

$$\mu^S(\mathbf{x}_s) \equiv \kappa^S(\mathbf{x}_s) = \varepsilon L \int_{-\infty}^{+\infty} [\bar{\mu}(\mathbf{x}_s, \tilde{n}) - \bar{\mu}(\mathbf{x}_s, 0)] d\tilde{n} \,. \quad (16.3\text{-}26)$$

This equality of interfacial viscosities may be viewed as providing an indirect theoretical confirmation of the observed fact (cf. Figure 2.5-2) that experimentally measured values of μ^s and κ^s are of similar orders of magnitude.

Validation of the symmetry condition.

In the preceding analysis it has been asserted that the term $\mathbf{I}^*_s \cdot \mathbf{D} \cdot \mathbf{n}^*\mathbf{n}^*$ in Eq. (16.3-6) does not contribute to the macroscale constitutive equation for τ^s [owing to the restriction (16.3-17)]. This is

---

$$\lambda^S = \varepsilon L \int_{-\infty}^{+\infty} [\bar{\lambda}(\tilde{n}) - \bar{\lambda}(0)] d\tilde{n} \,,$$

[whose integrand $\bar{\lambda}(\tilde{n}) - \bar{\lambda}(0)$ is known to exhibit a maximum in the vicinity of $\tilde{n} = 0$] it follows that, in the $\varepsilon \to 0$ limit (de Bruijn 1958),

$$\lambda^S \approx \varepsilon L \sqrt{2\pi} \frac{[\bar{\lambda}(0) - \bar{\lambda}(0)]}{\sqrt{-\bar{\lambda}(0)^{-1} (d^2\bar{\lambda} / d\tilde{n}^2)_0}}$$

obtains. This relation clearly illustrates the preceding textual remarks, namely, that the 'height' of the difference $\bar{\lambda}(0) - \bar{\lambda}(0)$ must be $O(1/\varepsilon)$ for the existence of a sensible [i.e. $O(1)$] surface-excess field λ^S. The above formula may be rewritten in terms of the dimensional independent variable n, as

$$\boxed{\lambda^S \approx \sqrt{2\pi} \frac{[\bar{\lambda}(0) - \bar{\lambda}(0)]}{\sqrt{-\bar{\lambda}(0)^{-1} (d^2\bar{\lambda} / dn^2)_0}} \,.}$$

This provides an alternative, asymptotic formulation of the generic relation (15.6-18), and may be employed with formulas such as (16.1-2), (16.3-3), (16.3-19) or (16.3-20) for the purpose of calculating macroscale interfacial quantities from limited microscale data.

demonstrated in the following to be a consequence of the microscale linear momentum equation

$$\rho \frac{\mathrm{D}\mathbf{v}}{\mathrm{D}t} = \mathbf{F} + \nabla \cdot \mathbf{P}. \qquad (16.3\text{-}27)$$

[Note that it is convenient here to use the conventional material-convected derivative $\mathrm{D}/\mathrm{D}t \equiv \partial/\partial t\,)_{\mathbf{x}} + \mathbf{v} \cdot \nabla$ [rather than Eq. (15.4-7)] within the interfacial transition region owing to the assumed material nature of the interface (whence the convected parent coordinate surface q_0^3 is itself material). Thus, the interfacial fluid particles convect together with the interface.]

Within the interfacial region, the tangential component of Eq. (16.3-27) can, with the use of Eq. (16.3-1), be written in the form

$$\frac{1}{\varepsilon L}\left(\mathbf{n}^* \cdot \frac{\partial \boldsymbol{\tau}}{\partial \tilde{n}}\right) \cdot \mathbf{I}_s^* = \rho \frac{\mathrm{D}\mathbf{v}}{\mathrm{D}t} \cdot \mathbf{I}_s^*$$

$$- \mathbf{F} \cdot \mathbf{I}_s^* + \nabla_s^* p - \nabla_s^* \cdot \left(\boldsymbol{\tau} \cdot \mathbf{I}_s^*\right). \qquad (16.3\text{-}28)$$

Upon assuming all of the terms on the RHS to be of $O(1)$, this equation furnishes the scaling

$$\mathbf{n}^* \cdot \frac{\partial \boldsymbol{\tau}}{\partial \tilde{n}} \cdot \mathbf{I}_s^* = O(\varepsilon). \qquad (16.3\text{-}29)$$

Application of the order-of-magnitude relation (15.5-33) yields

$$\frac{\partial}{\partial \tilde{n}}\left(\mathbf{n}^* \cdot \boldsymbol{\tau} \cdot \mathbf{I}_s^*\right) = O(\varepsilon), \qquad (16.3\text{-}30)$$

as the appropriate scaling of the linear momentum equation within the interfacial transition region. Since $\boldsymbol{\tau}$ is symmetric, the identity

$$\mathbf{n}^* \cdot \boldsymbol{\tau} \cdot \mathbf{I}_s^* = \mathbf{I}_s^* \cdot \boldsymbol{\tau} \cdot \mathbf{n}^* \qquad (16.3\text{-}31)$$

is applicable. Upon employing Eqs. (15.5-23)–(15.5-27), the above two relations combine to yield

$$\widetilde{\left(\mathbf{I}_s^* \cdot \boldsymbol{\tau} \cdot \mathbf{n}^*\mathbf{n}^*\right)} = \lim_{\bar{n} \to 0+} \overline{\left(\mathbf{I}_s^* \cdot \boldsymbol{\tau} \cdot \mathbf{n}^*\mathbf{n}^*\right)}_{\mathrm{I}} = \lim_{\bar{n} \to 0-} \overline{\left(\mathbf{I}_s^* \cdot \boldsymbol{\tau} \cdot \mathbf{n}^*\mathbf{n}^*\right)}_{\mathrm{II}}.$$

$$(16.3\text{-}32)$$

Since the inner and outer limits in the latter are seen to be identical, the term $\mathbf{I}_s^* \cdot \boldsymbol{\tau} \cdot \mathbf{n}^*\mathbf{n}^*$ will not contribute to the surface-excess stress tensor $\boldsymbol{\tau}^s$ upon inserting the corresponding term in Eq. (16.3-6) into Eq. (16.3-4). Hence, $\boldsymbol{\tau}^s \cdot \mathbf{n}^* = \mathbf{0}$, as asserted in Eq. (16.3-17). Q.E.D.

16.4 A Microscale Example of an Expanding Surfactant-Adsorbed Bubble

This section re-examines the expanding bubble problem of Chapter 4 (see example 2 of §4.4). In contrast with the purely macroscale approach to this problem in Chapter 4, here, a continuous, three-dimensional *microscale* point of view is adopted for the radially-expanding interfacial fluid region. By now pursuing in detail the asymptotic procedures outlined in §15.5 for the microscale interpretation of macroscale surface-excess fields, and in §15.7 for the derivation of the generic surface-excess transport law (17.7-20), the result (4.4-33) will be alternatively derived here, but now with a more fundamental basis.

Upon supposing the inhomogeneous microfluid to be incompressible throughout the interfacial region and in the contiguous bulk fluid phases, the kinematical condition (4.4-31) may be applied to describe the microscale velocity field; explicitly,

$$\mathbf{v} = \mathbf{i}_r \frac{Q}{4\pi r^2}, \tag{16.4-1}$$

where Q is a time-dependent function (see Figure 4.4-3) and $\mathbf{i}_r$ the unit radial spherical-polar vector drawn from the bubble center. The three-dimensional stress tensor for the interfacial transition region and surrounding bulk phases will, for simplicity, be assumed isotropic and Newtonian; thus, since Eqs. (4.1-23) and (4.1-24) are thereby valid, the interfacial domain linear momentum condition reduces to (cf. Table 4.1-3)

$$\frac{\partial p}{\partial r} = F_r - \frac{Q}{\pi r^3}\left(\frac{\partial \mu}{\partial r}\right). \tag{16.4-2}$$

(As in Chapter 4, inertial terms have been neglected in the above. Retention of the body force term F_r in this relation corresponds to the action of internal forces arising from physicochemical gradients existing within the highly inhomogeneous interfacial transition region.) An expression for $p(r)$ may be obtained by direct integration of the above from $r = r$ to $r = \infty$.

In order to compare the resulting expression with the comparable result of §4.4, we must multiply Eq. (16.4-2) by M^* prior to integrating. This owes to the scaling relation (15.3-10) between areal elements defined on different r-coordinate surfaces, and ultimately permits the assignation of surface-excess quantities deriving from Eq. (16.4-2) to a single r-surface; namely, the parent surface, which will be designated as the spherical surface $r \equiv a$ [cf. Eqs. (15.6-7)–(15.6-16)]. For spherical geometry,

$$H^* = -1/r, \tag{16.4-3}$$

whence it follows from Eq. (15.3-11) that

$$M^* = (r/a)^2 . \tag{16.4-4}$$

Multiply Eq. (16.4-2) by (16.4-4) and subsequently integrate over the normal interfacial coordinate r to obtain [with $n \equiv r - a$; cf. Eq. (15.3-7)]

$$p\left(1+\frac{n}{a}\right)^2\Bigg|_{-n_2}^{n_1} - \frac{2}{a}\int_{-n_2}^{n_1} p\left(1+\frac{n}{a}\right)dn$$

$$= \int_{-n_2}^{n_1} F_r\left(1+\frac{n}{a}\right)^2 dn - \frac{Q}{\pi a^3}\left(\frac{\mu}{1+\frac{n}{a}}\right)\Bigg|_{-n_2}^{n_1}$$

$$- \frac{Q}{\pi a^4}\int_{-n_2}^{n_1} \frac{\mu}{\left(1+\frac{n}{a}\right)^2} dn \qquad (16.4\text{-}5)$$

for the microscale linear-momentum transport relation, corresponding in this particular case to Eq. (15.7-9) (upon multiplying the above by an areal element A on the parent surface $r=a$). Here, n_1 and n_2 represent arbitrary integration limits lying within the interfacial region.

Subtract from the above equation the corresponding macroscale linear momentum equations, valid over the entire range n_1 to n_2 [cf. Eq. (15.7-12)], namely

$$\bar{p}\left(1+\frac{n}{a}\right)^2\Bigg|_{-n_2}^{n_1} - \frac{2}{a}\int_{-n_2}^{n_1} \bar{p}\left(1+\frac{n}{a}\right)dn$$

$$= -\frac{Q}{\pi a^3}\left(\frac{\bar{\mu}}{1+\frac{n}{a}}\right)\Bigg|_{-n_2}^{n_1} - \frac{Q}{\pi a^4}\int_{-n_2}^{n_1} \frac{\bar{\mu}}{\left(1+\frac{n}{a}\right)^2} dn - \frac{Q}{\pi a^3}\|\bar{\mu}\| + \|\bar{p}\|,$$

$$(16.4\text{-}6)$$

and subsequently pass to the $\varepsilon \rightarrow 0$ limit [as in Eq. (15.7-11)], to eventually obtain

$$[\bar{p}_{\text{I}}(0+) - \bar{p}_{\text{II}}(0-)] - \varepsilon L \frac{2}{a}\int_{\tilde{n}=-\infty}^{+\infty} (\tilde{p} - \bar{p})\,d\tilde{n}$$

$$= -\varepsilon L \frac{Q}{\pi a^4}\int_{\tilde{n}=-\infty}^{+\infty} (\tilde{\mu} - \bar{\mu})d\tilde{n} - \frac{Q}{\pi a^3}[\bar{\mu}_{\text{I}}(0+) - \bar{\mu}_{\text{II}}(0-)]. \qquad (16.4\text{-}7)$$

In the above limit, the integral containing the force term F_r vanishes by its nature as deriving from an *internal* (physicochemical) body force. Use of the surface-excess definitions (16.3-3), (16.3-21) and (16.3-26) thereby yields the relation

$$\|\bar{p}\| + \frac{2\sigma}{a} = -\frac{Q}{\pi a^3}\|\bar{\mu}\| - \frac{Q}{\pi a^4}\kappa^s, \tag{16.4-8}$$

which, upon neglect of the gas-phase viscosity ($\bar{\mu}_{\text{II}} = 0$), achieves a form identical with the previously derived macroscale result, Eq. (4.4-33).

This example demonstrates the viability of independently approaching interfacial transport phenomena from either the macro- (two-dimensional) or micro- (three-dimensional) points of view. The advantage of the former lies in the fact that detailed knowledge of the microphysical nature of the interfacial transition zone need not be known for purposes of achieving a macroscale, phenomenological understanding of interfacial behavior. The microscale point of view, on the other hand, with its (possibly restrictive) need of more detailed *a priori* knowledge of the physical nature of the diffuse, highly inhomogeneous interfacial region, is able to furnish detailed *microscale* predictions of (three-dimensional) interfacial transport phenomena—in addition, of course, to arriving at macroscale (two-dimensional) predictions of interfacial behavior identical with that derived from the purely macroscale description of the phenomena.

16.5 Summary

In this chapter, expressions for the surface-excess linear (16.1-16) and angular (16.2-8) momentum equations have been derived, as mechanical examples of the generic surface-excess balance equation (15.7-20). These macroscale relations are generally applicable to the case of *nonmaterial* fluid interfaces; in the limit of *material* interfaces, they reduce to expressions previously cited in Chapter 4, namely Eqs. (4.2-3) and (4.2-7). The Boussinesq-Scriven form of the surface-excess interfacial stress tensor (16.3-23) has also been derived from the microscale asymptotic theory, upon assuming the microscale fluid to be an incompressible and transversely isotropic linear fluid; cf. Eq. (16.3-6). In addition to reproducing the phenomenological form of the Boussinesq-Scriven equation, our asymptotic approach also yielded expressions for the phenomenological interfacial viscosity coefficients appearing therein in terms of comparable (strongly spatially inhomogeneous) microscale phenomenological viscosity coefficients existing within the interfacial transition region. Thereby, expressions were obtained for the following interfacial properties in terms of quadratures of the microscale data: (i) interfacial tension, (16.3-3); (ii) interfacial shear viscosity, (16.3-21); and (iii) interfacial dilatational viscosity [cf. Eqs. (16.3-19) and (16.3-25).

Micro- and macro-approaches to interfacial hydrodynamics have also been confirmed to agree in the macroscale limit by focusing upon a simple illustrative problem pertaining to the expansion of a spherical gas bubble within a viscous Newtonian liquid.

Appendix 16.A

The Mass-Average Velocity of a Nonmaterial Interface

The mass-average velocity $\mathbf{V}$, say, at a point $\mathbf{x}_s$ of a *nonmaterial* interface will be *defined* as the velocity vector

$$\mathbf{V} \overset{\text{def}}{=} \frac{1}{\rho^s}(\rho \mathbf{v})^s, \qquad (16.\text{A-}1)$$

wherein the surface-excess terms appearing on the right-hand side of the above are defined in Eqs. (16.1-2) and (16.1-3). Below, we derive the explicit representation (16.1-5) set forth for $\mathbf{V}$.

Consider the microscale continuity equation

$$\frac{\delta \rho}{\delta t} + (\mathbf{v} - \mathbf{u}_n^*) \cdot \nabla \rho + (\nabla \cdot \mathbf{v})\rho = 0, \qquad (16.\text{A-}2)$$

which employs the parent coordinate-convected time-derivative (15.4-7). As a matter of focus, within the interfacial transition region, the above equation may be written alternatively in the form

$$(\mathbf{v} - \mathbf{u}_n^*) \cdot \mathbf{n}^* \frac{\partial \rho}{\partial \tilde{n}} + \rho\, \mathbf{n}^* \cdot \frac{\partial \mathbf{v}}{\partial \tilde{n}} + \varepsilon L\left(\frac{\delta \rho}{\delta t} + \rho \nabla_s^* \cdot \mathbf{v}\right) = 0.$$

(16. A- 3)

The mass density ρ will be assumed to vary sharply within the interfacial region relative to the normal coordinate n, such that

$$\frac{1}{\bar{\rho}} \frac{\partial \rho}{\partial \tilde{n}} \equiv O(1), \qquad (16.\text{A-}4)$$

and relatively weakly in transverse directions, such that

$$\frac{L}{\bar{\rho}} \nabla_s^* \rho \equiv O(1).$$

In addition, we shall limit ourselves to circumstances such that

$$\frac{L}{U} \nabla_s^* \mathbf{v} = O(1), \qquad (16.\text{A-}5)$$

which encompass all conceivable nonpathological *physical* situations. As a consequence of these assumptions, it follows from Eq. (16.A-3) that within the interfacial transition region,

$$\frac{1}{U} \frac{\partial \mathbf{v}}{\partial \tilde{n}} \equiv O(1). \qquad (16.\text{A-}6)$$

With use of Eq. (15.5-33),

$$\frac{\partial \mathbf{n}^*}{\partial \tilde{n}} = \varepsilon \frac{\partial \mathbf{n}^*}{\partial n} \equiv O(\varepsilon), \qquad (16.\text{A-}7)$$

whence Eq. (16.A-3) yields the scaling

$$\frac{\partial}{\partial \tilde{n}}[\rho \mathbf{n}^* \cdot (\mathbf{v} - \mathbf{u}^*)] \equiv O(\varepsilon). \qquad (16.A\text{-}8)$$

Integration of the latter expression produces the relation

$$\rho \mathbf{n}^* \cdot (\mathbf{v} - \mathbf{u}^*) = c(\mathbf{x}_s) + O(\varepsilon), \qquad (16.A\text{-}9)$$

valid within the interfacial transition region. Here, $c(\mathbf{x}_s)$ is a function which is independent of n. Form the limit of the above equation as $\varepsilon \to 0$ with $\tilde{n}$ fixed to obtain

$$\tilde{\rho}(\mathbf{x}_s, \tilde{n})\mathbf{n} \cdot [\tilde{\mathbf{v}}(\mathbf{x}_s, \tilde{n}) - \mathbf{u}(\mathbf{x}_s)] = c(\mathbf{x}_s), \quad (16.A\text{-}10)$$

in which Eq. (15.5-27) has been employed. Matching conditions (15.5-21) and (15.5-22) permit us to conclude from the above that

$$\bar{\rho}_\alpha(\mathbf{x}_s, 0)\mathbf{n} \cdot [\bar{\mathbf{v}}_\alpha(\mathbf{x}_s, 0) - \mathbf{u}(\mathbf{x}_s)] = c(\mathbf{x}_s) \qquad (\alpha = \text{I, II}).$$

$$(16.A\text{-}11)$$

Subtraction of the latter from Eq. (16.A-10) thereby gives

$$\tilde{\rho}(\tilde{n})\mathbf{n} \cdot \tilde{\mathbf{v}}(\tilde{n}) = \bar{\rho}(0)\mathbf{n} \cdot \bar{\mathbf{v}}(0) + [\tilde{\rho}(\tilde{n}) - \bar{\rho}(0)]\mathbf{n} \cdot \mathbf{u}, \qquad (16.A\text{-}12)$$

in which the explicit dependence of the indicated arguments upon $\mathbf{x}_s$ has been suppressed.

In addition to the kinematical assumption (16.A-5), we will further suppose that the transverse velocity field,

$$\mathbf{v}_s \equiv \mathbf{I}_s^* \cdot \mathbf{v} \qquad (16.A\text{-}13)$$

is a macroscale quantity, thus obeying the scaling relation

$$\frac{1}{U}\frac{\partial \mathbf{v}_s}{\partial \tilde{n}} = O(1). \qquad (16.A\text{-}14)$$

The above, together with Eqs. (15.5-23)–(15.5-27), enables the following assertion:

$$\begin{aligned}\tilde{\mathbf{v}}_s(\mathbf{x}_s, \tilde{n}) &= \mathbf{v}_s(\mathbf{x}_s, 0) \\ &= \lim_{\bar{n} \to 0+} \bar{\mathbf{v}}_{s_{\text{I}}}(\mathbf{x}_s, \bar{n}) \\ &= \lim_{\bar{n} \to 0-} \bar{\mathbf{v}}_{s_{\text{II}}}(\mathbf{x}_s, \bar{n}). \qquad (16.A\text{-}15)\end{aligned}$$

Upon utilizing Eqs. (16.A-15), (16.1-6), (16.A-12) and (15.5-27) together with Eq. (16.A-7), the surface-excess linear momentum density expression (16.1-3) may finally be written as

$$(\rho \mathbf{v})^s = \varepsilon L \int_{\tilde{n}=-\infty}^{+\infty} [\tilde{\rho}(\tilde{n})\mathbf{n} \cdot \tilde{\mathbf{v}}(\tilde{n}) - \bar{\rho}(0)\mathbf{n} \cdot \bar{\mathbf{v}}(0)] d\tilde{n}$$

$$+ \varepsilon L \int_{\tilde{n}=-\infty}^{+\infty} [\tilde{\rho}(\tilde{n})\tilde{\mathbf{v}}_s(\tilde{n}) - \bar{\rho}(0)\bar{\mathbf{v}}_s(0)] d\tilde{n}$$

$$\equiv \rho^s \mathbf{u} \cdot \mathbf{n}\mathbf{n} + \rho^s \mathbf{v}_s. \qquad (16.\text{A-}16)$$

According to the definition of $\mathbf{V}$ provided by Eq. (16.A-1), the above relation yields

$$\mathbf{V} \equiv \mathbf{u} \cdot \mathbf{n}\mathbf{n} + \mathbf{v}_s. \qquad (16.\text{A-}17)$$

Q.E.D.

Questions for Chapter 16

16.1 *Derivation of the surface-excess internal energy equation*

i. Show that the surface-excess total energy equation is given by the following expression for a material interface [i.e. one satisfying the kinematical condition (16.1-9)] whose *microscale*, three-dimensional pressure tensor is symmetric [e.g. the microfluid tensor considered in §16.3; cf. Eqs. (16.3-1) and (16.3-6)]:

$$\frac{\partial U^s}{\partial t} + \frac{1}{2}\frac{\partial}{\partial t}(\rho^s \mathbf{v}^o \cdot \mathbf{v}^o)$$
$$+ \nabla_s \cdot \left(\mathbf{v}^o U^s + \tfrac{1}{2}\rho^s \mathbf{v}^o \cdot \mathbf{v}^o + \mathbf{q}^s - \mathbf{P}^s \cdot \mathbf{v}^o\right)$$
$$- \mathbf{F}^s \cdot \mathbf{v}^o + \mathbf{n} \cdot \left\| \bar{\mathbf{q}} - \bar{\mathbf{P}} \cdot \bar{\mathbf{v}} \right\| = 0. \qquad (\text{a})$$

For this purpose, choose $\psi \equiv U + (1/2)\rho\mathbf{v}\cdot\mathbf{v}$ in Eq. (15.7-20), employ the assignations of Table 3.5-1 appropriate to this choice, and use the following equivalences, each of which are applicable to the case of a material fluid interface (whose precursor microscale, three-dimensional stress tensor is symmetric):

$$(\rho \mathbf{v} \cdot \mathbf{v})^s \equiv \rho^s \mathbf{v}^o \cdot \mathbf{v}^o,$$
$$(\mathbf{v} \cdot \mathbf{P})^s = (\mathbf{P} \cdot \mathbf{v})^s \equiv \mathbf{P}^s \cdot \mathbf{v}^o,$$
$$(\mathbf{F} \cdot \mathbf{v})^s \equiv \mathbf{F}^s \cdot \mathbf{v}^o.$$

ii. Show that the surface-excess mechanical energy equation for such a material interface is given by the expression

$$\frac{1}{2}\frac{\partial}{\partial t}(\rho^s \mathbf{v}^o \cdot \mathbf{v}^o) + \nabla_s \cdot \left(\tfrac{1}{2}\rho^s \mathbf{v}^o \cdot \mathbf{v}^o - \mathbf{P}^s \cdot \mathbf{v}^o\right)$$
$$+ \mathbf{P}^s : \nabla_s \mathbf{v}^o - \mathbf{F}^s \cdot \mathbf{v}^o - \mathbf{n} \cdot \left\| \bar{\mathbf{P}} \cdot \bar{\mathbf{v}} \right\| = 0.$$
$$(\text{b})$$

For this purpose, dot multiply Eq. (16.1-15) [after noting the kinematical simplification (16.1-9)] with the material interfacial velocity $\mathbf{v}^o$, and employ the identity

$$(\nabla_s \cdot \mathbf{P}^s) \cdot \mathbf{v}^o \equiv \nabla_s \cdot (\mathbf{P}^s \cdot \mathbf{v}^o) - \mathbf{P}^s : \nabla_s \mathbf{v}^o .$$

iii. Subtract Eq. (b) from Eq. (a) and use the definition (3.4-11), to obtain the expression

$$\frac{D_s U^s}{Dt} + U^s \nabla_s \cdot \mathbf{v}^o + \nabla_s \cdot \mathbf{q}^s - \mathbf{P}^s : \nabla_s \mathbf{v}^o = -\mathbf{n} \cdot \|\bar{\mathbf{q}}\| , \qquad \text{(c)}$$

which constitutes the surface-excess internal energy equation for a material interface. This constitutes the internal energy portion of the *total* energy equation (a).

16.2 *Derivation of the surface-excess entropy equation*

i. Derive the surface-excess entropy equation for a material interface (16.1-9) by choosing $\psi \equiv S$ in Eq. (15.7-20). In the derivation, employ the assignations of Table 3.5-1 appropriate to this choice and use Eq. (3.4-11) to thus obtain

$$\rho^s \frac{D_s S^s}{Dt} + S^s \nabla_s \cdot \mathbf{v}^o + \frac{1}{T} \nabla_s \cdot \mathbf{q}^s - \frac{1}{T^2} \mathbf{q}^s \cdot \nabla_s T - \Phi^s = -\mathbf{n} \cdot \left\| \frac{1}{T} \bar{\mathbf{q}} \right\| . \qquad \text{(a)}$$

16.3 *Derivation of the surface-excess entropy inequality*

i. Utilize the definitions

$$\hat{U}^s \overset{\text{def}}{=} \frac{1}{\rho^s} U^s ,$$

$$\hat{S}^s \overset{\text{def}}{=} \frac{1}{\rho^s} S^s ,$$

to show [by employing Eq. (16.1-8) together with Eq. (16.1-9), Eq. (c) of Question 16.1 and Eq. (a) of Question 16.2] that the following alternative forms of the surface-excess internal energy and entropy equations obtain:

$$\rho^s \frac{D_s \hat{U}^s}{Dt} + \nabla_s \cdot \mathbf{q}^s - \mathbf{P}^s : \nabla_s \mathbf{v}^o = -\mathbf{n} \cdot \|\bar{\mathbf{q}}\| , \qquad \text{(a)}$$

$$\rho^s \frac{D_s \hat{S}^s}{Dt} + \frac{1}{T} \nabla_s \cdot \mathbf{q}^s - \frac{1}{T^2} \mathbf{q}^s \cdot \nabla_s T - \Phi^s = -\mathbf{n} \cdot \left\| \frac{1}{T} \bar{\mathbf{q}} \right\| . \qquad \text{(b)}$$

ii. The surface thermodynamic relation (Adamson 1982)

$$d\hat{U}^s = Td\hat{S}^s + \sigma d\hat{A},$$

where $\hat{A} \equiv 1/\rho^s$ denotes the specific surface (of a material interface), may be used with Eq. (16.1-8) to obtain the relation (valid strictly for material interfaces)

$$\rho^s \frac{D_s\hat{U}^s}{Dt} = T\frac{D_s\hat{S}^s}{Dt} + \sigma \mathbf{I}_s : \nabla_s \mathbf{v}^o .$$

Substitute the above into Eq. (b), and subtract the result from (a) to show that

$$\Phi^s \equiv \frac{1}{T}\boldsymbol{\tau}^s : \nabla_s \mathbf{v}^o - \frac{1}{T^2}\mathbf{q}^s \cdot \nabla_s T \geq 0 \qquad \text{(c)}$$

is the surface-excess dissipation function. [Note Eq. (16.3-2).]

iii. For isothermal systems, show for the case of a pure shear field, namely

$$\mathbf{D}_s : \mathbf{I}_s = 0,$$

that Eq. (c) above requires satisfaction of the following inequalities imposed upon the interfacial shear viscosity μ^s of a Newtonian interface, Eq. (16.3-24):

$$\mu^s \geq 0.$$

iv. Again, for isothermal systems, show for the case of a pure dilatation, namely

$$\mathbf{D}_s = D_s \mathbf{I}_s,$$

that Eq. (c) above requires satisfaction of the inequality

$$\kappa^s \geq 0.$$

imposed upon the interfacial dilatational viscosity κ^s of a Newtonian interface, Eq. (16.3-24).

16.4 Utilizing Eqs. (16.4-3) and (15.3-11), derive Eq. (16.4-4) for the scale factor M^* appropriate to familial spherical coordinate surfaces.

16.5 *Curvature dependence of interfacial tension*

i. Choose Q=0 in Eqs. (16.4-5) and (16.4-6), and let

$$F_r \equiv -\frac{\partial \phi}{\partial r}$$

represent the (conservative) interfacial body force density and

$$p_t \equiv p + \phi$$

the tangential pressure component: show that, in place of Eq. (16.4-7), there now obtains the expression

$$\|\bar{p}\| = \frac{2}{a} \lim_{\varepsilon \to 0} \varepsilon L \int_{-\frac{n_2}{\varepsilon L}}^{\frac{n_1}{\varepsilon L}} (\bar{p} - p_t)\left(1 + \varepsilon L \frac{\tilde{n}}{a}\right) d\tilde{n} .$$

Despite the fact that the dimensionless term involving n/a in the above integral is asymptotically small relative to unity, it is maintained in most classical theories of capillarity (see Gibbs 1957, Buff 1956). [Recall, as demonstrated in Question 15.6 (see also §15.2), that the surface-excess quadrature appearing on the right-hand side of the above equation is independent of the nonphysical linear dimension L *only* in the asymptotic limit $\varepsilon \equiv l/L \to 0$, with the curvature radius a satisfying the scaling $a/L = O(1)$.]

ii. Noting that for spherical geometry $H = -1/a$ and $K = 1/a^2$, and upon maintaining the asymptotically small term in the above integral, demonstrate that the preceding relation may be expressed alternatively as [see Eq. (22) of Buff (1956)]

$$\|\bar{p}\| = -2H\gamma - (4H^2 - K)\gamma_1,$$

wherein

$$\gamma \overset{\text{def}}{=} \lim_{\varepsilon \to 0} \varepsilon L \int_{-\frac{n_2}{\varepsilon L}}^{\frac{n_1}{\varepsilon L}} (\bar{p} - p_t)(1 - \varepsilon L 2H\tilde{n})\, d\tilde{n},$$

$$\gamma_1 \overset{\text{def}}{=} \lim_{\varepsilon \to 0} \varepsilon L \int_{-\frac{n_2}{\varepsilon L}}^{\frac{n_1}{\varepsilon L}} (\bar{p} - p_t)\varepsilon L \tilde{n}\, d\tilde{n},$$

corresponding respectively to Buff's (1956) definition of interfacial tension, and the *bending stress* (Murphy 1966). (Observe that in the asymptotic limit $\varepsilon \to 0$ these respectively reduce to their classical forms both $\gamma \to \sigma$ [cf. Eq. (16.3-3)] and $\gamma_1 \to 0$.)

iii. Show from the above that

$$\frac{\partial \gamma}{\partial(-2H)} = \gamma_1,$$

which provides an expression for the 'curvature dependence' of the interfacial tension. Explain why there is no curvature dependence of interfacial tension according to the asymptotic theory outlined in this text.

*Additional Reading for Chapter 16**

§16.1 Conservation of Linear Momentum

Scriven, L.E. 1960 Dynamics of a fluid interface. Equations of motion for Newtonian surface fluids. *Chem. Engng Sci.* **12**, 98–108.

---

* See the footnote accompanying the ***Additional Reading for Chapter 15***.

Slattery, J.C. 1964 Surfaces–I. Momentum and moment of momentum balances for moving surfaces. *Chem. Engng Sci.* **19**, 379–385.

§16.3 Constitutive Law for the Surface-Excess Pressure Tensor

Deemer, A.R. & Slattery, J.C. 1978 Balance equations and structural models for phase interfaces. *Int. J. Multiphase Flow* **4**, 171–192.

Aubert, J.H. 1983 Interfacial properties of dilute polymer solutions. *J. Colloid Interface Sci.* **96**, 135–149.

17

"'Would your Lordship indicate or explain to me in what direction is the Third Dimension, unknown to me?'

"'I come from it . . . It is above and down below.'

"'My Lord means seemingly that it is Northward and Southward.'

"'I mean nothing of the kind. I mean a direction in which you cannot look, because you have no eye in your side.'

". . . 'An eye in my side! . . . Your Lordship jests.'

" . . . 'now prepare to receive proof positive of the truth of my assertions. You cannot indeed see more than one of my sections or Circles, at a time; for you have no power to raise your eye out of the plane of Flatland; but you can at least see that, as I rise in Space, so my sections become smaller. See now, I will rise . . .'

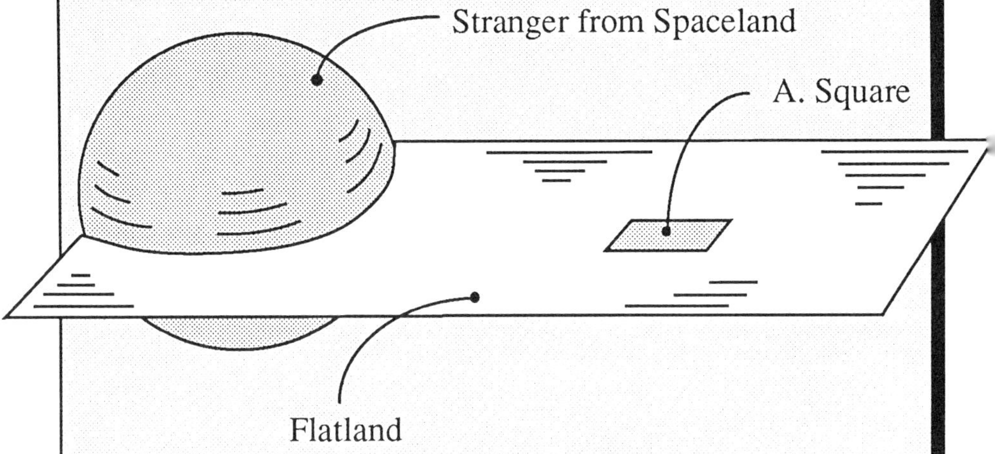

"'Fool! Madman! Irregular!,' I exclaimed.

"'Ha! Is it come to this?' thundered the Stranger: 'then meet your fate: out of your Plane you go. Once twice, thrice!

"'Tis done!'"

Edwin A. Abbott (1885)

CHAPTER 17

Surface-Excess Transport of Species

The generic surface-excess transport equation (15.7-20) is applied in this chapter to the interfacial transport of chemical species, particularly surfactant species. [This macroscale equation provides the generalization of Eq. (5.2-2) to *nonmaterial* fluid interfaces.] Concomitantly, integral formulations are provided for the surface-excess, macroscale fields appearing therein in terms of integrations of comparable *microscale* fields over the interfacial transition region. These surface-excess field formulations permit the Fickian constitutive *assumption* (5.2-3) for the surface-excess species flux-density vector $\mathbf{j}_i^s$ to be rationally derived from a detailed micromechanical model of the three-dimensional interfacial region. The latter model is based, *inter alia*, upon the plausible assumption that Fick's law applies at the microscale—albeit in an anisotropic form. By assuming that local (microscale) equilibrium prevails, explicit linear macroscale constitutive relations are also derived for adsorption and bulk-phase partitioning phenomena, including explicit expressions for the macroscale phenomenological coefficients appearing in these constitutive equations in terms of quadratures of the specified *microscale* data over the transition region. These adsorption formula serve to provide a quantitative physical interpretation of the coefficients appearing in Langmuir-type adsorption-isotherm models [cf. Eq. (5.3-2)], whereas the bulk-phase partition formula derived furnishes a direct interpretation of Eq. (5.2-1).

The final section of this chapter provides a summary of interfacial transport conditions for both material and nonmaterial interfaces. In turn, this compilation serves to review the general macroscale boundary conditions derived in Parts I and II. Also summarized are the surface-excess microscale quadrature formulations derived for the interfacial phenomenological coefficients appearing in these equations.

17.1 Conservation of Species

Upon making reference to the explicit physical identifications made in Table 3.5-1 for the continuous three-dimensional fields pertinent to species transport, and subsequently employing the surface-excess definitions (15.6-18), (15.6-36) and (15.7-19), the generic interfacial conservation equation (15.7-20) applied to the species mass density ($\psi \equiv \rho_i$) field adopts the form

$$\frac{\partial \rho_i^s}{\partial t} - 2Hu_n \rho_i^s + \nabla_s \cdot \left[(\rho_i \mathbf{v})^s + \mathbf{I}_s \cdot \mathbf{j}_i^s \right] - R^s$$
$$= -\mathbf{n} \cdot \left\| (\bar{\mathbf{v}} - \mathbf{u}) \bar{\rho}_i + \bar{\mathbf{j}}_i \right\|, \quad (17.1\text{-}1)$$

in which the following surface-excess densities appear:

Surface-excess species nonconvective flux vector density [cf. Eq. (15.6-36)*]:

$$\boxed{\mathbf{j}_i^s(\mathbf{x}_s) \stackrel{\text{def}}{=} \varepsilon L \int_{-\infty}^{+\infty} \mathbf{I}_s \cdot [\tilde{\mathbf{j}}_i(\mathbf{x}_s, \tilde{n}) - \bar{\mathbf{j}}_i(\mathbf{x}_s, 0)] d\tilde{n};} \quad (17.1\text{-}2)$$

Surface-excess species convective flux vector density:

$$(\rho_i \mathbf{v})^s (\mathbf{x}_s) \stackrel{\text{def}}{=} \varepsilon L \int_{-\infty}^{+\infty} [\tilde{\rho}_i \tilde{\mathbf{v}}(\mathbf{x}_s, \tilde{n}) - \bar{\rho}_i \bar{\mathbf{v}}(\mathbf{x}_s, 0)] d\tilde{n}; \quad (17.1\text{-}3)$$

Surface-excess reaction rate:

$$\boxed{R_i^s(\mathbf{x}_s) \stackrel{\text{def}}{=} \varepsilon L \int_{-\infty}^{+\infty} \left[\tilde{R}_i(\mathbf{x}_s, \tilde{n}) - \bar{R}_i(\mathbf{x}_s, 0) \right] d\tilde{n};} \quad (17.1\text{-}4)$$

Surface-excess species mass density:

$$\boxed{\rho_i^s(\mathbf{x}_s) \stackrel{\text{def}}{=} \varepsilon L \int_{-\infty}^{+\infty} [\tilde{\rho}_i(\mathbf{x}_s, \tilde{n}) - \bar{\rho}_i(\mathbf{x}_s, 0)] d\tilde{n}.} \quad (17.1\text{-}5)$$

* Note that $\mathbf{j}$ and $\mathbf{j}_i$ have distinctly different physical meanings. The former symbol (which appears *only* in Chapter 15) refers to a net areal flux vector or tensor field of some generic property $\mathcal{P}$, whereas the latter denotes specifically the species flux density vector, as measured relative to the mass-average velocity; that is, $\mathbf{j}_i$ refers to a *diffusive* flux of species i [see the footnote immediately preceding Eq. (5.1-2)].

In Eq. (17.1-1), employ Eq. (16.1-4) (upon replacing the mass density ρ with the species density ρ_i), and observe that

$$\mathbf{I}_s \cdot \mathbf{j}_i^s = \mathbf{j}_i^s,$$

to obtain

$$\boxed{\frac{\partial \rho_i^s}{\partial t} - 2Hu_n \rho_i^s + \nabla_s \cdot \left(\rho_i^s \mathbf{v}_s + \mathbf{j}_i^s\right) - R_i^s = -\mathbf{n} \cdot \left\| (\bar{\mathbf{v}} - \mathbf{u})\bar{\rho}_i + \bar{\mathbf{j}}_i \right\|.}$$

(17.1- 6)

This represents the surface-excess species transport equation for a nonmaterial interface. When Eq. (16.1-14) holds, this relation reduces to Eq. (5.2-2), applicable to material interfaces. Furthermore note upon summing Eq. (17.1-6) over all components i present at the macroscale interface, that the surface-excess mass balance equation (16.1-1) is recovered.

17.2 Derivation of the Surfactant Adsorption Isotherm, the Bulk-Phase Partitioning Relation, and the Constitutive Equation Governing the Surface-Excess Species Flux Vector

Equation (17.1-6) supplies the general jump boundary condition relating the respective normal components $\mathbf{n} \cdot \bar{\mathbf{j}}_i^{\alpha}$ $(\alpha = 1,2)$ of the macroscale species flux vector $\bar{\mathbf{j}}_i^{\alpha}$ on either side of a nonmaterial fluid interface. In order to apply this equation to any specific physical system, macroscale constitutive equations for R_i^s and $\mathbf{j}_i^s$ are needed. Such equations may be developed from the defining relations (17.1-4) and (17.1-2), once microscale constitutive models for the comparable three-dimensional quantities R_i and $\mathbf{j}_i$ are adopted. In addition, appropriate adsorption isotherms, e.g. Eqs. (5.3-1) and (5.3-2), relating bulk and surface-excess surfactant densities are required. [These also jointly provide the corresponding bulk-phase partitioning relation, Eq. (5.2-1)].

A macroscale constitutive equation for $\mathbf{j}_A^s$ is derived in the following analysis for an inert, nonreactive surfactant species A. The precursor microscale constitutive equation for $\mathbf{j}_A$ will be assumed to possess a transversely isotropic Fick's law form [cf. Eqs. (17.2-16) and (17.2-17)] for the diffusive contribution. In addition, a conservative microscale physicochemical 'adsorption' force $\mathbf{F}^{\phi}(\mathbf{x})$ is imagined to act upon the surfactant molecules in proximity to the interface. Ultimately [cf. Eq.

(17.2-33)], the interfacial Fick's law relation (5.2-3) is found to result, however, now with a microscale interpretation [cf. Eq. (17.2-16)] of the surface diffusion coefficient D^s appearing therein.

Our analysis will focus upon the case of *diffusion-controlled* surfactant transport at an interface (cf. §5.4). Thus, local microscale equilibrium is assumed to exist in the vicinity of the interface, leading eventually to the deduction of a linear, macroscale, equilibrium, adsorption isotherm of the Henry's-law type [cf. Eq. (5.3-2)], together with a microscale interpretation [cf. Eq. (17.2-10b)] of the Henry's law coefficient K_a appearing therein.

We begin with a consideration of the nature of microscale body forces acting upon surfactant species within the interfacial region.

Macro-Surface-Excess and Micro-Adsorptive Forces at an Interface

The volumetric microscale body-force density $\mathbf{F}(\mathbf{x})$ acting within the interfacial region and surrounding bulk fluid regions may generally be decomposed into the sum

$$\mathbf{F}(\mathbf{x}) = \mathbf{F}^{\phi}(\mathbf{x}) + \mathbf{F}^{e}(\mathbf{x}), \qquad (17.2\text{-}1)$$

where $\mathbf{F}^{\phi}(\mathbf{x})$ denotes an (internal, 'short range')* physicochemical body force which acts preferentially upon surfactant species in the neighborhood of the interface due to the amphoteric nature of such species, whereas $\mathbf{F}^{e}(\mathbf{x})$ denotes the (external, 'long range') body-force density, if any, (e.g. deriving from a gravitational or electromagnetic potential) acting upon all species, including the solvent.

Equivalently, the body-force per unit volume may be expressed in the form

$$\mathbf{F}(\mathbf{x}) = \rho\hat{\mathbf{F}} \equiv \sum_{i=1}^{N} \rho_i \hat{\mathbf{F}}_i\,, \qquad (17.2\text{-}2)$$

where $\hat{\mathbf{F}}$ denotes the body-force per unit fluid mass, and $\hat{\mathbf{F}}_i$ denotes the comparable body-force density acting upon the ith species in an N-component solution. Restricting our attention to the case of a *binary* mixture composed of a surfactant A, say, within a solvent, for which (internal and external) body forces act only upon the surfactant molecules, it follows from the preceding relation that

$$\mathbf{F}(\mathbf{x}) = \rho_A \hat{\mathbf{F}}_A. \qquad (17.2\text{-}3)$$

---

* By the phrase "internal" is meant, *inter alia*, that

$$\int_{-L_{II}}^{L_I} \mathbf{F}^{\phi}(\mathbf{x}_s, n)dn = 0,$$

that is, the *net* force exerted on the fluid is identically zero.

This latter relation is assumed to hold within the interfacial region, $\tilde{n} \equiv O(1)$, and as well from the interface, namely $\bar{n} \equiv O(1)$.

According to Eq. (17.2-1), the surfactant-specific body-force density $\hat{\mathbf{F}}_A$ may be expressed as the sum

$$\hat{\mathbf{F}}_A(\mathbf{x}) = \hat{\mathbf{F}}_A^{\phi}(\mathbf{x}) + \hat{\mathbf{F}}_A^{e}(\mathbf{x}). \tag{17.2-4}$$

Owing to the fact that the physicochemical adsorptive force $\hat{\mathbf{F}}_A^{\phi}$ vanishes far from the interface, it follows from Eqs. (15.5-12) and (15.5-13) that in the outer bulk fluid domains,

$$\left(\hat{\mathbf{F}}_A^{\phi}\right)_{\mathrm{I}} = \mathbf{0}, \qquad \left(\hat{\mathbf{F}}_A^{\phi}\right)_{\mathrm{II}} = \mathbf{0}. \tag{17.2-5}$$

It will further be supposed that the external force contribution to the total force $\hat{\mathbf{F}}_A^{e}$ exhibits no sharp spatial gradients within the interfacial transition zone, explicitly,*

$$\frac{\partial \hat{\mathbf{F}}_A^{e}}{\partial \bar{n}} = O(1) \tag{17.2-6}$$

everywhere. It follows from the latter together with Eqs. (15.5-23)–(15.5-27) and Eq. (17.2-5) that the inner and outer limits of $\hat{\mathbf{F}}_A^{e}$ are, simultaneously, asymptotically identical, and equivalent to the total (macroscale) body force $\hat{\mathbf{F}}_A(\mathbf{x}_s, 0)$ acting upon the surfactant species at the interface; explicitly,

$$\begin{aligned} \hat{\mathbf{F}}_A^{e}(\mathbf{x}_s, n) &= \lim_{\bar{n} \to 0+} \overline{\left(\hat{\mathbf{F}}_A\right)_{\mathrm{I}}} \\ &= \lim_{\bar{n} \to 0-} \overline{\left(\hat{\mathbf{F}}_A\right)_{\mathrm{II}}} \\ &\equiv \hat{\mathbf{F}}_A(\mathbf{x}_s, 0) \qquad \text{for} \quad n = O(l). \end{aligned} \tag{17.2-7}$$

Upon employing the quadrature formula (16.1-14), it follows from the above that

* The parental surface-fixed coordinate system described in §15.3 is employed in this and the following sections.

$$\mathbf{F}^s = \varepsilon L \int_{-\infty}^{\infty} (\tilde{\mathbf{F}} - \overline{\mathbf{F}}) d\tilde{n},$$

$$= \varepsilon L \int_{-\infty}^{\infty} [\widetilde{(\rho_A \hat{\mathbf{F}}_A^{\phi})} - \overline{(\rho_A \hat{\mathbf{F}}_A^{\phi})}] d\tilde{n} + \varepsilon L \int_{-\infty}^{\infty} [\widetilde{(\rho_A \hat{\mathbf{F}}_A^{e})} - \overline{(\rho_A \hat{\mathbf{F}}_A^{e})}] d\tilde{n} \quad (17.2\text{-}8)$$

The first integral on the right-hand side of this equation vanishes owing to its nature as an *internal* body force [cf. footnote pertaining to Eq. (17.2-1)]. Hence, upon utilizing Eqs. (17.2-7) and (16.1-2), we obtain

$$\boxed{\mathbf{F}^s = \rho_A^s \overline{\hat{\mathbf{F}}_A}(\mathbf{x}_s, 0),} \quad (17.2\text{-}9)$$

as an explicit constitutive relation for the surface-excess body-force density vector in terms of the surface-excess mass density ρ_A^s of surfactant and the macroscale body force $\overline{\hat{\mathbf{F}}}_A(\mathbf{x}_s, 0)$ acting upon surfactant A at the interface.

Whereas the physicochemical force $\hat{\mathbf{F}}_A^{\phi}$ acting upon surfactant A within the interfacial region does not explicitly contribute to the surface-excess areal force density vector (17.2-9), it nevertheless plays a fundamental *implicit* role by strongly influencing surfactant transport within the interfacial region. This may be seen by consideration of the following expression for the force vector $\hat{\mathbf{F}}_A^{\phi}$ in terms of the dimensionless physicochemical potential $E(\mathbf{x}_s, n)$:

$$\hat{\mathbf{F}}_A^{\phi} = -\frac{RT}{M_A} \nabla E \,. \quad (17.2\text{-}10)$$

Here, R is the gas constant, M_A the molecular weight of surfactant A, and T the absolute temperature. The main features of this force are that: (i) it is nonzero only in the interfacial transition region, $n = O(\varepsilon)$ [cf. Eq. (17.2-5)]; and (ii) it is an *attractive* force, tending to cause surfactant molecules to accumulate in the vicinity of the macroscale interface, $\bar{n} = 0$. In general, E possesses a deep potential well within the interfacial transition region (Figure 17.2-1). This well accounts for the strong adsorption of the surfactant at the interface and, consequently, the existence of an experimentally observable (i.e. nonvanishing) value for the surface-excess density ρ_A^s. Far away from the interface, the adsorptive force vanishes, as in Eq. (17.2-5). Thus, the potential energy function E possesses the macroscale asymptotic form

$$\bar{E} = \begin{cases} \bar{E}_{\mathrm{I}} & \text{for } \bar{n} > 0, \\ \bar{E}_{\mathrm{II}} & \text{for } \bar{n} < 0, \end{cases} \quad (17.2\text{-}11)$$

where $\bar{E}_{\mathrm{I}}$ and $\bar{E}_{\mathrm{II}}$ are constants independent of both $\mathbf{x}_s$ and n.

Owing to the sharp variation of E in the normal direction within the interfacial region, the potential possesses the scaling

$$L\mathbf{n}^* \cdot \nabla E = O\left(\frac{1}{\varepsilon}\right), \tag{17.2-12}$$

whereas, in contrast, *transverse* variations of E within the interfacial region will be assumed to be of a macroscale magnitude; whence

$$L\nabla_s^* E = O(1). \tag{17.2-13}$$

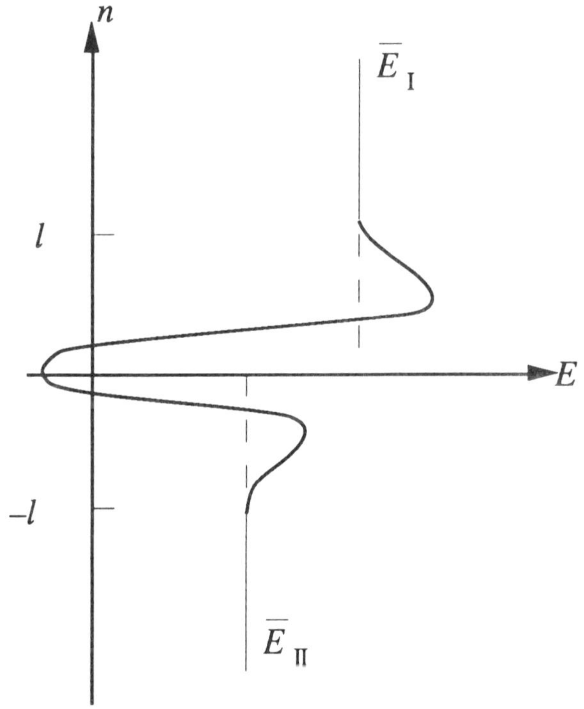

Figure 17.2-1 Variation in the adsorption potential E with respect to the normal distance n from the parent coordinate surface. This qualitative sketch displays the deep potential well within the interfacial transition region, as well as the possible energy barriers on either side this region.

Thus, upon rewriting Eq. (17.2-10) in terms of its normal and tangential components,

$$\hat{\mathbf{F}}_A^{\phi} = -\frac{RT}{M_A}\left[\frac{1}{\varepsilon L}\mathbf{n}^* \frac{\partial E}{\partial \tilde{n}} + \nabla_s^* E\right]. \tag{17.2-14}$$

It follows from Eqs. (17.2-12) and (17.2-13) that to leading order within the inner interfacial transition zone,

$$\hat{\mathbf{F}}_A \approx -\frac{RT}{M_A}\mathbf{n}^* \cdot \nabla E \,. \qquad (17.2\text{-}15)$$

Diffusion-Controlled Interfacial Surfactant Transport

Consider the transport of a nonreactive (R_A=0) surfactant species A whose microscale transport is governed by the constitutive equations given in the following three paragraphs.

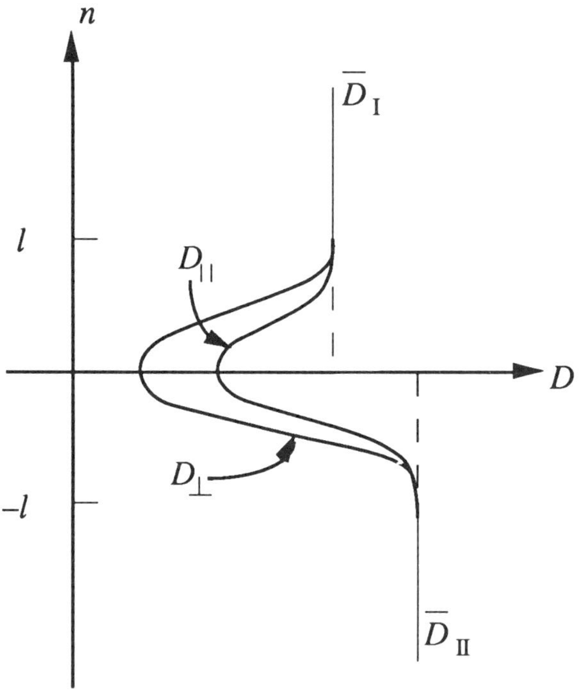

Figure 17.2-2 Variation in the normal and tangential microscale diffusivity coefficients with respect to normal distance n from the parent reference surface. This qualitative sketch displays the expected diffusivity minimum within the interfacial transition region. Both parallel and perpendicular diffusivity coefficients attain identical bulk-fluid values at sufficiently large distances from the interface;.

The microscale constitutive equation for the nonconvective flux of A, i.e. the flux relative to the mass-average velocity **v** [see the footnote preceding Eq. (5.1-2) for comments pertaining to this nonconvective flux, particularly noting the contribution attributed to 'forced diffusion'], is assumed to be given by the expression*

* This equation results from a dilute solution approximation for the usual species mass flux constitutive equation of a constant mass density binary fluid mixture composed of a

$$\mathbf{j}_A = -\mathbf{D}\cdot[(\mathbf{F}_A M_A / RT - \nabla\rho_A] \qquad (17.2\text{-}16)$$

at all points $\mathbf{x}$ of the microfluid, including the interfacial transition region. Here, $\mathbf{D}(\mathbf{x})$ is the diffusivity dyadic.

Within the interfacial transition region, the existence of short-range, anisotropic, intermolecular forces [attributable to the presence of steep mass density and/or species concentration gradients in the normal direction to the macroscale interface; cf. Eq. (17.2-10) and the discussion thereof] is assumed to endow the microscale fluid with axial symmetry about the normal vector $\mathbf{n}^*$. Thus, $\mathbf{D}$ is assumed to possess the transversely isotropic form

$$\mathbf{D}(\mathbf{x}_s, n) = \mathbf{I}^*_s D_\perp(\mathbf{x}_s, n) + \mathbf{n}^*\mathbf{n}^* D_{||}(\mathbf{x}_s, n). \qquad (17.2\text{-}17)$$

In general, $D_\perp$ and $D_{||}$, which respectively represent the diffusivity coefficients perpendicular and parallel respectively to the transversely isotropic symmetry axis $\mathbf{n}^*$, are each expected to attain minimum values within the interfacial transition region due to the strong surfactant retarding forces existing therein (Fig. 17.2-2). Far from the interface, in the two bulk-phase regions, both diffusivity components attain the same values; namely

$$\lim_{|\tilde{n}|\to\infty} D_{||} = \lim_{|\tilde{n}|\to\infty} D_\perp \equiv \bar{D}, \qquad (17.2\text{-}18)$$

where

$$\bar{D} = \begin{cases} \bar{D}_\mathrm{I} & \text{for } \bar{n} > 0, \\ \bar{D}_\mathrm{II} & \text{for } \bar{n} < 0. \end{cases} \qquad (17.2\text{-}19)$$

Thus, far from the interface the microscale diffusivity dyadic attains the isotropic form

$$\bar{\mathbf{D}} = \mathbf{I}\bar{D}. \qquad (17.2\text{-}20)$$

It will be supposed in what follows that within the interfacial region and to terms of the lowest-order in ε, the microscale surfactant mass density $\tilde{\rho}_A$ is given by an equilibrium Boltzmann distribution (justification for which is provided at the conclusion of this section), namely

$$\tilde{\rho}_A = C(\mathbf{x}_s)\exp[-\tilde{E}(\mathbf{x}_s, \tilde{n})], \qquad (17.2\text{-}21)$$

where the normalization coefficient $C(\mathbf{x}_s)$ is independent of n. The above expression is equivalent to assuming that the surfactant rapidly attains local thermodynamic equilibrium in the normal direction. This circumstance, as

---

solvent species and a surfactant species (A). In addition, the microscale equations of this section are similar to those developed by Brenner & Leal (1982) using a micromechanical model in which the surfactant molecules are considered to be solid, Brownian spheres of much larger radii than the 'thickness' l of the interfacial transition region.

was discussed in §5.4, corresponds to the case of *diffusion-controlled* surfactant transport, wherein surfactant molecules overcome potential energy barriers to adsorption at a rate that appreciably exceeds the rate of molecular diffusion towards the interface from the bulk fluids.

It will also be assumed in what follows that E possesses a relatively weak dependence upon $\mathbf{x}_s$ compared with that of C [cf. the transition from Eq. (17.2-12) to (17.2-13)]; explicitly,

$$\frac{C\,|\nabla_s E|}{|\nabla_s C|} << 1. \qquad (17.2\text{-}22)$$

Use of the matching conditions (15.5-21) and (15.5-22), together with the outer limiting form (17.2-11) for E, yields*

$$\lim_{\bar{n}\to 0+} \bar{\rho}_A^{\mathrm{I}}(\mathbf{x}_s, \bar{n}) = C(\mathbf{x}_s)\exp(-\bar{E}_{\mathrm{I}}), \qquad (17.2\text{-}23\,\mathrm{a})$$

$$\lim_{\bar{n}\to 0-} \bar{\rho}_A^{\mathrm{II}}(\mathbf{x}_s, \bar{n}) = C(\mathbf{x}_s)\exp(-\bar{E}_{\mathrm{II}}). \qquad (17.2\text{-}23\,\mathrm{b})$$

The surface-excess mass fraction for the surfactant species can now be determined. Substitution of the limiting forms (17.2-21) and (17.2-23) into (17.1-5) yields

$$\rho_A^s(\mathbf{x}_s) = \varepsilon L\, C(\mathbf{x}_s) \int_{-\infty}^{+\infty} \left\{ \exp\left[-\tilde{E}(\mathbf{x}_s, \tilde{n})\right] - \exp(-\bar{E}) \right\} d\tilde{n}\,.$$

(17.2-24)

Upon comparison with Eqs. (17.2-23), this provides a linear adsorption isotherm; explicitly,

$$\bar{\rho}_A^{\alpha} = K_a^{\alpha}\, \rho_A^s \qquad (\alpha = \mathrm{I, II}), \qquad (17.2\text{-}25)$$

in which

---

* Here, the relations $\widetilde{\exp(a)} = \exp(\tilde{a})$ and $\overline{\exp(a)} = \exp(\bar{a})$ have been employed. The validity of these relations can easily be proven. In particular, define $\alpha_1 = \ln(1 + \alpha\, /\, |\exp \tilde{a}|)$. Then $\alpha_1 > 0$ when $\alpha > 0$. Now, if $\tilde{a}$ exists, then there exists a $\delta > 0$ such that $|a(\varepsilon) - \tilde{a}| < \alpha_1$ for all $0 < \varepsilon < \delta$. Thus, $|\exp[a(\varepsilon)] - \exp \tilde{a}| < \alpha$ for all $0 < \varepsilon < \delta$, which proves the validity of the first relation. This same proof holds for the second relation as well.

$$\boxed{K_a^{\alpha} \stackrel{\text{def}}{=} \frac{\exp(-\bar{E}_{\alpha})}{\varepsilon L \int_{\tilde{n}=-\infty}^{+\infty} \left[\exp(-\tilde{E}) - \exp(-\bar{E})\right] d\tilde{n}}} \qquad (17.2\text{-}26)$$

is the (equilibrium) adsorption coefficient based upon the α-phase. Furthermore, upon division of Eq. (17.2-23a) by (17.2-23b), we obtain

$$\bar{\rho}_A^{\mathrm{I}}(0+) = K\bar{\rho}_A^{\mathrm{II}}(0-), \qquad (17.2\text{-}27)$$

with

$$\boxed{K \stackrel{\text{def}}{=} \exp\left(\bar{E}_{\mathrm{II}} - \bar{E}_{\mathrm{I}}\right),} \qquad (17.2\text{-}28)$$

the bulk-phase partition coefficient [cf. (5.2-1)].*

The appropriate form of the interfacial constitutive equation for the surface-excess mass flux $\mathbf{j}_A^s$ can be established via Eq. (17.1-2) from the above microscale model of the surfactant system as follows. Substitute Eq. (17.2-15) [cf. also Eq. (17.2-3)] into Eq. (17.2-16), thereby obtaining,

$$\mathbf{j}_A = -\mathbf{D} \cdot [\rho_A \mathbf{n}^* \cdot \nabla E - \nabla \rho_A]. \qquad (17.2\text{-}29)$$

Form the inner limit (15.5-7) of the above equation, and simplify the resulting expression using the product identity (15.5-19), together with the Boltzmann relation (17.2-21) and the inequality (17.2-22), to obtain

$$\tilde{\mathbf{j}}_A = -\tilde{\mathbf{D}} \cdot (\nabla C)\exp(-\tilde{E}). \qquad (17.2\text{-}30)$$

Similarly, application of the outer limiting operations (15.5-12) and (15.5-13), followed by subsequent simplification of the resulting expressions using Eqs. (15.5-18) and (17.2-23), and eventual formation of the limits $\bar{n} \to 0\pm$, produces the relations

$$\lim_{\bar{n} \to 0+} \bar{\mathbf{j}}_A^{\mathrm{I}} = -\bar{\mathbf{D}}_{\mathrm{I}}(0+) \cdot (\nabla C)\exp(-\bar{E}_{\mathrm{I}}), \qquad (17.2\text{-}31)$$

$$\lim_{\bar{n} \to 0-} \bar{\mathbf{j}}_A^{\mathrm{II}} = -\bar{\mathbf{D}}_{\mathrm{II}}(0-) \cdot (\nabla C)\exp(-\bar{E}_{\mathrm{II}}). \qquad (17.2\text{-}32)$$

Substitute the limiting forms (17.2-30), (17.2-31) and (17.2-32) into the integral microscale formulation (17.1-2) and subsequently simplify the

* Equations (17.2-24)–(17.2-28) may be compared with their counterparts [cf. Eqs. (15.2-4), (15.2-5), (15.2-17) and (15.2-19)], the latter presented from a phenomenological, macroscale point of view.

resulting expression using the order-of-magnitude scaling relation (17.2-22). This yields the interfacial Fick's law-type diffusion formula [cf. Eq. (5.2-3)]

$$\boxed{\mathbf{j}_A^s \equiv - D^s \nabla_s \rho_A^s} \qquad (17.2\text{-}33)$$

as the appropriate constitutive equation for the surface-excess surfactant mass flux vector, in which

$$\boxed{D^s(\mathbf{x}_s) \stackrel{\text{def}}{=} \frac{\int_{-\infty}^{+\infty} \{ \tilde{D}_{||}(\mathbf{x}_s, \tilde{n}) \exp[\tilde{E}(\mathbf{x}_s, \tilde{n})] - \bar{D}(\mathbf{x}_s, 0) \exp(-\bar{E}) \} d\tilde{n}}{\int_{-\infty}^{+\infty} \{ \exp[\tilde{E}(\mathbf{x}_s, \tilde{n})] - \exp(-\bar{E}) \} d\tilde{n}}} \qquad (17.2\text{-}34)$$

is the interfacial diffusion coefficient for the surfactant.

Validation of the Local Equilibrium Assumption, Eq. (17.2-21)

The microscale surfactant transport conservation equation [cf. Eq. (5.1-1)] may be expressed in terms of the parental surface-fixed coordinate system of §15.3 as

$$\frac{\delta \rho_A}{\delta t} + (\mathbf{v} - \mathbf{u}_n^*) \cdot \nabla \rho_A + (\nabla \cdot \mathbf{v}) \rho_A + \nabla \cdot \mathbf{j}_A = 0. \qquad (17.2\text{-}35)$$

Separation of the gradient operator into its normal and tangential components yields

$$\frac{1}{\varepsilon L}\left[(\mathbf{v} - \mathbf{u}^*) \cdot \mathbf{n}^* \frac{\partial \rho_A}{\partial \tilde{n}} + \mathbf{n}^* \cdot \left(\frac{\partial \mathbf{v}}{\partial \tilde{n}} \right) \rho_A + \mathbf{n}^* \cdot \frac{\partial \mathbf{j}_A}{\partial \tilde{n}} \right]$$
$$+ \left[\frac{\delta \rho_A}{\delta t} + (\mathbf{v} - \mathbf{u}_n^*) \cdot \nabla_s^* \rho_A + (\nabla_s^* \cdot \mathbf{v}) \rho_A + \nabla_s^* \cdot \mathbf{j}_A \right] = 0. \qquad (17.2\text{-}36)$$

Upon assuming that all terms appearing in the second bracket are $O(1)$, and subsequently applying the scaling relation (15.5-33) for $\mathbf{n}^*$ [see also the footnote below Eq. (15.7-5) regarding the appropriate scaling of $\mathbf{u}^*$] the preceding expression becomes

$$\frac{\partial}{\partial \tilde{n}}\left[\rho_A \mathbf{n}^* \cdot (\mathbf{v} - \mathbf{u}^*) + \mathbf{n}^* \cdot \mathbf{j}_A \right] = O(\varepsilon). \qquad (17.2\text{-}37)$$

Substitution of Eq. (17.2-29) into (17.2-37) yields the expression

$$\frac{D_\perp}{\varepsilon L}\left(\frac{\partial \rho_A}{\partial \tilde{n}} + \rho_A \frac{\partial E}{\partial \tilde{n}} \right) + \rho_A (\mathbf{v} - \mathbf{u}^*) \cdot \mathbf{n}^* = f(\mathbf{x}_s), \qquad (17.2\text{-}38)$$

valid within the interfacial transition region. Here, the function $f(\mathbf{x}_s)$, which represents the lowest-order contribution to the normal mass flux $\rho_A \mathbf{n}^* \cdot (\mathbf{v} - \mathbf{u}^*) + \mathbf{n}^* \cdot \mathbf{j}_A$ of species A across the interface, is independent of n. Examination of the lowest-order terms appearing in Eq. (17.2-20) yields the expression

$$\left(\frac{\partial \rho_A}{\partial \tilde{n}} + \rho_A \frac{\partial E}{\partial \tilde{n}} \right) = O(\varepsilon), \qquad (17.2\text{-}39)$$

valid within the interfacial region $n = (\varepsilon)$. Since, in this 'inner' region $\rho_A = \tilde{\rho}_A + O(\varepsilon)$, this relation confirms the Boltzmann form (17.2-21) for $\tilde{\rho}_A$ that was used in our derivation of the Fick's law surface diffusion relation (17.2-33).

17.3 Summary: Interfacial Boundary Conditions for Mass, Momentum and Species Surfactant Transport Along and Across Material and Nonmaterial Interfaces

This concludes our development of the asymptotic microscale theory of interfacial transport processes. The present section summarizes the (macroscale) interfacial mass, linear momentum and surfactant species boundary conditions thereby derived, together with a tabulation of surface-excess identifications appropriate to each such boundary condition. The summary (which incorporates results obtained from *both* Parts I and II), begins with the case of a material fluid interface, to which case most attention has been addressed in the literature. Thereafter, we discuss the necessary generalizations required for nonmaterial interfaces.

Summary of Interfacial Transport Boundary Conditions
I. Material Interface

<u>Mass</u>

(i) Continuity of velocity — Eq. (3.6-17) or, equivalently, Eq. (16.1-9)

<u>Hydrodynamics</u>

(ii) Interfacial stress boundary condition — Eq. (4.2-4)
Explicit form for Newtonian interface Eq. (4.2-20); see also Eq. (17.2-9)

<u>Surface-excess formulations of phenomenological coefficients</u>

Interfacial tension, σ — Eq. (16.3-3)
Interfacial shear viscosity, μ^s — Eq. (16.3-21)

	Interfacial dilatational viscosity, κ^s	Eq. (16.3-25)

Surfactant transport

(iii)	Interphase partition relation	Eq. (5.2-1)
	Surface-excess formulations of phenomenological coefficients	
	Partition coefficient, K	Eq. (17.2-28)
(iv)	Interfacial species flux condition	Eq. (5.2-2)
	Explicit form for Fickian interface	Eq. (5.2-6)
	Surface-excess formulations of phenomenological coefficients	
	Interfacial species mass density, ρ_i^s	Eq. (17.1-5)
	Interfacial reaction rate, R_i^s	Eq. (17.1-4)
	Interfacial species diffusivity, D_i^s	Eq. (17.2-34)
(v)	Equilibrium adsorption isotherm	
	Langmuir	Eq. (5.3-2)
	Frumkin	Eq. (5.3-1)
	Surface-excess formulations of phenomenological coefficients	
	Adsorption coefficient, K_a	Eq. (17.2-26)
(vi)	Nonequilibrium adsorption kinetic relation	
	Langmuir	Eq. (5.3-4)
	Frumkin	Eq. (5.3-5)
	Small departure from equilibrium	Eq. (5.3-6)
(vii)	Surface equation of state	
	Langmuir	Eq. (5.5-2)
	Frumkin	Eq. (5.5-1)
	Szyskowsky	Eq. (5.5-3)
	Small departure from equilibrium	Eq. (5.3-4)

II. Nonmaterial Interface

Mass

(i)	Continuity of velocity condition:	Replace Eq. (3.6-17) by: Continuity of tangential velocity, Eq. (16.1-6) Continuity of normal mass flux, Eq. (16.1-11)

Hydrodynamics

(ii)	Interfacial stress boundary condition:	Replace Eq. (4.2-4) by Eq. (16.1-16)

Surfactant transport

(iv) Interfacial species flux condition: Replace Eq. (5.2-2) by Eq. (17.1-6)

As inspection of the above boundary conditions reveals, one of the primary distinctions between material and nonmaterial interfaces lies the definition of 'surface velocity' for material vs nonmaterial cases. For material interfaces, the macroscale, mass-average velocities at the interface $\overline{\mathbf{v}}(\mathbf{x}_s, 0+)$ and $\overline{\mathbf{v}}(\mathbf{x}_s, 0-)$ are observed to be continuous; thus, for the material case, the bulk-phase, mass-average velocity 'at' the interface is regarded as the material surface velocity, $\mathbf{v}^o(\mathbf{x}_s, 0)$. For nonmaterial interfaces, the normal component of the bulk-phase, mass-average velocities on either side of the interface $\mathbf{n}\cdot\overline{\mathbf{v}}(\mathbf{x}_s, 0+)$ and $\mathbf{n}\cdot\overline{\mathbf{v}}(\mathbf{x}_s, 0-)$ may be discontinuous across the interface, and will generally differ (on either side of the interface) from the normal component of the interfacial velocity $\mathbf{u}_n$. The significance of this latter fact has been illuminated through examples (cf. Example 3 of §3.7, as well as Questions 15.3–15.5); it is further considered in Question 17.3 at the end of this chapter. The role (if any) of nonmaterial interfacial behavior upon interfacial rheological constitutive behavior has yet to be investigated and will not be further addressed here.

17.4 Summary

Chapter 17 has addressed the surface-excess species balance equation (17.1-6) (specialized to the case of binary surfactant transport) for nonmaterial fluid interfaces. This result generalizes the corresponding material interface result (5.2-2). A Fickian constitutive relation for the surface-excess species flux vector has been derived [cf. Eq. (17.2-33)] by employing the surface-excess asymptotic theory of Chapter 15: the three-dimensional, interfacial species flux vector was assumed in the derivation to possess a transversely-isotropic Fickian form [cf. Eqs. (17.2-16) and (17.2-17)]. In addition, by focusing attention upon the case of diffusion-controlled surfactant transport, a linear (Henry's law) adsorption constitutive relation was derived [cf. Eq. (17.2-25)] together with a corresponding linear bulk-phase partition relation [cf. Eq. (17.2-27)]. Integral microscale surface-excess formulations were provided for both the adsorption and partition coefficients appearing in the above cited relations.

Questions for Chapter 17

17.1 A surface-active species A undergoes a first-order, irreversible (microscale) chemical reaction in the interfacial region, in which the rate of reaction is dependent upon the normal distance n in accordance with the formula

$$R_A = k(n)\rho_A(n)$$

with

$$k(n) \to 0 \quad \text{as} \quad n \to \pm\infty.$$

What would be the appropriate (macroscale) definition of the interfacial reaction rate k^s in the following first-order, irreversible expression for the surface-excess areal reaction rate R_A^s:

$$R_A^s = k^s \rho_A^s.$$

17.2 The Boltzmann expression (17.2-21) was derived [cf. Eqs. (17.2-35)–(17.2-39)] on the basis of 'diffusion-controlled' surfactant transport.

i. By inspection of Eq. (17.2-38), what plausible physicochemical condition would have to hold for surfactant transport process at an interface *not* to be diffusion controlled?

ii. Would you expect a concentrated surfactant monolayer to exhibit diffusion-controlled or adsorption-controlled surfactant transport in nonequilibrium circumstances? Why?

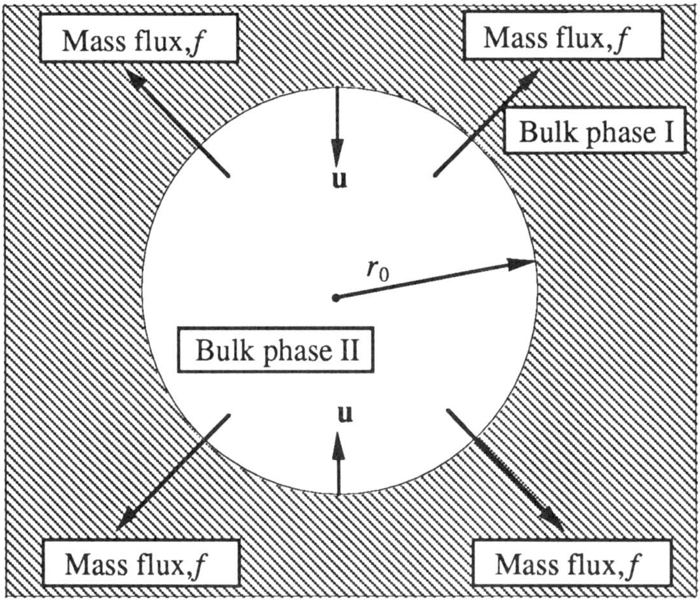

17.3 A spherical droplet of instantaneous radius r_0 containing within its interior an instantaneous total mass $M(t)$ of incompressible liquid phase II (mass density ρ_{II}) is being absorbed by the surrounding, incompressible liquid phase I (mass density $\bar{\rho}_I \neq \bar{\rho}_{II}$) at an areal rate

$$f = \bar{\rho}_{II}\mathbf{n} \cdot [\bar{\mathbf{v}}_{II}(0-) - \mathbf{u}]$$

i. Refer to Example 3 of §3.7 as well as Question 15.5 to determine the three pertinent velocities, namely, **u**, $\bar{\mathbf{v}}_{\mathrm{II}}(0-)$ and $\bar{\mathbf{v}}_{\mathrm{I}}(0+)$ as functions of the mass flux f.

ii. State in which case the interfacial velocity of phase I will be directed *inward*, toward the drop center, and in which case will it be directed outward (i.e. assuming $f > 0$).

*Additional Reading for Chapter 17**

§17.1 Conservation of Species

Deemer, A.R. & Slattery, J.C. 1978 Balance equations and structural models for phase interfaces. *Int. J. Multiphase Flow* **4**, 171–192.

§17.2 Derivation of Adsorption Isotherm, Surfactant Partitioning Relation and Constitutitive Relation for the Surface-Excess Species Flux Vector

Brenner, H. & Leal, L.G. 1978a A micromechanical derivation of Fick's law for interfacial diffusion of surfactant molecules. *J. Colloid Interface Sci.* **65**, 191–209.

* See the footnote accompanying the ***Additional Reading for Chapter 15***.

18

"I saw before me a vast multitude of small Straight lines . . . interspersed with other Beings still smaller and of the nature of lustrous points—all moving to and fro in one and the same Straight Line, and, as nearly as I could judge, with the same velocity.

"'Woman, what signifies this concourse, and this strange and confusing chirping, and this monotonous motion to and fro in one and the same Straight Line?'

"'I am no Woman,' replied the small Line: 'I am the Monarch of the World. But thou, whence intrudest thou into my realm of Lineland? . .' It seemed that the poor ignorant Monarch—as he called himself—was persuaded that the Straight Line which he called his Kingdom, and in which he passed his existence, constituted the whole of the world, and indeed, the whole of Space."

King: Exhibit to me, if you please, this motion from left to right.

I: Nay, that I cannot do, unless you could step out of your Line altogether.

King: Out of my Line? Do you mean out of my World? Out of Space?

I: Well, yes. Out of your World. Out of your Space. For your Space is not the true Space. True Space is a Plane; but your Space is only a Line.

King: If you cannot indicate this motion from left to right by yourself moving in it, then I beg you to describe it to me in words.

I: If you cannot tell your right side from your left, I fear that no words of mine can make my meaning clear to you. But surely you cannot be ignorant of so simple a distinction.

King: I do not in the least understand you.

Edwin A. Abbott (1885)

CHAPTER 18

A Micromechanical Line-Excess Theory of Equilibrium Line Tension

A three-dimensional, highly inhomogeneous, lineal transition region exists at the common junction of three bulk-fluid phases. To a macroscale observer—one whose characteristic length scale L is significantly larger than the characteristic linear dimension l of the inhomogeneity—this diffuse region will appear to be a one-dimensional line of discontinuity formed by the intersection of three (two-dimensional) interfaces separating the three bulk phases, as in Figure 18.0-1. Such a line of discontinuity is called a *three-phase contact line.** This chapter develops an *equilibrium*, micromechanical 'excess' theory of the macroscale behavior of such contact lines based upon asymptotic $O(l/L)$ scaling arguments applied to such lines, similar in conception to the comparable arguments of Chapter 15, there applied to surfaces. The analysis is limited to systems in a state of equilibrium. As yet, no comprehensive treatment exists for the nonequilibrium transport state.

The macroscopic equilibrium contact angles at which three fluid interfaces meet is determined by a balance of forces acting upon the contact line. In particular, for the case of a *circular* contact line, such as forms around a fluid lens at the common interface between three immiscible fluids (see Figure 18.0-1), the classical point-wise, contact-line force balance (known as the "Neumann" or "Neumann-Young" equation) adopts the form

$$\sigma^{\mathrm{III,\,I}} - \sigma^{\mathrm{II,\,III}} \cos \alpha - \sigma^{\mathrm{I,\,II}} \cos \beta = 0. \qquad (18.0\text{-}1)$$

Here, $\sigma^{\gamma,\delta}$ (hereafter abbreviated as $\sigma^{\gamma\delta}$) refers to the interfacial tension existing at the interface between the contiguous bulk phases γ and δ (with γ, δ = I, II, III); α and β represent the equilibrium contact angles shown in the figure.

---

* Contact lines have previously been encountered in the text in the context of foam and emulsion systems (cf. Figures 11.1-4 and 11.1-5, as well as the figures of §14.1).

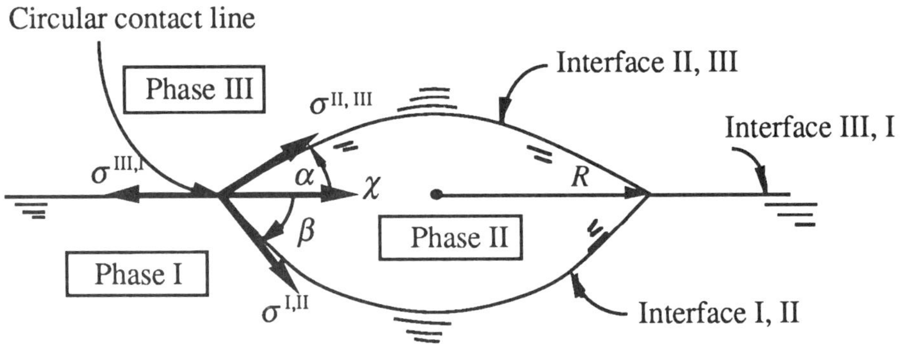

Figure 18.0-1 Areal (i.e. interfacial) and lineal forces acting at a circular three-phase contact line separating immiscible bulk phases I, II and III. The contact angles α and β are determined by the relative magnitudes of the three interfacial tensions $\sigma^{\gamma\delta}$ acting along the interfaces, as well as by the magnitude of the line tension χ, the latter acting tangentially to the contact line.

As first discussed by Gibbs [see Gibbs (1957)], and later proposed by Vesselovsky & Pertzov (1936), circumstances may exist for which the preceding equation is incomplete owing to the existence of an intrinsic *line tension*, χ, acting tangentially along the contact line. In such circumstances the Neumann equation for a circular contact line generalizes to the form

$$\sigma^{\text{III, I}} - \sigma^{\text{II, III}} \cos\alpha - \sigma^{\text{I, II}} \cos\beta - \frac{1}{R}\chi = 0, \qquad (18.0\text{-}2)$$

with R the (uniform) radius of curvature of the circular contact line.

The significance of the line tension in this latter *generalized Neumann equation* is quantified by the dimensionless group $\chi/R\sigma$ (with $\sigma \equiv \min \sigma^{\gamma\delta}$ a norm of the interfacial tension) whose magnitude has been a subject of considerable experimental controversy (Gershfeld & Good 1967, Torza & Mason 1971, Good & Koo 1979, Platikanov *et al.* 1980, Kralchevsky *et al.* 1986). Positive values of χ ranging from 10^3 to 10^{-6} dynes [as well as *negative* values (Good & Koo 1979)] have been indicated by experimental measurements of these and earlier observers. Current interest in line tension phenomena as a practical phenomenon, rather than as a laboratory curiosity, is based upon the assumed existence of experimentally reliable circumstances for which the dimensionless group $\chi/R\sigma$ is of the order of unity [as, for example, indicated by the higher range of χ values reported by Kralchevsky *et al.* (1986)].

In this, the final chapter of the text, a micromechanical, line-excess theory of equilibrium line tension is outlined.* This asymptotic theory,

* Other (primarily thermodynamic) analyses of equilibrium line tension may be found in the works of Buff & Saltsburg (1957), de Feijter & Vrij (1972) and Churaev *et al.* (1982).

which goes beyond earlier purely continuum-mechanical, phenomenological macroscale theories (by providing a *microscale* physicochemical basis for rationalizing line tension phenomena), corresponds in most respects to that outlined by Vignes-Adler & Brenner (1985), whose paper is referred to for several details not reproduced here. Our emphasis here is focused upon the unique *physics* of lineal domains (as opposed to the areal domains considered thus far in Part II), which accounts for our desire to burden the reader with as little mathematical detail as possible, and yet still provide a basic understanding of the phenomenon of line tension and its consequences. Differences between the methodology outlined in this chapter and that of Vignes-Adler & Brenner (1985) owe to our attempt herein to adhere more closely to the basic philosophy previously established in Chapter 15.

We begin in §18.1 with a microscale description of a three-phase transition zone, where all fields are regarded as being everywhere continuous—albeit changing steeply in a lateral direction relative to the 'axis' of the lineal domain. This is followed in §18.2 by an outline of a line-excess singular perturbation approach, pursuing analogies that exist between the present asymptotic line-excess theory and the comparable surface-excess theory of Chapter 15. In particular, following Vignes-Adler & Brenner (1985), we ultimately advance the view that the basic (macroscale) equations, such as Eq. (18.0-2), governing lineal phenomena may be viewed simply as *mathematical* matching conditions arising as a consequence of the singular nature of the *continuous* microscale fields in the $\varepsilon = l/L \to 0$ limit! Thus, line tension is viewed as a strictly macroscale phenomenon. In particular, in §18.3 generic definitions of line-excess properties are derived, from which analysis follows an equilibrium expression for the (macroscale) line tension in terms of steeply inhomogeneous microscale equilibrium pressure and (macroscale) surface tension fields. The subject of *curved* contact lines together with the corresponding development of the generalized Neumann equation, is taken up in §18.4.

18.1 Microscale Description

Referring to Figure 18.1-1a, suppose that a small quantity of immiscible fluid (II) [depicted as possessing a negative spreading coefficient: cf. Eq. (11.2-12)] is placed at the interface between two immiscible fluids (I and III). The whole system, assumed to exist in a state of thermodynamic equilibrium, may be regarded from the vantage point of some small microlength scale l (Figure 18.1-1b) as being a continuum, possessing continuously varying properties throughout the entire space occupied by the microscale fluid. Denote by $\mathbf{x}$ the position vector of an arbitrary point P with respect to an origin fixed at O. In addition, as in Figure 18.1-2a, let the position vector $\mathbf{x}_{\mathcal{L}}$ specify the location of a point $P_{\mathcal{L}}$ lying on a (generally curvilinear) line $\mathcal{L}$ somewhere within the contact-line transition zone. This line represents the common intersection of the three (mathematical) 'parent surfaces' identified with the three interfacial transition regions which meet in the vicinity of the contact line region (see Figure 18.2-1b). Eventually, this *parent line* will be identified with the (macroscale) contact line. [The latter 'physical' curve will itself be designated by the same symbol $\mathcal{L}$ (Figure 18.1-2).]*

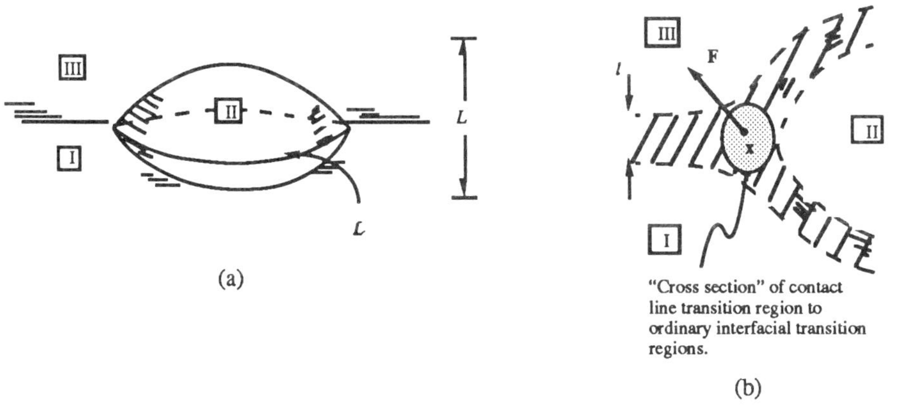

Figure 18.1-1 (a) Macroscale view of the circular contact line $\mathcal{L}$ existing at the triple junction between three immiscible phases. As a mnemonic, the phases are labeled I, II, III in order of decreasing density (the implicit presence of a gravity field being assumed so as to secure the configuration depicted in the figure). The radius of the circular lens (or thickness in the more general case) defines a characteristic *macroscopic* dimension L; (b) Microscale view of the contact line $\mathcal{L}$ in a meridian plane $\mathcal{P}$ of the axisymmetric lens configuration. The linear dimension l denotes a microlength scale; depicted from this scale are the three transition regions between the bulk phases I, II and III, as well as the contact line transition region.

At each point $\mathbf{x}$ of the three-dimensional microscale domain an *exact*, continuous thermodynamical pressure field $p(\mathbf{x})$ is assumed to exist. Viewed from a coarser scale L [i.e. the macroscale L ($>> l$), say, a characteristic linear dimension of the lens straddling the interface (Figure 18.1-1a)], this pressure field may no longer appear continuous. Rather, at this level of description, macroscale interfaces, representing mathematically singular surfaces, will *appear* to exist: this fact has been rigorously demonstrated in Chapter 15 using asymptotic perturbation methods. These surfaces constitute the interfaces existing between the macroscopically immiscible bulk phases, while the generally skew singular curve lying at the common intersection of

* This procedure of specifying a particular parent line ($\mathcal{L}$) within the contact-line transition zone is entirely analogous to our earlier identification in §15.3 of a parent surface (A) within the interfacial transition region. Owing to our focus here upon the static, equilibrium case, no need presently exists to introduce concepts such as 'line-fixed familial coordinates', or to consider the kinematical description of (possibly nonmaterial) lines. For this reason, many of the analogies between the surface and line theories that might otherwise have been invoked will not appear in the present chapter.

these three interfaces constitutes the so-called contact line, the latter being a strictly macroscale construct.

In addition to the pressure field $p(\mathbf{x})$, a continuous, microscale body-force vector density $\mathbf{F}(\mathbf{x})$ is assumed to act at each point $\mathbf{x}$ (Figure 18.1-1b) owing to internally-derived intermolecular interactions arising from the large physicochemical inhomogeneities present in the lineal transition region. The origin of this short-range physicochemical force is identical to that of the comparable body force existing within the neighboring interfacial transition zones (the latter having been been discussed in the first subsection of §17.2). Simultaneously, there may exist contributions to $\mathbf{F}$ owing to longer-ranged external (e.g. electromagnetic or gravitational) body forces acting upon the system.

For the sake of simplifying the subsequent analysis, attention will be restricted (until §18.4) to *rectilinear* (rather than *curvilinear*) contact lines, such as the (macroscale) contact line depicted in Figure 18.1-2. For such systems the internal microscale body force $\mathbf{F}(\mathbf{x})$ may be regarded as possessing a component $\mathbf{t}\cdot\mathbf{F} \equiv F_t$, say, parallel to the (unit tangent vector $\mathbf{t}$ in the) transverse direction within the contact-line region, and two other components (F_x, F_y) lying in the normal plane $\mathcal{P}(x,y)$ to $\mathcal{L}(t)$; here, the scalar t denotes the (tangential) distance within the contact line region measured in the transverse direction to $\mathcal{P}$. [Cartesian coordinates have been chosen such that $t \equiv z$, as illustrated in Figures 18.1-2a and 18.3-1. Nevertheless, we continue to employ the notation t rather than z, since comparable results to be obtained later will then apply equally well to *curvilinear* contact lines (§18.4), for which $t \neq z$; e.g. see Eq. (18.3-8).]* It suffices for purposes of the present analysis to assume the quintessential case, where F_t arises essentially from coarse-scale physicochemical gradients (i.e. fields varying on the macroscale L), whereas the forces (F_x, F_y) normal to $\mathcal{L}$ arise exclusively from physicochemical gradients varying on the microscale l. This physical assignation of orders of magnitude has important consequences in clarifying the subsequent exposition.

Our assumption that the system exists in a state of equilibrium requires that the hydrostatic equation,

$$-\nabla p + \mathbf{F} = \mathbf{0}, \tag{18.1-1}$$

prevail at each point $\mathbf{x}$. The above equation implicitly assumes that the local microscale pressure tensor may be represented as an isotropic pressure [cf. Eq. (18.3-10)], as was also assumed to be the case in §16.3; (note also the discussions of statistical and continuum interfacial approaches appearing at the conclusion of §15.1). Equation (18.1-1) is to be regarded as a physically *exact* description of the mechanical state of the system (lacking only

---

* That rectilinear and curvilinear contact lines can be treated on an equal footing derives from the fact that the eventual *asymptotic* theory of contact-line phenomena to be derived applies only to circumstances wherein the curvature of the contact line [cf. Eq. (18.3-9)] is a *macroscale* quantity; hence, in this sense, any curvilinear line can be regarded locally as rectilinear.

constitutive equations for the microscale pressure p, together with the intrinsic and extrinsic volumetric external force densities $\mathbf{F}$ in order to be complete), at the continuum microscale level.

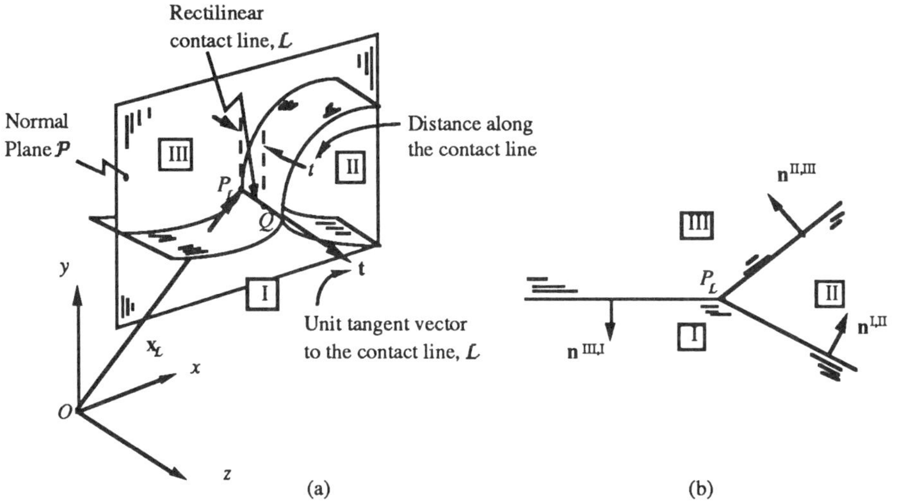

Figure 18.1-2 (a) Macroscale view of a *rectilinear* (rather than *curvilinear*), three-phase contact line. $\mathcal{L}$ denotes the contact line; t denotes the physical distance from a point $P_{\mathcal{L}}$ lying on the contact line to an arbitrary point Q also lying on $\mathcal{L}$; $\mathbf{t} \equiv \partial \mathbf{x}_{\mathcal{L}}/\partial t$ is the unit tangent vector along the contact line; $\mathcal{P}$ denotes the plane normal to $\mathcal{L}$ at $P_{\mathcal{L}}$. The position vector $\mathbf{x}_{\mathcal{L}}$ of an arbitrary point $P_{\mathcal{L}}$ on the contact line relative to a generic space-fixed origin O can be parameterized by the Cartesian coordinates $\mathbf{x}_{\mathcal{L}} \equiv (x,y,z)$. For the special case of the rectilinear contact line shown, these coordinates may be conveniently chosen such that the Cartesian coordinate $z \equiv t$ lies along $\mathcal{L}$. In that case, the plane $\mathcal{P}$ corresponds to the planar surface $z =$ const.; (b) Cross-sectional macroscale view of the contact line in a plane $\mathcal{P}$ normal to $\mathcal{L}$. Conventions: A general point lying on the interface formed by the common boundary of the α and β phases will be represented as $P^{\alpha\beta}$. For convenience, the three symbols (α, β, γ), in that order, will be taken to be a cyclic permutation of (I, II, III). The symbol $\mathbf{n}^{\alpha\beta} = -\,\mathbf{n}^{\beta\alpha}$ denotes a unit normal vector to the α–β interface, drawn in such a manner as to point from the α phase toward the β phase. By convention, the sense of the direction of the unit tangent vector $\mathbf{t}$ to the contact line will be such that a right-handed screw turned in the direction of the vector $\mathbf{n}^{\mathrm{I,II}}$ or $\mathbf{n}^{\mathrm{II,III}}$ or $\mathbf{n}^{\mathrm{III,I}}$ will advance in the $\mathbf{t}$ direction.

As in the previous surface-excess theory of §§15.5 and 15.6, matched asymptotic expansion methods will be employed in the subsequent integration and interpretation of macroscale results issuing from Eq. (18.1-1). It is important to bear in mind once again that, in a strict sense, such asymptotic methods are purely *mathematical*, and hence introduce no new (micro-) *physics* into the problem, beyond what is already explicitly and implicitly contained in Eq. (18.1-1). What may superficially appear to be new (macro-) physics is, in a strict sense, not physics but mathematics, since the same results could be arrived at formally by an applied mathematician, wholly unschooled in the physics of line tension. What may appear to be new macrophysics is merely the *physical interpretation* of the strictly mathematically-derived asymptotic results for the special scaling circumstances to which the mathematical analysis is limited. Adopting this purist stance vis-a-vis what is physical and what is mathematical, and insisting upon the maintenance of this distinction, does much to clarify the nature of subsequent operations.

18.2 Asymptotic Analysis

A generic surface-excess scheme was described in Chapter 15 (cf. §§15.5 and 15.6) via which all of the various macroscale mathematical (and concomitant macrophysical) entities arising therein (e.g. the 'interface' itself, surface-excess densities, etc.) were reconciled with the exact microscale physics. This reconciliation was effected within the context of singular perturbation theory. In what follows, the adaptation of those microphysical →macrophysical arguments to the present lineal geometric configuration is effected. The reader may refer to Vignes-Adler & Brenner (1985) for details of a closely related scheme.

Consider the contact region proximate to the common intersection of three immiscible bulk fluid phases. As in Figure 18.2-1, the diffuse microscale contact-line transition region between these bulk phases may be envisioned as a long, slender, circular cylindrical domain whose length is of order L and whose characteristic cross-sectional radius is of order l. So-called *slender-body theory* (Cox 1970) furnishes a convenient theoretical framework for the subsequent mathematical asymptotic analysis of contact-line phenomena. Inasmuch as the precise location of the parent line $\mathcal{L}$ is unimportant in the subsequent asymptotic theory (see Question 18.1), provided only that it lies somewhere within the lineal transition region, the symbol $\mathcal{L}$ will henceforth be understood to coincide with the center line of the cylinder.

Physical distances along $\mathcal{L}$, measured from an arbitrary point $P_{\mathcal{L}}$ (Figure 18.1-2a), will be denoted by t. The radial distance, measured from $P_{\mathcal{L}}$ (Figure 18.2-1b), perpendicular to $\mathcal{L}$, will be identified as the 'normal' coordinate R of the transition region. The physical interpretation of this coordinate (as coinciding with some physicochemical inhomogeneity possessing the steepest gradient within the transition region) is analogous to that ascribed to the normal coordinate n of the *interfacial* transition region, the latter coordinate being discussed in detail in §§15.3 and 15.5.

A generic, continuum, microscale, tensorial field $\lambda \equiv \lambda(\mathbf{x})$ (e.g. p or $\mathbf{F}$) may, according to the preceding remarks, be parameterized alternatively as

$$\lambda \equiv \lambda(\mathbf{x}_{\mathcal{L}}, R), \tag{18.2-1}$$

relative to the position $\mathbf{x}_{\mathcal{L}}$ along the parent line $\mathcal{L}$. Within each of the three respective bulk-fluid and interfacial regions, and for a fixed value of $\mathbf{x}_{\mathcal{L}}$, the field λ will generally change only gradually with the distance R; however, within the lineal transition region, λ may undergo rapid changes with R.

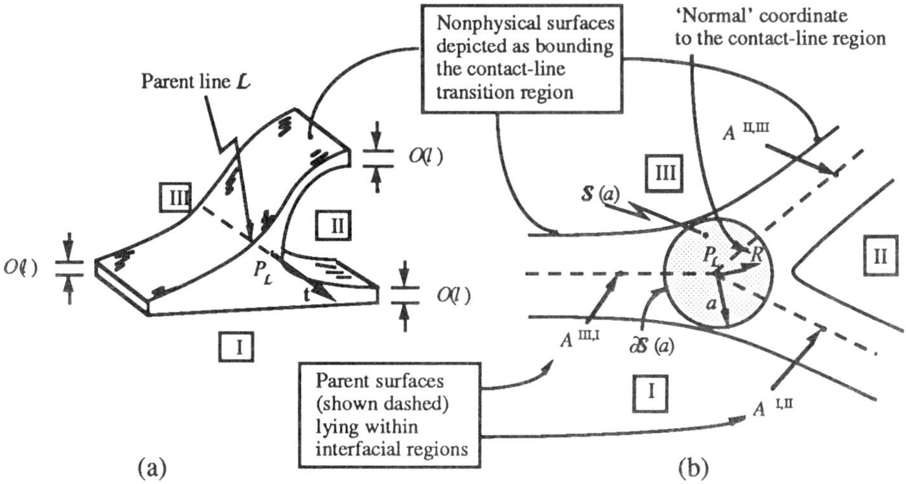

Figure 18.2-1 (a) Microscale view of the (rectilinear) contact-line transition region. At this microscale level of description, the three interfaces are diffuse transition regions possessing thicknesses of $O(l)$. (For the sake of clarity, these transition regions are depicted as existing between the three pairs of *mathematical* surfaces, I–II, II–III, and III–I); (b) Representation of the contact-line transition region in the plane $\mathcal{P}$ normal to t; $\mathcal{S}(a)$ is the cross-sectional domain (shown shaded) of the largest circular cylinder (whose radius is a) centered at $P_{\mathcal{L}}$, wherein all *three* bulk phases interact; the circle $\partial\mathcal{S}(a)$ bounds this cross section. Radial distances R measured from the cylinder center $P_{\mathcal{L}}$ are identified witht the 'normal' coordinate of the transition zone, transverse to the contact line. For convenience, the (planar) parent surfaces of the three interfacial regions may be described by meridian planes $\theta = \text{const} = \theta^{\alpha\beta}$, say, for the α–β surface ($0 \le \theta \le 2\pi$). For simplicity, the plane $\theta = 0$ (and, hence, $\theta = 2\pi$) may be conveniently chosen so as to coincide with one of the surfaces, say, $\theta^{\text{I,II}}$ for definiteness. With this choice, the I–II surface may be assigned either of the pair of values $\theta^{\text{I,II}} = 0$ or 2π, according to circumstances. The α phase then corresponds to the set of θ values, min $\theta^{\gamma\alpha} \le \theta \le$ max $\theta^{\alpha\beta}$, where min $\theta^{\text{I,II}} = 0$ and max $\theta^{\text{I,II}} = 2\pi$. For the other two parent surfaces, $\theta^{\text{II,III}}$ and $\theta^{\text{III,I}}$, max $\theta^{\beta\gamma} =$ min $\theta^{\beta\gamma} = \theta^{\beta\gamma}$.

The respective characteristic lengths l and L of the lineal and bulk-fluid regions allow definition of the nondimensional normal distance

$$\tilde{R} \overset{\text{def}}{=} R / l, \tag{18.2-2}$$

which is characteristic of *microscale* distances, and

$$\bar{R} = R / L \equiv \varepsilon\tilde{R} , \tag{18.2-3}$$

which is characteristic of *macroscale* distances. The characteristic length ratio

$$\varepsilon \equiv l / L << 1 \tag{18.2-4}$$

will be assumed in what follows to satisfy the indicated inequality.

The field λ may change significantly within the diffuse transition region of thickness l, i.e. with $O(1)$ changes in $\tilde{R}$. Thus, the functional form

$$\lambda = \lambda(\mathbf{x}_L, \tilde{R}; \varepsilon) \tag{18.2-5}$$

is a useful representation of Eq. (18.2-1) for those positions $\mathbf{x}_L$ lying within the lineal transition region. In this form, changes in the value of λ occurring over the microscale distance $|\tilde{R}| = O(1)$ can be distinguished. Those gradual changes in λ, if any, occurring within the bulk (and interfacial) regions will take place only over distances of the order of the macrolength L, i.e. over distances $|\tilde{R}| = O(\varepsilon^{-1})$.

Sensible changes in the value of λ within the bulk-fluid (and interfacial) regions, $|\bar{R}| = O(1)$, occur only over the much larger length scale $|\Delta\bar{R}| = O(1)$. The functional form

$$\lambda = \lambda(\mathbf{x}_L, \bar{R}; \varepsilon) \tag{18.2-6}$$

is thus a more useful representation than Eq. (18.2-5) for those positions $\mathbf{x}$ lying within these bulk-fluid (and interfacial) regions. In this form, changes in λ occuring over macroscale distances, $|\bar{R}| = O(1)$, can be distinguished. Any rapid changes in λ within the lineal transition region will occur over distances characterized by $|\bar{R}| = O(\varepsilon)$. As $\varepsilon \to 0$, such changes will appear to be *discontinuous* when the functional form (18.2-6) is used in a regular perturbation expansion scheme that utilizes ε as the small parameter.

The requisite asymptotic theory for lineal transition regions follows from the preceding remarks via arguments similar to those employed in the asymptotic theory of the interfacial transition region developed in §15.5. Thus, analogous to Eqs. (15.5-7), (15.5-12) and (15.5-13), one may define the leading-order terms of inner and outer perturbation expansions of the (exact) field density (18.2-1) as follows:

$$\tilde{\lambda}(\mathbf{x}_L, \tilde{R}) \stackrel{\text{def}}{=} \lim_{\substack{\varepsilon \to 0 \\ \tilde{R} \text{ fixed}}} \lambda(\mathbf{x}_L, \tilde{R}; \varepsilon), \tag{18.2-7}$$

representing the leading-order *inner limit* of the exact field density λ, and

$$\bar{\lambda}_{\mathrm{I}}(\mathbf{x}_L, \bar{R}) \stackrel{\text{def}}{=} \lim_{\substack{\varepsilon \to 0 \\ \bar{R} \text{ fixed}}} \lambda(\mathbf{x}_L, \bar{R}; \varepsilon) \quad (\min \theta^{\mathrm{III,I}} \le \theta \le 2\pi), \tag{18.2-8}$$

$$\bar{\lambda}_{\mathrm{II}}(\mathbf{x}_{\mathcal{L}}, \bar{R}) \stackrel{\text{def}}{=} \lim_{\substack{\varepsilon \to 0 \\ \bar{R} \text{ fixed}}} \lambda(\mathbf{x}_{\mathcal{L}}, \bar{R}; \varepsilon) \quad (0 \le \theta \le \max \theta^{\mathrm{II},\mathrm{III}}), \tag{18.2-9}$$

$$\bar{\lambda}_{\mathrm{III}}(\mathbf{x}_{\mathcal{L}}, \bar{R}) \stackrel{\text{def}}{=} \lim_{\substack{\varepsilon \to 0 \\ \bar{R} \text{ fixed}}} \lambda(\mathbf{x}_{\mathcal{L}}, \bar{R}; \varepsilon) \quad (\min \theta^{\mathrm{II},\mathrm{III}} \le \theta \le \max \theta^{\mathrm{III},\mathrm{I}}), \tag{18.2-10}$$

representing the respective leading-order *outer limits* of the exact field density λ in the three bulk phases I, II and III (see the caption to Figure 18.2-1 for a clarification of the above bounds upon θ). Explicit forms of the expressions (18.2-7)–(18.2-10) will subsequently appear in the generic line-excess definition (18.3-1), which definition—as in §15.6—may be obtained by contrasting micro- (l-scale) and macro- (L-scale) views of the total amount of the generic continuum property $\mathcal{P}$ (whose volumetric density is λ) contained within a control volume δV.

18.3 Line-Excess Properties

Our procedure for determining line-excess properties is based upon the identification of an infinitesimal control volume δV, such as depicted in Figure 18.3-1 (whose surface-excess counterpart is the control volume V depicted in Figure 15.6-1). Macro- and microscale perspectives pertaining to resolution of the total amount of λ in δV may be reconciled (analogous to §15.6) by defining the macroscale *line-excess lineal density field*, $\lambda^{\mathcal{L}}$, say (i.e. the excess amount of λ *per unit length*). Through this procedure, the lineal analog of Eq. (15.6-18) [and also of Eq. (15.6-36)] may be obtained. Leaving these (largely repetitive) details as exercises for the reader (see Questions 18.2 and 18.3), in the following we simply summarize the pertinent results of the generic line-excess matched asymptotic analysis of the three-phase transition zone described in §§18.1 and 18.2; thus, the line-excess quantity $\lambda^{\mathcal{L}}$ is shown to satisfy the expression:

$$\boxed{\lambda^{\mathcal{L}}(\mathbf{x}_{\mathcal{L}}) = \varepsilon^2 L^2 \int_0^\infty \int_0^{2\pi} \left[\tilde{\lambda}(\mathbf{x}_{\mathcal{L}}, \tilde{R}) - \tilde{\lambda}(\mathbf{x}_{\mathcal{L}}, 0)\right] \tilde{R}\, d\theta\, d\tilde{R} - \varepsilon L \sum_{\alpha\beta} \int_0^\infty \lambda^s_{\alpha\beta}(\mathbf{x}_{\mathcal{L}}, \tilde{R}^{\alpha\beta}) d\tilde{R}^{\alpha\beta},} \tag{18.3-1}$$

wherein

$$\tilde{\lambda}(\mathbf{x}_{\mathcal{L}},0) \equiv \begin{cases} \lim\limits_{\bar{R}\to 0} \bar{\lambda}_{\mathrm{I}}(\mathbf{x}_{\mathcal{L}}, \bar{R}) \ (\min \theta^{\mathrm{III,\,I}} \le \theta \le 2\pi), \\ \lim\limits_{\bar{R}\to 0} \bar{\lambda}_{\mathrm{II}}(\mathbf{x}_{\mathcal{L}}, \bar{R}) \ (0 \le \theta \le \max \theta^{\mathrm{II,\,III}}), \\ \lim\limits_{\bar{R}\to 0} \bar{\lambda}_{\mathrm{III}}(\mathbf{x}_{\mathcal{L}}, \bar{R}) \ (\min \theta^{\mathrm{II,\,III}} \le \theta \le \max \theta^{\mathrm{III,\,I}}) \end{cases} \quad (18.3\text{-}2\,\mathrm{a})$$

and

$$\sum_{\alpha\beta}\int_0^\infty \overset{s}{\lambda}_{\alpha\beta}(\mathbf{x}_{\mathcal{L}}, \tilde{R}^{\alpha\beta})d\tilde{R}^{\alpha\beta} \equiv \int_0^\infty \overset{s}{\lambda}_{\mathrm{I,\,II}}(\mathbf{x}_{\mathcal{L}}, \tilde{R}^{\mathrm{I,\,II}})d\tilde{R}^{\mathrm{I,\,II}}$$

$$+\int_0^\infty \overset{s}{\lambda}_{\mathrm{II,\,III}}(\mathbf{x}_{\mathcal{L}}, \tilde{R}^{\mathrm{II,\,III}})d\tilde{R}^{\mathrm{II,\,III}} + \int_0^\infty \overset{s}{\lambda}_{\mathrm{III,\,I}}(\mathbf{x}_{\mathcal{L}}, \tilde{R}^{\mathrm{III,\,I}})d\tilde{R}^{\mathrm{III,\,I}} \quad (18.3\text{-}2\,\mathrm{b})$$

Equation (18.3-1) expresses the line-excess lineal density field $\lambda^{\mathcal{L}}$ (at a point $\mathbf{x}_{\mathcal{L}}$ along the contact line in terms of the inner and outer limits ($\tilde{\lambda}$ and $\bar{\lambda}$, respectively) of the exact volumetric density field $\lambda(\mathbf{x})$ and the surface-excess density field $\overset{s}{\lambda}$, the latter being defined in Eq. (15.6-18).

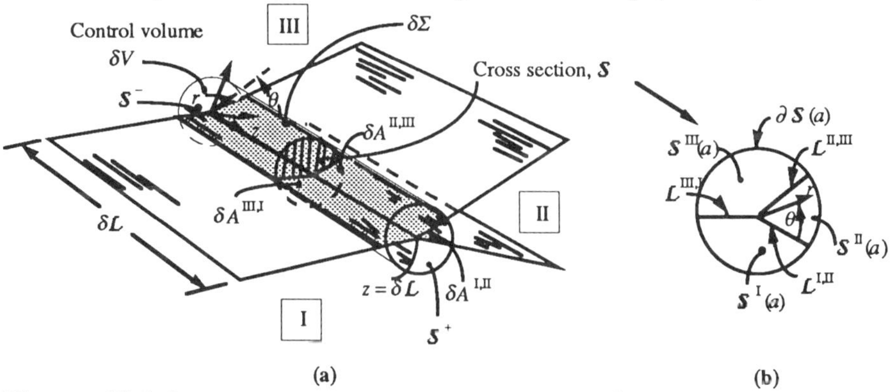

Figure 18.3-1 The circular cylindrical control volume δV, which plays a role in the calculation of line-excess properties, is bounded laterally by the open-ended cylindrical surface $\delta\Sigma$ and end caps S^+ and S^-. The circular cross-sectional domain $S(a)$, girdled by the circle $\partial S(a)$, is of radius a and contains at its center the parent line $\mathcal{L}$ whose length is $\delta\mathcal{L}$. $S^\gamma(\gamma =$ I, II, III) represents the pie-shaped region formed by that portion of γ phase lying inside of the circle. Areal domains $\delta A^{\alpha\beta}$ intersect $S(a)$ along the line segments $\mathcal{L}^{\alpha\beta}$ (i.e. $\mathcal{L}^{\alpha\beta}$ lies in the plane $\theta = \theta^{\alpha\beta}$ = const., and is of length a); $\mathbf{n}^{\alpha\beta}$ is a unit normal vector to the surface $\delta A^{\alpha\beta}$, as in Figure 18.1-2b.

Similarly (see Question 18.3), the matched asymptotic, line-excess definition of the point-flux density field $\mathbf{j}^{\mathcal{L}}(\mathbf{x}_{\mathcal{L}})$ [cf. its surface-excess analog Eq. (15.6-36)], representing the total efflux of a generic continuum property through a point $\mathbf{x}_{\mathcal{L}}$ on $\mathcal{L}$, is found to be given by the expression

$$\boxed{\begin{aligned}\mathbf{j}^{\mathcal{L}}(\mathbf{x}_{\mathcal{L}}) = \varepsilon^2 L^2 \int_0^{\infty}\int_0^{2\pi} \mathbf{I}_{\mathcal{L}} \cdot \left[\tilde{\mathbf{j}}(\mathbf{x}_{\mathcal{L}}, \tilde{R}) - \bar{\mathbf{j}}(\mathbf{x}_{\mathcal{L}}, 0)\right] d\theta \, \tilde{R}\, d\tilde{R} \\ - \varepsilon L \sum_{\alpha\beta} \int_0^{\infty} \mathbf{I}_{\mathcal{L}} \cdot \mathbf{j}^s_{\alpha\beta}(\mathbf{x}_{\mathcal{L}}, \tilde{R}^{\alpha\beta}) d\tilde{R}^{\alpha\beta},\end{aligned}} \qquad (18.3\text{-}3)$$

wherein

$$\bar{\mathbf{j}}(\mathbf{x}_{\mathcal{L}}, 0) \equiv \begin{cases} \lim_{\bar{R}\to 0} \bar{\mathbf{j}}_{\mathrm{I}}(\mathbf{x}_{\mathcal{L}}, \bar{R}) & (\min \theta^{\mathrm{III,I}} \le \theta \le 2\pi), \\ \lim_{\bar{R}\to 0} \bar{\mathbf{j}}_{\mathrm{II}}(\mathbf{x}_{\mathcal{L}}, \bar{R}) & (0 \le \theta \le \max \theta^{\mathrm{II,III}}), \\ \lim_{\bar{R}\to 0} \bar{\mathbf{j}}_{\mathrm{III}}(\mathbf{x}_{\mathcal{L}}, \bar{R}) & (\min \theta^{\mathrm{II,III}} \le \theta \le \max \theta^{\mathrm{III,I}}), \end{cases} \qquad (18.3\text{-}4)$$

and with the surface-excess (lineal) flux field $\mathbf{j}^s$ defined as in Eq. (15.6-36).

Appearing in Equation (18.3-3) is the lineal idemfactor $\mathbf{I}_{\mathcal{L}}$. This symmetric unit dyadic may be expressed in terms of the unit tangent vector $\mathbf{t}$ to the contact line as

$$\mathbf{I}_{\mathcal{L}} = \mathbf{t}\,\mathbf{t}, \qquad (18.3\text{-}5)$$

and possesses the important 'projection' properties that

$$\mathbf{I}_{\mathcal{L}} = \mathbf{I}_{\mathcal{L}} \cdot \mathbf{I} \qquad (18.3\text{-}6)$$

and

$$\mathbf{I}_{\mathcal{L}} = \mathbf{I}_{\mathcal{L}} \cdot \mathbf{I}_s. \qquad (18.3\text{-}7)$$

Here, the dyadics $\mathbf{I}$ and $\mathbf{I}_s$ are the unit spatial [cf. Eq. (A.1-10)] and surface [cf. Eq. (3.2-6)] idemfactors respectively. For later purposes note also that, as a consequence of the definition of the lineal gradient operator [cf. Eq. (3.2-11)], namely

$$\nabla_{\mathcal{L}} \overset{\text{def}}{=} \mathbf{I}_{\mathcal{L}} \cdot \nabla \equiv \mathbf{t}\frac{\partial}{\partial t}, \qquad (18.3\text{-}8)$$

there obtains, upon use of Eq. (15.3-8), the following lineal corollary to the surface identity (3.3-8):

$$\nabla_{\mathcal{L}} \cdot \mathbf{I}_{\mathcal{L}} = \kappa_{\mathcal{L}}\mathbf{p}, \qquad (18.3\text{-}9)$$

where $\mathbf{p}$ denotes the principal unit normal vector to the contact line (see the figure accompanying Question 18.5) and $\kappa_{\mathcal{L}}$ is the radius of curvature of the contact line. For the special case of a rectilinear contact line, $\kappa_{\mathcal{L}}=0$.

The line tension χ acting along $\mathcal{L}$ may be defined by considering the special case of the generic flux relation (18.3-3), for which $\mathbf{j} \equiv -\mathbf{P}$, with the latter (equilibrium) pressure tensor satisfying the following isotropic constitutive relation [cf. Eq. (4.1-13)]:

$$\mathbf{P}(\mathbf{x}_{\mathcal{L}}, \tilde{R}) = -\mathbf{I}\, p(\mathbf{x}_{\mathcal{L}}, \tilde{R}). \tag{18.3-10}$$

This yields [cf. Eq. (16.3-2)]

$$\mathbf{P}^s(\mathbf{x}_{\mathcal{L}}, \tilde{R}^{\alpha\beta}) = \mathbf{I}_s \sigma^{\alpha\beta}(\mathbf{x}_{\mathcal{L}}, \tilde{R}^{\alpha\beta}), \tag{18.3-11}$$

whence, together with Eqs. (18.3-6) and (18.3-7), Eq. (18.3-3) furnishes the macroscale expression

$$\boxed{\mathbf{P}^{\mathcal{L}}(\mathbf{x}_{\mathcal{L}}) = \mathbf{I}_{\mathcal{L}}\, \chi(\mathbf{x}_{\mathcal{L}}),} \tag{18.3-12}$$

in which we have defined

$$\boxed{\begin{aligned}\chi(\mathbf{x}_{\mathcal{L}}) \stackrel{\text{def}}{=} \varepsilon^2 L^2 \int_0^\infty \int_0^{2\pi} \left[\bar{p}(\mathbf{x}_{\mathcal{L}}, 0) - \tilde{p}(\mathbf{x}_{\mathcal{L}}, \tilde{R})\right] d\theta\, \tilde{R}\, d\tilde{R} \\ - \varepsilon L \sum_{\alpha\beta} \int_0^\infty \sigma^{\alpha\beta}(\mathbf{x}_{\mathcal{L}}, \tilde{R}^{\alpha\beta})\, d\tilde{R}^{\alpha\beta}\end{aligned}} \tag{18.3-13}$$

as the equilibrium line tension χ.*

Upon applying Eq. (18.3-1) to the choice $\lambda \equiv \mathbf{F}$, with $\mathbf{F}(\mathbf{x})$ the external body-force volumetric vector density, there obtains the following definition for the macroscale, line-excess, external body-force (lineal) vector density:

$$\begin{aligned}\mathbf{F}^{\mathcal{L}}(\mathbf{x}_{\mathcal{L}}) \stackrel{\text{def}}{=} \varepsilon^2 L^2 \int_0^\infty \int_0^{2\pi} \left[\tilde{\mathbf{F}}(\mathbf{x}_{\mathcal{L}}, \tilde{R}) - \bar{\mathbf{F}}(\mathbf{x}_{\mathcal{L}}, 0)\right] d\theta\, \tilde{R}\, d\tilde{R} \\ \varepsilon L \sum_{\alpha\beta} \int_0^\infty \mathbf{F}^s_{\alpha\beta}(\mathbf{x}_{\mathcal{L}}, \tilde{R}^{\alpha\beta})\, d\tilde{R}^{\alpha\beta},\end{aligned} \tag{18.3-14}$$

with the surface-excess body force $\mathbf{F}^s$ defined as in Eq. (16.1 14).

The preceding generic line-excess definitions (18.3-1) and (18.3-3) [together with the explicit relations (18.3-13) and (18.3-14)] merit the three essential clarifying remarks, discussed sequentially in each of the following paragraphs.

(i) The line-excess lineal density field $\lambda^{\mathcal{L}}(\mathbf{x}_{\mathcal{L}})$ will only be a macroscale quantity when either (or both) of the following conditions are satisfied [similar comments pertain to $\mathbf{j}^{\mathcal{L}}(\mathbf{x}_{\mathcal{L}})$]:

---

* More properly, χ should be referred to as the line-*excess* tension. However, we reluctantly bow in the matter to accepted terminology.

$$\tilde{\lambda}(\mathbf{x}_{\mathcal{L}}, \tilde{R}) - \bar{\lambda}(\mathbf{x}_{\mathcal{L}}, 0) \equiv O(\frac{1}{\varepsilon^2}) \quad \text{or} \quad \lambda^s(\mathbf{x}_{\mathcal{L}}, \tilde{R}^{\alpha\beta}) \equiv O(\frac{1}{\varepsilon}),$$

within the contact-line transition region, $\tilde{R} = O(1)$. The former condition shows the necessity of an inhomogeneity within the contact-line transition region that is $O(1/\varepsilon)$ larger than the comparable inhomogeneities within the three interfacial transition regions if line-excess densities are to be macroscopically discernable. In addition, it is evident from the latter condition that the (macroscale) surface-excess density contribution to the line-excess densities defined in both (18.3-1) and (18.3-3) will be negligibly small in most circumstances, insofar as the surface-excess densities λ^s and $\mathbf{j}^s$ are generally $O(1)$ quantities.

(ii) Only those fields in the immediate neighborhood of the contact line contribute to the value of the line-excess density $\lambda^{\mathcal{L}}(\mathbf{x}_{\mathcal{L}})$ [and, equivalently, $\mathbf{j}^{\mathcal{L}}(\mathbf{x}_{\mathcal{L}})$]. Indeed, denote $\mathcal{S}(a, a_\infty)$ an annular region (Figure 18.3-2) possessing very large radii a, a_∞ and bounding the control surface $\mathcal{S}(a)$ centered at the point $P_{\mathcal{L}}$, and consider the expression

$$\Delta\lambda^{\mathcal{L}}(a, a_\infty) = \iint_{\mathcal{S}(a,\, a_\infty)} (\tilde{\lambda} - \bar{\lambda})\, dS - \sum_{\alpha,\, \beta} \int_{\mathcal{L}^{\alpha\beta}(a,\, a_\infty)} \lambda^s\, dR^{\alpha\beta} \tag{18.3-15}$$

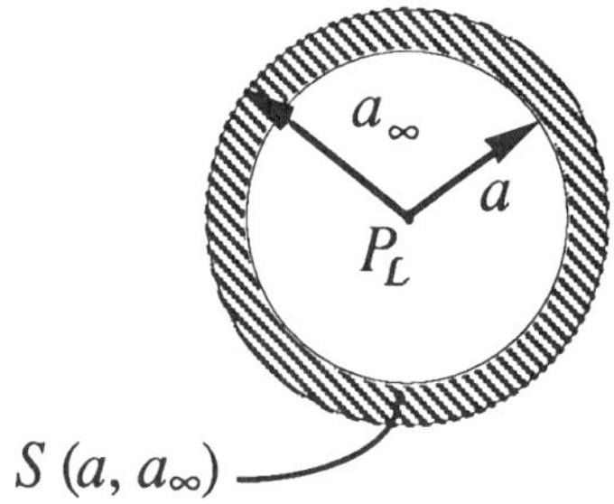

Figure 18.3-2 Annular domain in the plane $\mathcal{P}$ normal to the contact line and centered about the contact line.

for the incremental contribution to the value of the line-excess density $\lambda^{\mathcal{L}}(\mathbf{x}_{\mathcal{L}})$ made by this annular area. A fundamental aspect of the formulation (18.3-1) requiring emphasis is that in the macroscale limit $\varepsilon \to 0$, the quantity $\tilde{\lambda} - \bar{\lambda}(0)$ vanishes everywhere in the plane $\mathcal{P}$ which contains the surface $\mathcal{S}$, except near to the lines $\mathcal{L}^{\alpha\beta}$ (which lines, as shown in Figure 18.3-1, exist upon the interfaces $A^{\alpha\beta}$ across which interfaces $\bar{\lambda}$ is generally discontinuous). Since a and a_∞ are both very large, the surface element dS appearing in Eq. (18.3-15) may be conveniently approximated by the asymptotic relation

$$dS \approx dR^{\alpha\beta}\, dn^{\alpha\beta},$$

where $n^{\alpha\beta}$ is the distance measured normal to $\mathcal{L}^{\alpha\beta}$. Consequently, with use of Eq. (15.6-18),

$$\iint_{S\,(a,\, a_\infty)} (\mathcal{X} - \bar{\lambda})\, dS \approx \sum_{\alpha,\,\beta} \int_a^{a_\infty} dR^{\alpha\beta} \int_{-\infty}^{+\infty} (\mathcal{X} - \bar{\lambda})\, dn^{\alpha\beta}$$

$$\equiv \sum_{\alpha,\,\beta} \int_a^{a_\infty} dR^{\alpha\beta}\, \lambda^s, \tag{18.3-16}$$

whence, the right-hand side of Eq. (18.3-15) vanishes asymptotically, in the limit. Q.E.D.

(iii) In the definitions (18.3-1) and (18.3-3), the integration domains $\mathcal{S}$ and $\mathcal{L}^{\alpha\beta}$ were supposed infinite in extent. As a consequence of the preceding remark, it suffices for them to be merely of macroscopic size. This apparent vagueness in the precise location of the boundary delineating the transition region between the inner and outer expansions accords with the general ideas of matched asymptotic expansion methods — namely, the existence of an extended overlap region, where *both* expansions are asymptotically valid.

18.4 Hydrostatics of a Curved Contact Line

As demonstrated in Question 18.4,* the preceding line-excess quantities, namely χ in Eq. (18.3-13) and $\mathbf{F}^{\mathcal{L}}$ in Eq. (18.3-14), make their appearance in the hydrostatic lineal balance equation

$$\boxed{\mathbf{F}^{\mathcal{L}} + \nabla_L \cdot (\mathbf{I}_{\mathcal{L}}\, \chi) - \mathbf{t} \times \boldsymbol{\sigma} = \mathbf{0},} \tag{18.4-1}$$

in which $\boldsymbol{\sigma}$ is the vector

$$\boldsymbol{\sigma} = \mathbf{n}^{\mathrm{I,\,II}}\sigma^{\mathrm{I,\,II}} + \mathbf{n}^{\mathrm{II,\,III}}\sigma^{\mathrm{II,\,III}} + \mathbf{n}^{\mathrm{III,\,I}}\sigma^{\mathrm{III,\,I}}. \tag{18.4-2}$$

Equation (18.4-1) applies locally, at every point $\mathbf{x}_{\mathcal{L}}$ along the contact line.

Equation (18.4-1) represents the fundamental equation of hydrostatic equilibrium appropriate to a curvilinear contact line. It corresponds to the line analog of the standard equation of hydrostatic equilibrium for a curved interface [cf. Eq. (3.3-12)].

* See Vignes-Adler & Brenner (1985) for a general derivation of the relations provided in this section for curvilinear contact lines.

Upon utilizing Eq. (18.3-9), the following alternative formulation of Eq. (18.4-1) is obtained:

$$\mathbf{F}^{\mathcal{L}} + \nabla_L \chi + \kappa_{\mathcal{L}} \chi \mathbf{p} - \mathbf{t} \times \boldsymbol{\sigma} = \mathbf{0}. \tag{18.4-3}$$

As $\mathbf{t}$ is normal to the vectors $\mathbf{p}$ and $\mathbf{t} \times \mathbf{n}^{\alpha\beta}$, the assumption that the line-excess external force vector $\mathbf{F}^{\mathcal{L}}$ lies parallel to $\mathbf{t}$ causes Eq. (18.4-3) to split componentwise into the respective pair of relations

$$\boxed{\mathbf{F}^{\mathcal{L}} + \nabla_L \chi = \mathbf{0}} \tag{18.4-4}$$

and

$$\boxed{\mathbf{p}\kappa_{\mathcal{L}} \chi - \mathbf{t} \times \boldsymbol{\sigma} = \mathbf{0}.} \tag{18.4-5}$$

Equation (18.4-5) [in conjunction with Eq. (18.4-2)] may be recognized as the generalized Neumann's equation [cf. Eq. (18.0-2)], the lineal analog of the (areal) equilibrium normal stress boundary condition [cf. Eq. (4.2-20a) with $\mathbf{v}^o=\mathbf{0}$], whereas Eq. (18.4-4) is the lineal analog of the (areal) equilibrium tangential stress boundary condition [cf. Eq. (4.2-20b) with $\mathbf{v}^o=\mathbf{0}$]. Equation (18.4-4) points up the possibility of "line Marangoni effects" whenever χ varies along the contact line; for, according to Eq. (18.4-4), in the absence of a line-excess *external* force $\mathbf{F}^{\mathcal{L}}$ to balance this gradient, a state of disequilibrium, necessarily arises, posing the possibility of concomitant lineal Marangoni-type effects.

18.5 Summary

A matched-asymptotic, line-excess theory (albeit limited to equilibrium systems) appropriate to curvilinear three-phase contact regions has been outlined in this chapter by adopting mathematical perturbation techniques similar to those as employed in previous chapters of Part II for comparable surface-excess entities.

The generalized Neumann equation has been derived [cf. Eq. (18.4-5)]. This lineal boundary condition relates the line tension χ to discontinuities in surface tension at a curved contact line, in much the same way as Laplace's equation relates interfacial tension to discontinuities in pressure across curved interfaces. In addition, a lineal analog of the tangential stress boundary condition for equilibrium interfaces [cf. Eq. (18.4-4)] has been developed, the latter relating line tension gradients to the action of an external line-excess force. A line-excess definition of line tension has been provided in Eq. (18.3-13), and a definition of the line-excess force density vector in Eq. (18.3-14).

Possibilities for extending the line-excess theory to dynamical, nonequilibrium systems, and/or extensions of such 'excess' theories to novel concepts such as "point tension", etc. clearly exist (Vignes-Adler & Brenner

1985), such research thrusts being naturally founded upon the generic 'excess' formalism developed and utilized in this second part of the text.

Questions for Chapter 18

18.1 Use methods analogous to those employed in Eqs. (15.6-37)–(15.6-40) to demonstrate that the precise location of the reference line $\mathcal{L}$ in the contact line transition region is irrelevant [i.e. to $O(\varepsilon)$] in the definition (18.3-1) of the line-excess density $\lambda^{\mathcal{L}}$.

18.2 *Derivation of the line-excess lineal density* $\lambda^{\mathcal{L}}$: Making reference to Figure 18.3-1, let

$$\Lambda = \int_{\delta V} \lambda dV$$

represent the true amount of λ contained in the volume δV, and let

$$\bar{\Lambda} = \int_0^{\delta\mathcal{L}} dt \left[\int_{S^{\mathrm{I}}(a)} \bar{\lambda}_{\mathrm{I}} dS + \int_{S^{\mathrm{II}}(a)} \bar{\lambda}_{\mathrm{II}} dS + \int_{S^{\mathrm{III}}(a)} \bar{\lambda}_{\mathrm{III}} dS \right] + \sum_{\alpha\beta} \int_{\mathcal{L}^{\alpha\beta}} \lambda^{s}_{\alpha\beta} dR^{\alpha\beta}$$

$$\equiv \int_{\delta V} \bar{\lambda} dV ,$$

denote the amount that would be observed in δV by a macroscale observer.

i. Follow steps analogous to the surface-excess approach outlined in the scheme leading from Eqs. (15.6-2) to (15.6-9) (after noting the appreciably simpler geometry of the control volume δV in the present case) to derive an intermediate form of the lineal excess density (18.3-1) corresponding to Eq. (15.6-9).

ii. Qualitatively indicate the steps necessary to obtain the solution (18.3-1) from the results of (i).

18.3 *Derivation of the line-excess lineal density* $\mathbf{j}^{\mathcal{L}}$: Making reference to Figure 18.3-1, let

$$\mathbf{J} = \int_{\delta\Sigma \oplus S^{+} \oplus S^{-}} d\mathbf{S} \cdot \mathbf{j}$$

represent the true efflux of λ through the boundary of the volume δV, and let

$$\bar{\mathbf{J}} = \int_0^{\delta\mathcal{L}} dt \left[\int_{\partial S^{\mathrm{I}}(a)} d\mathbf{S} \cdot \bar{\mathbf{j}}_{\mathrm{I}} + \int_{\partial S^{\mathrm{II}}(a)} d\mathbf{S} \cdot \bar{\mathbf{j}}_{\mathrm{II}} + \int_{\partial S^{\mathrm{III}}(a)} d\mathbf{S} \cdot \bar{\mathbf{j}}_{\mathrm{III}} \right]$$
$$+ \int_{S^{+} \oplus S^{-}} d\mathbf{S} \cdot \bar{\mathbf{j}} + \sum_{\alpha\beta} \int_{\mathcal{L}^{\alpha\beta}} \mathbf{t} \cdot \mathbf{j}^{s}_{\alpha\beta}\, dR^{\alpha\beta}$$
$$\equiv \int_{\overline{\delta\Sigma} \oplus \overline{S^{+}} \oplus \overline{S^{-}}} d\mathbf{S} \cdot \bar{\mathbf{j}},$$

denote the comparable efflux observed by a macroscale observer.

i. Follow steps analogous to the surface-excess approach outlined from Eqs. (15.6-19) to (15.6-27) (noting the appreciably simpler geometry of the control volume δV in the present case) to derive an intermediate form of the point-flux density (18.3-3) corresponding to Eq. (15.6-27).

ii. Qualitatively indicate the steps necessary to obtain the solution (18.3-3) from the results of (i).

18.4 Using methods analogous to those developed in §15.7, compute the difference between the true balance of forces acting upon the control volume δV in Figure 18.3-1 and that which would be observed by a macroscale observer. Use this to establish the invariant line-excess boundary condition (18.4-1).

18.5 Referring to the accompanying figure, consider the plane curve C lying in the plane of this page, and let x be a Cartesian coordinate drawn along the tangent line to C at the point O, with the second Cartesian coordinate axis y drawn as indicated (in the concave sense shown).

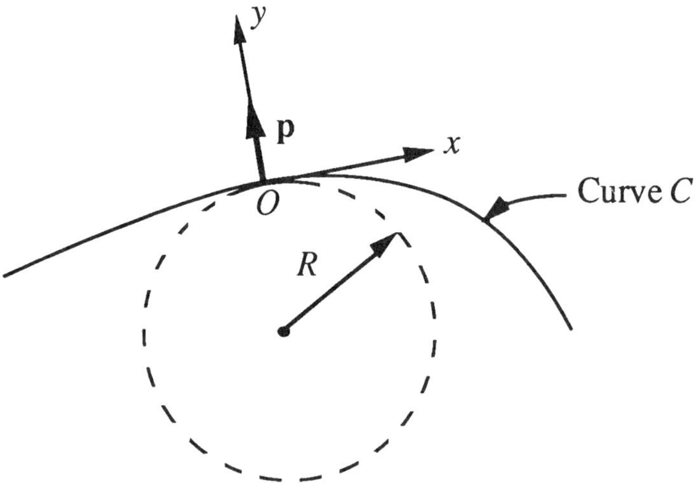

i. Following techniques similar to those outlined in Questions 3.8 and 3.9 for a surface, show that the equation for the curve C about the point O is

$$y = \frac{1}{2}\mathbf{rr}:(\nabla_1\nabla_1 f)_o + O(r^3),$$

where

$$\nabla_1 = \mathbf{i}_x\frac{\partial}{\partial x},$$

$$\mathbf{r} = x\,\mathbf{i}_x,$$

$$r = |\mathbf{r}|$$

and the general equation for the curve C is provided by

$$F(x, y) = y - f(x) = 0.$$

ii. Show that

$$y = \frac{1}{2}\mathbf{r}\cdot\mathbf{b}\cdot\mathbf{r} + O(r^3),$$

where the curvature dyadic $\mathbf{b}$ for the curve C (at O) is given by the expression

$$\mathbf{b} = -\nabla_{\mathcal{L}}\mathbf{p}$$

[analogous to the corresponding surface formula (3.2-10)].

iii. Utilizing

$$\mathbf{b} = \mathbf{i}_x\mathbf{i}_x\kappa_{\mathcal{L}},$$

show as a consequence that

$$y = \tfrac{1}{2}\kappa_{\mathcal{L}}x^2 + O(x^3)$$

iv. Referring to the figure, show that the name 'curvature' is warranted in the sense that

$$\kappa_{\mathcal{L}} = -\frac{1}{R},$$

where R is the radius of the circle tangent to C at O.

v. Given the one-dimensional curvature result

$$\mathbf{b} = -\nabla_{\mathcal{L}}\mathbf{p},$$

and the two-dimensional result

$$\mathbf{b} = -\nabla_s\mathbf{n},$$

would you expect the formula

$$\mathbf{b} = -\nabla_N\mathbf{q}$$

to hold for the curvature dyadic of an N-dimensional space whose unit normal vector $\mathbf{q}$ points from the 'concave' to the 'convex sides' of the N-dimensional space, into the (N+1)st dimension? If so, what is the definition of ∇_N and how many independent curvature components does the dyadic $\mathbf{b}$ possess in a curved N-dimensional space?

vi. The formulas summarized in part (v) above were derived for Euclidean (i.e. 'flat') spaces, namely a plane for which $\mathbf{b} = -\nabla_{\mathcal{L}}\mathbf{p}$ or the ordinary Euclidean 3-space of our 'world' (for which $\mathbf{b} = -\nabla_s\mathbf{n}$).

If we considered a curve drawn in a non-Euclidean space (e.g. a curve C drawn on the two-dimensional surface of a sphere, rather than on a two-dimensional plane), would you expect the 'invariant' formula $\mathbf{b} = -\nabla_{\mathcal{L}} \mathbf{p}$ to continue to hold? In particular, would the 'space' curve C, now be expected to contain elements of the curvature of the space in which it was embedded?

Additional Reading for Chapter 18

§18.4 Hydrostatics of a Curved Contact Line

Vignes-Adler, M. & Brenner, H. 1985 A micromechanical derivation of the differential equations of interfacial statics. III. Line Tension. *J. Colloid Interface Sci.* **103**, 11–43.

Appendix A

An elementary review of basic tensor analysis is provided in the first section of this appendix. In it we adopt the notation of Menzel (1961), whose book may be consulted for further details. Our focus here is upon a three-dimensional parameterization of space, relative to which the two-dimensional parameterizations of surfaces in §3.2 and Appendix B form close analogs. The vector-polyadic (Gibbsian) notation used in the text is briefly summarized in the following section: further elucidation may be found in the text of Wilson & Gibbs (1929). A useful tabulation of differential operators in general orthogonal curvilinear coordinates is provided in §A.3.

Readers more familiar (from their earlier transport processes courses) with Cartesian tensors than the so-called general or curvilinear tensors outlined in this appendix may initially find it unnecessary to peruse this material, arguing that Cartesian tensors are equally useful and, indeed, conceptually simpler. There is much pragmatic truth to this observation, at least for transport problems in ordinary three-dimensional space owing to its property of being flat (or Euclidean). However, the *two-dimensional spaces* characterizing *curved* interfaces are not flat (but are, rather, generally Riemannian), and Cartesian tensors are essentially useless. It is in addressing such problems that general tensors show their superiority over Cartesian tensors.

Appendix A is intended to provide an easy transition between these two schemes, since the reader familiar with Cartesian tensors will not have any difficulty in seeing their relation to general tensors in the familiar context of three dimensions. However, the real value of mastering general tensors does not make itself manifest until Appendix B (and Chapter 3, *et seq.*), which are concerned with *curved* interfaces, and for which Cartesian tensors are of no utility.

A.1 Basic Tensor Analysis

Base Vectors

Let the vector $\mathbf{x}$ denote a two-point vector field, specifying the position of a point P in three-dimensional Euclidean space relative to an origin O. Furthermore, let the trio $q^i \equiv (q^1, q^2, q^3)$, hereafter referred to as curvilinear coordinates of the point $\mathbf{x}$ (Happel & Brenner 1983—their Appendix A) represent specified functions of $\mathbf{x}$. Then,*

$$d\mathbf{x} = \frac{\partial \mathbf{x}}{\partial q^i} dq^i$$

$$\equiv \mathbf{g}_i \, dq^i \qquad \text{(A.1-1)}$$

provides a relation between the two-point (bilocal) differential displacement $d\mathbf{x}$ between neighboring points $\mathbf{x}$ and $\mathbf{x}+d\mathbf{x}$ and the comparable differential displacement dq^i of the coordinates q^i between these same neighboring points. The 'one-point' vectors

$$\mathbf{g}_i \overset{\text{def}}{=} \frac{\partial \mathbf{x}}{\partial q^i} \qquad \text{(A.1-2)}$$

comprise a system of base vectors, defined locally at the point P [$\equiv \mathbf{x} \equiv (q^1, q^2, q^3)$], 'induced' by the specific choice of the functional dependence of the q^i upon $\mathbf{x}$. The base vectors $\mathbf{g}_i$ are the three-dimensional analogs of the two-dimensional surface base vectors $\mathbf{a}_\alpha$ [cf. Eq. (3.2-2)].

A set of reciprocal base vectors [analogous to the reciprocal surface base vectors $\mathbf{a}^\alpha$ defined in Eq. (B.1-4b)] may be defined such that

$$\mathbf{g}^j \overset{\text{def}}{=} \frac{\partial q^j}{\partial \mathbf{x}}, \qquad \text{(A.1-3)}$$

and possess the property that

$$\mathbf{g}_i \cdot \mathbf{g}^j = \delta_i^j . \qquad \text{(A.1-4)}$$

Here,

$$\delta_i^j = \begin{cases} 1 & (i = j) \\ 0 & (i \neq j) \end{cases} \qquad \text{(A.1-5)}$$

is the three-dimensional Kronecker delta.

---

* As noted also in Appendix B, the Einstein summation convention for tensors is that upper and lower repeated dummy indices are to be summed over their respective ranges. Greek indices ($\alpha, \beta, \gamma, \ldots$) will always span the range (1, 2), whereas italic letters ($i, j, k, \ldots$) will generally span the range (1, 2, 3), though in certain n-dimensional problems the latter may be envisaged with a range (1, 2, 3, . . . , $n-1$, n). Thus, for example, Eq. (A.1-1) written out explicitly is $d\mathbf{x} = \mathbf{g}_1 dq^1 + \mathbf{g}_2 dq^2 + \mathbf{g}_3 dq^3$. Care should be taken to distinguish between (lightface) tensorial quantities, like g_{ij}, whose indices denote components, and (boldface) vector quantities, like $\mathbf{g}_i$, whose indices denote a particular vector.

A set of self-reciprocal basis vectors is defined by the equivalence of base and reciprocal base vectors, $\mathbf{g}_i = \mathbf{g}^i$ (i=1,2,3). Such self reciprocity occurs only for the case of *orthonormal* curvilinear coordinate systems (q^1, q^2, q^3) (see Question 3.19).

Covariant and Contravariant Vectors

A vector $\mathbf{v}$ may be expressed in a particular curvilinear coordinate system q^i in one of two equivalent ways, depending upon whether one chooses $\mathbf{g}_i$ or $\mathbf{g}^i$ as the basis vectors for the representation. Thus, the representation of $\mathbf{v}$ in terms of its *covariant* components v_i is

$$\mathbf{v} = \mathbf{g}^i v_i \equiv \mathbf{g}^1 v_1 + \mathbf{g}^2 v_2 + \mathbf{g}^3 v_3 . \qquad \text{(A.1-6)}$$

Alternatively, the representation of $\mathbf{v}$ in terms of its *contravariant* components v^i is

$$\mathbf{v} = \mathbf{g}_i v^i . \qquad \text{(A.1-7)}$$

Components of Polyadics

Polyadics are frequently encountered in the text, examples being the dyadic idemfactor $\mathbf{I}$, the pressure dyadic $\mathbf{P}$, the pseudotriadic ε, and the surface Riemann tetradic (§B.2). These invariant entities may be expressed in component form either in terms of the base vectors $\mathbf{g}_k$ and/or $\mathbf{g}^k$; more usually, in the case of orthogonal systems, the component representation is effected in terms of a set of orthonormal basis vectors $\mathbf{i}_k$ (as is generally done in this text, and discussed in §A.2).

By way of example, a dyadic $\mathbf{D}$ may be expressed in terms of its components in any of the following alternative forms:*

$$\begin{aligned} \mathbf{D} &= \mathbf{g}_i \mathbf{g}_j D^{ij} \\ &= \mathbf{g}^i \mathbf{g}^j D_{ij} \\ &= \mathbf{g}^i \mathbf{g}_j D_{i\cdot}^{\cdot j} \\ &= \mathbf{g}_i \mathbf{g}^j D_{\cdot j}^{i \cdot}, \end{aligned} \qquad \text{(A.1-8)}$$

where the D^{ij} are the *contravariant* components of $\mathbf{D}$, and D_{ij} its *covariant* components; $D_{i\cdot}^{\cdot j}$ and $D_{\cdot j}^{i\cdot}$ are its *mixed* components. Similarly, a triadic $\mathbf{T}$ may be expressed alternatively in any of the following forms:

$$\begin{aligned} \mathbf{T} &= \mathbf{g}_i \mathbf{g}_j \mathbf{g}_k T^{ijk} \\ &= \mathbf{g}^i \mathbf{g}^j \mathbf{g}^k T_{ijk}, \text{ etc.,} \end{aligned} \qquad \text{(A.1-9)}$$

and so on for all higher-order polyadics.

Idemfactor

The *dyadic idemfactor* **I** may be represented as

$$\mathbf{I} \equiv \mathbf{g}_i \mathbf{g}^i$$

$$= \mathbf{g}_j \mathbf{g}^i \delta_i^{\ j} . \qquad \text{(A.1-10)}$$

It possesses the important 'identity' property that

$$\mathbf{I} \cdot \mathbf{v} = \mathbf{v} \qquad \text{(A.1-11)}$$

for any vector **v**, which, in fact, constitutes the invariant definition of **I** (in the invariant sense of Gibbs).

Metric Tensor

The *metric tensor* is defined as [cf. Eq. (3.2-3)]

$$g_{ij} \overset{\text{def}}{=} \mathbf{g}_i \cdot \mathbf{g}_j \qquad \text{(A.1-12)}$$

which gives rise to the corresponding *metric dyadic*

$$\mathbf{g} \overset{\text{def}}{=} \mathbf{g}^i \mathbf{g}^j g_{ij} . \qquad \text{(A.1-13)}$$

As the order of the vectors appearing in the dot product in (A.1-12) is immaterial, it follows that $g_{ij}=g_{ji}$, whence the metric tensor is symmetric. Likewise, in (A.1-13), $\mathbf{g}^\dagger = \mathbf{g}$, whence the metric dyadic is also symmetric.

Analogous to Eq. (A.1-12), a conjugate metric tensor may be defined as

$$g^{ij} \overset{\text{def}}{=} \mathbf{g}^i \cdot \mathbf{g}^j \qquad \text{(A.1-14)}$$

together with the corresponding conjugate metric dyadic

$$\mathbf{G} = \mathbf{g}_i \mathbf{g}_j g^{ij} . \qquad \text{(A.1-15)}$$

The dyadics **G** and **g** may be contracted to provide

$$\mathbf{G} \cdot \mathbf{g} = g^{ij} \mathbf{g}_i \mathbf{g}_j \cdot \mathbf{g}^k \mathbf{g}^l g_{kl}$$

$$= g^{ij} \mathbf{g}_i \delta_j^k \mathbf{g}^l g_{kl}$$

$$= g^{ik} g_{kl} \mathbf{g}_i \mathbf{g}^l$$

$$= \delta_l^i \mathbf{g}_i \mathbf{g}^l$$

$$\equiv \mathbf{I} . \qquad \text{(A.1-16)}$$

The determinants of the metric and its conjugate are specified by the notation

$$g \overset{\text{def}}{=} \det \mathbf{g} \qquad \text{(A.1-17 a)}$$

$$G \overset{\text{def}}{=} \det \mathbf{G}, \qquad \text{(A.1-17 b)}$$

and their product yields

$$gG = 1, \qquad \text{(A.1-18)}$$

as follows upon forming the determinant of (A.1-16) and using the product rule, namely det(**A**·**B**)=(det **A**)(det **B**), for **A** and **B** any dyadics.

Cross Product of Two Vectors

The cross (or vector) product of two vectors **a** and **b** may be represented as

$$\mathbf{a} \times \mathbf{b} = \mathbf{g}_k \, \varepsilon^{ijk} \, a_i b_j, \tag{A.1-19}$$

where the unit alternator ε^{ijk} is a third-rank pseudotensor defined as

$$\varepsilon^{ijk} \overset{\text{def}}{=} \frac{e^{ijk}}{\sqrt{g}}, \tag{A.1-20}$$

with e^{ijk} the permutation symbol. Each component of the latter assumes the value +1 or −1 according as *ijk* is an even or odd permutation of 123, and is zero if any pair of indices of the trio *i j k* are equal.

The unit isotropic pseudotriadic, $\varepsilon = -\mathbf{I} \times \mathbf{I}$, possesses the representation

$$\varepsilon \equiv \mathbf{g}_i \mathbf{g}_j \mathbf{g}_k \, \varepsilon^{ijk}, \tag{A.1-21}$$

or, alternatively,

$$\varepsilon \equiv \mathbf{g}^i \mathbf{g}^j \mathbf{g}^k \, \varepsilon_{ijk}. \tag{A.1-22}$$

Dot Product of Two Vectors

The dot (or scalar) product of two vectors **a** and **b** may be expressed as

$$\begin{aligned}\mathbf{a} \cdot \mathbf{b} &= \mathbf{g}_i a^i \cdot \mathbf{g}^j b_j \\ &= \mathbf{g}_i \cdot \mathbf{g}^j \, a^i b_j \\ &= \delta_i^j \, a^i b_j \\ &= a^i b_i.\end{aligned} \tag{A.1-23}$$

This operation affords the following representation of the magnitude of a vector **v** in terms of the metric tensor:

$$|\mathbf{v}| \overset{\text{def}}{=} (\mathbf{v} \cdot \mathbf{v})^{1/2} \equiv \left(g_{ij} v^i v^j \right)^{1/2}. \tag{A.1-24}$$

Use of the so-called "nesting convention" of Chapman & Cowling (1961) (rather than that of Gibbs) permits the double-dot product of two dyads **ab** and **cd** to eventually be expressed as the scalar

$$\begin{aligned}\mathbf{ab:cd} &= (\mathbf{b}\cdot\mathbf{c})(\mathbf{a}\cdot\mathbf{d})\\ &= \left(\mathbf{g}_i b^i \cdot \mathbf{g}^j c_j\right)\left(\mathbf{g}_k d^k \cdot \mathbf{g}^l d_l\right)\\ &= \left(\mathbf{g}_i \cdot \mathbf{g}^j\, b^i c_j\right)\left(\mathbf{g}_k \cdot \mathbf{g}^l\, a^k d_l\right)\\ &= \left(\delta_i^j b^i c_j\right)\left(\delta_k^l a^k d_l\right)\\ &= \left(b^i c_i\right)\left(a^k d_k\right)\\ &\equiv a^k b^i c_i d_k . \end{aligned} \tag{A.1-25}$$

Orthogonal Curvilinear Coordinate Systems

The properties of orthogonal curvilinear coordinate systems are extensively summarized by Happel & Brenner (1983) as well as by many others [e.g. Reddy & Rasmussen (1983); Menzel (1961)]. For such systems $g_{ij} = 0$ $(i \neq j)$. In such circumstances one typically uses normalized or *unit* base vectors, namely

$$\begin{aligned}\mathbf{i}_i &\equiv \frac{\mathbf{g}_i}{|\mathbf{g}_i|} \qquad \text{(no sum on } i\text{)}\\ &= \frac{\mathbf{g}_i}{\sqrt{g_{ii}}}\\ &= h_i \mathbf{g}_i , \end{aligned} \tag{A.1-26}$$

with $|\mathbf{i}_i| = 1$ $(i=1,2,3)$, and where the *metrical coefficients* h_i (>0) are given by

$$h_i = \frac{1}{\sqrt{g_{ii}}} \qquad \text{(no sum on } i\text{)}. \tag{A.1-27}$$

The trio of unit vectors $(\mathbf{i}_1, \mathbf{i}_2, \mathbf{i}_3)$ form an orthonormal basis at each point $P \equiv (q^1, q^2, q^3)$, in the sense that

$$\mathbf{i}_j \cdot \mathbf{i}_k = \delta_{jk} \qquad (j,\ k = 1,\ 2,\ 3),$$

with δ_{jk} the covariant form of the Kronecker delta.

The unit vector $\mathbf{i}_j$ corresponds physically to the unit tangent vector $\mathbf{i}_j \equiv \delta\mathbf{x}/\delta l_j$ along the q^j-coordinate curve, where $\delta\mathbf{x}$ is the displacement along the q^j-coordinate curve and $\delta l_j \overset{\text{def}}{=} \mathbf{i}_j \cdot \delta\mathbf{x} \equiv \delta q^j / h_j$ (no sum on j), as follows from Eqs. (A.1-2) and (A.1-26). Thus,

$$\mathbf{i}_i = h_i \frac{\partial \mathbf{x}}{\partial q^i} \qquad \text{(no sum on } i\text{)}. \tag{A.1-28}$$

Since $h_j>0$, Eq. (A.1-28) contains the implicit convention that the unit tangent vectors $\mathbf{i}_j$ are always drawn in the direction of *increasing* q^j. Moreover, in applications, the orthonormal bases ($\mathbf{i}_1$, $\mathbf{i}_2$, $\mathbf{i}_3$) will always be ordered such as to form a *right-handed* system, i.e. $\mathbf{i}_1 \cdot \mathbf{i}_2 \times \mathbf{i}_3 = 1$.

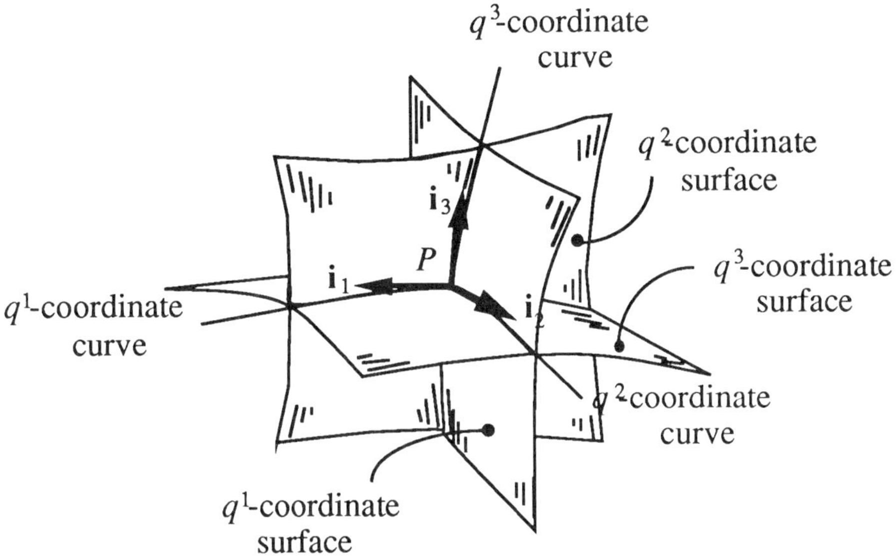

Figure A.1-1 Orthogonal coordinate surfaces q^i=constant (i=1,2,3) intersect at right angles. The respective coordinate curves arise from the intersection of pairs of coordinate surfaces. (Thus, the q^1-coordinate *curve* arises from the intersection of the q^2-and q^3-coordinate *surfaces*; only the coordinate q^3 varies along this curve.) These curves are mutually perpendicular.

Comparison of Eqs. (A.1-3), (A.1-4) and (A.1-26) yields an alternative physical interpretation of the orthonormal trio ($\mathbf{i}_1$, $\mathbf{i}_2$, $\mathbf{i}_3$), namely

$$\mathbf{i}_j = \frac{\nabla q_j}{h_j}. \qquad \text{(A.1- 29 a)}$$

Since the gradient vector ∇F is always perpendicular to the surface $F(\mathbf{x})$=constant, it follows from Eq. (A.1-28) that $\mathbf{i}_j$ is perpendicular to the q^j-coordinate surface (and points in the direction of increasing q^j since $h_j>0$). Upon forming the magnitude of both sides of (A.1-28) it follows from the unitary nature of $\mathbf{i}_j$ that

$$|h_i| = |\nabla q^i|. \qquad \text{(A.1- 29 b)}$$

Equation (A.1-29b) may be used to provide the following explicit representation [relative to a cartesian coordinate system (x, y, z)] of the metrical coefficients (or 'scale factors') h_i:

$$h_1 = \sqrt{\left(\frac{\partial q^1}{\partial x}\right)^2 + \left(\frac{\partial q^1}{\partial y}\right)^2 + \left(\frac{\partial q^1}{\partial z}\right)^2}, \quad h_2 = \sqrt{\left(\frac{\partial q^2}{\partial x}\right)^2 + \left(\frac{\partial q^2}{\partial y}\right)^2 + \left(\frac{\partial q^2}{\partial z}\right)^2},$$

$$h_3 = \sqrt{\left(\frac{\partial q^3}{\partial x}\right)^2 + \left(\frac{\partial q^3}{\partial y}\right)^2 + \left(\frac{\partial q^3}{\partial z}\right)^2}. \qquad \text{(A.1- 30)}$$

A.2 Polyadic Notation

As attention will be focused more-or-less exclusively in this text upon orthogonal curvilinear coordinate systems (rather than more general nonorthogonal systems), it is a matter of convenience that the unit base vectors $\mathbf{i}_i$ are consistently employed in the following representations rather than their more fundamental counterparts $\mathbf{g}_i$. Such representations greatly simplify the treatment of vectors and polyadics, as is made clear in the remaining sections of the appendix.

Component Representation of Polyadics in terms of Orthogonal Unit Basis Vectors

In this text, polyadics will frequently be represented in terms of their curvilinear components in a system of orthonormal base vectors ($\mathbf{i}_1$, $\mathbf{i}_2$, $\mathbf{i}_3$) for the reasons outlined in the preceding paragraph [see Question 3.21]. The notation conventionally adopted in such cases differs from the general tensorial notation of §A.1 in the following way: Only subscripted indices are employed, as in v_i [and not v^i, since the orthonormal unit vectors $\mathbf{i}_i$ constitute a self-reciprocal basis system ($\mathbf{i}_i = \mathbf{i}^i$)]; in this same context,* tensor indices will consistently be rendered as subscripts, as for example in the scalar $\mathbf{I}:\mathbf{D} = D_{ii}$ ($\equiv D_{11}+D_{22}+D_{33}$) and the dyadic idemfactor, $\mathbf{I} = \mathbf{i}_j\mathbf{i}_k\delta_{jk}$.

Explicitly, the representation of a vector $\mathbf{v}$ in terms of the orthonormal trio ($\mathbf{i}_1$, $\mathbf{i}_2$, $\mathbf{i}_3$) is

$$\mathbf{v} = \sum_{j=1}^{3} \mathbf{i}_j v_j \equiv \mathbf{i}_j v_j. \qquad \text{(A.2- 1)}$$

Likewise, a dyadic $\mathbf{D}$ is generally represented as

$$\mathbf{D} = \sum_{j=1}^{3}\sum_{k=1}^{3} \mathbf{i}_j \mathbf{i}_k D_{jk} \equiv \mathbf{i}_j \mathbf{i}_k D_{jk}, \qquad \text{(A.2- 2)}$$

and a triadic $\mathbf{T}$ as

---

* In a somewhat similar context, the orthogonal curvilinear coordinates themselves will often be written as (q_1, q_2, q_3) rather than (q^1, q^2, q^3).

$$\mathbf{T} = \sum_{j=1}^{3}\sum_{k=1}^{3}\sum_{l=1}^{3} \mathbf{i}_j \mathbf{i}_k \mathbf{i}_l T_{jkl} \equiv \mathbf{i}_j \mathbf{i}_k \mathbf{i}_l T_{jkl}\,. \qquad \text{(A.2-3)}$$

Symmetric and Antisymmetric Dyadics

A *symmetric* dyadic **S** is defined as a dyadic possessing the property that

$$\mathbf{S} = \mathbf{S}^{\dagger}, \qquad \text{(A.2-4)}$$

where † is the transposition operator. In component form this gives

$$S_{ij} = S_{ji}. \qquad \text{(A.2-5 a)}$$

Explicitly,

$$S_{12} = S_{21}, \quad S_{23} = S_{32}, \quad S_{13} = S_{31}. \qquad \text{(A.2-5 b)}$$

An *antisymmetric* dyadic **A** is defined by the property that

$$\mathbf{A} = -\,\mathbf{A}^{\dagger}, \qquad \text{(A.2-6)}$$

which gives, in component form,

$$A_{ij} = -\,A_{ji}. \qquad \text{(A.2-7 a)}$$

Explicitly,

$$A_{11} = 0, \quad A_{22} = 0, \quad A_{33} = 0,$$
$$A_{12} = -\,A_{21}, \quad A_{23} = -\,A_{32}, \; A_{13} = -\,A_{31}\,. \qquad \text{(A.2-7 b)}$$

A symmetric dyadic generally possesses *six* independent components and an antisymmetric dyadic *three* independent components, as is readily seen from the following matrix representations:

$$[S_{ij}] = \begin{bmatrix} S_{11} & S_{12} & S_{13} \\ S_{12} & S_{22} & S_{23} \\ S_{13} & S_{23} & S_{33} \end{bmatrix} \qquad \text{(A.2-8)}$$

and

$$[A_{ij}] = \begin{bmatrix} 0 & A_{12} & A_{13} \\ -\,A_{12} & 0 & A_{23} \\ -\,A_{13} & -\,A_{23} & 0 \end{bmatrix}. \qquad \text{(A.2-9)}$$

The three independent components of the antisymmetric dyadic **A** may be expressed in terms of the components of a *pseudovector* $\mathbf{A}_{\times}$, as in

$$\mathbf{A}_{\times} \equiv -\frac{1}{2}\,\varepsilon\!:\mathbf{A} \equiv \mathbf{i}_1 A_{23} + \mathbf{i}_2 A_{31} + \mathbf{i}_3 A_{12}\,. \qquad \text{(A.2-10)}$$

Any dyadic **D** may be expressed as a sum of a symmetric and an antisymmetric dyadic:

$$\mathbf{D} = \mathbf{S} + \mathbf{A}, \qquad \text{(A.2-11)}$$

in which

$$\mathbf{S} \equiv \frac{1}{2}(\mathbf{D} + \mathbf{D}^{\dagger}) \qquad \text{(A.2-12)}$$

is symmetric, and

$$\mathbf{A} \equiv \frac{1}{2}(\mathbf{D} - \mathbf{D}^{\dagger}) \tag{A.2-13}$$

is antisymmetric.

A.3 Differential Operators in Orthogonal Curvilinear Coordinates

The following is a listing of certain useful vector differential operators in general orthogonal curvilinear coordinates (q^1, q^2, q^3). Additional general representations, particularly for surface differential operators, are provided in §B.5. Scale factors are provided in §B.1 for planar, circular-cylindrical and spherical-polar coordinates.

Of particular value are the following formulas for the nine possible derivatives $\partial \mathbf{i}_j/\partial q^k$ $(j, k=1,2,3)$ of the unit basis vectors [see Reddy & Rasmussen (1982) or Appendix A of Happel & Brenner (1983)], derived from Eq. (A.1-28):

$$
\begin{aligned}
\frac{\partial \mathbf{i}_1}{\partial q^1} &= -\mathbf{i}_2 h_2 \frac{\partial}{\partial q^2}\left(\frac{1}{h_1}\right) - \mathbf{i}_3 h_3 \frac{\partial}{\partial q^3}\left(\frac{1}{h_1}\right), \\
\frac{\partial \mathbf{i}_1}{\partial q^2} &= \mathbf{i}_2 h_1 \frac{\partial}{\partial q^1}\left(\frac{1}{h_2}\right), \\
\frac{\partial \mathbf{i}_1}{\partial q^3} &= \mathbf{i}_3 h_1 \frac{\partial}{\partial q^1}\left(\frac{1}{h_3}\right), \\
\frac{\partial \mathbf{i}_2}{\partial q^1} &= \mathbf{i}_1 h_2 \frac{\partial}{\partial q^2}\left(\frac{1}{h_1}\right), \\
\frac{\partial \mathbf{i}_2}{\partial q^2} &= -\mathbf{i}_3 h_3 \frac{\partial}{\partial q^3}\left(\frac{1}{h_2}\right) - \mathbf{i}_1 h_1 \frac{\partial}{\partial q^1}\left(\frac{1}{h_2}\right), \\
\frac{\partial \mathbf{i}_2}{\partial q^3} &= \mathbf{i}_3 h_2 \frac{\partial}{\partial q^2}\left(\frac{1}{h_3}\right), \\
\frac{\partial \mathbf{i}_3}{\partial q^1} &= \mathbf{i}_1 h_3 \frac{\partial}{\partial q^3}\left(\frac{1}{h_1}\right), \\
\frac{\partial \mathbf{i}_3}{\partial q^2} &= \mathbf{i}_2 h_3 \frac{\partial}{\partial q^3}\left(\frac{1}{h_2}\right), \\
\frac{\partial \mathbf{i}_3}{\partial q^3} &= -\mathbf{i}_1 h_1 \frac{\partial}{\partial q^1}\left(\frac{1}{h_3}\right) - \mathbf{i}_2 h_2 \frac{\partial}{\partial q^2}\left(\frac{1}{h_3}\right);
\end{aligned} \tag{A.3-1}
$$

Gradient operator, ∇ (see Question 3.25)

$$\nabla = \mathbf{i}_1 h_1 \frac{\partial}{\partial q^1} + \mathbf{i}_2 h_2 \frac{\partial}{\partial q^2} + \mathbf{i}_3 h_3 \frac{\partial}{\partial q^3}; \tag{A.3-2}$$

Laplacian, ∇^2 (see Question 3.26)

$$\nabla^2 = h_1 h_2 h_3\left[\frac{\partial}{\partial q^1}\left(\frac{h_1}{h_2 h_3}\frac{\partial}{\partial q^1}\right) + \frac{\partial}{\partial q^2}\left(\frac{h_2}{h_3 h_1}\frac{\partial}{\partial q^2}\right) + \frac{\partial}{\partial q^3}\left(\frac{h_3}{h_1 h_2}\frac{\partial}{\partial q^3}\right)\right]. \tag{A.3-3}$$

Then, for any general vector function

$$\mathbf{v} = \mathbf{i}_1 v_1 + \mathbf{i}_2 v_2 + \mathbf{i}_3 v_3, \tag{A.3-4}$$

the operations given below may be derived from Eqs. (A.3-2) and (A.3-3):

Divergence of a vector, $\nabla\cdot\mathbf{v}$

$$\nabla\cdot\mathbf{v} = h_1 h_2 h_3\left[\frac{\partial}{\partial q^1}\left(\frac{v_1}{h_2 h_3}\right) + \frac{\partial}{\partial q^2}\left(\frac{v_2}{h_3 h_1}\right) + \frac{\partial}{\partial q^3}\left(\frac{v_2}{h_1 h_2}\right)\right]; \tag{A.3-5}$$

Curl of a vector, $\nabla\times\mathbf{v}$

$$\begin{aligned}\nabla\times\mathbf{v} = {} & \mathbf{i}_1 h_2 h_3\left[\frac{\partial}{\partial q^2}\left(\frac{v_3}{h_3}\right) - \frac{\partial}{\partial q^3}\left(\frac{v_2}{h_2}\right)\right] \\ & + \mathbf{i}_2 h_3 h_1\left[\frac{\partial}{\partial q^3}\left(\frac{v_1}{h_1}\right) - \frac{\partial}{\partial q^1}\left(\frac{v_3}{h_3}\right)\right] \\ & + \mathbf{i}_3 h_1 h_2\left[\frac{\partial}{\partial q^1}\left(\frac{v_2}{h_2}\right) - \frac{\partial}{\partial q^2}\left(\frac{v_1}{h_1}\right)\right]\end{aligned} \tag{A.3-6}$$

Laplacian of a vector, $\nabla^2\mathbf{v}$

$$\begin{aligned}\nabla^2\mathbf{v} = \mathbf{i}_1\Bigg[& \nabla^2 v_1 - \frac{v_1}{h_1}\nabla^2 h_1 \\ & + \frac{v_1}{h_1} h_1 \frac{\partial}{\partial q^1}(\nabla^2 q_1) + \frac{v_2}{h_2} h_1 \frac{\partial}{\partial q^1}(\nabla^2 q_2) + \frac{v_3}{h_3} h_1 \frac{\partial}{\partial q^1}(\nabla^2 q_3) \\ & - 2h_1^2\frac{\partial h_1}{\partial q^1}\frac{\partial}{\partial q^1}\left(\frac{v_1}{h_1}\right) - 2h_2^2\frac{\partial h_1}{\partial q^2}\frac{\partial}{\partial q^1}\left(\frac{v_2}{h_2}\right) - 2h_3^2\frac{\partial h_1}{\partial q^3}\frac{\partial}{\partial q^1}\left(\frac{v_3}{h_3}\right) \\ & + h_1\frac{\partial h_1^2}{\partial q^1}\frac{\partial}{\partial q^1}\left(\frac{v_1}{h_1}\right) + h_1\frac{\partial h_2^2}{\partial q^1}\frac{\partial}{\partial q^2}\left(\frac{v_2}{h_2}\right) + h_1\frac{\partial h_3^2}{\partial q^1}\frac{\partial}{\partial q^3}\left(\frac{v_3}{h_3}\right)\Bigg] \\ & + \ldots \text{ (cycl.)};\end{aligned} \tag{A.3-7}$$

Divergence of a dyadic, ∇·D

$$\nabla \cdot \mathbf{D} = \mathbf{i}_1\Bigg[h_1 h_2 h_3\left\{\frac{\partial}{\partial q^1}\left(\frac{D_{11}}{h_2 h_3}\right) + \frac{\partial}{\partial q^2}\left(\frac{D_{21}}{h_3 h_1}\right) + \frac{\partial}{\partial q^3}\left(\frac{D_{31}}{h_1 h_2}\right)\right\}$$
$$+ h_1 h_1 D_{11}\frac{\partial}{\partial q^1}\left(\frac{1}{h_1}\right) + h_1 h_2 D_{12}\frac{\partial}{\partial q^2}\left(\frac{1}{h_1}\right) + h_1 h_3 D_{13}\frac{\partial}{\partial q^3}\left(\frac{1}{h_1}\right)$$
$$- h_1 h_1 D_{11}\frac{\partial}{\partial q^1}\left(\frac{1}{h_1}\right) + h_1 h_2 D_{22}\frac{\partial}{\partial q^1}\left(\frac{1}{h_1}\right) + h_1 h_3 D_{33}\frac{\partial}{\partial q^1}\left(\frac{1}{h_1}\right)\Bigg]$$
$$+ \ldots \text{(cycl.)}. \tag{A.3- 8}$$

In the latter two relations, the $\mathbf{i}_2$ and $\mathbf{i}_3$ components are formed in an entirely analogous way to the $\mathbf{i}_1$ component shown.

Appendix B

Material is provided in this appendix pertaining to subjects presented in Part I (primarily Chapters 3 to 5) of the text. Section B.1 provides an explicit consideration of surface geometry in semi-orthogonal curvilinear coordinates.

B.1 Surface Geometry in Semi-Orthogonal Curvilinear Coordinates

Explicit formulas outlined in §3.2 for the key geometrical properties of a surface are here further developed in the context of *semi-orthogonal* curvilinear components (q^1, q^2, q^3). The terminology 'semi-orthogonality' is introduced here to denote three-dimensional curvilinear coordinate systems for which two of the three curvilinear coordinates [namely the *surface* coordinate pair (q^1, q^2)] need not be orthogonal to one another, whereas the third coordinate (namely the *normal* coordinate q^3) is, however, orthogonal to the other two. Explicitly, in the notation of Appendix A, $\mathbf{g}_3 \cdot \mathbf{g}_1 = 0$ and $\mathbf{g}_3 \cdot \mathbf{g}_2 = 0$, whereas $\mathbf{g}_1 \cdot \mathbf{g}_2 \neq 0$, in general; when the latter inequality becomes an equality, namely $\mathbf{g}_1 \cdot \mathbf{g}_2 = 0$, the semi-orthogonal system then becomes a fully orthogonal one. The latter are discussed at length in Appendix A. Reasons for distinguishing semi-orthogonal systems from the more general class of nonorthogonal systems lie in the fact that whereas in many—if not most—physical applications the interface may be conveniently parameterized as a q^3-coordinate surface in space within the context of a *fully* orthogonal curvilinear coordinate system (q^1, q^2, q^3), circumstances may nevertheless be envisioned (e.g. a deforming fluid interface) in which the surface coordinates (q^1, q^2) would not necessarily be orthogonal to one another (although they might both be orthogonal to the q^3-coordinate curve). Observe that the surface geometrical formulas presented in §3.2 are appropriate to such nonorthogonal surface coordinate pairs (q^1, q^2).

Ultimately, the interface is to be regarded as embedded within three-dimensional space, and for these purposes it is convenient to choose the third coordinate (i.e. the q^3 coordinate) as representing a coordinate normal to the surface; as such, the unit surface normal **n** at a surface point $P_s \equiv P_s(q^1, q^2)$ also represents the tangent of the q^3-coordinate curve at P_s (see Figure B.1-3). According to the definition of the unit surface normal [cf. Eq. (3.2-5)], the q^3-coordinate curve is locally orthogonal to the surface, with the latter representing a q^3=constant coordinate surface (since q^3 does not vary over this surface).

In this section the surface geometry of the q^3 coordinate surfaces in a semi-orthogonal coordinate system (q^1, q^2, q^3) is further developed, supplementing the presentation of §3.2. At the conclusion of this section, relations for certain fundamental surface geometrical properties are expressed in an explicit form in fully orthogonal coordinates, as most calculations will eventually involve such an orthogonal parameterization of the surface. [Orthogonal curvilinear coordinates (see Appendix A) encompass many, if not most, of the standard geometrical surface shapes encountered in classical hydrodynamic problems; included in this class are cartesian, spherical and circular cylindrical surfaces, and possibly small deformations from these shapes.]

Throughout the text, Einstein's summation convention is followed in tensorial relations [such as in Eq. (3.2-7), or Eqs. (B.1-5) through (B.1-8) below], wherein all dummy indices repeated in both superscript (contravariant) and subscript (covariant) positions are to be summed over the range of their indices (i.e. α=1,2); for example, $A_\alpha B^\alpha \equiv A_1 B^1 + A_2 B^2$. Nonrepeated indices are called *free* indices, and are allowed to assume any value within the index range. The reader should be careful to distinguish between (lightface) quantities such as $a^{\beta\alpha}$, $a_{\alpha\beta}$ and δ_α^β, which are tensors, whose indices denote *components*, and (boldface) quantities such as $\mathbf{a}_\alpha$, $\mathbf{a}^\beta$, which are vectors, whose indices identify a particular vector. Relations also arise between vector or scalar quantities [cf. Eq. (3.2-12)] wherein the summation convention is to be suspended. In these cases the phrase "no sum on α" is generally appended to the relation.

Curvilinear Surface Coordinates

As discussed in the preceding subsection, the essential conceptual simplification employed in this text is to regard three-dimensional space as parameterized by the semi-orthogonal curvilinear coordinate system (q^1, q^2, q^3), wherein the (macroscale) *interface* coincides with a particular (i.e. *parent*) q^3-coordinate surface (see Figure B.1-3; see also §15.3). As such, in the special but important case where the surface coordinates (q^1, q^2) are mutually orthogonal, the geometrical description of the interface reduces to the limiting case of general three-dimensional orthogonal curvilinear coordinates (q^1, q^2, q^3) (cf. Appendix A). This limiting case point of view is developed in the following paragraphs of this appendix, initially by focusing upon the *intrinsic* surface coordinates (q^1, q^2) themselves, as well as certain of their consequent geometrical properties. Later, the surface is viewed *extrinsically* as a particular q^3-coordinate surface in three-dimensional space, whereby the respective curvatures of the q^1 and q^2 coordinate curves are explicitly developed in terms of a set of orthonormal basis vectors and spatial derivatives in the third dimension, q^3 [cf. Eqs. (B.1-21) and (B.1-22)].

Imagine, as depicted in Figure B.1-1, the specification of a point P_s on a surface by the two-point position vector $\mathbf{x}_s$ relative to a fixed origin O. Explicitly, the surface point P_s may be regarded as specified by a function $F^0(x, y, z)=0$, where (x, y, z) are a set of cartesian coordinates, as pursued in Question 3.8. Let the coordinates q^1 and q^2 be independent functions of position within this surface, such that

$$q^\alpha = q^\alpha(\mathbf{x}_s) \qquad (\alpha = 1, 2)\,. \tag{B.1-1}$$

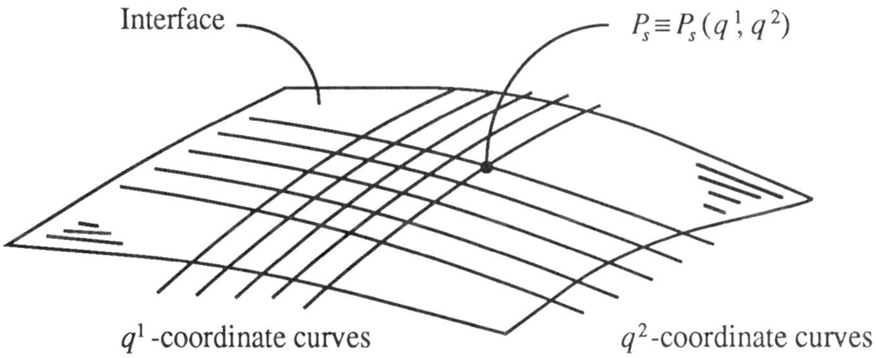

Figure B.1-1 The parameterization of an interface by (q^1, q^2)-curvilinear coordinate curves, not necessarily orthogonal. The surface point P_s corresponds to the pair of coordinate values (q^1, q^2).

These relations may be expressed alternatively as

$$q^\alpha = f^\alpha(x, y, z),$$

or

$$F^\alpha \equiv q^\alpha - f^\alpha(x, y, z) = 0,$$

such that the following three relations hold between the space coordinates (x, y, z) and the surface coordinates (q^1, q^2):

$$F^0(x, y, z; q^1, q^2) = 0,$$

$$F^1(x, y, z; q^1, q^2) = 0,$$

$$F^2(x, y, z; q^1, q^2) = 0.$$

In regions where the Jacobian determinant

$$\frac{\partial (F^0, F^1, F^2)}{\partial (x, y, z)} = \begin{vmatrix} \frac{\partial F^0}{\partial x} & \frac{\partial F^0}{\partial y} & \frac{\partial F^0}{\partial z} \\ \frac{\partial F^1}{\partial x} & \frac{\partial F^1}{\partial y} & \frac{\partial F^1}{\partial z} \\ \frac{\partial F^2}{\partial x} & \frac{\partial F^2}{\partial y} & \frac{\partial F^2}{\partial z} \end{vmatrix} \neq 0 \qquad \text{(B.1-2)}$$

is nonzero (Aris 1962), the system of equations (B.1-1) may be solved simultaneously for the position vector $\mathbf{x}_s$ to obtain

$$\mathbf{x}_s = \mathbf{x}_s(q^1, q^2). \qquad \text{(B.1-3)}$$

According to Eq. (B.1-3), assigning numerical values to the pair (q^1,q^2) serves to specify the position vector $\mathbf{x}_s$: that is, the choice locates the surface point $P_s \equiv P_s(q^1, q^2)$ on the (q^3=constant coordinate) surface. The set of two numbers (q^1, q^2) may, therefore, be regarded as the *curvilinear coordinates* of a point lying on the surface, which is itself situated somewhere in space. These coordinates have a simple geometric interpretation (cf. Appendix A).

Whereas in Cartesian coordinates (x, y, z), the respective differentials (dx, dy, dz) correspond to distances measured along each of the three cartesian coordinates, the analogous differentials (dq^1, dq^2) of the surface curvilinear coordinates (q^1,q^2) do not necessarily possess the same interpretation as representing physical distances. In particular, differential displacements (dq^1, dq^2) of the curvilinear coordinates (q^1,q^2) are related to a differential displacement of the surface position vector $d\mathbf{x}_s$ by the relation*

$$dq^\alpha = \frac{\partial q^\alpha}{\partial \mathbf{x}_s} \cdot d\mathbf{x}_s$$
$$\equiv \mathbf{a}^\alpha \cdot d\mathbf{x}_s, \qquad \text{(B.1-4 a)}$$

where the 'one-point' vectors

---

* Appearing in Eqs. (B.1-4) is the surface gradient operator, defined as

$$\frac{\partial}{\partial \mathbf{x}_s} \equiv \nabla_s .$$

The left-hand notation for this operator is symbolic since the vector $\mathbf{x}_s$ appears in the denominator rather than in the numerator, where vectors normally appear. However, this notation may be given a quite literal interpretation [as in Eq. (B.1-4a)] by defining $\partial / \partial \mathbf{x}_s$ to be the vector operator possessing the property that

$$d\mathbf{x}_s \cdot \frac{\partial \mathbf{f}}{\partial \mathbf{x}_s} \overset{\text{def}}{=} d\mathbf{f}$$

for $\mathbf{f}(\mathbf{x}_s)$ a scalar, vector or polyadic surface field of any rank; here, $d\mathbf{f}$ represents the change in this field between the neighboring surface points $\mathbf{x}_s$ and $\mathbf{x}_s + d\mathbf{x}_s$.

$$\mathbf{a}^{\beta} = \frac{\partial q^{\beta}}{\partial \mathbf{x}_s} \qquad (\beta = 1, 2). \qquad \text{(B.1-4 b)}$$

comprise a system of reciprocal surface base vectors [cf. Eq. (3.2-2)].

Reciprocal Base Vectors

As noted in §3.2, the reciprocal base vectors (B.1-4b) possess the reciprocity property that

$$\mathbf{a}_{\alpha} \cdot \mathbf{a}^{\beta} = \delta_{\alpha}^{\beta} .$$

The surface unit tensor δ_{α}^{β} (which is a mixed, second-rank tensor) may be viewed here as being closely related to the dyadic surface idemfactor $\mathbf{I}_s$, defined in Eq. (3.2-6). Thus,

$$\mathbf{I}_s = \mathbf{a}^{\alpha}\mathbf{a}_{\beta}\delta_{\alpha}^{\beta} \qquad \text{(B.1-5 a)}$$

or, equivalently

$$\mathbf{I}_s = \mathbf{a}_{\beta}\mathbf{a}^{\alpha}\delta_{\alpha}^{\beta} , \qquad \text{(B.1-5 b)}$$

since $\mathbf{I}_s$ is symmetric.

In tensor algebra manipulations the surface Kronecker delta δ_{α}^{β} facilitates the exchanging of indices, as illustrated by use of Eq. (B.1-4a):

$$\begin{aligned} dq^{\beta} &= \frac{\partial q^{\beta}}{\partial \mathbf{x}_s} \cdot d\mathbf{x}_s \\ &= \frac{\partial q^{\beta}}{\partial \mathbf{x}_s} \cdot \left(\frac{\partial \mathbf{x}_s}{\partial q^{\alpha}} dq^{\alpha} \right) \\ &= \frac{\partial q^{\beta}}{\partial q^{\alpha}} dq^{\alpha} \\ &= \delta_{\alpha}^{\beta}\, dq^{\alpha} \end{aligned}$$

the third step following from an application of the chain rule together with the identity $\partial \mathbf{x}_s / \partial \mathbf{x}_s = \mathbf{I}_s$ (see Question 3.14) and the last step following as a consequence of the relation $\partial q^{\beta} / \partial q^{\alpha} = \delta_{\alpha}^{\beta}$ (see Question 3.15), which is an obvious consequence of the definition of the curvilinear coordinates q^{γ} together with the fact that δ_{α}^{β} is the surface Kronecker delta, as in Eq. (3.2-7).

Raising or lowering of tensorial indices may be accomplished by use of the surface metric defined in Eq. (3.2-3); thus,

$$\mathbf{a}_\alpha = a_{\alpha\beta}\,\mathbf{a}^\beta, \tag{B.1-6 a}$$

$$\mathbf{a}^\beta = a^{\beta\alpha}\,\mathbf{a}_\alpha, \tag{B.1-6 b}$$

where, similar to Eq. (3.2-3),

$$a^{\alpha\beta} \stackrel{\text{def}}{=} \mathbf{a}^\alpha \cdot \mathbf{a}^\beta. \tag{B.1-7}$$

Analogous to the tensor representations (B.1-5) of the dyadic idemfactor is the comparable representation of the dyadic surface alternator [cf. Eq. (3.2-9)]

$$\boldsymbol{\varepsilon}_s = \mathbf{a}^\alpha\,\mathbf{a}^\beta\,\varepsilon_{\alpha\beta}, \tag{B.1-8 a}$$

where

$$\varepsilon_{\alpha\beta} = \sqrt{a}\,e_{\alpha\beta}, \tag{B.1-8 b}$$

with $e_{\alpha\beta}$ the second-order permutation symbol, given by

$$e_{11} = e_{22} = 0, \quad e_{12} = -\,e_{21} = 1. \tag{B.1-8 c}$$

and [cf. Eq. (3.2-4)] $a = a_{11}a_{22} - a_{12}a_{21}$.

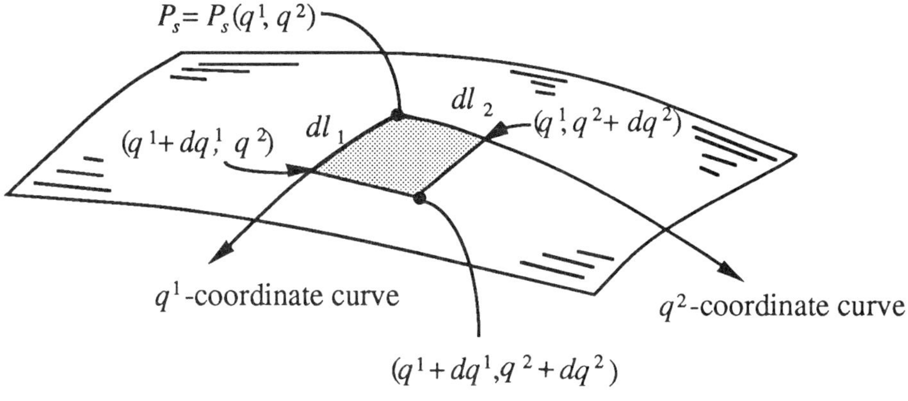

Figure B.1-2 Differential length elements dl_1 and dl_2 respectively measured along the q^1 and q^2 coordinate curves.

Metrical Coefficients for an Orthogonal Surface Coordinate System

Let dl (> 0) represent the scalar displacement between neighboring interfacial points $\mathbf{x}_s$ and $\mathbf{x}_s + d\mathbf{x}_s$, such that

$$dl = |d\mathbf{x}_s|. \tag{B.1-9}$$

Since, from (B.1-9), $dl^2 = d\mathbf{x}_s \cdot d\mathbf{x}_s$, it therefore follows from Eqs. (3.2-2) and (3.2-3) that

$$dl^2 = a_{\alpha\beta}\, dq^\alpha\, dq^\beta, \tag{B.1-10a}$$

which defines the fundamental metrical relationship between distance and coordinate displacement dq^γ on a Riemannian surface, as subsequently discussed in §B.2. In matrix form this relation is

$$dl^2 = (\, dq^1\; dq^2) \begin{pmatrix} a_{11} & a_{12} \\ a_{21} & a_{22} \end{pmatrix} \begin{pmatrix} dq^1 \\ dq^2 \end{pmatrix}. \tag{B.1-10b}$$

In the remainder of this appendix, attention is confined to the special case of an *orthogonal* surface pair (q^1, q^2). In this case, Eq. (B.1-10b) becomes

$$dl^2 = (\, dq^1\; dq^2) \begin{pmatrix} a_{11} & 0 \\ 0 & a_{22} \end{pmatrix} \begin{pmatrix} dq^1 \\ dq^2 \end{pmatrix}, \tag{B.1-10c}$$

where the vanishing of the cross-diagonal components arises from the orthogonal nature of the surface coordinate system [see Eq. (3.2-17)]. Thus, it follows that in important circumstances,

$$dl^2 = dl_1^{\,2} + dl_2^{\,2}, \tag{B.1-10d}$$

with

$$dl_1 = \frac{dq^1}{h_1}, \qquad dl_2 = \frac{dq^2}{h_2}, \tag{B.1-10e}$$

where the *scale factor* (or scalar *metrical coefficient*) $h_\alpha(q^1, q^2)$, representing the ratio of coordinate displacement dq^α to physical length dl_α along the q^α-coordinate curve, is defined as

$$h_1 = \frac{1}{\sqrt{a_{11}}} \equiv \frac{1}{|\mathbf{a}_1|}, \qquad h_2 = \frac{1}{\sqrt{a_{22}}} \equiv \frac{1}{|\mathbf{a}_2|}. \tag{B.1-11}$$

(Note that h_α is a scalar, not a first-rank tensor as might be otherwise suggested by the appearance of the subscript index.)

Unit Orthogonal Surface Base Vectors

For *orthogonal* systems the surface base vectors $\mathbf{a}_\alpha$ are conveniently expressed in terms of the comparable *unit* base vectors $\mathbf{i}_\alpha$ according to the relations [cf. Eq. (3.2-19)]

$$\mathbf{i}_1 = h_1 \mathbf{a}_1, \qquad \mathbf{i}_2 = h_2 \mathbf{a}_2. \tag{B.1-12}$$

Hence, upon employing Eqs. (B.1-6) and (B.1-11) it follows that

$$\mathbf{i}_1 = \frac{\mathbf{a}^1}{h_1}, \qquad \mathbf{i}_2 = \frac{\mathbf{a}^2}{h_2}. \tag{B.1-13}$$

[The reciprocal unit vectors $\mathbf{i}^\alpha$ are clearly identical to the base vectors $\mathbf{i}_\alpha$ themselves, as the vector pair $(\mathbf{i}_1, \mathbf{i}_2)$ is *self-reciprocal*.]

From Eqs. (B.1-13) and (B.1-4b) it follows that the orthogonal curvilinear basis vectors $\mathbf{i}_\alpha$ may be expressed as

$$\mathbf{i}_{\alpha} = \frac{1}{h_{\alpha}}\left(\frac{\partial q^{\alpha}}{\partial \mathbf{x}}_{s}\right)$$
$$\equiv \frac{\nabla_{s} q^{\alpha}}{h_{\alpha}} \quad \text{(no sum on } \alpha\text{)}. \tag{B.1-14}$$

Equation (B.1-14) [representing the two-dimensional counterpart of Eq. (A.1-28)] reveals that the unit basis vector $\mathbf{i}_{\alpha}$ may be physically interpreted as being the unit tangent vector to the q^{α}-coordinate curve, as depicted in Figure B.1-2. Scalar multiplication of the above relation with the unit vector $\mathbf{i}_{\beta}$ furnishes the following pair of relations, upon utilizing Eq. (3.2-3) [cf. Eqs. (A.1-29)]:

$$\nabla_{s} q^{\alpha} \cdot \nabla_{s} q^{\alpha} = h_{\alpha}^{2} \quad \text{(no sum on } \alpha\text{)} \tag{B.1-15}$$

and

$$\nabla_{s} q^{\alpha} \cdot \nabla_{s} q^{\beta} = 0 \quad (\alpha \neq \beta)\,. \tag{B.1-16}$$

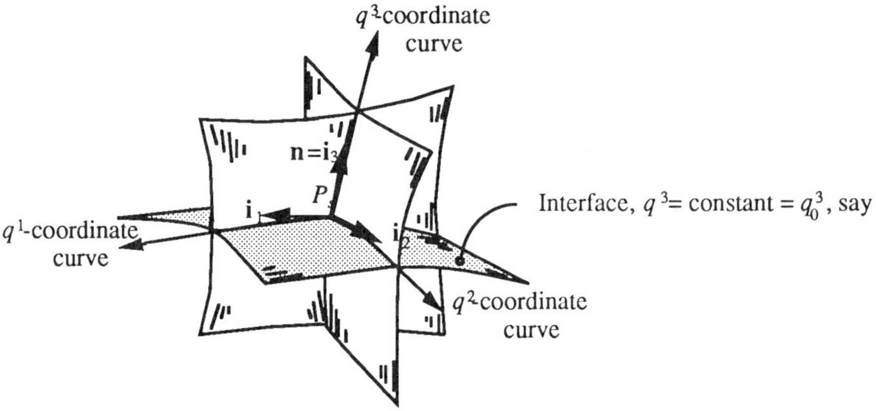

Figure B.1-3 Coordinate curves, surfaces and unit basis vectors in an orthogonal parameterization of the interface in three-dimensional space. The *parent* coordinate surface q^3=constant coincides with the macroscale interface (see §15.3 for further elaboration). The unit basis vectors ($\mathbf{i}_1$, $\mathbf{i}_2$, $\mathbf{i}_3$) are those existing at the point P_s (q^1, q^2, q^3) located on the interface (cf. Fig. A.1-1.)

The preceding orthogonal surface parameterization is simply the two-dimensional portion of the three-dimensional orthogonal curvilinear component parameterization of space considered in Appendix A. In this context, the q^3-coordinate curve intersects the interface orthogonally, as depicted in Figure B.1-3.

Surface Vector-Dyadic Invariants

In right-handed orthogonal curvilinear coordinates, the surface idemfactor, Eq. (3.2-6), may be expressed as

$$\mathbf{I}_s = \mathbf{i}_1\mathbf{i}_1 + \mathbf{i}_2\mathbf{i}_2, \tag{B.1-17}$$

whereas the dyadic surface alternator, Eq. (3.2-9), adopts the form

$$\boldsymbol{\varepsilon}_s = \mathbf{i}_1\mathbf{i}_2 - \mathbf{i}_2\mathbf{i}_1. \tag{B.1-18}$$

The unit surface normal, Eq. (3.2-5), is simply

$$\mathbf{n} = \mathbf{i}_1 \times \mathbf{i}_2 \equiv \mathbf{i}_3, \tag{B.1-19}$$

where the triad of orthonormal vectors ($\mathbf{i}_1$, $\mathbf{i}_2$, $\mathbf{i}_3$), form a right-handed coordinate basis; explicitly, $\mathbf{i}_1 \cdot \mathbf{i}_2 \times \mathbf{i}_3 = 1$. Derivatives of these orthogonal unit base vectors are furnished in Eq. (A.3-1). The unit normal vector to the interface is $\mathbf{n}$ ($\equiv \mathbf{i}_3$).

Upon using Eqs. (B.1-19) and (A.3-1) in conjunction with the definition (3.2-10), the surface curvature dyadic $\mathbf{b}$ may be expressed as

$$\mathbf{b} = -\mathbf{i}_1\mathbf{i}_1 h_3 h_1 \frac{\partial}{\partial q^3}\left(\frac{1}{h_1}\right)\bigg|_{q^3 = q^3_0} - \mathbf{i}_2\mathbf{i}_2 h_3 h_2 \frac{\partial}{\partial q^3}\left(\frac{1}{h_2}\right)\bigg|_{q^3 = q^3_0} . \tag{B.1-20}$$

Comparison of this equation with (3.2-12a) furnishes the following representations of the principal curvatures $\kappa(q^1, q^2)$ ($\alpha = 1,2$) expressed in terms of operations in orthogonal curvilinear coordinates on the surface q^3=constant= q^3_0, say:

$$\kappa_1 = -\, h_3 h_1 \frac{\partial}{\partial q^3}\left(\frac{1}{h_1}\right)\bigg|_{q^3 = q^3_0} , \tag{B.1-21}$$

$$\kappa_2 = -\, h_3 h_2 \frac{\partial}{\partial q^3}\left(\frac{1}{h_2}\right)\bigg|_{q^3 = q^3_0} . \tag{B.1-22}$$

Equation (A.1-30) is employed below to determine the curvature dyadic $\mathbf{b}$, as well as the mean and total curvatures (H and K respectively), for planar, circular cylindrical, and spherical surfaces, using the preceding relations.

Planar Surface (q^1=x, q^2=y) and q^3=z

The extrinsic parameterization of the planar surface depicted in Figure B.1-4 corresponds to the choice z=constant=z_0, say. It follows trivially from Eq. (A.1-30) that

$$h_x = h_y = 1, \tag{B.1-23}$$

as well as

$$h_z = 1. \tag{B.1-24}$$

Hence, from Eqs. (B.1-20)-(B.1-22), together with (3.2-13) and (3.2-14), the planar surface is characterized by the curvature parameters:

$$\mathbf{b} = \mathbf{0}, \qquad H = 0, \qquad K = 0. \tag{B.1-25}$$

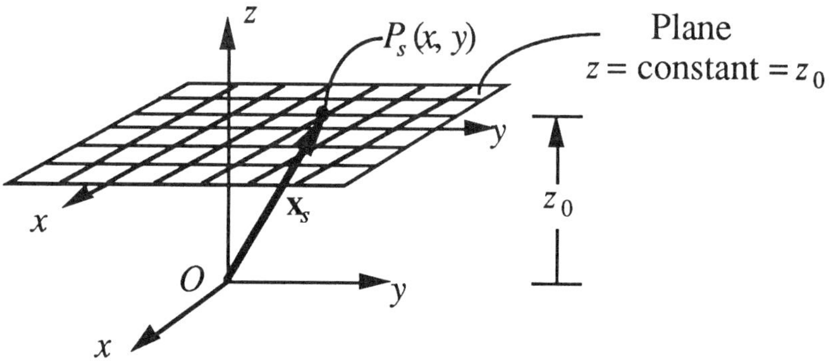

Figure B.1-4 Extrinsic parameterization of a planar surface showing the x and y coordinate lines (i.e. 'axes') ruled on the surface.

Cylindrical Surface ($q^1=\phi$, $q^2=z$) **and** $q^3=R$

The extrinsic parameterization of the circular cylindrical surface depicted in Figure B.1-5, corresponds to the choice R=constant=R_0, say. The surface coordinates (ϕ, z) on this cylinder are defined in relation to the cartesian system (x, y, z) shown, by the relations

$$x = R_0 \cos\,\phi, \quad y = R_0 \sin\,\phi, \tag{B.1-26}$$

together with

$$z = z\,.$$

From Eq. (A.1-30), the respective scale factors for this cylindrical surface are

$$h_\phi = \frac{1}{R_0}, \quad h_z = 1, \tag{B.1-27}$$

together with

$$h_R = 1. \tag{B.1-28}$$

From Eqs. (B.1-20)-(B.1-22), together with Eqs. (3.2-13) and (3.2-14), the cylindrical surface is characterized by the curvature parameters*

---

* Note that the definition of the mean curvature H provided by Eq. (3.2-13) renders $H(q^1, q^2)$ a *negative* scalar, when the unit surface normal $\mathbf{n}(q^1, q^2)$ points in the direction of

$$\mathbf{b} = -\mathbf{i}_\phi \mathbf{i}_\phi \frac{1}{R_0}, \qquad H = -\frac{1}{2R_0}, \qquad K = 0. \qquad \text{(B.1-29)}$$

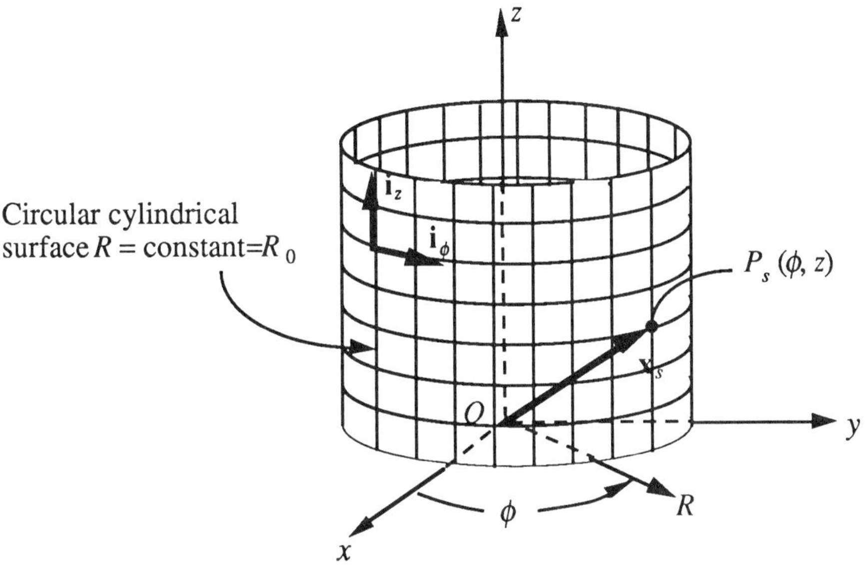

Figure B.1-5 Extrinsic parameterization of a circular cylindrical surface, showing the ϕ and z coordinate lines ruled on the surface.

Spherical Surface $(q^1=\theta,\ q^2=\phi)$ **and** $q^3=r$

The extrinsic parameterization of the spherical surface depicted in Figure B.1-6 corresponds to the choice r=constant=r_0, say. The surface coordinates (θ, ϕ) on this sphere are defined in relation to the cartesian system (x, y, z) shown, by the relations

$$x = r_0 \sin\theta \cos\phi, \quad y = r_0 \sin\theta \sin\phi, \qquad \text{(B.1-30)}$$

together with

$$z = r_0 \cos\theta. \qquad \text{(B.1-31)}$$

The scale factors for the spherical surface of Figure B.1-6 are

$$h_\theta = \frac{1}{r_0}, \quad h_\phi = \frac{1}{r_0 \sin\theta}, \qquad \text{(B.1-32)}$$

---

the convex side of the surface [at the surface point (q^1, q^2)]. This explains the negative values of H in Eqs. (B.1-29) and (B.1-34).

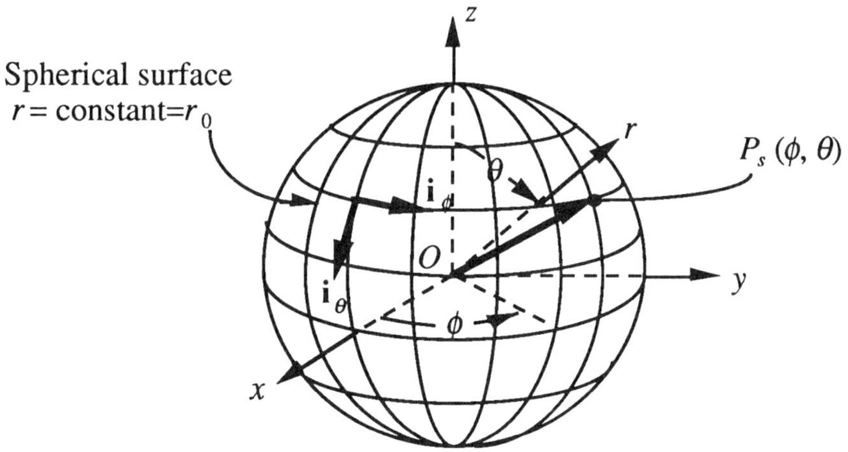

Figure B.1-6 Extrinsic parameterization of a spherical surface showing the θ and ϕ coordinate lines ruled on the surface.

together with

$$h_r = 1. \tag{B.1-33}$$

From Eqs. (B.1-20) through (B.1-22), in conjunction with Eqs. (3.2-13) and (3.2-14), it follows that for the spherical surface:

$$\mathbf{b} = -(\mathbf{i}_\theta \mathbf{i}_\theta + \mathbf{i}_\phi \mathbf{i}_\phi)\frac{1}{r_0}, \qquad H = -\frac{1}{r_0}, \qquad K = \frac{1}{r_0^2}. \tag{B.1-34}$$

Bibliography

Abbot, E.A. 1952 *Flatland. A Romance of Many Dimensions*. New York: Dover.

Adam, N.K. 1948 *The Physics and Chemistry of Surfaces*. London: Oxford University Press.

Adamson, A.W. 1982 *Physical Chemistry of Surfaces*, 4th edn. New York: Wiley.

Adler, P. M. & Brenner, H. 1985 Spatially periodic suspensions of convex particles in linear shear flows: I. Description and kinematics. *Int. J. Multiphase Flow* **11**, 361–385.

Adler, P. M., Zuzovsky, M. & Brenner, H. 1985 Spatially periodic suspensions of convex particles in linear shear flows: II. Rheology. *Int. J. Multiphase Flow* **11**, 387–417.

Adler, P.M., Nadim, A. & Brenner, H. 1990 Suspension rheology. *Adv. Chem. Eng.* **15**, 1–72.

Agrawal, M.L. & Neuman, R.D. 1988a Surface diffusion in monomolecular films. I. *J. Colloid Interface Sci.* **121**, 355–365.

Agrawal, M.L. & Neuman, R.D. 1988b Surface diffusion in monomolecular films. II. Experiment and theory. *J. Colloid Interface Sci.* **121**, 356–380.

Agrawal, S.K. & Wasan, D.T. 1979 The effect of interfacial viscosities on the motion of drops and bubbles. *Chem. Engng J.* **18**, 215–223.

Allen, R.F. 1988 The mechanics of splash. *J. Colloid Interface Sci.* **134**, 309–316.

Aris, R. 1962 *Vectors, Tensors, and the Basic Equations of Fluid Mechanics*. Englewood Cliffs, New Jersey: Prentice Hall.

Aubert, J.H. 1983 Interfacial properties of dilute polymer solutions. *J. Colloid Interface Sci.* **96**, 135–149.

Barber, A.D. & Hartland, S. 1976 The effects of surface viscosity on the axisymmetric drainage of planar liquid films. *Can. J. Chem. Engng.* **54**, 279–284.

Barthes, B.D. & Acrivos, A. 1973 Deformation and burst of a liquid droplet freely suspended in a linear shear flow. *J. Fluid Mech.* **61**, 1–21.

Barton, K.D. & Subramanian, R.S. 1989 The migration of liquid drops in a vertical temperature gradient. *J. Colloid Interface Sci.* **133**, 214–222.
Batchelor, G.K. 1967 *An Introduction to Fluid Mechanics*. London: Cambridge University Press.
Becher, P. 1965 *Emulsions: Theory and Practice*. New York: Reinhold.
Becher, P. 1983 *Encyclopedia of Emulsion Technology*. New York: Marcel Dekker.
Bénard, H. 1901 Les tourbillons cellulaires dans une nappe liquide transportant de la chaleur par convection régime permanent. *Ann. Chim. Phys.* **23**, 62–144.
Bendure, R.L. 1971 Dynamic surface tension determination with the maximum bubble pressure method. *J. Colloid Interface Sci.* **35**, 238–248.
Berg, J.C. 1988 The effect of surface-active agents on distillation processes. In *Surfactants in Chemical/Process Engineering*, vol. 28 (eds. D.T. Wasan, M.E. Ginn & D.O. Shah), pp. 29–76. New York: Marcel Dekker.
Bikerman, J.J. 1948 *Surface Chemistry for Industrial Research*. New York: Academic Press.
Bikerman, J.J. 1958 *Surface Chemistry*, 2nd edn. New York: Academic Press.
Bikerman, J.J. 1973 *Foams*. New York: Springer-Verlag.
Bird, R.B., Stewart, W.E. & Lightfoot, E.N. 1960 *Transport Phenomena*. New York: Wiley.
Bird, R.B., Armstrong, R.C. & Hassager, O. 1977 *Dynamics of Polymeric Liquids: Volume I Fluid Mechanics*. New York: Wiley.
Biswas, B. & Haydon, D.A. 1963 Rheology of some interfacial adsorbed films of macromolecules I. Elastic and creep phenomena. *Proc. Roy. Soc.* **A271**, 296–305.
Blakey, B.C. & Lawerence, A.S.C. 1954 *Discuss. Faraday Soc.* **18**, 268.
Block, M.J. 1956 Surface tension as the cause of Bénard cells and deformation in a liquid film. *Nature* **178**, 650–651.
Blumenthal, R., Changeux, J.G. & Lefever, R. 1970 Membrane exciteability and dissipative instabilities. *J. Membrane Biol.* **2**, 351.
Bohr, N. 1909 Determination of the surface tension of water by the method of jet vibration. *Phil. Trans.* **A209**, 281–317.
Borwanker, R.P. & Wasan, D.T. 1983 The kinetics of adsorption of surface-active agents at gas-liquid surfaces. *Chem. Engng Sci.* **38**, 1637–1649.
Boussinesq, M.J. 1913a Sur l'existance d'une viscosite' superficielle, dans la mince couche de transition separant un liquide d'une autre fluide contigu. *Ann. Chim. Phys.* **29**, 349–357.
Boussinesq, M.J. 1913b The application of the formula for surface viscosity to the surface of a slowly falling droplet in the midst of a large unlimited amount of fluid which is at rest and possesses a smaller specific gravity. *Ann. Chim. Phys.* **29**, 357.
Boyd, J., Parkinson, C. & Sherman, P. 1972 Factors affecting emulsion stability and HLB concept. *J. Colloid Interface Sci.* **41**, 359–370.

Brand, L. 1947 *Vector and Tensor Analysis*. New York: Wiley.
Brenner, H. 1970 Rheology of a dilute suspension of dipolar spherical particles in an external field. *J. Colloid Interface Sci.* **32**, 141–158.
Brenner, H. 1979 A micromechanical derivation of the differential equations of interfacial statics. *J. Colloid Interface Sci.* **68**, 422–439.
Brenner, H. & Leal, L.G. 1982 Conservation and constitutive equations for adsorbed species undergoing surface diffusion and convection at a fluid-fluid interface. *J. Colloid Interface Sci.* **88**, 136–184.
Brenner, H. 1984 Antisymmetric stress induced by the rigid body rotation of dipolar suspended particles. Vortex Flows. *Int. J. Engng Sci.* **22**, 645–682.
Bretherton, F.P. 1961 The motion of long bubbles in tubes. *J. Fluid Mech.* **12**, 166–188.
Briley, P.B., Deemer, A.R. & Slattery, J.C. 1976 Blunt knife-edge and disk surface viscometers. *J. Colloid Interface Sci.* **56**, 1–18.
Brochard, F. & Lennon, J.F. 1975 Frequency spectrum of flicker phenomena in erythrocytes. *J. Phys.* **36**, 1035–1047.
Brown, R.C. 1940 Surface tension. *Phys. Soc. Rept. Progress Physics* **7**, 180–194.
Brown, A.G., Thuman W.C. & McBain, J.W. 1953 The surface viscosity of detergent solutions as a factor in foam stability. *J. Colloid Sci.* **8**, 491–507.
Buff, F.P. & Saltsburg, H. 1957 Curved fluid interfaces II. The generalized Neumann formula. *J. Chem. Phys.* **26**, 23–31.
Buff, F.P. 1955 Curved fluid interfaces I. The generalized Gibbs-Kelvin equation. *J. Chem. Phys.* **23**, 419–427.
Buff, F.P. 1960 The theory of capillarity. In *Handbuch der Physik*, vol. X, pp. 281–304. Berlin: Springer-Verlag.
Burkholder, H.C. & Berg, J.C. 1974a Effect of mass transfer on laminar jet breakup: Part I. Liquid jets in gases. *AIChE J.* **20**, 863–872.
Burkholder, H.C. & Berg, J.C. 1974b Effect of mass transfer on laminar jet breakup: Part I. Liquid jets in liquids. *AIChE J.* **20**, 872–880.
Burrill, K.A. & Woods, D.R. 1973 Film shapes for deformable drops at liquid-liquid interfaces II. Mechanisms of film drainage. *J. Colloid Interface Sci.* **42**, 35–51.
Burton, A.L., Anderson, W.L. & Andrews, R.V. 1968 Quantitative studies on flicker phenomena in erythrocytes. *Blood* **32**, 819.
Byrne, D. & Earnshaw, J.C. 1979 Photo correlation spectroscopy of liquid interfaces II. Monolayers of fatty acids. *J. Phys. D* **12**, 1145–1157.
Camp, M. & Durham, K. 1955 The foaming of sodium laurate solutions—factors influencing foam stability. *J. Phys. Chem.* **59**, 993–997.
Campanelli, J.R. & Cooper, D.G. 1989 Interfacial viscosity and the stability of emulsions. *Can. J. Chem. Eng.* **67**, 851–855.
Carless, J.E. & Hallworth, G. 1966 Viscosity of oil-water interfaces and emulsion stability. *Chem. Ind. London* **30**, 66.
Caskey, J.A. & Barlage, W.B. 1971 An improved experimental technique for determining dynamic surface tension of water and surfactant solutions. *J. Colloid Interface Sci.* **35**, 46–52.

Chaffey, C.E. & Brenner, H. 1967 A second-order theory for shear deformation of drops. *J. Colloid Sci.* **24**, 258–268.

Chandrasekhar, S. 1961 *Hydrodynamic and Hydromagnetic Stability.* Oxford: Clarendon Press.

Chapman, S. & Cowling, T.G. 1961 *The Mathematical Theory of Nonuniform Gases,* 2nd ed. Cambridge: Cambridge University Press.

Choi, S.J. & Schowalter, W.R. 1975 Rheological properties of non-dilute suspensions of deformable particles. *Phys. Fluids* **18**, 420–427.

Chu, X.L. & Velarde, M.G. 1989 Transverse and longitudinal waves induced and sustained by surfactant gradients at liquid-liquid interfaces. *J. Colloid Interface Sci.* **131**, 471–484.

Churaev, N.V., Starov, V.M. & Derjaguin, B.V. 1982 The shape of the transition core between a thin film and a bulk liquid and the line tension. *J. Colloid Interface Sci.* **89**, 16–24.

Clint, J.H., Neustadter, E.L. & Jones, T.J. 1981 Dynamic interfacial phenomena related to EOR. *Dev. Pet. Sci.* **13**, 135–148.

Clunie, J.S., Goodman, J.F. & Ingram, B.T. 1971 Thin liquid films. *Surface & Colloid Sci.* **3**, 167–239.

Cockbain, E.G. & McRoberts, T.S. 1953 The stability of elementary emulsion drops and emulsions. *J. Colloid Sci.* **8**. 440–451.

Cockbain, E.G. 1956 The adsorption of serum albumin and sodium dodecyl sulfate at emulsion interfaces. *J. Colloid Sci.* **11**, 575–584.

Cole, J.D. 1968 *Perturbation Methods in Applied Mechanics.* Waltham, Massachusetts: Blaisdell Publishing Co.

Courant, R. & Robbins, H. 1941 *What is Mathematics?* New York: Oxford University Pres.

Cox, R.G. 1969 Deformation of a drop in a general time-dependent fluid flow. *J. Fluid Mech.* **37**, 601.

Cox, R.G. 1970 Motion of long slender bodies in a viscous fluid I. General theory. *J. Fluid Mech.* **44**, 4.

Criddle, D.W. & Meader, A.L. Jr. 1955 Viscosity and elasticity of oil surfaces and oil-water interfaces. *J. Appl. Phys.* **26**, 838–842.

Cumper, C.W.N. & Alexander, A.E. 1950 The surface chemistry of proteins. *Trans. Faraday Soc.* **46**, 235–253.

Cumper, C.W.N. 1953 The stabilization of foams by proteins. *Tans. Faraday Soc.* **49**, 1360–1369.

Dagan, Z. & Pismen, L.M. 1984 Marangoni waves induced by a multistable chemical reaction on thin liquid films. *J. Colloid Interface Sci.* **99**, 215–225.

Dahler, J.S. & Scriven, L.E. 1963 Theory of structured continua I. General consideration of angular momentum and polarization. *Proc. Roy. Soc.* **A275**, 504–527.

Davies, J.T. 1957 *Proc. 2nd Int. Congr. Surf. Act.* **1**, 220.

Davies, J.T. & Rideal, E.K. 1963 *Interfacial Phenomena*, 2nd. edn. New York: Academic Press.

Davis, A.M. & O'Neill, M.E. 1979 The slow rotation of a sphere submerged in a fluid with a surfactant surface layer. *Int. J. Multiphase Flow* **5**, 413–425.

Davis, H.T. & Scriven, L.E. 1982 Stress and structure in fluid interfaces. *Adv. Chem. Phys.* **49**, 357-454.

Davis, H.T. 1987 Kinetic theory of flow in strongly inhomogeneous fluids. *Chem. Eng. Commun.* **58**, 413–430.

DeBruijn 1958 *Asymptotic Methods in Analysis*. Amsterdam: North Holland Publishing Co.

Deemer, A.R. & Slattery, J.C. 1978 Balance equations and structural models for phase interfaces. *Int. J. Multiphase Flow* **4**, 171–192.

de Feijter, J.A. & Vrij, A. 1972 Transition regions: line tensions and contact angles in soap films. *J. Electroanal. Chem. Interfacial Electrochem.* **37**, 9.

Defay, R. & Hommelen, J.R. 1958 I Measurement of dynamic surface tensions of aqueous solutions by the oscillating jet method. *J. Colloid Interface Sci.* **13**, 553–564.

Defay, R. & Hommelen, J.R. 1959 II Measurement of dynamic surface tensions of aqueous solutions by the falling meniscus method. *J. Colloid Interface Sci.* **14**, 401–410.

Defay, R. & Prigogine, I. 1966 *Surface Tension and Adsorption*. New York: Wiley.

Derjaguin, B.V. 1955 Definition of the concept of and magnitude of the disjoining pressure and its role in the statics and kinetics of thin layers of liquids. *Collid J.* (*USSR*) **17**, 191–197.

Derjaguin, B.V. & Gutop, Y.V. 1965 The thermodynamics and stability of free films. *Colloid J.* (*USSR*) **27**, 298–305.

Derjaguin, B.V. & Kussakov, M.M. 1937 Experimental study of solvation of surfaces as applied to the mathematical theory of the stability of lyophilic colloids. Izv. AN SSR, ser. khim. **6**, 1119–1152.

Derjaguin, B.V. & Landau, L. 1941 Theory of the stability of strongly charged lyophobic sols and of the adhesion of strongly charged particles in solutions of electrolytcs. *Acta Physicochim.* (*USSR*) **14**, 633–662.

Derjaguin, B.V. & Titiyevskaya, A.S. 1953 The repulsive action of frecly liquid films and its role in the stability of foams. *Collid J.* (*USSR*) **15**, 431–443.

Derjaguin, B.V. & Titievskaya, A.S. 1957 Static and kinetic stability of free films and froths. *Proc. 2nd. Int. Cong. Surf. Act.* **1**, 211–223.

Dervichian, D.G. & Joly, M. 1939 *J. Phys. Radium* **10**, 375.

Dijkstra, H.A. 1990 The Coupling of Marangoni and Capillary Instabilities in an Annular Thread of Liquid. *J. Colloid Interface Sci.* **136**, 151–159.

Djabbarah, N.F. & Wasan, D.T. 1982 Relationship between surface viscosity and surface composition of adsorbed surfactant films. *I&EC Fundamentals* **21**, 27–31.

Dumais, J.F. 1980 Two and three-dimensional interfacial dynamics. *Physica* **104A**, 143–180.

Edwards, D.A. 1987 *Surface Rheology*. Ph.D. Thesis, Illinois Institute of Technology, Chicago, Illinois.
Edwards, D.A., Brenner, H. & Wasan, D.T. 1989 On a relation between foam and surface dilatational viscosities. *J. Colloid Interface Sci.* **130**, 266–270
Edwards, D.A. & Wasan, D. T. 1990 Foam dilatational rheology I. Dilatational viscosity. *J. Colloid Interface Sci.* **139**, 479–487.
Einstein, A. 1901 Folgerungern aus den kapillaritatsercheinungen. *Ann. Phys.* **4**, 513–523.
Einstein, A. 1905 Die von der molekularkinetischen Theorie der warme geforderte bewegung von in ruhenden flussigkeiten suspendierken teilchen. *Ann. Phys.* **17**, 549–560.
Einstein, A. 1906 Eine neue bestimmung der molekuldimensionen. *Ann. Phys.* **29**, 289–306.
Einstein, A. 1956 *Investigations on the Theory of Brownian Movement.* New York: Dover.
Eliassen, J.D. 1963 *Interfacial Mechanics.* Ph.D. Thesis, Universtity of Minnesota, Minneapolis, Minnesota.
Entov, V.M. & Yarin, A.L. 1984 The dynaimcs of thin liquid jets in air. *J. Fluid Mech.* **140**, 91–111.
Eriksson, J.C. & Toshev, B.V. 1986 On the mechanical equilibrium of a soap film and its adjacent meniscus. *Coll. Polym. Sci.* **264**, 807–811.
Evans, R. 1979 Nature of the vapor-liquid interface and other topics in the statistical mechanics of nonuniform classical fluids. *Adv. Phys.* **28**, 143–200.
Ewers, W.E. & Sack, R.A. 1954 The surface viscosity of soluble films. *Australian J. Chem.* **7**, 40–54.
Faraday, M. 1851 *Phil. Trans. Roy. Soc. London* **141**, 7.
Felderhof, B. 1968 Dynamics of free liquid films. *J. Chem. Physics*, **49**, 44–68.
Ferry, J.D. (1958) *Rheol. Theory Appl.* **2**, 442
Feuillebois, F. 1989 Thermocapillary migration of two equal bubbles parallel to the line of centers. *J. Colloid Interface Sci.* **131**, 267–274.
Flummerfelt, R.W. 1980 Effects of dynamic interfacial properties on drop deformation and orientation in shear and extensional flow fields. *J. Colloid Interface Sci.* **76**, 330–349.
Frankel, N.A. & Acrivos, A. 1970 A constitutive equation for a dilute emulsion. *J. Fluid Mech.* **44**, 65–75.
Franklin, B. 1773 Personal letter to William Bronrigg, as quoted in *Benjamin Franklin*. Literary Classics of the United States, Inc., New York (1987).
Frenkel, J. 1955 *Kinetic Theory of Liquids*. New York: Dover.
Frumkin, A. 1925 Die kapillarkurve der hohren fettsauren und die zustadsgleichung der oberflachenschicht. *Z. Phys. Chemie* **116**, 466–480.
Frumkin, A. 1938 *Zhur. Phys. Khim.* **12**, 337.
Frumkin, A.N. & Levich, V.G. 1947 *Zhur. Fiz. Khim.* **21**, 1183.

Gaestel, C. 1967 Breaking mechanism of cationic bitumen emulsions. *Chem. Ind.* **6**, 221.

Gaines, G.L. 1966 *Insoluble Monolayers at Liquid-Gas Interfaces*. New York: Interscience Publishers.

Gardner, J.W., Addison, J.V. & Schechter, R.S. 1978 A constitutive equation for a viscoelastic interface. *AIChE J.* **24**, 400–405.

Gardner, J.W. & Schechter, R.S. 1976 Evaluation of surface rheological models. *Colloid Interface Sci.* **4**, 98–115.

Garrett, P.R. & Joos, P. 1976 Dynamic dilatational surface properties of submicellar multicomponent surfactant solutions II. Thermodynamics of adsorption and comparison with experiment. *J. Chem. Soc., Faraday Trans.* **72**, 2174–2184.

Garrett, P.R. & Ward, D.R. 1988 A reexamination of the measurement of dynamic surface tensions using the maximum bubble pressure method. *J. Colloid Interface Sci.* **132**, 475–490.

Geffen, T.M. 1973 Oil production to expect from known technology. *Oil Gas J.* 66.

Gershfeld, N.L. & Good, R.J. 1967 Line tension and penetration of a cell membrane by an oil drop. *J. Theor. Biol.* **17**, 246.

Gibbs, J.W. 1957 *The Collected Works of J. Willard Gibbs*, Vol. I. New Haven, Connecticut: Yale University Press.

Giordano, R.M. & Slattery, J.C. 1983 Effect of interfacial viscosities upon displacement in capillaries with special application to tertiary oil recovery. *AIChE J.* **29**, 483–502.

Gladden, G.P. & Neustadter, E.L. 1972 Oil-water interfacial viscosity and crude oil emulsion stability. *J. Inst. Pet. London* **58**, 351.

Good, R.J. & Koo, M.N. 1979 Effect of drop size on contact angle. *J. Colloid Interface Sci.* **71**, 283–292.

Goodrich, F.C. 1981a Surface viscosity as a capillary excess transport property. In *The Modern Theory of Capillarity* (eds. F.C. Goodrich and A.I. Rusanov). Berlin: Akademie-Verlag.

Goodrich, F.C. 1981b The theory of capillary excess viscosities. *Proc. Roy. Soc. Lond.* A**374**, 341-370.

Goodrich, F.C. & Allen, L.H. 1972 The theory of absolute surface shear viscosity. *J. Colloid Interface Sci.* **40**, 329–336.

Goodrich, F.C. & Chatterjee, A.K. 1970 The theory of absolute surface shear viscosity II. The rotating disk problem. *J. Colloid Interface Sci.* **34**, 36–42.

Goodrich, F.C., Allen, L.H. & Poskanzer, A. 1975 A new surface viscometer of high sensitivity I. Theory. *J. Colloid Interface Sci.* **52**, 201–212.

Gorodetskaya, A.V. 1949 Speed of rise of bubbles in water and water solutions with large Reynolds number. (Russ.) *Zhur. Fiz. Khim.* **23**, 71–77.

Gouda, J.H. & Joos, P. 1975 Application of longitudinal wave theory to describe interfacial instability. *Chem. Eng. Sci.* **30**, 521–528.

Graham, D.E. & Phillips, M.C. 1976 In *Proc. Symp. on Foams*, (ed. R.J. Akers), p. 237. London: Academic Press.

Graham, D.E. 1976 PhD Thesis, Council of National Academic Awards, England.

Gumerman, R.J. & Homsy, G. 1975 Stability of uniformly driven flows with application to convection driven by surface tension. *J. Fluid Mech.* **68**, 191–207.

Haber, S. & Hetsroni, G. 1971 Hydrodynamics of a drop submerged in an unbounded arbitrary velocity field in the presence of surfactants. *Appl. Sci. Res.* **25**, 215–233.

Hadamard, J.S. 1911 Mouvement permenent lent d'une sphére liquide et visqueuse dans un liquide visqueux. *Comp. Rend. Acad. Sci.* (Paris) **152**, 1735–1738.

Hajiloo, A., Ramamohan, T.R. & Slattery, J.C. 1987 Effect of interfacial viscosity on the stability of a liquid thread. *J. Colloid Interface Sci.* **117**, 384–393.

Haines, W.B. 1930 Studies in the physical properties of soil: V. The hysteresis effect in capillary properties and the modes of mosture distribution associated therewith. *J. Agric. Sci.* **20**, 97.

Hansen, R.S. & Mann, S.A. Jr. 1964 Propagation characteristics of capillary ripples I. Theory of velocity dispersion and amplitude attenuation of plane capillary waves on viscoelastic films. *J. Appl. Phys.* **35**, 152.

Hansen, R.S. 1961 Diffusion and the kinetics of adsorption of aliphatic acids and alcohols at the water-air interface. *J. Colloid Sci.* **16**, 549–560.

Hansen, R.S. and Ahmad, J. 1971 Waves at interfaces. In *Progress in Surface and Membrane Science*, vol. 4, 5, (eds. J.F. Danielli, M.D. Rosenberg and D.A. Cadenhead), pp. 1–55. New York: Academic Press.

Happel, J. & Brenner, H. 1983 *Low Reynolds Number Hydrodynamics.* Dordrecht, Netherlands: Nijhof.

Hard, S. & Lofgren, H. 1977 Elasticity and viscosity measurement of monomolecular propyl stearate films at air-water interfaces using laser-scattering techniques. *J. Colloid Interface Sci.* **60**, 529–539.

Hard, S. & Neuman, R.D. 1981 Laser light scattering measurements of viscoelastic monomolecular films. *J. Colloid Interface Sci.* **83**, 315–334.

Hard, S. & Neuman, R.D. 1987 Viscoelasticity of monomolecular films: A laser light-scattering study. *J. Colloid Interface Sci.* **120**, 15–29.

Harkins, W.D. & Meyers, R.J. 1937 Viscosity of monomolecular films. *Nature* **140**, 465.

Hariri, H.H., Nadim, A. & Borhan, A. 1990 Effect of inertia on the thermocapillary velocity of a drop. Submitted to *Phys. Fluids.*

Hartland, S. 1967 Coalescence of a liquid drop at a liquid-liquid interface III. Film rupture. *Trans. Inst. Chem. Engrs.* **45**, T102.

Hedge, M.G. & Slattery, J.C. 1971a Capillary waves at a gas-liquid interface. *J. Colloid Interface Sci.* **35**, 183–203.

Hedge, M.G. & Slattery, J.C. 1971b Studying nonlinear surface behavior with the deep channel surface viscometer. *J. Colloid Interface Sci.* **35**, 593–600.

Heller, J.P. 1968 The drying through the top surface of a vertical porous column. *Soil Sci. Soc. Am. Proc.* **32**, 778.

Hennenberg, M., Sorensen, T.S. & Sanfeld, A. 1977 Deformational instability of a plane interface with transfer of matter. Part I. Non-oscillatory critical states with a linear concentration profile. *J. Chem. Soc. Trans. Faraday II* **73**, 48–66.

Hill, T.L. 1959 Exact definition of quasi-thermodynamical point functions in statistical mechanics. *J. Chem. Phys.* **30**, 1521–1523.

Hirasaki, G.J. & Lawson, J.B. 1983 Mechanisms of foam flow in porous media—apparent viscosity in smooth capillaries. Presented at the 58th Annual Technical Conference and Exhibition in San Franscisco, CA.

Hooke, R. 1672 Communication to the Royal Society. See T. Birch, *History of the Royal Society.* A Millard, London (1757).

Huang, D. 1986 *Enhanced Oil Recovery: Foams, Emulsions and Immiscible Flooding.* PhD thesis, Illinois Institute of Technology, Chicago, Illinois.

Ibanez, J.L. & Velarde, M.F. 1977 Hydromechanical stability of an interface between two immiscible liquids—role of Langmuir-Henshelwood saturation law. *J. Physique* **38**, 1479–1483.

Irving, J.H. & Kirkwood, J.G. 1950 The statistical mechanical theory of transport processes: IV. The equations of hydrodynamics. *J. Chem. Phys.* **18**, 817–829.

Ito, K., Sauer, B.B., Skarlupka, R.J., Sano, M. & Hyuk, Y. 1990 Dynamic interfacial properties of poly (ethylene oxide) and polystyrene at toluene/water interface. *Langmuir J.* **6**, 1379–1388.

Ivanov, I.B. & Dimitrov, D.S. 1974 Hydrodynamics of thin liquid films. Effect of surface viscosity on thinning and rupture of foam films. *Coll. Polym. Sci.* **252**, 982–990.

Ivanov, I.B. & Jain, R.K. 1979 In *Dynamics and Stability of Fluid Interfaces*, (ed. T.S. Sorensen), p. 120. New York: Springer-Verlag.

Ivanov, I.B. & Toshev, B.V. 1975 Thermodynamics of thin liquid films II. Film thickness and its relation to the surface tension and the contact angle. *Coll. Polym. Sci.* **253**, 593–599.

Jain, R.K. & Ivanov, I.B. 1980 Thinning and rupture of ring-shaped films. *J. Chem. Soc. Faraday Trans. II* **76**, 250–266.

Jain, R.K. & Maldarelli, C. 1983 Stability of thin viscoelastic films with applications to biological membrane deformation. *Ann. New York Acad. Sci.*, **1**, 89–102.

Jain, R.K.& Ruckenstein, E. 1974 *Faraday Trans. II* **70**, 132.

Jain, R.K.& Ruckenstein, E. 1976 Stability of stagnant viscous films on a solid surface. *J. Colloid Interface Sci.* **54**, 108–116.

Jain, R.K., Maldarelli & Ruckenstein, E. 1978 *AIChE Symp. Ser. Biorhel.* **74**, 120.

Jiang, T.S., Chen, J.D. & Slattery, J.C. 1983 Nonlinear interfacial stress-deformation behavior measured with several interfacial viscometers. *J. Colloid Interface Sci.* **96**, 7–19.

Johannes, W. & Whitaker, S. 1965 Thinning of soap films—The effect of surface viscosity. *J. Phys. Chem.* **69**, 1471–1477.

Joly, M. 1956 Non-Newtonian surface viscosity. *J. Colloid Interface Sci.* **11**, 519–531.

Joly, M. 1964, Surface viscosity. In *Recent progress in surface science*, vol. I (eds. J.F. Danielli, K.G.A. Pankhurst, and A.C. Riddiford), pp. 1–50. New York: Academic Press.

Jones, T.J., Durham, K., Evans, W.P. & Camp, M. 1957 *Proc. 2nd Int. Congr. Surf. Act.* **1**, 225.

Jones, T.J., Neustadter, E.L. & Whittingham, K. 1978 Water-in-crude-oil emulsion stability and emulsion destabilization by chemical demulsifiers. *J. Can. Pet. Tech.* **17**, 100–108.

Joos, P. & Rillaerts, E. 1981 Theory on the determination of the dynamic surface tension with the drop volume and maximum bubble pressure methods. *J. Colloid Interface Sci.* **79**, 96–100.

Kao, R.L., Edwards, D.A., Wasan, D.T. & Chen, E. 1991a Measurement of the interfacial dilatational viscosity at high rates of interfacial expansion using the maximum bubble pressure method I. Gas-liquid surface. *J. Colloid Interface Sci.*(To appear.)

Kao, R.L., Edwards, D.A. & Wasan, D.T. 1991b Measurement of the interfacial dilatational viscosity at high rates of interfacial expansion using the maximum bubble pressure method II. Liquid-liquid surface. *J. Colloid Interface Sci.* (To appear.)

Kaplun, S. 1967 *Fluid Mechanics and Singular Perturbations* (eds. P.A. Lagerstrom, L.N. Howard, C.S. Liu). New York: Academic Press.

Karim, S.M. & Rosenhead, L. 1952 The second coefficient of viscosity of liquids and gases. *Reviews of Modern Physics* **24**, 108–116.

Keller, J.B., Rubinov, S.J. & Tu, Y.Q. 1973 Spatial Instability of a Jet. *Phys. Fluids* **16**, 2052–2055.

Kevorkian, J. & Cole, J.D. 1981 *Perturbation Methods in Applied Mathematics*. New York: Springer-Verlag.

Khan, S.A. 1985 *Rheology of Large Gas Fraction Foams*. PhD thesis, Massachusetts Institute of Technology, Cambridge, Massachusetts.

Khan, S.A. 1987 Foam rheology: Relation between elongation and shear deformations in high gas fraction foams. *Rheol. Acta* **26**, 78–84.

Khan, S.A. & Armstrong, R.C. 1987a Rheology of foams: I. Theory for dry foams. *J. Non-Newtonian Fluid Mech.* **22**, 1–22.

Khan, S.A. & Armstrong, R.C. 1987b Rheology of foams: II. Effects of polydispersivity and liquid viscosity for foams having gas graction approaching unity. *J. Non-Newtonian Fluid Mech.* **25**, 61–92.

Kim, H.S. & Subramanian, R.S. 1989 Thermocapillary migration of a droplet with insoluble surfactant I. Surfactant cap. *J. Colloid Interface Sci.* **127**, 417–428.

King, A. 1941 Some factors governing the stability of oil-in-water emulsions. *Trans. Faraday Soc.* **37**, 168–180.

Kirkwood, J.G. & Buff, F.P. 1949 The statistical mechanical theory of surface tension. *J. Chem. Phys.* **17**, 338–343.
Kitchener, J.A. 1964 Foams and free liquid films. In *Recent Progress in Surface Science*, vol. I, (eds. J.F. Danielli, K.G.A. Pankhurst, and A.C. Riddiford), pp. 51–90, New York: Academic Press.
Kosinski, W. 1986 *Field Singularities and Wave Analysis in Continuum Mechanics*. Chichester: Ellis Howard Limited.
Kott, A.T., Gardner, J.W., Schechter, R.S. & DeGroot, W. 1974 The elasticity of pulmonary lung surfactant. *J. Colloid Interface Sci.* **47**, 265–266.
Kralchevsky, P. & Ivanov, I.B. 1988 In *Thin Liquid Films*, New York: Marcel Dekkar.
Kralchevsky, P.A., Nikolov, A.D. & Ivanov, I.B. 1986 Film and line tension effects on the attachment of particles to an interface IV. Experimental studies with bubbles in solutions of dodecyl sodium sulfate. *J. Colloid Interface Sci.* **112**, 132–143.
Krantz, W.B. & Zollars, R.L. 1976 The linear hydrodynamic stability of film flow down a vertical cylinder. *AIChE J.* **22**, 930–934.
Kraynik, A.M. 1981 Rheological aspects of thermoplastic foam extrusion. *Polym. Eng. Sci.* **21**, 80–85.
Kraynik, A.M. 1987 Foam rheology: The linear viscoelastic response of a spatially peridoic model. *Proc. Can. Congr. Appl. Mech., 11th, Edmonton* **2**, B2–3.
Kraynik, A.M. 1988 Foam flows. *Ann. Rev. Fluid Mech.* **20**, 325–357.
Kraynik, A.M. & Hansen, M.G. 1986 Foam and emulsion rheology: A quasistatic model for large deformations of spatially periodic cells. *J. Rheol.* **30**, 409–439.
Kraynik, A.M. & Hansen, M.G. 1987 Foam rheology: A model of viscous phenomena. *J. Rheol.* **31**, 175–205.
Lamb, H. 1945 *Hydrodynamics*. New York: Dover.
Landau, L. & Lifshitz, E.M. 1960 *Fluid Mechanics*. Reading: Addison-Wesley.
Lane, A.R. & Ottewill, R.H. 1976 The preparation and properties of Bitumen emulsions stabilized by cationic surface-active agents. In *Theory and Practive of Emulsion Technology*, (ed. A.L. Smith), pp. 157–177. New York: Academic Press.
Langevin, D. 1981 Light-scattering of monolayer viscoelasticity. *J. Colloid Interface Sci.* **80**, 412–425.
Langevin, D. & Griesmar, C. 1980 Light-scattering study of fatty acid monolayers. *J. Phys. D* **13**, 1189–1199.
Langmuir, I. 1917 The constitution and fundamental properties of solids and liquids II. Liquids. *J. Am. Chem. Soc.* **39**, 1848–1906.
Langmuir, I. 1918 The adsorption of gases on plane surfaces of glass, mica and platinum. *J. Am. Chem. Soc.* **40**, 1361–1403.
Langmuir, I. 1931 Experiments with oil on water *J. Chemical Education* **8**, 850–862.
Langmuir, I. 1936 Two-dimensional gases, liquids and solids. *Science* **84**, 378–386.

Laplace, P.S. 1806 *Mechanique Celeste*, Supplement to Book 10.
Lawerence, A.S.C. 1952 Emulsions and films. *Nature* **170**, 232–234.
Lebedev, A.A. 1916 *Zhur. Russ. Fiz. Khim.* **48**.
Lemlich, R. 1972 *Adsorptive Bubble Separation Techniques*. Academic Press: New York.
Leung, R., Hou, M.J. & Shah, D.O. 1988 Microemulsions: formation, structure, properties, and novel applications. In *Surfactants in Chemical/Process Engineering*, Vol. 28 (eds. D.T. Wasan, M.E. Ginn & D.O. Shah), pp. 315–367. New York: Marcel Dekker.
Levich, V.G. 1941 *Zh. Eksperim. Theor. Fiz.* **11**, 340.
Levich, V.G. 1962 *Physicochemical Hydrodynamics*.Englewood Cliffs, New Jersey: Prentice Hall.
Levich, V.G. 1969 Surface-tension-driven phenomena. In *Annual Review of Fluid Mechanics*, vol. 1, pp. 283–316. California: Annual Reviews Inc.
Li, D.L. & Slattery, J.C. 1988 Measuring nonlinear surface stress-deformation behavior for aqueous solutions of dodecyl sodium sulfate and dodecyl alcohol. *J. Colloid Interface Sci.* **125**, 190–197.
Liebermann, L.N. 1949 The second viscosity of liquids. *Physical Review* **75**, 1415–1422.
Lifshutz, N., Hegde, M.G. & Slattery, J.C. 1971 Knife-edge surface viscometers. *J. Colloid Interface Sci.* **37**, 73.
Lindsay, K.A. & Straughan, B. 1978 A thermodynamic viscous interface theory and associated stability problems. *Arch. Rational Mech. Anal.* **59**, 307–326.
Lissant, K.J. 1966 The geometry of high-internal-phase-ratio emulsions. *J. Colloid Interface Sci.* **22**, 462–468.
Lissant, K.J. 1974 *Emulsions and Emulsion Technology: Part I*. Surfactant Science Series vol. 6. New York: Marcel Dekker.
Loewenthal, M. 1931 Tears of strong wine. *Phil. Mag.* **12**, 462–472.
Lovett, R., DeHaven, P.W., Vieceli, J.J. & Buff, F.P. 1972 Generalized van der Waals theory for surface tension and interfacial width. *J. Chem. Phys.* **58**, 1880–1885.
Low, A. R. & Brunt, D. 1925 Instabilities of viscous fluid motion. *Nature* **115**, 299–301.
Lu, H.L. & Apfel, R.E. 1990 Quadrupole oscillations of drops for the study of dynamic interfacial properties. *J. Colloid Interface Sci.* **134**, 245–255.
Lucassen, J. 1968 Longitudinal capillary waves. *Trans. Faraday Soc.* **64**, 2221–2234.
Lucassen, J. & Hansen, R.S. 1966 Damping of waves on monolayer-covered surfaces I. Systems with negligible surface dilatational viscosity. *J. Colloid Interface Sci.* **22**, 32–44.
Lucassen, J. & Hansen, R.S. 1967 Damping of waves on monolayer-covered surfaces II. Influence of bulk-to-surface diffusional interchange on ripple characteristics. *J. Colloid Interface Sci.* **23**, 319–328.

Lucassen, J., Hollway, F. and Buckingham, J.H. 1978 Surface properties of mixed solutions of poly-l-lysine and sodium dodecyl sulfate II. Dynamic surface properties. *J. Colloid Interface Sci.* **67**, 432–440.
Lucassen, J. & van den Tempel, M. 1972 Longitudinal waves on visco-elastic surfaces. *J. Colloid Interface Sci.* **41**, 491–498.
Lucassen, J., van den Temple, M. Vrig, A & Hesselink, M. 1970 Waves in thin liquid films. *Proc. Konkl. Ned. Akad. Wet.* **B73**, 109–135.
Lucassen-Reynders, E.H. 1976 Adsorption of surfactant monolayers at gas/liquid and liquid/liquid interfaces. In *Progress in surface and membrane science* (eds. D.A. Cadenhead and J.F. Danielli), pp. 253–359. New York: Marcel Dekker.
Lucassen-Reynders, E.H. 1981 Adsorption at fluid interfaces. In *Anionic surfactants* (ed. E.H. Lucassen-Reynders), pp. 1–54. Marcel Dekker: New York.
Lunkenheimer, K., Hartenstein, C., Miller, R. & Wantke, K.D. 1984 Investigations of the method of the radially oscillating bubble. *Colloids and Surfaces* **8**, 271–288.
Lyklema, J. & Mysels, K.J. 1965 A study of double layer repulsion and van der Waals attraction in soap films. *J. Am. Chem. Soc.* **87**, 2539.
Maldarelli, C, Jain, R.K., Ivanov, I. & Ruckenstein, E. 1980 Stability of symmetric and unsymmetric thin liquid films to short and long wavelength perturbations. *J. Colloid Interface Sci.* **78**, 118–143.
Maldarelli, C. & Jain, R.K. 1982a The linear, hydrodynamic stability of an interfacially perturbed, transversely isotropic, thin, planar viscoelastic film I. General formulation and a derivation of the dispersion equation. *J. Colloid Interface Sci.* **90**, 233–261.
Maldarelli, C. & Jain, R.K. 1982b The linear, hydrodynamic stability of an interfacially perturbed, transversely isotropic, thin, planar viscoelastic film II. Extension of the theory to the study of the onset of small-scale cell membrane motions. *J. Colloid Interface Sci.* **90**, 263–276.
Malhotra, A.K. & Wasan, D.T. 1987 Effects of surfactant adsorption-desorption kinetics and interfacial rheological properties on the rate of drainage of foam and emulsion films. *Chem. Eng. Commun.* **55**, 95–128.
Malhotra, A.K. & Wasan, D.T. 1988 Interfacial rheological properties of absorbed surfactant films with applications to emulsion and foam stability. In *Thin Liquid Films*, (ed. I.B. Ivanov). New York: Marcel Dekker.
Malysa, K., Lunkenheimer, K., Miller, R. and Hartenstein, C. 1981 Surface elasticity and frothability of n-octanol and n-octanoic acid solutions. *Colloids and Surfaces* **3**, 329–338.
Manev, E., Scheludko, A. & Exerowa, D. 1974 Effect of surfactant concentration on critical thickness of liquid films. *Colloid Polym. Sci.* **252**, 586–593.
Manev, E.D., Sazdanova, S.V. & Wasan, D.T. 1984 Emulsion and foam stability—the effect of film size on film drainage. *J. Colloid Interface Sci.* **97**, 591–594.

Mann, J.A. & Hansen, R.S. 1963 Propagation characteristics of capillary ripples II. Instrumentation for measurement of ripple velocity and amplitude. *J. Colloid Sci.* **18**, 757–771.
Mannheimer, R.J. 1969 Surface rheological properties of foam stabilizers in nonaqueous liquids. *AIChE J.* **15**, 88.
Mannheimer, R.J. & Burton, R.A. 1970 A theoretical estimation of viscous-interaction effects with a torsional (knife-edge) surface viscometer. *J. Colloid Interface Sci.* **32**, 73–80.
Mannheimer, R.J. & Schechter, R.S. 1967 An analysis for Bingham plastic surface flow in a canal viscometer (viscous-traction type). *J. Colloid Interface Sci.* **25**, 434–437.
Mannheimer, R.J. & Schechter, R.S. 1970a An improved apparatus and analysis for surface rheological measurements. *J. Colloid Interface Sci.* **32**, 195–211.
Mannheimer, R.J. & Schechter, R.S. 1970b Shear-dependent surface rheological measurements of foam stabilizers in nonaqueous liquids. *J. Colloid Interface Sci.* **32**, 212–224.
Mannheimer, R.J. & Schechter, R.S. 1970c The theory of interfacial viscoelastic measurement by the viscous-traction method. *J. Colloid Interface Sci.*, **32**, 225–241.
Marangoni, C.G.M. 1871 Uber die ausbreitung der tropfen einer flussigkeit auf der oberfluche einer anderen. *Ann. Physik. (Poggendorff)* **3**, 337–354.
Maru, H.C. & Wasan, D.T. 1979 Dilatational viscoelastic properties of fluid interfaces II. Experimental study. *Chem. Eng. Sci.* **34**, 1295–1307.
Mavrovouniotis, G. 1989 *Transport of Interfaces containing Surface-Active Substances*. PhD Thesis. Massachusetts Institute of Technology, Cambridge, Massachussetts.
Mavrovouniotis, G. & Brenner, H. 1991 A micromechanical investigation of interfacial transport processes: I. Interfacial conservation equations. (Submitted for publication.)
Mcbain, J.W. & Robinson, J.V. 1949 NACA Tech. Note, 1844.
McConnell, A.J. 1957 *Applications of Tensor Analysis*. New York: Dover.
Melrose, J.C. 1968 Thermodynamic aspects of capillarity. *Ind. Eng. Chem.* **60**, 53–70.
Melrose, J.C. 1970 Interfacial phenomena as related to oil recovery mechanism. *Can. J. Chem. Eng.* **48**, 638.
Menzel, D.H. 1961 *Mathematical Physics*. New York: Dover.
Merritt, R.M. & Subramanian, R.S. 1989 Migration of a gas bubble normal to a plane horizontal surface in a vertical temperature gradient. *J. Colloid Interface Sci.* **131**, 514–525.
Miller, C.A. 1978 Stability of interfaces. In *Surface and Colloid Science*, vol. 10 (ed. E. Matijevic), pp. 227–293. New York: Plenum.
Miller, E.E. & Miller, R.D. 1956 Physical theory for capillary flow phenomena, *J. Appl. Phys.* **27**, 324.
Miller, C.A. & Neogi, P. 1985 *Interfacial Phenomena*. Surfactant Science Series, vol. 17. New York: Marcel Dekker.

Minssaiux, L. 1974 Oil displacement by foams in relation to their physical properties in porous media. *J.Petrol.Tech.*, 100–108.
Moeckel, G.P. 1975 Thermodynamics of an interface. *Arch. Rat. Mech. Anal.* **57**, pp. 255–280.
Morrell, J.C. & Egloff, G. 1931 *J. Colloid Chem.* **3**, 503.
Morse, P.M. & Feschbach, H. 1953 *Methods of Theoretical Physics, Part I.* New York: McGraw-Hill.
Motomura, K. & Matamura, R. 1963 Viscoelasticity of linear polymer monolayers. *J. Colloid Sci.* **18**, 295.
Mouquin, H. & Rideal, E.K. 1927 The rigidity of solid unimolecular films. *Proc. Roy. Soc.* **A114**, 690–697.
Murdoch, P.G. & Leng, D.E. 1971 The mathematical formulation of hydrodynamic film thinning and its application to colliding drops suspended in a second liquid—II. *Chem. Engng Sci.* **26**, 1881–1892.
Murphy, C.L. 1966 *Thermodynamics of Low Tension and Highly Curved Interfaces*. Ph.D. Thesis, University of Minnesota, Minneapolis.
Mysels, K.J. 1964 Soap films and some problems in surface and colloid chemistry. *J. Phys. Chem.* **68**, 3441–3448.
Mysels, K.J. 1986 Improvements in the maximum bubble pressure method of measuring surface tension. *Langmuir J.* **2**, 428–432.
Mysels, K.J., Shinoda, S. & Frankel, S. 1959 *Soap Films — A Study of Their Thinning and a Bibliography*. New York: Pergamon.
Nayfeh, A. H. 1973 *Perturbation methods*. New York: Wiley.
Newton, I. 1704 *Optiks*, Book II, Part I, observation 17. London: Smith and Watford.
Nicolson, G.L. 1974 Interactions of lectins with animal-cell surfaces. *Int. Rev. Cytol.* **39**, 89–190.
Nicolson, G.L. 1976 Control of receptors on normal and tumor cells. I. Cytoplasmic influence over cell surface components. *Biochim. Biophys. Acta*, **457**, 57–108.
Nikolov, A.D. & Wasan, D.T. 1989 Ordered micelle structuring in thin liquid films formed from anionic surfactant solutions. I. Experimental. *J. Colloid Interface Sci.*, **133**, 1–12.
Oh, S.G. & Slattery, J.C. 1978 Disk and biconical interfacial viscometers. *J. Colloid Int. Sci.*, **67**, 516–525.
Oldroyd, J.G. 1955 The effect of interfacial stabilizing films on the elastic and viscous properties of emulsions. *Proc. Roy. Soc. London* **A232**, 567–577.
Ono, S. & Kondo, S. 1960 Molecular theory of surface tension in liquids. *Encyclopedia of Physics*, vol. 10 (ed. S. Flugge), pp. 134–280. Berlin: Springer.
Orell, A. 1961 PhD Thesis, University of Illinois, Urbana.
Osborne, M.F.M. 1968 *Kolloid Z.* **2**, 150.
Overbeek, J. 1960 Black soap films. *J. Phys. Chem.* **64**, 1178–1183.
Pacetti, S.D. 1985 *Structural Modeling of Foam Rheology*. MS thesis, Univ. Houston, Houston, Texas.
Pasternak, A. 1976 *J. Theor. Biol.* **58**, 365.

Pearson, J.R.A. 1958 On convection cells induced by surface tension *J. Fluid Mech.* **4**, 489–500.

Perri, J.M. 1953 In *Foams: Theory and Industrial Applications*, (ed. J.J. Bikerman), pp. 195–211. New York: Reinhold.

Phillips, W.J., Graves,R.W. & Flumerfelt, R.W. 1980 Droplet dynamics in shear fields. Role of dynamic interfacial effects. *J. Colloid Interface Sci.* **76**, 350–370.

Pintar, A.J., Israel, A.B. & Wasan, D.T. 1971 Interfacial shear viscosity phenomena in solutions of macromolecules *J. Colloid Interface Sci.* **37**, 52–67.

Plateau, J.A.F. 1869 Experimental and theoretical researches into the figures of equilibrium of a liquid mass without weight VIII. Researches into the causes upon which the easy development and the persistence of liquid films depend.— On the superficial tension of liquids. — On a new principle relating to the surfaces of liquid. *Phil. Mag.* **38**, 445–455.

Plateau, J.A.F. 1872 Stratique experimentale aux seules forces moleculaires. *Bull. Acad. Belg., Ser. 2* **34**, 404–410.

Platikanov, D., Nedjalkov, M. & Scheludko, A. 1980 Line tension of Newton black films I. Determination by the critical bubble method. *J. Colloid Interface Sci.* **75**, 612–619.

Platikanov, D., Panajotov, I. & Scheludko, A. 1966 Elasticity of adsorption films in solutions of surface-active substances. *Abh. Dt. Akad. Wiss. Berlin, Kl. Chemie* **66**, 773–782.

Pliny AD77 *Natural History*, Book II. cvi. See Loeb Classical Library. Harvard University Press, Cambridge (1979).

Porter, K., Prescott, D. & Frye, J. 1973 Changes in surface morphology of Chinese hamster overy cells during a cell cycle. *J. Cell Biology* **57**, 815–836.

Poskanzer, A.M. & Goodrich, F.C. 1975a Surface viscosity of sodium dodecyl sulfate solutions with and without added dodecanol. *J. Physical Chemistry* **79**, 2122–2126.

Poskanzer, A.M. & Goodrich, F.C. 1975b A new surface viscometer of high sensitivity II. Experiments with stearic acid monolayers. *J. Colloid Interface Sci.* **52**, 213–221.

Prandtl, L. & Tietjens, O.G. 1934 *Funamentals of Hydro- and Aeromechanics*. New York: Dover.

Princen, H.M. 1983 Rheology of foams and highly concentrated emulsions: I. Elastic properties and yield stress of a cylindrical model system. *J. Colloid Interface Sci.* **91**, 160–175.

Princen, H.M. 1985 Rheology of foams and highly concentrated emulsions: II. Experimental study of the yield stress and wall effects for concentrated oil-in-water emulsions. *J. Colloid Interface Sci.* **105**, 150–171.

Princen, H.M., Aronson, M.P. & Moser, J.C. 1980 Highly concentrated emulsions: II. Real systems. The effect of film thickness and contact angle on the volume fraction in creamed emulsions. *J. Colloid Interface Sci.* **75**, 246–270.

Princen, H.M., Cazabat, A.M., Cohen Stuart, M.A, Heslot, F. & Nicolet, S. 1988 Instabilities during wetting processes: Wetting by tensioactive liquids. *J. Colloid Interface Sci.* **126**, 84–102.

Princen, H.M. & Levinson, P. 1987 The surface area of Kelvin's minimal tetrakaidecahedron: the ideal foam cell? *J. Colloid Interface Sci.* **120**, 172–175.

Prud'homme, R.K. 1981 Foam Flow. Presented at the Annual Meeting of the Society of Rheology, Louisville, Ky.

Prud'homme, R.K. & Bird, R.B. 1978 The dilatational properties of suspensions of gas bubbles in incompressible Newtonian and non-Newtonian fluids. *J. Non-Newtonian Fluid Mech.* **3**, 261–279.

Radoev, B.P., Dimitrov, D.S. & Ivanov, I.B. 1974 Hydrodynamics of thin liquid films. Effect of the surfactant on the rate of thinning. *Coll. Polym. Sci.* **252**, 50–55.

Ramamohan, R.R. & Slattery, J.C. 1984 Effects of surface viscoelasticity in the entrapment and displacement of residual oil. *Chem. Eng. Commun.* **26**, 241–263.

Rayleigh, Lord 1878 On the instability of jets. *Proc. London Math. Soc.* **10**, 4–13.

Rayleigh, Lord 1879 On the capillary phenomena of jets. *Proc. Roy. Soc.* (London) **29**, 71–97.

Rayleigh, Lord 1916 On convection currents in a horizontal layer of fluid when the higher temperature is on the under side. *Phil. Mag.* **32**, 529–546.

Reddy, J.N. & Rasmussen, M.L. 1982 *Advanced Engineering Analysis.* New York: Wiley.

Reynolds, O. 1886 On the theory of lubrication and its application to Mr. Beauchamp Tower's experiments, including an experimental determination of the viscosity of olive oil. *Phil. Trans. Roy. Soc.* (London) **A177**, 157–234.

Robb, I.D. 1982 *Microemulsions.* New York: Plenum Press.

Roberts, K., Axberg, C. & Osterlund, R. 1977 Effect of spontaneous emulsification of defoamers on foam prevention. *J. Colloid Interface Sci.* **62**, 264–272.

Ronis, D., Bedeaux, D. & Oppenheim, I. 1978 Derivation of dynamical equations for a system with an interface I. General theory. *Physica* **90A**, 487–506.

Ronis. D. & Oppenheim, I. 1983 Derivation of dynamical equations for a system with an interface II. The gas-liquid interface. *Physica* **117A**, 317–354.

Ross, S. 1967 Mechanisms of foam stabilization and anti-foaming action. *Chem. Eng. Prog.* 41–47.

Ross, S. 1980 Foams. In *Encyclopedia of Chemical Technology*, vol. II (3rd edn.), pp. 127–145. New York: Wiley.

Rowlinson, J.S. & Widom, B. 1982 *Molecular Theory of Capillarity.* Oxford: Clarendon Press.

Rumscheidt, F.D. & Mason, S.G. 1961 Particle motions in sheared suspensions XI. Internal circulations in fluid droplets. (Experimental.) *J. Colloid Interface Sci.* **16**, 210–237.

Rusanov, A. I. 1967 Thermodynamics of films 5. Three-phase equilibria involving films. *Colloid J. (USSR)* **29**, 118–188.

Rusanov, A. & Krotov, V. 1979 Gibbs elasticity of liquid films, threads and foams. In *Progress in Surface and Membrane Science*, vol. 13, pp. 415–524. New York: Academic Press.

Rybczynski, W. 1911 *Bull. Intern. Acad. Sci.* Cracovie (ser. A).

Sanfeld, A. & Steinchen, A. 1975 Coupling between a transconformation surface-reaction and hydrodynamic motion. *Biophys. Chem.* **3**, 99–106.

Sanfeld, A., Steinchen, A., Hennenberg, M. Bisch, P.M., Van Lamsweerde-Gallez, D. & Dalle-Vedove, W. 1979 In *Dynamics and Stability of Fluid Interfaces*, (ed. T.S. Sorensen), p. 168. New York: Springer-Verlag.

Sche, S. & Fijnaut, H.M. 1978 Dynamics of thin free liquid films stabilized with ionic surfactants. *Surface Sci.* **76**, 186–202.

Scheludko, A.D. 1963 Theorie der flotation. *Kolloid Z.* **191**, 52–58.

Scheludko, A.D. 1967 Thin liquid films. *Adv. Colloid Interface Sci.* **1**, 391–464.

Scheludko, A, Chakarov, V. & Toshev, B. 1981 Water condensation on hexadecane and linear tension. *J. Colloid Interface Sci.* **82** 1, 83.

Schofield, P. & Henderson, J.R. 1982 Statistical mechanics of inhomogeneous fluids. *Proc. R. Soc. Lond.* **A379**, 231–246.

Schowalter, W.R., Chaffey, C.E. & Brenner, H. 1968 Rheological behavior of a dilute emulsion. *J. Colloid Interface Sci.* **26**, 152–160.

Schulman, J.H. & Cockbain, E.G. 1940 Molecular interactions of oil-water interfaces. Part I. Molecular complex formation and the stability of oil-in-water emulsions. *Trans. Faraday Soc.* **36**, 651–661.

Schwartz, A.M., Perry, J.W. & Berch, J. 1977 *Surface Active Agents and Detergents*. New York: Krieger.

Schwartz, L.W. & Princen, H.M. 1987 A theory of extensional viscosity for flowing foams and concentrated emulsions. *J. Colloid Interface Sci.* **118**, 201–211.

Scott, J.A.N. 1976 A general description of the breaking process of cationic Bitumen emulsions contacted with mineral aggregates. In *Theory and Practice of Emulsion Technology*, (ed. A.L. Smith), pp. 179–200. New York: Academic Press.

Scriven, L.E. 1960 Dynamics of a fluid interface. *Chem. Eng. Sci.* **12**, 98–108.

Scriven, L.E. & Sternling, C.V. 1964 On cellular convection driven by surface-tension gradients: effects of mean surface tension and surface viscosity. *J. Fluid Mech.* **19**, 321–340.

Shahin, G.T. 1986 *The Stress Deformation Interfacial Rheometer*. PhD Thesis, Department of Chemical Engineering, University of Pennsylvania, State College, Pennsylvania.

Shail, R. 1978 The torque on a rotating disk in the surface of a liquid with an adsorbed film. *J. Engng Math.* **12**, 59–76.
Shail, R. 1979 The slow rotation of an axisymmetric solid submerged in a fluid with a surfactant surface layer—I. The rotating disk in a semi-infinite fluid. *Int. J. Multiphase Flow* **5**, 169–183.
Shail, R. 1983 On the slow motion of a solid submerged in a fluid with a surfactant surface film. *J. Engng Math.* **17**, 239–256.
Shail, R. & Gooden, D.K. 1981 The slow rotation of an axisymmetric solid submerged in a fluid with a surfactant surface layer—II. The rotating solid in a bounded fluid. *Int. J. Multiphase Flow* **7**, 245–260.
Sherman, P. 1953 Studies in water-in-oil emulsions III. The properties of interfacial films of sorbiton sequioleate. *J. Colloid Interface Sci.* **8**, 35–37.
Sherman, P. 1968 *Emulsion Science*. New York: Academic Press.
Sherwood, T.K. & Wei, J.C. 1957 Interfacial phenomena in liquid extraction. *Ind. Eng. Chem.* **49**, 1030–1034.
Silvey, A. 1916 The fall of mercury droplets in a viscous medium. *Phys. Rev.* **7**, 106–111.
Slattery, J.C. 1964a Surfaces—I. Momentum and moment-of-momentum balances for moving surfaces. *Chem. Engng Sci.* **19**, 379–385.
Slattery, J.C. 1964b Surfaces—II. Kinematics of diffusion in a heterogeneous surface. *Chem. Engng Sci.* **19**, 453–455.
Slattery, J.C. 1967 General balance equation for a phase interface. *I&EC Fundam.* **6**, 108–115.
Slattery, J.C. 1974 Interfacial effects in the entrapment displacement of residual oil. *AIChE J.* **20**, 1145.
Slattery, J.C. 1979 Interfacial effects in the displacement of residual oil by foam. *AIChE J.* **25**, 283.
Slattery, J.C. 1981 *Momentum, Energy and Mass Transfer in Continua.* Huntington, New York: Krieger.
Slattery, J.C. 1990 *Interfacial Transport Phenomena*. New York: Springer-Verlag.
Slattery, J.C., Chen, J., Thomas, C.P. & Fleming, P.D. 1980 Spinning drop interfacial viscometer. *J. Colloid Interface Sci.* **73**, 483–499.
Smith, F.I. P. 1970 Stability of liquid film flow down an inclined plane with oblique air flow. *Phys. Fluids* **13**, 1693.
Smoluchowski, M. 1916 Drei vortrage uber diffusion Brownshe molekular und koagulation von kolloidteilchen. *Physik Z.* **17**, 557–573.
Snik, A., Kouijzer, W., Keltjens, J. and Houkes, Z. 1983 The measurement of dilatational properties of liquid surfaces. *J. Colloid Interface Sci.* **93**, 301–306.
Sohl, C.H., Miyano, K. and Ketterson, J.B. 1978 Novel technique for dynamic surface tension and viscosity measurements at liquid-gas interfaces. *Rev. Sci. Instrum.* **49**, 1464–1469.
Somasundaran, P., Danitz, M. & Mysels, K.J. 1974 A new apparatus for measurements of dynamic interfacial properties. *J. Colloid Interface Sci.* **48**, 410–416.

Somasundaran, P. & Ramachandran, R. 1988 Surfactants in flotation. In *Surfactants in Chemical/Process Engineering*, vol. 28 (eds. D.T. Wasan, M.E. Ginn & D.O. Shah), pp. 195–235. New York: Marcel Dekker.

Sonntag, H. & Strenge, K. 1972 *Coagulation and Stability of Disperse Systems*. New York: Halstead Press.

Sorensen, T.S., Hennenberg, M., Steinchen, A. & Sanfeld, A. 1976 Chemical and hydrodynamical analysis of the stability of a spherical interface. *J. Colloid Interface Sci.* **56**, 191–205.

Srivastava, S.N. & Haydon, D.A. 1964 *Proc. 4th Int. Congr. Surf. Act.*, **3**, 221.

Steinchen, A.& Sanfeld, A. 1977 Thermodynamic stability criteria for surfaces. *Chemical Phys.* **1**, 156.

Sternling, C.V. & Scriven, L.E. 1959 Interfacial turbulence: Hydrodynamic instability and the Marangoni effect. *AIChE J.* **5**, 514–523.

Stokes, G.G. 1851a On the effect of the internal friction of fluids on the motion of a pendulum. *Trans. Cambr. Phil. Soc.* **1**, 104–106.

Stokes, G.G. 1851b On the colours of thick plates. *Trans. Cambr. Phil. Soc.* **1**, 110–111.

Stokes, V.K. 1984 *Theories of Fluids with Microstructure*. New York: Springer-Verlag.

Tachibana, T., Inokuchi, K. & Inokuchi, T. 1958 The gelation of high polymers at the air-water interface. *Kolloid Z.*, **167**, 141–146.

Taubman, A.B. & Koretskii, A.F. 1958 The mechanism of the demulsifying action of solid emulsifiers. *Dan. SSR.* **118**, 991.

Taylor, G.I. 1932 The viscosity of a fluid containing small drops of another fluid. *Proc. Roy. Soc.* (London) **A138**, 41–48.

Taylor, G.I. 1934 The formation of emulsions in definable fields of flow. *Proc. Roy. Soc.* (London) **A146**, 501–523.

Taylor, G.I. 1954 The two coefficients of viscosity for an incompressible fluid containing air bubbles. *Proc. R. Soc.* (London) **A226**, 34–39.

Taylor, T.D. & Acrivos, A. 1964 On the deformation and drag of a falling viscous drop at low Reynolds number. *J. Fluid Mech.* **18**, 466–476.

Thiessen, D. 1966 Flusige filme unter stoff uber angsbedingungen. *Z. Phys. Chem.* **232**, 27–38.

Thompson, W. (Lord Kelvin) 1855 On certain curious motions observable at the surfaces of wine and other alcoholic liquors. *Phil. Mag* **10**, 330–333.

Thompson, W. (Lord Kelvin) 1871 The influence of wind on waves in water supposed frictionless. *Phil. Mag.* **42**, 368–374.

Thompson, W. (Lord Kelvin) 1887 On the division of space with minimum partitional area. *Philos. Mag.* **24**, 503–514.

Ting, L. 1984 *Dynamic Interfacial Properties of Surfactant Systems*. PhD Thesis, Illinois Institute of Technology, Chicago, Illinois.

Ting, L., Wasan, D.T., Miyano, K. & Xu, S.Q. 1984 Longitudinal surface waves for the study of dynamic properties of surfactant systems I. Gas-liquid interface. *J. Colloid Interface Sci.* **102**, 248.

Ting, L., Wasan, D.T. & Miyano, K. 1985 Longitudinal surface waves for the study of dynamic properties of surfactant systems II. Liquid-liquid interface. *J. Colloid Interface Sci.* **107**, 345–354.

Tisza, L. 1942 Supersonic absorption and Stokes' viscosity relation *Physical Review* **61**, 531–535.

Tolman, R.C. 1948 Consideration of the Gibbs theory of surface tension. *J. Chem. Phys.* **16**, 758–774.

Torrey, S. 1984 *Emulsions and Emulsifier Applications.* Park Ridge, New Jersey: Noyes Data Corporation.

Torza, S. & Mason, S.G. 1971 Effect of line tension on 3-phase liquid interactions. *Kolloid Z. Z. Polym.*, **246**, 593.

Toshev, B.V. & Ivanov, I.B. 1975 Thermodynamics of thin liquid films. I. Basic relations and conditions of equilibrium. *Colloid and Polymer Sci.* **253**, 558–565.

Trapeznikov, A.A. 1957 The influence of the concentration of an electrolyte solution on the exchange sorption by fatty acid crystals. *Dan. SSSR* **64**, 1281–1283.

Traykov, T.T., Manev, E.D. & Ivanov, I.B. 1977 Hydrodynamics of thin liquid films. Experimental investigation of the effect of surfactants on the drainage of emulsion films. *Int. J. Multiphase Flow* **3**, 485–494.

Truesdell, C. & Toupin, R. 1960 In *Handbuch der Physik,* vol III/1 (ed. S. Flugge), p. 533.

Tschoegl, N.W. 1958 Elastic moduli in monolayers. *J. Colloid Sci.* **13**, 500–507.

Tschoegl, N.W. & Alexander, A.E. 1960 The surface chemistry of wheat gluten II. Measurement of surface viscoelasticity. *J. Colloid Sci.* **15**, 168–182.

Turner, A. & Gurnee, E.F. 1958 *Rheol. Theory Appl.* **1**, 408.

Valentini, J.E., Thomas, W.R., Sevenhysen, P., Jiang, T.S., Liu, Y. & Yen, S.C. 1991 Materials and interfaces. Role of dynamic surface tension in slide coating. *Ind. Eng. Chem.* **30**, 453–461.

Van Oss, C.J. 1978 Phagocytosis as a surface phenomenon. *Ann. Rev. Microbiol.* **32**, 19–39.

van den Tempel, M. 1957 *Proc. 2nd Int. Congr. Surf. Act.* **1**, 439.

van der Waals, J.D. & Kohnstamm, P. 1908 *Lehrbuch der Thermodynamik.*, vol. I. Leipzig: Mass and van Suchtelen.

Van Dyke, M. 1964 *Perturbation Methods in Fluid Mechanics.* London: Academic Press.

Velarde, M.G., Ibanes, J.L., Sorensen, T.S., Sanfeld, A. & Hennenberg, M. 1977 *Proceedings Levich Conference*, Oxford.

Verwey, E.J.W. & Overbeek, J.Th.G. 1948 *Theory of the Stability of Lyophobic Colloids.* Amsterdam: Elsevier.

Vesselovsky, V.S. & Pertzov, V. 1936 *Zh. Fiz. Khim.* **8**, 245.

Vignes-Adler, M. & Brenner, H. 1985 A micromechanical derivation of the differential equations of interfacial statics III. Line tension. *J. Colloid Interface Sci.* **103**, 11–43.

Vold, R.D. & Vold, M.J. 1983 *Colloid & Interface Chemistry.* London: Addison-Wesley.

von Szyskowsky, B. 1908 Experimentalle stdien uber kapullare eigenshaften der wasserigen losurgen von fettsauren. *Z. Physik Chem.* **64**, 385–414.

Vrij, A. 1966 Possible mechanism for spontaneous rupture of free liquid films. *Discuss. Faraday Soc.* **42**, 23.

Vrij, A., Hesselink, F., Lucassen J. & van den Tempel, M. 1970 Waves in liquid films. Two symmetrical modes in very thin films and film rupture *Proc. Kon. Ned. Akad. Wet. B* **73**, 124.

Wantke, K.D., Miller, R. & Lunkenheimer, K. 1976 Theory of diffusion at an oscillatory bubble surface. *Abh. Akad. Wiss. DDR, Originalbertr.* Berlin, E. Ger.

Wasan, D.T., Gupta, L. & Vora, M.K. 1971 Interfacial shear viscosity at fluid-fluid interfaces. *AIChE J.* **17**, 1287–1295.

Wasan, D.T. & Mohan, V. 1977 Interfacial rheological properties of fluid interfaces containing surfactants. In *Improved Oil Recovery by Surfactant and Polymer Flooding* (eds. D.O. Shah and R.S. Schechter), pp. 161–201, New York: Academic Press.

Wasan, D.T., McNamara, J.J., Shah, S.M., Sampath, K. & Aderangi, N. 1979 Role of coalescence phenomena and interfacial rheological properties in enhanced oil recovery. *J. Rheol.* **23**, 181–207.

Wasan, D.T., Nikolov, A.D., Huang, D.D. & Edwards, D.A. 1988 Foam stability: Effects of oil and film stratification. In *Surfactant-Based Mobility Control* (ed. Duane H. Smith), pp. 136-162. Washington D.C.: American Chemical Society.

Weaire, D. 1989 A note on the elastic behavior of ordered hexagonal froths. *Phil. Mag.* **60**, 27–30.

Weatherburn, C.E. 1930 *Differential Geometry of Three Dimensions*, vol. I. London: Cambridge University Press.

Wei, J.C. 1955 Ph.D. Thesis, Massachusetts Institute of Technology, Cambridge, Massachusetts.

Wei, L., Schmidt, W. & Slattery, J.C. 1974 Measurement of the surface dilatational viscosity. *J. Colloid Interface Sci.* **48**, 1–8.

Wei, L.Y. & Slattery, J.C. 1976 Experimental study of nonlinear surface stress-deformation behavior with the deep-channel surface viscometer. *Colloid and Interface Science* **4**, 399–420.

Wendel, H., Gallez, D. & Bisch, P.M. 1981 On the dynamical stability of fluid dielectric films. *J. Colloid Interface Sci.* **84**, 1–7.

Whitaker, S. 1964 Effect of surface active agents on the stability of falling liquid films. *Ind. Eng. Chem. Fundamentals* **3**, 132.

Williams, M.B. & Davis, S.H. 1982 Nonlinear theory of film rupture. *J. Colloid Interface Sci.* **90**, 220–228.

Willingham, M.C. & Pastan, I. 1975 Cyclic AMP modulates microvillus formation and aggiutonability in tranformed and normal mouse fibroblasts. *Proc. Nat. Acad. Sci.* USA, **72**, 1263–1267.

Wilson, E.B. & J.W. Gibbs 1929 *Vector Analysis*. New York: Dover.

Woods, D.R. & Burrill, K.A. 1972 Stability of emulsions *J. Electroan. Chem.* **37**, 191.

Worthington, A.M. 1876 On the forms assumed by droplets of liquids falling vertically on a horizontal plane. *Proc. Roy. Soc.* (London) **25**, 261–271.

Yarin, A.L. 1979 Stability of a jet of viscoelastic liquid in the presence of a mass flux at its surface. *J. Eng. Phys.*. **37**, 904–910.

Yih, S.M. & Seagrave, R.C. 1978 Hydrodynamic stability of thin liquid films flowing down an inclined plane with accompanying heat transfer and interfacial shear. *AIChE J.* **24**, 803–810.

Young, N. O., Goldstein, J.S. & Block, M.J. 1959 The motion of bubbles in a vertical temperature gradient. *J. Fluid Mech.*, **6**, 350–356.

Young, T. 1805 An essay on the cohesion of fluids. *Phil. Trans. Roy. Soc. (London)* **5**, 65–87.

Zapryanov, Z., Malhotra, A.K., Aderangi, N. & Wasan, D.T. 1983 Emulsion stability: An analysis of the effects of bulk and interfacial properties on film mobility and drainage rate. *Int. J. Multiphase Flow* **9**, 105–129.

Zuzovsky, M., Adler, P.M. & Brenner, H. 1983 Spatially periodic suspensions of convex particles in linear shear flows: III. Dilute arrays of spheres suspended in Newtonian fluids. *Phys. Fluids* **26**, 1714–1723.

Author Index

Subject Index

Π extensive time-rate of production of generic physical property within fluid volume, Eq. (3.5-1)

π intensive time-rate of production of generic physical property at a point of fluid (field density: per unit volume), Eq. (3.5-4)

π^s surface-excess time-rate of production of a generic physical property at a point on the interface (field density: per unit area), Eq. (3.6-11a)

ρ bulk-phase mass density, Eq. (4.1-1)

ρ_ι bulk-phase species mass density, Eq. (5.1-1)

ρ^s surface-excess mass density, Eq. (4.2-1)

ρ^s_ι surface-excess species mass density, Eq. (5.2-2)

σ interfacial tension, Eq. (4.2-15a)

τ bulk-phase viscous stress tensor, Eq. (4.1-13)

τ^s surface-excess viscous stress tensor, Eq. (4.2-15)

ϕ azimuthal angle in spherical polar coordinates measured from the x-direction (see Figure B.1-6)

Φ bulk-phase energy dissipation density (see Table 3.6-1)

Φ^s surface-excess energy dissipation density (see Table 3.7-1)

Φ^s surface-excess energy dissipation density (see Table 3.7-1)

Φ extensive diffusive flux of generic physical property *out* of the bounding surface to a fluid volume, Eq. (3.5-1)

ϕ intensive diffusive flux of generic physical property (field density: per unit time per unit area), Eq. (3.5-6)

ϕ^s surface-excess diffusive flux of generic physical property (field density: per unit time per unit length), Eq. (3.6-10)

Ψ extensive amount of generic physical property within a fluid volume, Eq. (3.5-1)

ψ intensive amount of generic physical property at a point of fluid (field density: per unit volume), Eq. (3.5-2)

ψ^s surface-excess amount of generic physical property at a point on the interface (field density: per unit area), Eq. (3.6-7)

Z extensive rate of supply of generic physical property within a fluid volume, Eq. (3.5-1)

ζ intensive rate of supply of generic physical property at a point of fluid (field density: per unit volume), Eq. (3.5-5)

ζ^s surface-excess rate of supply of generic physical property at a point on the interface (field density: per unit area), Eq. (3.6-11b)

Subscripts

α covariant index of a surface vector or tensor, (=1, 2)

β covariant index of a surface vector or tensor, (=1, 2)

γ covariant index of a surface vector or tensor, (=1, 2)